Environmental Management Technologies

Environmental Management Technologies: Challenges and Opportunities details the environmental problems posed by the various types of toxic organic and inorganic pollutants discharged from both natural and anthropogenic activities and their toxicological effects in environments, humans, animals, and plants. This book also highlights the recent advanced and innovative methods for the effective degradation and bioremediation of organic pollutants, heavy metals, dyes, etc. from the environment for sustainable development.

Features of the book:

- Provides state-of-the-art information on pollutants, their sources, and deleterious impacts on the environment
- Elucidates the recent updates on Emerging Pollutants (EPs) in pharmaceutical waste and personal care products
- Discusses the various physico-chemical, biological, and combination treatment systems for sustainable development
- Details recent research findings in the area of environmental waste management and their future challenges and opportunities

Environmental Management Technologies

Challenges and Opportunities

Edited by
Pankaj Chowdhary, Vineet Kumar,
Sunil Kumar, and Vishvas Hare

CRC Press is an imprint of the
Taylor & Francis Group, an **informa** business

Cover image: © Shutterstock

First edition published 2023
by CRC Press
6000 Broken Sound Parkway NW, Suite 300, Boca Raton, FL 33487-2742

and by CRC Press
4 Park Square, Milton Park, Abingdon, Oxon, OX14 4RN

CRC Press is an imprint of Taylor & Francis Group, LLC

First edition published by CRC Press 2023

ISBN: 9781032145617 (hbk)
ISBN: 9781032358598 (pbk)
ISBN: 9781003239956 (ebk)

DOI: 10.1201/9781003239956

Typeset in Times New Roman
by Newgen Publishing UK

Contents

Preface......ix
Acknowledgments......xi
Editors' Biographies......xiii
Notes on Contributors......xvii

PART I Pollutants in the Environment

Chapter 1 Emerging Environmental Pollutants: Current and Future Challenges......3
Fazila Younas, Natasha, Irshad Bibi, Muhammad Mahroz Hussain, Muhammad Shahid, and Nabeel Khan Niazi

Chapter 2 Study on the Occurrence and Detection of Non-Steroidal Anti-Inflammatory Drugs in Wastewater Treatment Plants......23
Arun Kumar Thalla and K.T.N. Greeshma

Chapter 3 Organochlorine and Organophosphate Pesticides and Emerging Pollutants in the Ganga River System: An Overview......49
Leena Singh and Aryaman Singh

Chapter 4 Sources, Spread, and Surveillance of Antimicrobial Resistance: A Global Concern......69
Sneha Suresh, Preethy Chandran, and Joseph Kadanthottu Sebastian

Chapter 5 Black Trail in Blue Oceans......83
Cristina Carapeto

Chapter 6 Health Hazards of Food Allergens and Related Safety Measures......99
Sonia Morya, Nishi Singh, and Chinaza Godswill Awuchi

PART II Remediation and Management of Pollutants

Chapter 7 The Role of Microbes in Environmental Contaminants' Management......117
Pravin Khaire, Vajramma Boggala, Akshay Mamidi, and Tanaji Narute

Chapter 8 Microbial Bioformulation Technology for Applications in Bioremediation......155
Jia May Chin and Adeline Su Yien Ting

Chapter 9 The Role of Microbes in the Degradation of Plastics and Directions Toward Greener Bioplastic 181

A.K. Priya, D. Balaji, J. Vijayaraghavan, J. Thivya, and R. Anand

Chapter 10 Microalgae: The Role of Phycoremediation in Treated Chrome Sludge from the Electroplating Industry and in Biomass Production 201

M. Muthukumaran

Chapter 11 Environmental Sensing and Detection of Toxic Heavy Metals by Metal Organic Frameworks-Based Electrochemical Sensors 215

Komal Rizwan and Muhammad Bilal

Chapter 12 Beneficial Functions of Vermiwash and Vermicompost for Sustainable Agriculture 229

Ankeet Bhagat, Sumit Singh, Kasahun Gudeta, Siddhant Bhardwaj, and Sartaj Ahmad Bhat

Chapter 13 Phytoremediation of Mine Tailings 243

Biju P. Sahariah, Tanushree Chatterjee, Jyoti K. Choudhari, Mukesh K. Verma, Anandkumar J., and Jyotsna Choubey

Chapter 14 Advanced Functional Approaches of Nanotechnology in Food and Nutrition 257

Sonia Morya, Chinaza Godswill Awuchi, and Farid Menaa

Chapter 15 The Role of Nanotechnology in Insect Pest Management 273

S.A. Dwivedi and Lelika Nameirakpam

Chapter 16 Polymer Nanocomposites for Wastewater Treatment 295

Adnan Khan, Sumeet Malik, Nisar Ali, Muhammad Bilal, Yong Yang, Mohammed Salim Akhter, Vineet Kumar, and Hafiz M.N. Iqbal

Chapter 17 Ohmic Heating as an Advantageous Technology for the Food Industry: Prospects and Applications 307

Sonia Morya, Chinaza Godswill Awuchi, Pankaj Chowdhary, Suneel Kumar Goyal, and Farid Menaa

Chapter 18 Nanoparticles Synthesis from Kitchen Waste: Opportunities, Challenges, and Future Prospects 329

Deepa and Raunak Dhanker

Chapter 19 Artificial Intelligence-Based Smart Waste Management for the Circular Economy 341

Alaa El Din Mahmoud and Omnya Desokey

Index 359

Preface

Environmental Management Technologies: Challenges and Opportunities describes the potential role of recent technologies to combat environmental concerns. This book covers the various roles of biological as well as physico-chemical management technologies for environmental sustainability. Rapid industrialization and intensive increases in agricultural activities have caused a reduction of soil quality, fertility, and also many adverse impacts on the environment, which is a serious concern globally. Pollutants are pesticides and herbicides, non-halogenated compounds, phthalates, additives to plastics, pharmaceuticals, personal care, etc. In recent times we have become more technologically advanced with scientific knowledge and socio-economic awareness and people are becoming more concerned about their surrounding environment to make it clean and green. Various ongoing research studies have focused on pollutants' removal or degradation by physical, chemical, and biological methods, but these pollutants still remain in the environment. Despite previously available research data on various pollutants, their nature, behavior, and threats to the ecosystem and human health implication are still not well known.

This book details the environmental problems posed by antibiotics, including the various types of toxic environmental pollutants discharged from both natural and anthropogenic activities and their toxicological effects on the environment, humans, animals, and plants. This book also highlights the recent advanced and innovative methods for the useful degradation and bioremediation of organic pollutants, heavy metals, dyes, etc. in wastewater. Also, this book covers a wide range of topics: environmental microbiology, biotechnology, nanotechnology, green chemistry, environmental science, environmental engineering, etc. Environmental degradation is one of the most challenging phenomena due to the increasing rate of industrialization globally. Traditional methods like biological degradation, coagulation, and membrane filtration technology have played a major role in the treatment of toxic organic pollutants. To date, nanotechnology and several bioremediation processes have been adopted as a modern technique for the removal of water contaminants as they have potential advantages, such as low cost, reusability, and are highly efficient in removing and recovering the pollutants. In nanotechnology, advanced oxidation processes (AOP) and photocatalytic methods are a more advanced method for the purification of contaminated water.

Besides these processes, some other methods are also used, such as physical adsorption processes, chemical oxidation with the use of ozone, hydrogen peroxide, or chlorine. In nanotechnology, the nanostructured materials are used due to their exceptionally high surface area, showing much higher efficiency and faster adsorption rates in water treatment. A variety of efficient, low-cost and eco-friendly nanomaterials with unique functionalities have been proposed for potential applications in the detoxification of industrial effluents, groundwater, surface water, and drinking water. Therefore, this book also serves as an invaluable source of basic knowledge on recent developments in management technologies to combat these threats.

Acknowledgments

The editors would like to express their sincere thanks to the contributors for submitting their work in a timely and proper manner. The editors are also thankful to the reviewers for evaluation and their valuable suggestions and comments to enhance the book's quality. Dr. Chowdhary acknowledges the Council of Scientific and Industrial Research (CSIR), New Delhi, India for Research Associate (RA) work. Dr. Chowdhary also acknowledges the support received from his family, especially his parents (Ram Chandra and Malti Devi). Further, the editors also acknowledge the cooperation received from CRC Taylor & Francis Group, for their guidance to finalize this book.

Editors' Biographies

Pankaj Chowdhary is Founder and President of the Society for Green Environment (SGE) in New Delhi, India. Currently he is working as a Postdoctoral Fellow in the CSIR-Indian Institute of Toxicology Research. He received his post-graduation degree (2011) in Biotechnology from Deen Dayal Upadhyaya Gorakhpur University, Uttar Pradesh, India. Afterwards, he gained his PhD (2018) in Microbiology from Babasaheb Bhimrao Ambedkar University, Lucknow, Uttar Pradesh, India. During his PhD, his work has mainly focused on the role of ligninolytic enzyme-producing bacterial strains in the decolorizing and degradation of coloring compounds from distillery wastewater. His main research areas are microbial biotechnology, biodegradation, and bioremediation of environmental contaminants in industrial wastewaters, metagenomics, biofuel, and bioenergy production. He is the editor of three books: *Emerging and Eco-friendly Approaches for Waste Management*, *Contaminants and Clean Technologies*, and *Microorganisms for Sustainable Environment and Health*. He is also the author of two books: *New Technologies for Reclamation of Industrial Wastewater* and *Recent Advances in Distillery Waste Management for Environmental Safety*. He has published more than 70 research publications, including research/review papers and two book reviews in national and international peer-reviewed journals of high impact factor, published by Springer, Elsevier, the Royal Society of Chemistry (RSC), Taylor & Francis Group and Frontiers. He has also published many book chapters and magazine articles on the biodegradation and bioremediation of industrial pollutants. He has presented many posters/papers at national and international conferences. He has also served as a reviewer for various national and international journals. He is a life member of the Association of Microbiologists of India (AMI), the Indian Science Congress Association (ISCA) Kolkata, India, and the Biotech Research Society, India (BRSI).

Dr. Vineet Kumar is presently working as an Assistant Professor in the Department of Basic and Applied Sciences, School of Engineering and Sciences at GD Goenka University, Gurugram, Haryana, India. Prior to joining GD Goenka University, Dr. Kumar served the various reputed institutions in India like CSIR-National Environmental Engineering Research Institute (NEERI), Maharashtra, India; Guru Ghasidas Vishwavidyalaya (A Central University), Bilaspur, India; Jawaharlal Nehru University, Delhi; Dr. Shakuntala Misra National Rehabilitation University Lucknow, India, etc. He received his MSc and MPhil. degree in Microbiology from Ch. Charan Singh University, Meerut, India. Subsequently, he gained his PhD (2018) in Environmental Microbiology from Babasaheb Bhimrao Ambedkar University, Lucknow, India. Dr. Kumar's research interests include: bioremediation, phytoremediation, metagenomics, wastewater treatment, environmental monitoring, waste management; bioenergy, and biofuel production. Currently, his research mainly focuses on the development of integrated and sustainable treatment techniques that can help in minimizing or eliminating hazardous waste in the environment. He has published 20 research articles in reputed international journals, written four proceeding papers, 38 book chapters, and authored/edited over 14 books on the different aspects of phytoremediation, bioremediation, wastewater treatment, omics, genomics, and metagenomics, published by CRC Press, Elsevier Science, Springer Nature, Wiley, and the Royal Society of Chemistry. His recently published book is *Recent Advances in Distillery Waste Management for Environmental Safety* (CRC Press). He has presented several papers at national and international conferences. He is an active member of numerous scientific societies, including the Microbiology Society, UK, and the Indian Science Congress Association, India. Dr. Kumar has been a guest editor and reviewer for many prestigious international journals, including *Frontiers in Microbiology, Environmental Research, Chemosphere, Journal of Basic Microbiology*; *International Journal of Environmental Science* and *Technology*; *CLEAN-Soil, Air, Water*, etc. He is the founder of the Society for Green Environment, India.

Sunil Kumar is a well-rounded researcher with more than 20 years of experience in leading, supervising and undertaking research in the broad field of environmental engineering and science with a focus on solid and hazardous waste management. His primary area of expertise is solid waste management (municipal solid waste, electronic waste, biomedical waste, etc.) in a wide range of environmental topics, including contaminated sites, EIA, and wastewater treatment. He has contributed extensively to these fields and has a citation of 8825, the h-index of 44, and the i10-index of 168 (Google Scholar). His contributions since the inception of CSIR-National Environmental Engineering Research Institute (NEERI), India, in 2000 include 300 refereed journal publications, 5 books and 40 book chapters, 10 edited volumes and numerous project reports to various governmental and private, local and international academic/research bodies. He is the Associate Editor of peer-reviewed journals of international repute, e.g. *Environmental Chemistry Letters, International Journal of Environmental Science & Technology, ASCE, Journal of Hazardous, Toxic and Radioactive Waste*. He also serves on the editorial board of *Bioresource Technology*. He has completed more than 22 research projects as PI with 17 (7 awarded) PhD and 17 MPhil/MTech thesis/dissertations under his supervision. The list of his collaborations is long and includes key Indian universities, such as IIT Kharagpur, IIT Delhi, and IIT Mumbai and prestigious regional institutes, such as the Asian Institute of Technology (AIT) and Kasetsart University in Bangkok, Hong Kong Baptist University, as well as universities in the US (Columbia, Texas A&M), the University of Calgary, Canada, and Europe (UN University Dresden, and the University of Uppsala, Sweden). He has contributed immensely to the advancement of environmental engineering/science fields in India both regionally and internationally as an expert committee member for the revision of Solid Waste Management Rules, as an NGT member for Solid Waste Rules, organizing workshops/conferences and delivering invited papers at both Indian and international venues. Dr. Kumar has achieved recognition and was awarded the prize of Outstanding Scientist in 2011 and 2016 at CSIR-NEERI for Scientific Excellence in the field of Research & Development in Solid Waste Management. Dr. Kumar also won a most prestigious award from the Alexander von Humboldt-Stiftung, Bonn, Germany, as a Senior Researcher for Developing a Global Network and Excellence for More Advanced Research and Technology Innovation.

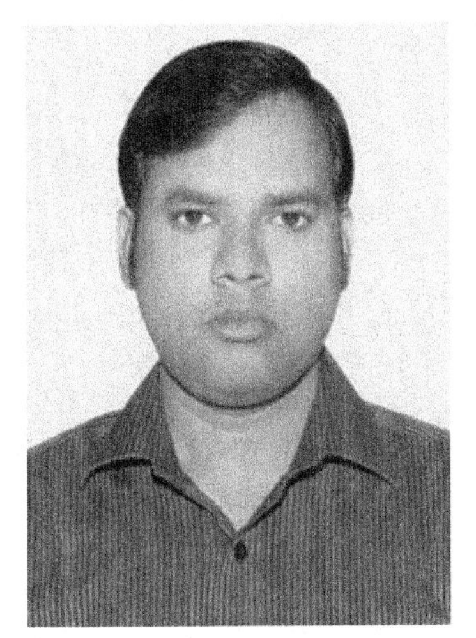

Vishvas Hare is currently working as Vice-President of the Society for Green Environment (SGE) in New Delhi. He gained his PhD (2018) in Microbiology from Babasaheb Bhimrao Ambedkar University (A Central University), Lucknow, Uttar Pradesh, India. He did his post-graduation degree in Medical Microbiology at Choudhary Charan Singh University Meerut. His PhD topic was "Studies on the Arsenic Affected Paddy Grown Area of Uttar Pradesh and its Remediation Approach". The main focus of his research is the remediation of arsenic from paddy and other crop fields. He has published many research and review articles in peer-reviewed journals of high impact. He has also published many international book chapters and a book review on the biodegradation and bioremediation of industrial pollutants. He has qualified in the ICAR NET in Microbiology and GATE in Life Science. He is also a life member of the Academy of the Association of Microbiologists of India.

Contributors

Mohammed Salim Akhter, Department of Chemistry, College of Science, University of Bahrain, Bahrain.

Nisar Ali, Key Laboratory of Regional Resource Exploitation and Medicinal Research, Faculty of Chemical Engineering, Huaiyin Institute of Technology, Huaian, Jiangsu Province, China.

R. Anand, Department of Electrical and Electronics Engineering. Nehru Institute of Engineering and Technology, Coimbatore, Tamil Nadu, India.

J. Anandkumar, National Institute of Technology, Raipur, Chattishgarh, India.

Chinaza Godswill Awuchi, School of Natural and Applied Sciences, Kampala International University, Kampala, Uganda.

D. Balaji, Department of Mechanical Engineering, KPR Institute of Engineering and Technology, Coimbatore, Tamil Nadu, India.

Ankeet Bhagat, Department of Zoology, Guru Nanak Dev University, Amritsar, Punjab, India.

Siddhant Bhardwaj, Government Degree College (Boys), Kathua, J&K, India.

Sartaj Ahmad Bhat, River Basin Research Center, Gifu University, Yanagido, Gifu, Japan.

Irshad Bibi, Institute of Soil and Environmental Sciences, University of Agriculture Faisalabad, Faisalabad, Pakistan.

Muhammad Bilal, School of Life Science and Food Engineering, Huaiyin Institute of Technology, Huaian, China.

Vajramma Boggala, Division of Plant Pathology, Indian Agricultural Research Institute, New Delhi, India.

Cristina Carapeto, Department of Science and Technology. Universidade Aberta, Portugal.

Preethy Chandran, School of Environmental Studies, Cochin University of Science and Technology, Kalamassery, Kochi, Kerala, India.

Tanushree Chatterjee, Raipur Institute of Technology, Raipur, Chattishgarh, India.

Jia May Chin, School of Science, Monash University Malaysia, Jalan Lagoon Selatan, Selangor, Malaysia.

Jyotsna Choubey, Raipur Institute of Technology, Raipur, Chattishgarh, India.

Jyoti K. Choudhari, Chhattisgarh Swami Vivekan and Technical University (CSVTU), Bhilai, Chattishgarh, India; Raipur Institute of Technology, Raipur, Chattishgarh, India.

Deepa, School of Engineering and Sciences, Department of Basic and Applied Sciences, GD Goenka University, Gurugram, Haryana, India.

Omnya Desokey, Environmental Sciences Department, Faculty of Science, Alexandria University, Alexandria, Egypt.

Raunak Dhanker, School of Engineering and Sciences, Department of Basic and Applied Sciences, GD Goenka University, Gurugram, Haryana, India.

S.A. Dwivedi, Department of Entomology, School of Agriculture Lovely Professional University, Punjab, India.

Suneel Kumar Goyal, Department of Farm Engineering, Institute of Agricultural Sciences, Banaras Hindu University, Varanasi, Uttar Pradesh, India.

K.T.N. Greeshma, Department of Civil Engineering, NITK Surathkal, Karnataka, India.

Kasahun Gudeta, Shoolini University Biotechnology and Management Sciences, School of Biological and Environmental Sciences, Solan, Himachal Pradesh, India; Department of Biology, Adama Science and Technology University, Adama, Ethiopia.

Muhammad Mahroz Hussain, Institute of Soil and Environmental Sciences, University of Agriculture Faisalabad, Faisalabad, Pakistan.

Hafiz M.N. Iqbal, School of Engineering and Sciences, Tecnológico de Monterrey, Monterrey, Mexico.

Pravin Khaire, Department of Plant Pathology and Agricultural Microbiology, PGI, MPKV, Rahuri, India.

Adnan Khan, Institute of Chemical Sciences, University of Peshawar, Khyber Pakhtunkhwa, Pakistan.

Vineet Kumar, Waste Re-processing Division, CSIR-National Environmental Engineering Research Institute (CSIR-NEERI), Nehru Marg, Nagpur, Maharashtra, India.

Alaa El Din Mahmoud, Environmental Sciences Department, Faculty of Science, Alexandria University, Alexandria, Egypt; Green Technology Group, Faculty of Science, Alexandria University, Alexandria, Egypt.

Sumeet Malik, Institute of Chemical Sciences, University of Peshawar, Khyber Pakhtunkhwa, Pakistan.

Akshay Mamidi, Department of Genetics and Plant Breeding, PJTSAU, Rajendranagar T.S., India.

Farid Menaa, Departments of Internal Medicine, Nanomedicine, and Advanced Technologies, Fluorotronics-CIC, La Jolla, San Diego, CA, USA.

Sonia Morya, Department of Food Technology and Nutrition, School of Agriculture, Lovely Professional University, Punjab, India.

M. Muthukumaran, PG and Research Department of Botany, Ramakrishna Mission Vivekananda College, Chennai, Tamil Nadu, India.

Lelika Nameirakpam, Department of Entomology, School of Agriculture Lovely Professional University Punjab, India.

Tanaji Narute, Department of Plant Pathology and Agricultural Microbiology, PGI, MPKV, Rahuri, India.

Natasha, Department of Environmental Sciences, COMSATS University Islamabad, Vehari Campus, Vehari, Pakistan.

Nabeel Khan Niazi, Institute of Soil and Environmental Sciences, University of Agriculture Faisalabad, Faisalabad, Pakistan.

A.K. Priya, Department of Civil Engineering, KPR Institute of Engineering and Technology, Coimbatore, Tamil Nadu, India.

Komal Rizwan, Department of Chemistry, University of Sahiwal, Sahiwal, Pakistan.

Biju P. Sahariah, Chhattisgarh Swami Vivekanand Technical University (CSVTU), Bhilai, Chattishgarh, India.

Joseph Kadanthottu Sebastian, Department of Life Sciences, CHRIST, Bangalore, Karnataka, India.

Muhammad Shahid, Department of Environmental Sciences, COMSATS University Islamabad, Vehari Campus, Vehari, Pakistan.

Aryaman Singh, NALSAR University of Law, Hyderabad, Telangana, India.

Leena Singh, Environmental Sciences, Kirori Mal College, University of Delhi, Delhi, India.

Nishi Singh, National Dairy Research Institute, Karnal, India.

Sumit Singh, Department of Zoology, Guru Nanak Dev University, Amritsar, Punjab, India.

Sneha Suresh, School of Environmental Studies, Cochin University of Science and Technology, Kalamassery, Kochi, Kerala, India.

Arun Kumar Thalla, Department of Civil Engineering, NITK Surathkal, Karnataka, India.

J. Thivya, Department of Civil Engineering, University College of Engineering Dindigul, Tamil Nadu, India.

Adeline Su Yien Ting, School of Science, Monash University Malaysia, Jalan Lagoon Selatan, Selangor, Malaysia.

Mukesh K. Verma, Chhattisgarh Swami Vivekanand Technical University (CSVTU), Bhilai, Chattishgarh, India; National Institute of Technology, Raipur, Chattishgarh, India.

J. Vijayaraghavan, Department of Civil Engineering, University College of Engineering Ramanathapuram, Tamil Nadu, India.

Yong Yang, Key Laboratory of Regional Resource Exploitation and Medicinal Research, Faculty of Chemical Engineering, Huaiyin Institute of Technology, Huaian, Jiangsu Province, China.

Fazila Younas, Institute of Soil and Environmental Sciences, University of Agriculture Faisalabad, Faisalabad, Pakistan,

Part I

Pollutants in the Environment

1 Emerging Environmental Pollutants

Current and Future Challenges

Fazila Younas,[1] Natasha,[2] Irshad Bibi,[1] Muhammad Mahroz Hussain,[1] Muhammad Shahid,[2] and Nabeel Khan Niazi[1,]*
[1]Institute of Soil and Environmental Sciences, University of Agriculture Faisalabad, Faisalabad, Pakistan
[2]Department of Environmental Sciences, COMSATS University Islamabad, Vehari, Pakistan
*Corresponding author Email: nabeelkniazi@gmail.com; nabeel.niazi@uaf.edu.pk

CONTENTS

1.1 Introduction 3
1.2 Characteristics, Behavior and Fate of Emerging Pollutants 4
1.3 The Presence of Emerging Pollutants in the Aquatic Environment 7
1.4 Transport and Bioaccumulation of Emerging Pollutants 12
1.5 Environmental and Health Risks of Emerging Pollutants 12
1.6 Challenges and Limitations 14
1.7 Conclusion 15
References 15

1.1 INTRODUCTION

Emerging pollutants are chemicals and compounds that have recently been identified as harmful to the environment and human health. In addition, many of these emerging pollutants are not subject to any national or international legislation and therefore pose a greater risk to life. These include various compounds such as industrial additives, steroids, antibiotics, drugs, endocrine disruptors, hormones, chemicals, as well as microplastics (Kumar et al. 2006; Chandra and Kumar 2017a, 2017b; Gavrilescu et al. 2015; Thomaidis et al. 2012). It has been reported that the worldwide demand for synthetic chemicals surged from 1 million tons to 400 million tons per year between 1930 and 2000. According to statistics given by EUROSTAT in 2013, environmentally hazardous substances accounted for more than 5% of overall chemical output between 2002 and 2011. More than 70% of these compounds have major environmental and human health impacts and because there is no legislation, their use is continuously increasing (Gavrilescu et al. 2015; Mishra et al. 2021).

The European Aquatic Environment Agency has classified over 700 EPs compounds, their metabolites, and transformation products (Geissen et al. 2015). In the environment, EPs originate from various sources that may eventually affect groundwater and surface water (Figure 1.1).

These sources can be divided into two types: (1) point sources, i.e., primarily urban, hospital, and industrial sources; and (2) diffuse sources of contamination. Important examples of both sources include industrial wastewater, municipal sewage treatment plants, disinfection by-products, buried

DOI: 10.1201/9781003239956-2

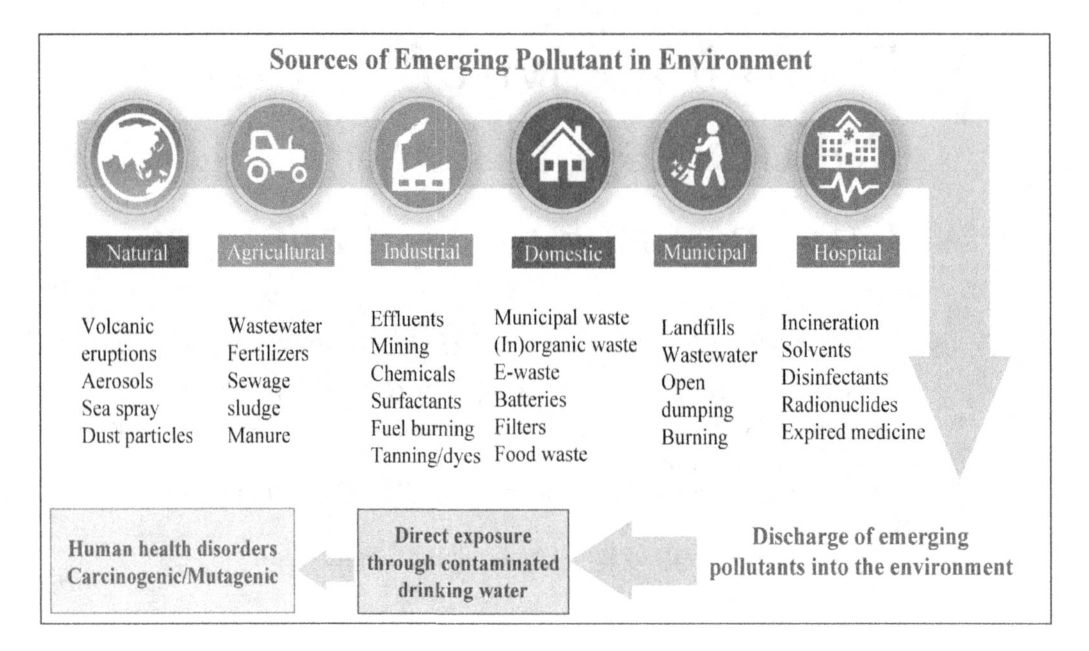

FIGURE 1.1 Possible sources of emerging pollutants (EPs) in the environment.

septic tanks, resource extraction, waste disposal sites, agriculture runoff from manure and biosolids, storm water, and urban runoff and leakage from the reticulated urban sewerage system (Table 1.1).

Many studies have been conducted because there are vast numbers of EPs, establishing priority lists, studying consumption, predicting their environmental concentrations as well as pharmacological, ecotoxicological, and physicochemical data. However, their impact on the environment, especially the health impacts and the ecotoxicological effects related to their presence, is still unknown. Also, the extent of their removal from water and wastewater treatment plants is still an unknown parameter. The inefficient removal of EPs from contaminated water poses a serious environmental issue. Because of the potential effect of these compounds on aquatic life and human health, as well as a lack of understanding about their behavior in the environment and a lack of analytical and sampling tools, action at several levels is lacking but urgently needed.

The aim of this chapter is to propose a concept that demonstrates the need for EPs monitoring systems, assesses their future fate and risk assessment tools, and policy requirements as a foundation for long-term water resource management. It also focuses on novel and sophisticated techniques and technology to monitor, avoid and reduce environmental and health concerns, as well as the future difficulties of lowering EPs' influence on the environment.

1.2 CHARACTERISTICS, BEHAVIOR AND FATE OF EMERGING POLLUTANTS

Point (primarily urban, hospital and industrial) or diffuse (agricultural) pollution may create emerging contaminants. Emerging pollutants migrate from sources to sinks (water bodies) which are regulated by physicochemical features, such as sorption, volatility, persistence, polarity, half-life, and the capacity to interact with other environmental components (Farré et al. 2010; Geissen et al. 2010). The EPs from municipal or industrial wastewater treatment facilities are released directly into rivers, endangering the aquatic and terrestrial ecosystems. These pollutants may undergo a different transformation that may result in the release of toxic intermediate metabolites and secondary products, thus contaminating the environment (Xie et al. 2013; Kumar et al. 2020; Kumar, Kaushal

TABLE 1.1
Environmental Contaminants

Contaminant	Description	References
Pesticides	Any substance used to kill, repel, or control certain forms of plant or animal life that are considered to be pests. These are substances or mixtures of substances that are mainly used in agriculture or in public health protection programs. Pesticides can contaminate soil, water, turf, and other vegetation. In addition to killing insects or weeds, pesticides can be toxic to a host of other organisms including birds, fish, beneficial insects, and non-target plants. Pesticides can cause short-term adverse health effects, called acute effects, as well as chronic adverse effects that can occur months or years after exposure. Examples of acute health effects include stinging eyes, rashes, blisters, blindness, nausea, dizziness, diarrhea, and death. Removal of pesticides is possible through various ways. Removal of pesticides is possible through various physical, chemical, and biological methods.	(Giri et al. 2021; Hu et al. 2015; Saleh et al. 2020; Syafrudin et al. 2021)
Pharmaceuticals	Biologically active compounds designed to interact with specific physiological pathways of targeted organisms Extensive use in human and veterinary medicine. Includes compounds such as antibiotics, analgesics, antidepressants, anti-diabetics, among others. Continuous entry into the aquatic environment, low concentrations, impacting water supply in ecosystems and human health. Majority are excreted without metabolizing after application, reaching the sewage. Cause risk to non-targeted organisms and act as additional stressors on ecosystems There is no complete remediation of pharmaceuticals with water treatment technologies	(Branchet et al. 2021; Sadutto et al. 2021; Vargas-Berrones et al. 2020)
Personal care products (PCPs)	Product applied or used on the human body for personal hygiene, personal grooming, or for beautification Released into the environment through domestic and commercial sources, including landfills of urban solids, disposal of household wastes, animals, effluent of water and sewage treatment plants, drug manufacturing processes, pharmaceutical companies PPCPs are frequently inserted into bodies of humans and animals and are subjected to metabolic transformations. However, they do not undergo the complete metabolism, and are excreted into the environment. Hundreds of chemicals, including phthalates, formaldehyde and lead have been discovered in cosmetic and personal care products Some of the chemicals in personal care and cosmetic products have been linked to health problems, including cancer and infertility.	(Keerthanan et al. 2021; Nohynek et al. 2010; Panico et al. 2019; Tasho and Cho 2016)
Endocrine disrupters (EDCs)	Compounds or mixtures of compounds that interfere with hormone action EDCs contribute to numerous disease aetiologias including infertility, neurodevelopmental problems, other reproductive disorders, and hormone-sensitive cancers like breast and prostate cancers	(Gao et al. 2020; Ho et al. 2021)

(*continued*)

TABLE 1.1 (Continued)
Environmental Contaminants

Contaminant	Description	References
	Cause developmental malformations, interference with reproduction, increased cancer risk; and disturbances in the immune and nervous system function Can be removed to some extent through photocatalysis and biodegradation	
Polycyclic aromatic hydrocarbons (PAHs)	Class of chemicals that occur naturally in coal, crude oil, and gasoline PAHs are formed during domestic and industrial food processing like roasting, toasting, drying, grilling, frying, baking, and barbecuing. Major sources of PAHs include residential heating, coal gasification and liquefying plants, carbon black, coal-tar pitch and asphalt production, coke and aluminum production, catalytic cracking towers and related activities in petroleum refineries as well as motor vehicle exhaust. Exposure to PAHs causes eye irritation and breathing passages irritation. Blood and liver abnormalities can develop from skin contact with the liquid form and from breathing in PAHs. Several PAHs are considered to be cancer-causing chemicals. Removal of PAHs is possible mainly through sorption process	(Chen and Liao 2006; Grmasha et al. 2020; Lamichhane et al. 2016)
Food additives	Substances that are added to food to maintain or improve the safety, freshness, taste, texture, or appearance of food Additives are not necessarily bad. Consuming small amounts of additives may be safe, but the health risks add up if you rely heavily on processed foods. A diet rich in processed foods is linked to allergic reactions to food additives are usually mild, skin irritation, intestinal upset, some breathing problems. Furthermore, it can cause chronic diseases such as obesity, high blood pressure, heart disease, and cancer. Biodegradation of food additives is possible.	(Gatidou et al. 2020; WHO 2018; Wu et al. 2021)
Fire retardants	Chemicals that are applied to materials to prevent the start or slow the growth of fire. Flame retardants of concern include organo-halogen and organophosphate chemicals such as polybrominated diphenyl ethers (PBDEs) and chlorinated tris (TDCPP). Some of the flame retardants remain persistent in the environment for years. They can also bioaccumulate, or build up in people and animals over time. Exposure to flame retardant can cause thyroid disruption, memory and learning problems, delayed mental and physical development, lower IQ, advanced puberty, and reduced fertility Removal of flame retardants is possible through various techniques	(Brown and Cordner 2011; Gross 2013; He et al. 2020; Rahman et al. 2001; Zhang and Zhang 2012)

et al. 2021; Kumar, Shahi et al. 2021; Kumar, Singh et al. 2021; Kumar and Chandra 2020). These EPs enter the environment through air transfusion, runoff, erosion, or leaching that ultimately spread the EPs across rural regions and represent potential human and environmental health problems, but scientific research and understanding of their prevalence in water resources and wastewater, as well as their routes and accumulation in the environment have still not been fully investigated (UNESCO

2021). Environmental, water quality, and wastewater discharge standards do not apply to most EPs. As a result, there is an urgent need for studies to enhance scientific understanding and implement appropriate analytical, legislative, and mitigation plans to overcome EPs contamination (ibid.). If the quantity of these chemicals in the aquatic environment exceeds critical levels, the quality of wastewater effluents will suffer dramatically, and the reuse of treated wastewater will be restricted. As a result, their persistence in the environment and half-life should also be explored along with their harmful effects on the aquatic environment and human health. Their pathways are unclear and mitigation strategies that should be explored to devise a plan of mitigation and legislative strategies to prevent EPs contamination in aquatic and terrestrial environments are lacking.

1.3 THE PRESENCE OF EMERGING POLLUTANTS IN THE AQUATIC ENVIRONMENT

Several microbiological and chemical agents that are not typically classified as contaminants may be found in various environmental areas, mainly because of their persistence in long-distance transportation. Waste and sewage produced by industrial, agricultural, or municipal operations are increasing, thus correlating them with the origin's EPs and paths of entry into the food chain (Ngo et al. 2020). Chemical micropollutants are often formed when organic molecules decompose, resulting in the build-up of persistent metabolites, or when items such as pharmaceuticals are discharged into the environment (Srensen et al. 2007). Biological contaminants in drinking water production and distribution may include changes in people's health. Surface and groundwater leaching, as well as health issues, may result from a change in agricultural techniques to intensify agriculture and the use of manure or sludge on farms (Goel 2006; Houtman 2010).

Pesticides may still be found in surface and groundwater (Table 1.2), even though some have been phased out and replaced with more eco-friendly alternatives (Gavrilescu 2005; Stuart et al. 2012).

The residues of pesticides that are biologically active and harmful have reportedly been found in soil and water (Clausen et al. 2007; Lapworth and Gooddy 2006). Other environmental contaminants with distinct chemical structures and characteristics are harmful to the human endogenous hormone system. Endocrine disruptors (EDCs) are contaminants that are poorly recorded and monitored, and there is limited knowledge regarding their incidence and environmental effects. A detailed literature review was carried out in order to assess the content of various EDCs in water systems that are classified as EPs (Table 1.3).

Hospital/medical wastes are also a major source of concern due to their nature. The presence of many pharmaceutical medications in the water system (Table 1.4) may result in different environmental issues by contaminating water bodies, hence impacting human health. Other goods, such as flame retardants, commonly used industrial chemicals (bisphenol A), and UV filters, may have an influence on human health, in addition to the EPs described above (Gavrilescu 2009; Preda et al. 2012). As a result, the existence of EPs and the intermediates of their degradation are the focus of substantial investigation (Preda et al. 2012), but there is limited available data on their toxicology symptoms.

Despite the fact that studies and overviews of EP sources, occurrence, environmental behavior, and fate have been published (Deblonde et al. 2011; Lapworth and Gooddy 2006; Stuart et al. 2012), the route of these pollutants from sources to receptors remains an important topic for further study. This is owing to a lack of knowledge, mostly due to the complexity of the target compound's physical and chemical characteristics, as well as the complexity of the environmental variables, such as EPs' unanticipated behavior in the air, water, or soil (Gavrilescu et al. 2015). The EP route, particularly from people and animals to environmental components, has been extensively investigated (Gavrilescu et al. 2015; Pal et al. 2010).

TABLE 1.2
Concentration of Pesticide Concentration Surface/Groundwater

Pesticide name	Water sampling area	Water source	Pesticide concentration	Reference
Propiconazole	Tanjung Karang, Selangor, Malaysia	Surface water	4493.1 ng/L	(Elfikrie et al. 2020)
Pymetrozine	Tanjung Karang, Selangor, Malaysia	Surface water	1.3 ng/L	(Elfikrie et al. 2020)
Acetamiprid	Dutch market, the Netherlands	Surface water	1.1 µg/L	(Sjerps et al. 2019)
Thiamethoxam	Dutch market, the Netherlands	Groundwater	0.4 µg/L	(Sjerps et al. 2019)
Propazine	Gaza	Groundwater	1.5 µg/L	(Shomar et al. 2006)
Atrazine,	Gaza	Groundwater	3.5 µg/L	(Shomar et al. 2006)
Atrazine-desisopropyl	Gaza	Groundwater	1.2 µg/L	(Shomar et al. 2006)
Simazine	Gaza	Groundwater	2.3 µg/L	(Shomar et al. 2006)
Glyphosate	California	Perennial or seasonal ponds	1.1 µg/L	(Smalling et al. 2012)
HCB	Rosetta and Damiatta branches of the Nile River	Surface water	0.195–0.240µg/L	(Abbassy et al. 1999)
Lindane	Rosetta and Damiatta branches of the Nile River	Surface water	0.286–0.352 µg/L	(Abbassy et al. 1999)
DDE	Rosetta and Damiatta branches of the Nile River	Surface water	0.035–0.067µg/L	(Abbassy et al. 1999)
DDD	Rosetta and Damiatta branches of the Nile River	Surface water	0.019–0.033µg/L	(Abbassy et al. 1999)
DDT	Rosetta and Damiatta branches of the Nile River	Surface water	0.024–0.031µg/L	(Abbassy et al. 1999)
Aroclor 1254	Rosetta and Damiatta branches of the Nile River	Surface water	0.390–0.70µg/L	(Abbassy et al. 1999)
Aroclor 1260	Rosetta and Damiatta branches of the Nile River	Surface water	0.166–0.330 µg/L	(Abbassy et al. 1999)
Dichlorvos	China	Surface water	17.8 ng/L	(Gao et al. 2009)
DDE	South Litani region in South Lebanon	Surface and groundwater	100 ng/L	(Youssef et al. 2015)
Demeton,	China	Surface water	35.4 ng/L	Gao et al. 2009)
pirimiphos-methyl	South Litani region in South Lebanon	Surface and groundwater	300.87 ng/L	(Youssef et al. 2015)
Metalaxyl	Beijing	Groundwater	89.58 ng/L	(Zhang et al. 2021)
Nicosulfuron	Washington	Surface water	3.0 µg/L	(Battaglin et al. 2009)
Glyphosate	Riley Spring, Washington	Surface water	328 µg/L	(Battaglin et al. 2009)
Carbendazim	Wujin District northwest of Taihu Lake, China	Surface water	508 ng/L	(Zhou et al. 2020)
Imidacloprid	China	Surface water	438 ng/L	(Zhou et al. 2020)
Fluometuron	Lake Vistonis Basin, Greece	Surface water	0.088 µg/L	(Papadakis et al. 2015)
Lambdacyhalothrin	Lake Vistonis Basin, Greece	Surface water	0.041 µg/L	(Papadakis et al. 2015)
Alphamethrin	Lake Vistonis Basin, Greece	Surface water	0.168 µg/L	(Papadakis et al. 2015)
Fluometuron	Lake Vistonis Basin, Greece	Surface water	317.6 µg/L	(Papadakis et al. 2015)
Dichlorvos	Shangyu, Zhejiang province, China	Surface water	0.01–5.63 µg/L	(Chen et al. 2016)
DDT	China	Surface water	14.6 ng/L	(Gao et al. 2008)
Heptachlor epoxide	China	Surface water	10 ng/L	(Gao et al. 2008)

TABLE 1.2 (Continued)
Concentration of Pesticide Concentration Surface/Groundwater

Pesticide name	Water sampling area	Water source	Pesticide concentration	Reference
Tetradifon	Massa river estuary, Morocco	Surface water	160,404ng/L	(Agnaou et al. 2018)
Bupirimate	Massa river estuary, Morocco	Surface water	62,644 ng/L,	Agnaou et al. 2018)
Dichlofluanide	Massa river estuary, Morocco	Surface water	56,219 ng/L	(Agnaou et al. 2018)
Lindane	Massa river estuary, Morocco	Surface water	46,134 ng/L.	(Agnaou et al. 2018)
Aldrin	Lake in the Sembrong Lake Basin in Malaysia	Surface water	23ng/L,	(Sharip et al. 2017)
δ-BHC	Lake in the Sembrong Lake Basin in Malaysia	Surface water	43.2 ng/L	(Sharip et al. 2017)
Heptachlor	Lake in the Sembrong Lake Basin in Malaysia	Surface water	50.4 ng/L	(Sharip et al. 2017)
Chlorpyrifos	Urbanized Linggi River, Negeri Sembilan, Malaysia	Surface water	0.0275 µg/L	(Zainuddin et al. 2020)
Diazinon	Urbanized Linggi River, Negeri Sembilan, Malaysia	Surface water	0.0328 µg/L	Zainuddin et al. 2020)
Quinalphos	Urbanized Linggi River, Negeri Sembilan, Malaysia	Surface water	0.0362 µg/L	(Zainuddin et al. 2020)
Diazinon EPTC malathion	Salton Sea Basin, California	Surface water	940– 3,830ng/L	(LeBlanc and Kuivila 2008)
Chlorpyrifos	Langat River, Selangor, Malaysia	Surface water	0.0202 µg/L	(Wee et al. 2016)
Quinalphos	Langat River, Selangor, Malaysia	Surface water	0.0178 µg/L	(Wee et al. 2016)
Diazinon	Langat River, Selangor, Malaysia	Surface water	0.0094 µg/L	(Wee et al. 2016)

TABLE 1.3
Occurrence of Different Endocrine Disrupters (EDCs) in Water

Names	Areas	Sources	Concentrations	References
Perchlorate	Kamkaria River, India	Surface water	0.97µg/L	(Kumar 2020)
Perchlorate	Kamkaria River, India	Surface water	1.03µg/L	(Kumar 2020)
Perchlorate	Sabarmati River, India	Surface water	65µg/L	(Kumar 2020)
Perchlorate	Sabarmati River, India	Surface water	1.14µg/L	(Kumar 2020)
Chlorophenols	Southern Finland	Drinking water	70 to 140µg/L	(Lampi et al. 1990)
Chlorophenols	Southern Finland	Groundwater	56,000 to 190,000µg/L	(Lampi et al. 1990)
Polychlorinated dibenzo-p-dioxins and furans	Taiwan	Fresh water	0.001 to 0.265pg/L	(Thuan et al. 2011)
Dioxins	Russia	Town's rinking water	28.4–74.1pg/L	(Revich et al. 2001)
Atrazine	Ontario, Canada	Surface water	<0.1 to 3.91µg/L	(Revich et al. 2001)
Metolachlor	Ontario, Canada	Surface water	<0.1 to 1.83µg/L	(Revich et al. 2001)
Atrazine	China and Hong Kong	Tap water	21.0ng/L	(Wang et al. 2020)

(continued)

TABLE 1.3 (Continued)
Occurrence of Different Endocrine Disrupters (EDCs) in Water

Names	Areas	Sources	Concentrations	References
Atrazine	Vojvodina Province	Groundwater	0.198μg/L	(Pucarević et al. 2002)
Deethylatrazine	Vojvodina Province	Groundwater	0.116μg/L	(Pucarević et al. 2002)
Deisopropylatrazine	Vojvodina Province	Groundwater	0.043μg/L	(Pucarević et al. 2002)
Deethyldeisopropylatrazine	Vojvodina Province	Groundwater	0.077μg/L	(Pucarević et al. 2002)
Bisphenol A	Austria	Groundwater	0.013 to 0.021μg/L	(Brueller et al. 2018)
Bisphenol A	Germany	Surface water	0.0005 to 0.41μg/L	(Fromme et al. 2002)
Bisphenol A	Germany	Sewage water	0.018 to 0.702μg/L	(Fromme et al. 2002)
Di2-ethylexyl phthalate, DEHP	Germany	Surface water	0.33 to 97.8μg/L	(Fromme et al. 2002)
Di2-ethylexyl phthalate, DEHP	Germany	Sewage water	1.74 to 182μg/L	(Fromme et al. 2002)
Polychlorinated biphenyls, PCBs	Bay of Bengal coast, Bangladesh	Coastal surface water	32.17 to 160.7ng/L	(Habibullah-Al-Mamun et al. 2019)
Polychlorinated biphenyls, PCBs	Bay of Bengal coast, Bangladesh	Coastal surface water	46.45 to 199.4ng/L	(Habibullah-Al-Mamun et al. 2019)
Polychlorinated biphenyls, PCBs	China	Surface water	0.2 to 985.2ng/L	(Han and Currell 2017)
PAHs	China	Surface water	11.84 to 393.12ng/L	(Li et al. 2017)
PAHs	China	Groundwater	8.57 to 402.84ng/L	(Li et al. 2017)
Polychlorinated biphenyls, PCBs	River Chenab, Pakistan	Surface water	7.7 to110ng/L	(Eqani et al. 2012)
Polychlorinated biphenyls, PCBs	River Chenab, Pakistan	Surface water	13 to 99ng/L	(Eqani et al. 2012)
Perchlorate	India	Drinking water	0.1μg/L	(Kannan et al. 2009)
Perchlorate	India	Groundwater	1.0μg/L	(Kannan et al. 2009)
Perchlorate	India	Bottled water	<0.02μg/L	(Kannan et al. 2009)
Perchlorate	India	Surface water	0.05μg/L	(Kannan et al. 2009)
Perchlorate	India	Rain water	<0.02μg/L	(Kannan et al. 2009)
Perchlorate	Colorado River, California	Surface water	non-detect to 9μg/L	(Tikkanen 2006)
Perchlorate ions	France	Groundwater	0.7 to 33μg/L	(Cao et al. 2018)
Perchlorate ions	France	Surface water	<0.5 to 10.2μg/L	(Cao et al. 2018)
Perchlorate	Great Lakes Basin	Surface water	0.2μg/L	
Perchlorate	Upper Tone River and its Tributary, India	Drinking water	93.19μg/L	(Backus et al. 2005)
Chlorate	Usui River, India	Drinking water	3692.07μg/L	(Sijimol et al. 2016)
Perchlorate	Tianjin, China	Rain water	0.35 to 27.3ng/L	(Qin et al. 2014)
Perchlorate	China	Bottled drinking water	0.037 to 2.013μg/L	(Shi et al. 2007)

TABLE 1.4
Pharmaceutical Drugs' Concentration in Water in Different Areas Around the Globe

Drug name	Area	Water type	Concentration	References
Ibuprofen	Pakistan	Wastewater	10700 ng/L	(Vergeynst et al. 2015)
Benzoylecgonine	Brazil	Drinking water	10–1019 ng/L	(Campestrini and Jardim 2017)
Norfloxacin	Brazil	Urban water	8–18 ng/L	(Torres et al. 2017)
Acetaminophen	United States	Groundwater	1890 ng/L	(Deo and Halden 2013)
Caffeine	United States	Groundwater	290 ng/L	(Deo and Halden 2013)
Carbamazepine	United States	Groundwater	420 ng/L	(Deo and Halden 2013)
Ciprofloxacin	United States	Groundwater	45 ng/L	(Deo and Halden 2013)
Codeine	United States	Groundwater	214 ng/L	(Deo and Halden 2013)
Ibuprofen	South Africa	Drinking water	19.2 µg/L	(Madikizela and Chimuka 2017)
Atorvastatin	Italy	Surface water	0.015 µg/L	(Ferrari et al. 2011)
Ranitidine	Italy	Surface water	0.005 µg/L	(Ferrari et al. 2011)
Propanolol	Italy	Surface water	0.001 µg/L	(Ferrari et al. 2011)
Carbamazepine	Italy	Surface water	0.012 µg/L	(Ferrari et al. 2011)
Metronidazole	Italy	Surface water	0.005 µg/L	(Ferrari et al. 2011)
Paracetamol	Italy	Surface water	0.174 µg/L	(Ferrari et al. 2011)
Furosemide	Italy	Surface water	0.013 µg/L	(Ferrari et al. 2011)
Nimesulide	Italy	Surface water	0.008 µg/L	(Ferrari et al. 2011)
Bezafibrate	Brazil	Wastewater	94.4 ng /L	(Queiroz et al. 2012))
Diclofenac	Brazil	Wastewater	99.9 ng /L	(Queiroz et al. 2012
Sulfamethoxazole	Brazil	Wastewater	13.0 ng /L	(Queiroz et al. 2012)
Trimethoprim	Brazil	Wastewater	61.5 ng /L	(Queiroz et al. 2012)
Ibuprofen	Pakistan	Wastewater	1673 µg/L	(Ashfaq et al. 2017)
Diclofenac	Pakistan	Wastewater	836 µg/L	(Ashfaq et al. 2017)
Naproxen	Pakistan	Wastewater	464 µg/L	(Ashfaq et al. 2017)
Ibuprofen	Belgium	Surface water	84.6 µg/L	(Matongo et al. 2015)
Lorazepam	Spain	Watershed	562 ng/L	(Esteban et al. 2012)
Venlafaxine	Spain	Watershed	44 ng/L	(Esteban et al. 2012)
Carbamazepine	Argentina	Wastewater	0.2–2.3µg/L	(Elorriaga et al. 2013)
Caffeine	Argentina	Wastewater	0.9–44.2 µg/L	(Elorriaga et al. 2013)
Ibuprofen	Argentina	Wastewater	0.4–13 µg/L	(Elorriaga et al. 2013)
Atenolol	Argentina	Wastewater	0.2–1.7 µg/L	(Elorriaga et al. 2013)
Diclofenac	Argentina	Wastewater	0.03–1.2 µg/L	(Elorriaga et al. 2013)
Ibuprofen	Europe	Groundwater	395 ng/L	(Loos et al. 2010)
Clotrimazole	England	River Tyne	20 ng/L	(Roberts and Thomas 2006)
Propranolol	England	River Tyne	61 ng/L	(Roberts and Thomas 2006)
Dextropropoxyphene	England	River Tyne	8 ng/L	(Roberts and Thomas 2006)
Tamoxifen	England	River Tyne	212 ng/L	(Roberts and Thomas 2006)
Ibuprofen	England	River Tyne	352 ng/L	(Roberts and Thomas 2006)
Erythromycin	England	River Tyne	4 ng/L	(Roberts and Thomas 2006)
Trimethoprim	England	River Tyne	9 ng/L	(Roberts and Thomas 2006)
Aspirin	South Africa	Msunduzi River	25345 ng/L	(Agunbiade and Moodley 2016)
Ibuprofen	South Africa	Msunduzi River	689 ng/L	(Agunbiade and Moodley 2016)
Diclofenac	South Africa	Msunduzi River	8174 ng/L	(Agunbiade and Moodley 2016)
Cocaine	Brazil	Drinking water	6–62 ng/L	(Campestrini and Jardim 2017)

1.4 TRANSPORT AND BIOACCUMULATION OF EMERGING POLLUTANTS

Emerging pollutants may migrate to water bodies through a variety of direct and indirect pathways from point to non-point sources, as well as bioaccumulate via food webs, posing a threat to human health. According to Dsikowitzky et al. (2018) the Ciliwung River (the biggest city-bound river in Jakarta) yearly transfers roughly 5–17 tons of the assessed pollutants (including EPs) into the Java Sea, based on a mass flux method assessment. In the air system, plastic polymers operate as passive samplers, concentrating hydrophobic organic pollutants by sorption or interactions, which may then be transmitted to other systems, such as the marine environment.

Littoral plastics trash, which may be thought of as passive samplers, has recently attracted a lot of attention. The EPs are taken up via adsorbing or reacting with microplastics, then transported from the littoral to the ocean, altering the marine plankton habitat (León et al. 2018). In 2018, there were several reports of emerging contaminants in wildlife. In Hong Kong, for example, marine bivalves collected near point source urban wastewater and landfill leachate effluent discharges included numerous kinds of pharmaceuticals, insecticides, and phosphorus-based flame retardants (Burket et al. 2018). They discovered high levels of target pollutants. The results highlight the need to pay closer attention to the possible dangers of ingesting aquatic products (Fu et al. 2018). The appearance of diverse EPs in aquatic animals implies that they may pose ecological hazards and, as a result, have an impact on human health through the food webs. Few studies have been carried out on the processes of transformation and EP metabolites in organisms (Koppel et al. 2017; Tang et al. 2019). In future, their impact on the environment (ecotoxicological and health risks) should be monitored.

1.5 ENVIRONMENTAL AND HEALTH RISKS OF EMERGING POLLUTANTS

The fundamental purpose of risk assessment is to preserve not only the biological community in the aquatic environment, but also the health of the people who are exposed to water directly or indirectly (for example, consumption of drinking water, dietary intake of irrigated food, aquatic animal consumption). The EPs' concentration in nanograms per liter (ng/L) has negative impacts on human and aquatic ecology, such as hormonal abnormalities, genetic toxicity, carcinogenicity, endocrine disorders, and immunotoxicity (Miraji et al. 2016). Most EPs' long-term consequences on the environment and human health are still uncertain and/or difficult to predict, and there is a lack of knowledge of their behavior and associated hazards/environmental dangers (Figure 1.2).

Existing research on the dangers of EPs in the environment, particularly in water, has considered aquatic flora and fauna (fish, algae, water fleas), as well as human health concerns, particularly in water recovery and reuse systems. However, due to the low concentrations of these contaminants in the liquid stream, gathering data on toxicity characteristics and human exposure is a major research undertaking (Lin et al. 2020; Ma et al. 2018). The US Environmental Protection Agency (USEPA) and the World Health Organization (WHO) have issued recommendations in this regard, which include methodologies for assessing the hazards of EPs to human health and determining exposure to different environmental variables (Lin et al. 2020). Protocols and procedures for studying chemical dangers were created in the European Union in 1980, and medications were examined in the early 1990s (Vasilachi et al. 2021). One of the most significant breakthroughs in the permitted registration and assessment of chemicals is the Registration, Evaluation, Authorisation and Restriction of Chemicals (REACH) legislation (ibid.).

The sorts of contaminants in wastewater and surface water were the subject of research on ecological and human health issues (Hang et al. 2016). Furthermore, several investigations have identified the build-up of EPs in sediments, which are often thought of as EP sinks (Panahandeh et al. 2018). Furthermore, if the sediment-forming particles travel and get suspended in a moving liquid stream (for example, during a flood), the sediment might become a source of EP exposure to aquatic plants and wildlife. Benthic organisms may take advantage of EPs linked with solid particles. If the

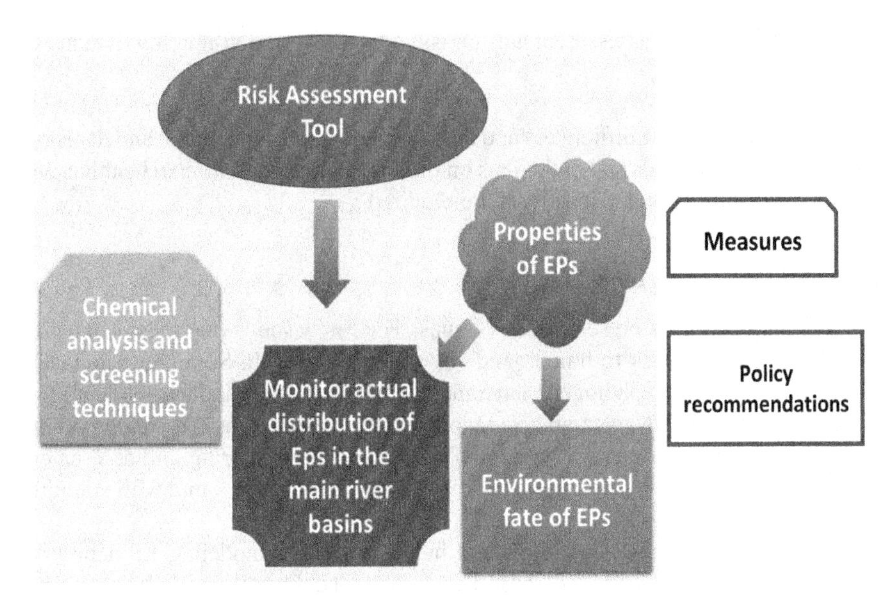

FIGURE 1.2 Management concept of EPs in the environment.

Source: Modified from Nassar and Younis (2019).

bioaccumulation is high enough, it will result in acute and chronic exposure, as well as the spread of increased nutrient levels (Dsikowitzky et al. 2020). Emerging pollutants may be converted into more harmful compounds in river water, which are adsorbed on sediments and deposited or carried to neighboring water bodies containing sensitive or protected flora and fauna, or seep down into the groundwater (Al Aukidy et al. 2012).

New studies are required to detect and estimate the acute and chronic doses of EPs in water to evaluate the danger of EPs from various sources (Naidu et al. 2016; Sanchez and Egea 2018). The ratio between the predicted environmental concentration (PEC) and the predicted no effect concentration (PNEC) of a given individual drug is now used in Environmental Risk Assessment (ERA) (Riva et al. 2019). The development of physiologically based pharmacokinetic (PBPK) models based on the physiological biotransformation of EP-affected entities has recently allowed the description of EP absorption, distribution, metabolism, and excretion (Geissen et al. 2015). However, most of the time, determining the route from the source to the receptor is challenging. Considering the interaction between various EPs and the environment, as well as the hazards associated with them, particularly for novel drugs, is problematic since this route is influenced by a variety of circumstances connected to the substance itself, including the source of EP exposure (danger), the surrounding environment, and the treatment options.

Furthermore, assessing and analyzing the hazards posed by EPs mixes require new methodologies to represent reality as accurately as possible, because if just one component or each chemical in the mixture is studied independently, the mixture's toxicity may be affected (Rivera-Jaimes et al. 2018). In the literature, two types of problems have been solved in this regard (Vasilachi et al. 2021):

- *evaluating the toxicity of the mixture*, where the results are only applicable to the mixture and cannot be extrapolated to other exposure scenarios;
- *evaluating the components of the mixture*, where the results can be interpreted in two ways: first, by considering the accumulative toxicity of the components or, second, by considering the independent effects of each component that can be extrapolated to other exposure scenarios,

thus comprehensive risk assessment aids in risk reduction decision-making (Naidu et al. 2016; Dsikowitzky et al. 2020).

Furthermore, to overcome these difficulties and meet the demands of the public and decision-makers regarding the effect of chemicals and pathogens on the environment and human health, scientifically proven and novel techniques and instruments are required.

1.6 CHALLENGES AND LIMITATIONS

Monitoring and analyzing EPs present unique issues. The first issue is that these contaminants are seldom monitored, posing a risk to human and environmental health. Some new impacts, such as antibiotic resistance in aquatic pathogens have made EPs monitoring and assessment increasingly more critical in recent years (Nawaz and Sengupta 2019). However, as technology advances and new compounds are used to remove or regulate pollutants, the monitoring and assessment of EPs have progressed as well. Some of the general obstacles and possibilities found while monitoring and analyzing EPs are listed below.

It is critical to organize knowledge and build helpful tools for modeling EP activity in water, soil, and air mediums. The selection of adsorbents and flocculants, membrane selection, dosing processes, system set-up, mixing conditions, and other optimization tactics are required to build a safe monitoring and removal mechanism that can manage the load (Maaz et al. 2019; Qu et al. 2013). The cost-effectiveness and identification of pollutant-specific treatments are hampered by a lack of comparison between various mechanisms to remove EPs and between conventional and emergency techniques. The peculiarities of certain locales may also play a role in defining the limits of the treatment techniques of EPS (Krauss et al. 2019). The necessary study should also take into account the constraints of a given area, as well as the regulatory requirements for the safe disposal of EPs.

It's also crucial to comprehend EP conversion pathways and products. EP monitoring operations have different regional and temporal dimensions depending on their properties and transport mediums, such as organisms, water, or air (McGuire and McDonnell 2006). Modeling tools for transportation process verification, calibration, and comprehension are important and valuable, however on a broader scale, such techniques are constrained by computation and data needs (Aalizadeh et al. 2019).

Implementing the United Nations Sustainable Development Goals (such as access to safe drinking water) is a difficult endeavor since it needs a wide range of policy and scientific assistance at all levels of strategy development (Aravindakumar et al. 2019). It does, however, offer an opportunity for nations to perform comparable appraisals and have a shared purpose. Modeling contaminant movement across biological and non-biological systems may assist in reducing these restrictions. To maximize their accumulation and toxicity, scientists must model the water flow via plant and animal systems. The synergistic impact of testing EPs, and the initial phase analysis of their life cycle, will help to avoid future threats to human and ecological health before any conversion products are generated.

Except for pesticides, the modeling framework for the fate of EPs in the soil-water ecosystem is underdeveloped and has to be improved. The aim is to track the movement and final destination of different EPs in the watershed from all potential sources (urban, industrial, agricultural). EPs are carried from the diffusion source to the sink (water body) through the soil-water systems (McGuire and McDonnell 2006).

The use of PBPK models has exploded in recent years. Fàbrega et al. (2014) employed the PBPK models to represent the biological distribution (intake, distribution, metabolism, and excretion) of chemical compounds in animals and humans. On the other hand, increased attempts have been made to evaluate the potential negative health impacts of pollutant combinations rather than single substances (Nadal et al. 2006). As an initial stage in analyzing chemical aggregation, substance comparisons are occasionally carried out using various priority methods (von der Ohe et al. 2011;

Fàbrega et al. 2013), which are often based on ranking and scoring systems. To estimate the dangers of metabolites and transformation products in comparison to their parent substances, prioritization approaches are required. As a result, they are critical in calculating the health and environmental dangers of chemical combinations. They will serve as the foundation for the development of novel methodologies to assess the cumulative risk of numerous stressors, such as chemical and physical/biological combinations.

1.7 CONCLUSION

The European Aquatic Environment Agency has classified over 700 EPs compounds, their metabolites, and transformation products. More than 70% of these compounds have a major environmental and human health impact and, because there is no governing legislation, their use is increasing. Only a small proportion of these EPs, as well as their influence on the environment and human health, have been thoroughly investigated. The extensive use of a broad range of steroids, antibiotics, drugs, endocrine disruptors, hormones, chemicals, as well as pharmaceuticals and personal care items, has resulted in increased levels of contamination in aquatic and terrestrial environment. As a result, it is determined that different treatment methods for the removal of EPs from environment must be improved as soon as possible. Because of the potential effect of these compounds on aquatic life and human health, as well as a lack of understanding about their behavior in the environment and a lack of analytical and sampling tools, action at several levels is urgently required. To estimate the toxicity of EPs and their transformation products in comparison to their parent substances, prioritization approaches are required. As a result, they will be critical in calculating the health and environmental effects of mixtures of EPs. They will serve as the foundation for the development of novel methodologies for assessing the cumulative risk of numerous stressors, such as chemical and physical/biological combinations.

REFERENCES

Aalizadeh, R., Nika, M.-C., and Thomaidis, N.S. (2019). Development and application of retention time prediction models in the suspect and non-target screening of emerging contaminants. *Journal of Hazardous Materials* 363, 277–285.

Abbassy, M.S., Ibrahim, H.Z., and El-Amayem, M.M.A. (1999). Occurrence of pesticides and polychlorinated biphenyls in water of the Nile river at the estuaries of Rosetta and Damiatta branches, north of delta, Egypt. *Journal of Environmental Science and Health, Part B* 34, 255–267.

Agnaou, M., Nadir, M., Alla, A.A., Bazzi, L., El Alami, Z., and Moukrim, A.J.E. (2018). The occurrence and spatial distribution of pesticides in sea water of the Agadir bay (South of Morocco). *Journal of Materials and Environmental Sciences* 13, 14.

Agunbiade, F.O. and Moodley, B. (2016). Occurrence and distribution pattern of acidic pharmaceuticals in surface water, wastewater, and sediment of the Msunduzi River, Kwazulu-Natal, South Africa. *Environmental Toxicology and Chemistry* 35, 36–46.

Al Aukidy, M., Verlicchi, P., Jelic, A., Petrovic, M., and Barcelò, D. (2012). Monitoring release of pharmaceutical compounds: Occurrence and environmental risk assessment of two WWTP effluents and their receiving bodies in the Po Valley, Italy. *Science of the Total Environment* 438, 15–25.

Aravindakumar, C.T., Aravind, U.K., and Kallenborn, R. (2019). Foreword to the Special Issue on the International Conference on Water: From pollution to purification. *Environmental Chemistry* 16, 561–562.

Ashfaq, M., Khan, K.N., Rehman, M.S.U., Mustafa, G., Nazar, M.F., Sun, Q., Iqbal, J., Mulla, S.I., and Yu, C.-P. (2017). Ecological risk assessment of pharmaceuticals in the receiving environment of pharmaceutical wastewater in Pakistan. *Ecotoxicology and Environmental Safety* 136, 31–39.

Backus, S., Klawuun, P., Brown, S., D'sa, I., Sharp, S., Surette, C., and Williams, D. (2005). Determination of perchlorate in selected surface waters in the Great Lakes Basin by HPLC/MS/MS. *Chemosphere* 61, 834–843.

Battaglin, W.A., Rice, K.C., Focazio, M.J., Salmons, S., and Barry, R.X. (2009). The occurrence of glyphosate, atrazine, and other pesticides in vernal pools and adjacent streams in Washington, DC, Maryland, Iowa, and Wyoming, 2005–2006. *Environmental Monitoring and Assessment* 155, 281–307.

Branchet, P., Arpin-Pont, L., Piram, A., Boissery, P., Wong-Wah-Chung, P., and Doumenq, P. (2021). Pharmaceuticals in the marine environment: What are the present challenges in their monitoring? *Science of the Total Environment* 766, 142644.

Brown, P. and Cordner, A. (2011). Lessons learned from flame retardant use and regulation could enhance future control of potentially hazardous chemicals. *Health Affairs* 30, 906–914.

Brueller, W., Inreiter, N., Boegl, T., Rubasch, M., Saner, S., Humer, F., Moche, W., ... Brezinka, C. (2018). Occurrence of chemicals with known or suspected endocrine disrupting activity in drinking water, groundwater and surface water, Austria 2017/2018. *Die Bodenkultur: Journal of Land Management, Food and Environment* 69, 155–173.

Burket, S.R., Sapozhnikova, Y., Zheng, J., Chung, S.S., and Brooks, B.W. (2018). At the intersection of urbanization, water, and food security: Determination of select contaminants of emerging concern in mussels and oysters from Hong Kong. *Journal of Agricultural and Food Chemistry* 66, 5009–5017.

Campestrini, I. and Jardim, W.F. (2017). Occurrence of cocaine and benzoylecgonine in drinking and source water in the São Paulo State region, Brazil. *Science of the Total Environment* 576, 374–380.

Cao, F., Jaunat, J., Ollivier, P., Cancès, B., Morvan, X., Hubé, D., Devos, A., ... Pannet, P. (2018). Sources and behavior of perchlorate ions (ClO_4^-) in chalk aquifer of Champagne-Ardenne, France: Preliminary results. *Proceedings of the International Association of Hydrological Sciences* 379, 113–117.

Chandra, R. and Kumar, V. (2017a). Detection of *Bacillus* and *Stenotrophomonas* species growing in an organic acid and endocrine-disrupting chemicals rich environment of distillery spent wash and its phytotoxicity. *Environmental Monitoring and Assessment* 189, 26. https://doi.org/10.1007/s10661-016-5746-9.

Chandra, R. and Kumar, V. (2017b). Detection of androgenic-mutagenic compounds and potential autochthonous bacterial communities during in-situ bioremediation of post methanated distillery sludge. *Frontiers in Microbiology* 8, 87. https://doi.org/10.3389/fmicb.2017.00887.

Chen, H., Zhu, J., Li, Z., Chen, A., and Zhang, Q. (2016). The occurrence and risk assessment of five organophosphorus pesticides in river water from Shangyu, China. *Environmental Monitoring and Assessment* 188, 614.

Chen, S.-C. and Liao, C.-M. (2006). Health risk assessment on humans exposed to environmental polycyclic aromatic hydrocarbons pollution sources. *Science of the Total Environment* 366, 112–123.

Clausen, L., Arildskov, N.P., Larsen, F., Aamand, J., and Albrechtsen, H.-J. (2007). Degradation of the herbicide dichlobenil and its metabolite BAM in soils and subsurface sediments. *Journal of Contaminant Hydrology* 89, 157–173.

Deblonde, T., Cossu-Leguille, C., and Hartemann, P. (2011). Emerging pollutants in wastewater: A review of the literature. *International Journal of Hygiene and Environmental Health* 214, 442–448.

Deo, R.P. and Halden, R.U. (2013). Pharmaceuticals in the built and natural water environment of the United States. *Water* 5, 1346–1365.

Dsikowitzky, L., Crawford, S.E., Nordhaus, I., Lindner, F., Irianto, H.E., Ariyani, F., and Schwarzbauer, J. (2020). Analysis and environmental risk assessment of priority and emerging organic pollutants in sediments from the tropical coastal megacity Jakarta, Indonesia. *Regional Studies in Marine Science* 34, 101021.

Dsikowitzky, L., Van der Wulp, S., Ariyani, F., Hesse, K., Damar, A., and Schwarzbauer, J. (2018). Transport of pollution from the megacity Jakarta into the ocean: Insights from organic pollutant mass fluxes along the Ciliwung River. *Estuarine, Coastal and Shelf Science* 215, 219–228.

Elfikrie, N., Ho, Y.B., Zaidon, S.Z., Juahir, H., and Tan, E.S.S. (2020). Occurrence of pesticides in surface water, pesticides' removal efficiency in drinking water treatment plant and potential health risk to consumers in Tengi River Basin, Malaysia. *Science of the Total Environment* 712, 136540.

Elorriaga, Y., Marino, D.J., Carriquiriborde, P., and Ronco, A.E. (2013). Human pharmaceuticals in wastewaters from urbanized areas of Argentina. *Bulletin of Environmental Contamination and Toxicology* 90, 397–400.

Eqani, S.A-M.A.S., Malik, R.N., Katsoyiannis, A., Zhang, G., Chakraborty, P., Mohammad, A., and Jones, K.C. (2012). Distribution and risk assessment of organochlorine contaminants in surface water from River Chenab, Pakistan. *Journal of Environmental Monitoring* 14, 1645–1654.

Esteban, S., Valcárcel, Y., Catalá, M., and Castromil, M.G. (2012). Psychoactive pharmaceutical residues in the watersheds of Galicia (Spain). *Gaceta Sanitaria* 26, 457–459.

Fàbrega, F., Kumar, V., Schuhmacher, M., Domingo, J.L., and Nadal, M. (2014). PBPK modeling for PFOS and PFOA: Validation with human experimental data. *Toxicology Letters* 230, 244–251.

Fàbrega, F., Marquès, M., Ginebreda, A., Kuzmanovic, M., Barceló, D., Schuhmacher, M., Domingo, J.L., and Nadal, M. (2013). Integrated risk index of chemical aquatic pollution (IRICAP): Case studies in Iberian rivers. *Journal of Hazardous Materials* 263, 187–196.

Farré, M., Pérez, S., Gajda-Schrantz, K., Osorio, V., Kantiani, L., Ginebreda, A., and Barceló, D. (2010). First determination of C60 and C70 fullerenes and N-methylfulleropyrrolidine C60 on the suspended material of wastewater effluents by liquid chromatography hybrid quadrupole linear ion trap tandem mass spectrometry. *Journal of Hydrology* 383, 44–51.

Ferrari, F., Gallipoli, A., Balderacchi, M., Ulaszewska, M.M., Capri, E., and Trevisan, M. (2011). Exposure of the main Italian river basin to pharmaceuticals. *Journal of Toxicology* epub 19 September 2011.

Fromme, H., Küchler, T., Otto, T., Pilz, K., Müller, J., and Wenzel, A. (2002). Occurrence of phthalates and bisphenol A and F in the environment. *Water Research* 36, 1429–1438.

Fu, L., Lu, X., Tan, J., Wang, L., and Chen, J. (2018). Multiresidue determination and potential risks of emerging pesticides in aquatic products from Northeast China by LC–MS/MS. *Journal of Environmental Sciences* 63, 116–125.

Gao, J., Liu, L., Liu, X., Lu, J., Zhou, H., Huang, S., Wang, Z., and Spear, P.A. (2008). Occurrence and distribution of organochlorine pesticides—lindane, p,p′-DDT, and heptachlor epoxide—in surface water of China. *Environment International* 34, 1097–1103.

Gao, J., Liu, L., Liu, X., Zhou, H., Lu, J., Huang, S., and Wang, Z. (2009). The occurrence and spatial distribution of organophosphorous pesticides in Chinese surface water. *Bulletin of Environmental Contamination and Toxicology* 82, 223–229.

Gao, X., Kang, S., Xiong, R., and Chen, M. (2020). Environment-friendly removal methods for endocrine disrupting chemicals. *Sustainability* 12, 7615.

Gatidou, G., Vazaiou, N., Thomaidis, N.S., and Stasinakis, A.S. (2020). Biodegradability assessment of food additives using OECD 301F respirometric test. *Chemosphere* 241, 125071.

Gavrilescu, M. (2005). Fate of pesticides in the environment and its bioremediation. *Engineering in Life Sciences* 5, 497–526.

Gavrilescu, M. (2009). Behaviour of persistent pollutants and risks associated with their presence in the environment—integrated studies. *Environmental Engineering and Management Journal* 8, 1517–1531.

Gavrilescu, M., Demnerová, K., Aamand, J., Agathos, S., and Fava, F. (2015). Emerging pollutants in the environment: Present and future challenges in biomonitoring, ecological risks and bioremediation. *New Biotechnology* 32, 147–156.

Geissen, V., Mol, H., Klumpp, E., Umlauf, G., Nadal, M., Van der Ploeg, M., Van de Zee, S.E., and Ritsema, C.J. (2015). Emerging pollutants in the environment: A challenge for water resource management. *International Soil and Water Conservation Research* 3, 57–65.

Geissen, V., Ramos, F.Q., Bastidas-Bastidas, P.d.J., Díaz-González, G., Bello-Mendoza, R., Huerta-Lwanga, E., and Ruiz-Suárez, L.E. (2010). Soil and water pollution in a banana production region in tropical Mexico. *Bulletin of Environmental Contamination and Toxicology* 85, 407–413.

Giri, B.S., Geed, S., Vikrant, K., Lee, S.S., Kim, K.-H., Kailasa, S.K., Vithanage, M., ... Singh, R.S. (2021). Progress in bioremediation of pesticide residues in the environment. *Environmental Engineering Research* 26, 77–100.

Goel, P. (2006). *Water Pollution: Causes, Effects and Control*. New York: New Age International.

Grmasha, R.A., Al-Sareji, O.J., Salman, J.M., and Hashim, K.S. (2020). Polycyclic aromatic hydrocarbons (PAHs) in urban street dust within three land-uses of Babylon governorate, Iraq: Distribution, sources, and health risk assessment. *Journal of King Saud University—Engineering Sciences*. DOI:10.1016/j.jksues.2020.11/002.

Gross, L. (2013). Flame retardants in consumer products are linked to health and cognitive problems. Washington Post, April 15.

Habibullah-Al-Mamun, M., Ahmed, M.K., Islam, M.S., Tokumura, M., and Masunaga, S. (2019). Occurrence, distribution and possible sources of polychlorinated biphenyls (PCBs) in the surface water from the Bay of Bengal coast of Bangladesh. *Ecotoxicology and Environmental Safety* 167, 450–458.

Han, D. and Currell, M.J. (2017). Persistent organic pollutants in China's surface water systems. *Science of the Total Environment* 580, 602–625.

Hang, C., Zhang, B., Gong, T., and Xian, Q. (2016). Occurrence and health risk assessment of halogenated disinfection byproducts in indoor swimming pool water. *Science of the Total Environment* 543, 425–431.

He, W., Song, P., Yu, B., Fang, Z., and Wang, H. (2020). Flame retardant polymeric nanocomposites through the combination of nanomaterials and conventional flame retardants. *Progress in Materials Science* 114, 100687.

Ho, V., Pelland-St-Pierre, L., Gravel, S., Bouchard, M., Verner, M.-A., and Labrèche, F. (2021). Endocrine disruptors: Challenges and future directions in epidemiologic research. *Environmental Research*, 111969.

Houtman, C.J. (2010). Emerging contaminants in surface waters and their relevance for the production of drinking water in Europe. *Journal of Integrative Environmental Sciences* 7, 271–295.

Hu, R., Huang, X., Huang, J., Li, Y., Zhang, C., Yin, Y., Chen, Z., ... Cui, F. (2015). Long- and short-term health effects of pesticide exposure: A cohort study from China. *PloS One* 10, e0128766.

Kannan, K., Praamsma, M.L., Oldi, J.F., Kunisue, T., and Sinha, R.K. (2009). Occurrence of perchlorate in drinking water, groundwater, surface water and human saliva from India. *Chemosphere* 76, 22–26.

Keerthanan, S., Jayasinghe, C., Biswas, J.K., and Vithanage, M. (2021). Pharmaceutical and Personal Care Products (PPCPs) in the environment: Plant uptake, translocation, bioaccumulation, and human health risks. *Critical Reviews in Environmental Science and Technology* 51, 1221–1258.

Koppel, N., Rekdal, V.M., and Balskus, E.P. (2017). Chemical transformation of xenobiotics by the human gut microbiota. *Science* 356 (6344).

Krauss, M., Hug, C., Bloch, R., Schulze, T., and Brack, W. (2019). Prioritising site-specific micropollutants in surface water from LC-HRMS non-target screening data using a rarity score. *Environmental Sciences Europe* 31, 1–12.

Kumar, M. (2020). Runoff from firework manufacturing as major perchlorate source in the surface waters around Diwali in Ahmedabad, India. *Journal of Environmental Management* 273, 111091.

Kumar, S., Mukherjee, S.N., Ghosh, S., and Ray, R. (2006). A study for evaluation of contaminant transport characteristics through fine-grained soil. *Water Environment Research* 78, 2261–2267. https://doi.org/10.2175/106143005X78645.

Kumar, V. and Chandra, R. (2020s. Bioremediation of melanoidins containing distillery waste for environmental safety. In Bharagava, R. and Saxena, G. (Eds.), *Bioremediation of Industrial Waste for Environmental Safety*. Singapore: Springer. https://doi.org/10.1007/978-981-13-3426-9_20

Kumar, V., Kaushal, A., Singh, K., and Shah, M.P. (2021). Phytoaugmentation technology for phytoremediation of environmental pollutants: Opportunities, challenges and future prospects. In Kumar, V., Saxena, G., and Shah, M.P. (Eds.), *Bioremediation for Environmental Sustainability: Approaches to Tackle Pollution for Cleaner and Greener Society.* Oxford: Elsevier, pp. 329–381. https://doi.org/10.1016/B978-0-12-820 318-7.00016-2

Kumar, V., Shahi, S.K., Ferreira, L.F.R., Bilal, M., Biswas, J.K., and Bulgariu, L. (2021). Detection and characterization of refractory organic and inorganic pollutants discharged in biomethanated distillery effluent and their phytotoxicity, cytotoxicity, and genotoxicity assessment using *Phaseolus aureus L.* and *Allium cepa L. Environmental Research* 201, 111551. https://doi.org/10.1016/j.envres.2021.111551.

Kumar, V., Singh, K., Shah, M.P., and Kumar, M. (2021). Phytocapping: An eco-sustainable green technology for cleaner environment. In Kumar, V., Saxena, G., and Shah, M.P. (Eds.), *Bioremediation for Environmental Sustainability: Approaches to Tackle Pollution for Cleaner and Greener Society.* Oxford: Elsevier, pp. 481–491. https://doi.org/10.1016/B978-0-12-820318-7.00022-8.

Kumar, V., Thakur, I.S., and Shah, M.P. (2020). Bioremediation approaches for pulp and paper industry wastewater treatment: Recent advances and challenges. In Shah, M.P. (Ed.), *Microbial Bioremediation & Biodegradation.* Singapore: Springer, pp. 1–48. https://doi.org/10.1007/978-981-15-1812-61.

Lamichhane, S., Bal Krishna, K.C., and Sarukkalige, R. (2016). Polycyclic aromatic hydrocarbons (PAHs) removal by sorption: A review. *Chemosphere* 148, 336–353.

Lampi, P., Vartiainen, T., Tuomisto, J., and Hesso, A. (1990). Population exposure to chlorophenols, dibenzo-p-dioxins and dibenzofurans after a prolonged ground water pollution by chlorophenols. *Chemosphere* 20, 625–634.

Lapworth, D. and Gooddy, D. (2006). Source and persistence of pesticides in a semi-confined chalk aquifer of southeast England. *Environmental Pollution* 144, 1031–1044.

LeBlanc, L.A. and Kuivila, K.M. (2008). Occurrence, distribution and transport of pesticides into the Salton Sea Basin, California, 2001–2002. In *The Salton Sea Centennial Symposium*. Cham: Springer, pp. 151–172.

León, V.M., García, I., González, E., Samper, R., Fernández-González, V., and Muniategui-Lorenzo, S. (2018). Potential transfer of organic pollutants from littoral plastics debris to the marine environment. *Environmental Pollution* 236, 442–453.

Li, J., Li, F., and Liu, Q. (2017). PAHs' behavior in surface water and groundwater of the Yellow River estuary: Evidence from isotopes and hydrochemistry. *Chemosphere* 178, 143–153.

Lin, X., Xu, J., Keller, A.A., He, L., Gu, Y., Zheng, W., Sun, D., ... Huang, X. (2020). Occurrence and risk assessment of emerging contaminants in a water reclamation and ecological reuse project. *Science of the Total Environment* 744, 140977.

Loos, R., Locoro, G., Comero, S., Contini, S., Schwesig, D., Werres, F., Balsaa, P., ... Blaha, L. (2010). Pan-European survey on the occurrence of selected polar organic persistent pollutants in ground water. *Water Research* 44, 4115–4126.

Ma, X.Y., Li, Q., Wang, X.C., Wang, Y., Wang, D., and Ngo, H.H. (2018). Micropollutants' removal and health risk reduction in a water reclamation and ecological reuse system. *Water Research* 138, 272–281.

Maaz, M., Yasin, M., Aslam, M., Kumar, G., Atabani, A., Idrees, M., Anjum, F., ... Khan, A.L. (2019). Anaerobic membrane bioreactors for wastewater treatment: Novel configurations, fouling control and energy considerations. *Bioresource Technology* 283, 358–372.

Madikizela, L.M. and Chimuka, L. (2017). Occurrence of naproxen, ibuprofen, and diclofenac residues in wastewater and river water of KwaZulu-Natal Province in South Africa. *Environmental Monitoring and Assessment* 189, 1–12.

Matongo, S., Birungi, G., Moodley, B., and Ndungu, P. (2015). Pharmaceutical residues in water and sediment of Msunduzi River, KwaZulu-Natal, South Africa. *Chemosphere* 134, 133–140.

McGuire, K.J. and McDonnell, J.J. (2006). A review and evaluation of catchment transit time modeling. *Journal of Hydrology* 330, 543–563.

Miraji, H., Othman, O.C., Ngassapa, F., and Mureithi, E. (2016). Research trends in emerging contaminants on the aquatic environments of Tanzania. *Scientifica* Epub 22 February 2016.

Mishra, R., Singh, E., Kumar, A., and Kumar, S. (2021). Application of remote sensing for assessment of change in vegetation cover and the subsequent impact on climatic variables. *Environmental Science and Pollution Research*, 1–13. https://doi.org/10.1007/s11356-021-13563-9.

Nadal, M., Kumar, V., Schuhmacher, M., and Domingo, J.L. (2006). Definition and GIS-based characterization of an integral risk index applied to a chemical/petrochemical area. *Chemosphere* 64, 1526–1535.

Naidu, R., Espana, V.A.A., Liu, Y., and Jit, J. (2016). Emerging contaminants in the environment: Risk-based analysis for better management. *Chemosphere* 154, 350–357.

Nassar, H.N. and Younis, S.A. (2019). From priority contaminants to emerged threat: Risk and occurrence-based analysis for better water management strategies in present and future. In Nassar, H.N. and Younis, S.A. (Eds.), *Nano and Bio-Based Technologies for Wastewater Treatement*. Wiley Online Library, pp. 41–103.

Nawaz, T. and Sengupta, S. (2019). Contaminants of emerging concern: Occurrence, fate, and remediation. In Ahuja, S. (Ed.), *Advances in Water Purification Techniques*. Oxford: Elsevier, pp. 67–114.

Ngo, H.H., Vo, H.N.P., Guo, W., Chen, Z., Liu, Y., and Varjani, S. (2020). Sustainable management and treatment technologies for micro-pollutants in wastewater. In Kataki, R. et al. (Eds.), *Current Developments in Biotechnology and Bioengineering*. Oxford: Elsevier, pp. 1–22.

Nohynek, G.J., Antignac, E., Re, T., and Toutain, H. (2010). Safety assessment of personal care products/cosmetics and their ingredients. *Toxicology and Applied Pharmacology* 243, 239–259.

Pal, A., Gin, K.Y.-H., Lin, A.Y.-C., and Reinhard, M. (2010). Impacts of emerging organic contaminants on freshwater resources: Review of recent occurrences, sources, fate and effects. *Science of the Total Environment* 408, 6062–6069.

Panahandeh, M., Mansouri, N., Khorasani, N., Karbassi, A., and Riazi, B. (2018). A study of pollution in sediments from Anzali wetland with Geo-accumulation Index and ecological risk assessment. *Environmental Engineering & Management Journal* 17, 2255–2262.

Panico, A., Serio, F., Bagordo, F., Grassi, T., Idolo, A., De Giorgi, M., Guido, M., Congedo, M., and De Donno, A. (2019). Skin safety and health prevention: An overview of chemicals in cosmetic products. *Journal of Preventive Medicine and Hygiene* 60, E50.

Papadakis, E.-N., Tsaboula, A., Kotopoulou, A., Kintzikoglou, K., Vryzas, Z., and Papadopoulou-Mourkidou, E. (2015). Pesticides in the surface waters of Lake Vistonis Basin, Greece: Occurrence and environmental risk assessment. *Science of the Total Environment* 536, 793–802.

Preda, C., Ungureanu, M.C., and Vulpoi, C. (2012). Endocrine disruptors in the environment and their impact on human health. *Environmental Engineering & Management Journal* 11, 1697–1706.

Pucarević, M., Šovljanski, R., Lazić, S., and Marjanović, N. (2002). Atrazine in groundwater of Vojvodina Province. *Water Research* 36, 5120–5126.

Qin, X., Zhang, T., Gan, Z., and Sun, H. (2014). Spatial distribution of perchlorate, iodide and thiocyanate in the aquatic environment of Tianjin, China: Environmental source analysis. *Chemosphere* 111, 201–208.

Qu, X., Alvarez, P.J., and Li, Q. (2013). Applications of nanotechnology in water and wastewater treatment. *Water Research* 47, 3931–3946.

Queiroz, F., Brandt, E., Aquino, S., Chernicharo, C., and Afonso, R. (2012). Occurrence of pharmaceuticals and endocrine disruptors in raw sewage and their behavior in UASB reactors operated at different hydraulic retention times. *Water Science and Technology* 66, 2562–2569.

Rahman, F., Langford, K.H., Scrimshaw, M.D., and Lester, J.N. (2001). Polybrominated diphenyl ether (PBDE) flame retardants. *Science of the Total Environment* 275, 1–17.

Revich, B., Aksel, E., Ushakova, T., Ivanova, I., Zhuchenko, N., Klyuev, N., Brodsky, B., and Sotskov, Y. (2001). Dioxin exposure and public health in Chapaevsk, Russia. *Chemosphere* 43, 951–966.

Riva, F., Zuccato, E., Davoli, E., Fattore, E., and Castiglioni, S. (2019). Risk assessment of a mixture of emerging contaminants in surface water in a highly urbanized area in Italy. *Journal of Hazardous Materials* 361, 103–110.

Rivera-Jaimes, J.A., Postigo, C., Melgoza-Alemán, R.M., Aceña, J., Barceló, D., and de Alda, M.L. (2018). Study of pharmaceuticals in surface and wastewater from Cuernavaca, Morelos, Mexico: Occurrence and environmental risk assessment. *Science of the Total Environment* 613, 1263–1274.

Roberts, P.H. and Thomas, K.V. (2006). The occurrence of selected pharmaceuticals in wastewater effluent and surface waters of the lower Tyne catchment. *Science of the Total Environment* 356, 143–153.

Sadutto, D., Andreu, V., Ilo, T., Akkanen, J., and Picó, Y. (2021). Pharmaceuticals and personal care products in a Mediterranean coastal wetland: Impact of anthropogenic and spatial factors and environmental risk assessment. *Environmental Pollution* 271, 116353.

Saleh, I.A., Zouari, N., and Al-Ghouti, M.A. (2020). Removal of pesticides from water and wastewater: Chemical, physical and biological treatment approaches. *Environmental Technology & Innovation* 19, 101026.

Sanchez, W. and Egea, E. (2018). *Health and Environmental Risks Associated with Emerging Pollutants and Novel Green Processes*. Berlin: Springer.

Sharip, Z., Hashim, N., and Suratman, S.J. (2017). Occurrence of organochlorine pesticides in a tropical lake basin. *Environmental Monitoring and Assessment* 189, 1–13.

Shi, Y., Zhang, P., Wang, Y., Shi, J., Cai, Y., Mou, S., and Jiang, G. (2007). Perchlorate in sewage sludge, rice, bottled water and milk collected from different areas in China. *Environment International* 33, 955–962.

Shomar, B.H., Müller, G., and Yahya, A. (2006). Occurrence of pesticides in groundwater and topsoil of the Gaza Strip. *Water, Air, & Soil Pollution* 171, 237–251.

Sijimol, M., Mohan, M., and Dineep, D. (2016). Perchlorate contamination in bottled and other drinking water sources of Kerala, southwest coast of India. *Energy, Ecology and Environment* 1, 148–156.

Sjerps, R.M.A., Kooij, P.J.F., van Loon, A., and Van Wezel, A.P. (2019). Occurrence of pesticides in Dutch drinking water sources. *Chemosphere* 235, 510–518.

Smalling, K.L., Orlando, J.L., Calhoun, D., Battaglin, W.A., and Kuivila, K.. (2012). Occurrence of pesticides in water and sediment collected from amphibian habitats located throughout the United States, 2009–2010. *San Francisco Estuary and Watershed Science* 5(1).

Sørensen, S.R., Holtze, M.S., Simonsen, A., and Aamand, J. (2007). Degradation and mineralization of nanomolar concentrations of the herbicide dichlobenil and its persistent metabolite 2, 6-dichlorobenzamide by *Aminobacter spp.* isolated from dichlobenil-treated soils. *Applied and Environmental Microbiology* 73, 399–406.

Stuart, M., Lapworth, D., Crane, E., and Hart, A. (2012). Review of risk from potential emerging contaminants in UK groundwater. *Science of the Total Environment* 416, 1–21.

Syafrudin, M., Kristanti, R.A., Yuniarto, A., Hadibarata, T., Rhee, J., Al-Onazi, W.A., Algarni, T.S., Almarri, A.H., and Al-Mohaimeed, A.M. (2021). Pesticides in drinking water: A review. *International Journal of Environmental Research and Public Health* 18, 468.

Tang, Y., Yin, M., Yang, W., Li, H., Zhong, Y., Mo, L., Liang, Y., Ma, X., and Sun, X. (2019). Emerging pollutants in water environment: Occurrence, monitoring, fate, and risk assessment. *Water Environment Research* 91, 984–991.

Tasho, R.P. and Cho, J.Y. (2016). Veterinary antibiotics in animal waste, its distribution in soil and uptake by plants: A review. *Science of the Total Environment* 563, 366–376.

Thomaidis, N.S., Asimakopoulos, A.G., and Bletsou, A. (2012). Emerging contaminants: A tutorial mini-review. *Global NEST Journal* 14, 72–79.

Thuan, N.T., Tsai, C.L., Weng, Y.M., Lee, T.Y., and Chang, M.B. (2011). Analysis of polychlorinated dibenzo-p-dioxins and furans in various aqueous samples in Taiwan. *Chemosphere* 83, 760–766.

Tikkanen, M.W. (2006). Development of a drinking water regulation for perchlorate in California. *Analytica Chimica Acta* 567, 20–25.

Torres, N.H., de Salles Pupo, M.M., Ferreira, L.F.R., Maranho, L.A., Américo-Pinheiro, J.H.P., Vilca, F.Z., de Hollanda, L.M., and Tornisielo, V.L. (2017). Spatial and seasonal analysis of antimicrobials and toxicity tests with *Daphnia magna*, on the sub-basin of Piracicaba river, SP, Brazil. *Journal of Environmental Chemical Engineering* 5, 6070–6076.

UNESCO (2021). Emerging pollutants in water and wastewater. United Nations Educational, Scientific and Cultural Organization. https://en.unesco.org/emergingpollutantsinwaterandwastewater.

Vargas-Berrones, K., Bernal-Jácome, L., Díaz de León-Martínez, L., and Flores-Ramírez, R. (2020). Emerging pollutants (EPs) in Latin America: A critical review of under-studied EPs, case of study—Nonylphenol. *Science of the Total Environment* 726, 138493.

Vasilachi, I.C., Asiminicesei, D.M., Fertu, D.I., and Gavrilescu, M. (2021). Occurrence and fate of emerging pollutants in water environment and options for their removal. *Water* 13, 181.

Vergeynst, L., Haeck, A., De Wispelaere, P., Van Langenhove, H., and Demeestere, K. (2015). Multi-residue analysis of pharmaceuticals in wastewater by liquid chromatography—magnetic sector mass spectrometry: Method quality assessment and application in a Belgian case study. *Chemosphere* 119, S2–S8.

von der Ohe, P.C., Dulio, V., Slobodnik, J., De Deckere, E., Kühne, R., Ebert, R.-U., Ginebreda, A., … Brack, W. (2011). A new risk assessment approach for the prioritization of 500 classical and emerging organic microcontaminants as potential river basin specific pollutants under the European Water Framework Directive. *Science of the Total Environment* 409, 2064–2077.

Wang, A., Hu, X., Wan, Y., Mahai, G., Jiang, Y., Huo, W., Zhao, X., … Xia, W. (2020). A nationwide study of the occurrence and distribution of atrazine and its degradates in tap water and groundwater in China: Assessment of human exposure potential. *Chemosphere* 252, 126533.

Wee, S.Y., Omar, T.F.T., Aris, A.Z., and Lee, Y. (2016). Surface water organophosphorus pesticides concentration and distribution in the Langat river, Selangor, Malaysia. *Exposure and Health* 8, 497–511.

WHO (2018). Food additives. Available at: www.who.int/news-room/fact-sheets/detail/food-additives.

Wu, L., Zhang, C., Long, Y., Chen, Q., Zhang, W., and Liu, G. (2021). Food additives: From functions to analytical methods. *Critical Reviews in Food Science and Nutrition*, 1–21. DOI:10.1080/10408398.2021.1929823.

Xie, Q., Chen, J., Zhao, H., Qiao, X., Cai, X., and Li, X. (2013). Different photolysis kinetics and photo-oxidation reactivities of neutral and anionic hydroxylated polybrominated diphenyl ethers. *Chemosphere* 90, 188–194.

Youssef, L., Younes, G., Kouzayha, A., and Jaber, F.J. (2015). Occurrence and levels of pesticides in South Lebanon water. *Chemical Speciation and Bioavailability* 27, 62–70.

Zainuddin, A.H., Wee, S.Y., and Aris, A.Z. (2020). Occurrence and potential risk of organophosphorus pesticides in urbanised Linggi River, Negeri Sembilan, Malaysia. *Environmental Geochemistry and Health* 42, 3703–3715.

Zhang, C.-C. and Zhang, F.-S. (2012). Removal of brominated flame retardant from electrical and electronic waste plastic by solvothermal technique. *Journal of Hazardous Materials* 221, 193–198.

Zhang, Y., Qin, P., Lu, S., Liu, X., Zhai, J., Xu, J., Wang, Y., … Wan, Z. (2021). Occurrence and risk evaluation of organophosphorus pesticides in typical water bodies of Beijing, China. *Environmental Science and Pollution* 28, 1454–1463.

Zhou, Y., Wu, J., Wang, B., Duan, L., Zhang, Y., Zhao, W., Wang, F., … Xu, D. (2020). Occurrence, source and ecotoxicological risk assessment of pesticides in surface water of Wujin District (northwest of Taihu Lake), China. *Environmental Pollution* 265, 114953.

2 Study on the Occurrence and Detection of Non-Steroidal Anti-Inflammatory Drugs in Wastewater Treatment Plants

*Arun Kumar Thalla** *and K.T.N. Greeshma*
Department of Civil Engineering, NITK Surathkal, Karnataka, India
*Corresponding author Email: arunkumar@nitk.edu.in

CONTENTS

2.1 Introduction 23
2.2 Detection and Analysis of NSAIDs 24
  2.2.1 Target Drugs and Their Mechanism of Action 24
  2.2.2 Sample Preparation Methods for Analyzing NSAIDs in Wastewater 25
  2.2.3 Sample Analysis Methods for NSAIDs in Wastewater 28
2.3 Occurrence of NSAIDS in WWTPs 29
  2.3.1 Occurrence of NSAIDs in WWTPs: the Indian context 29
    2.3.1.1 The South Zone 35
    2.3.1.2 The North Zone 36
    2.3.1.3 The Central Zone, the East Zone, the West Zone, and the North-east Zone 36
  2.3.2 Environmental Risk Assessment 37
2.4 Removal of NSAIDS in WWTPs and Other Advanced Technologies 37
  2.4.1 Conventional WWTPs and Treatment Technologies 37
  2.4.2 Constructed Wetlands 41
  2.4.3 Adsorption 41
  2.4.4 Advanced Oxidation Processes 42
2.5 Conclusion 43
References 43

2.1 INTRODUCTION

Water is one of the most valuable resources on Earth and it is essential to all known forms of life. Some 71% of the Earth's surface is covered by water, of which the ocean holds about 96.5%. But only 2.5% contributes fresh water. Massive amounts of wastewater are discharged as a result of increasing urbanization and industrialization, which significant reduced fresh water availability. The aim of wastewater treatment is to ensure that wastewater can be disposed of properly, without endangering public health or contaminating water sources. Since a wastewater treatment plant (WWTP) is such a complex and dynamic system, it must be properly operated and controlled in

DOI: 10.1201/9781003239956-3

order to protect human and environmental health. The performance of any WWTP is influenced by the quality of the raw and treated effluent. Sequential combinations of physical, chemical and biological units are used in WWTP to obtain the prescribed effluent standards set by regulatory agencies. Usually physical processes come under the primary treatment system. But it alone will not be sufficient to remove organic materials. Then comes the secondary treatment. Organic material is metabolized by bacteria in the secondary treatment. Based on the required effluent water quality, tertiary treatment units are also included in wastewater treatment plants.

Despite treatment, some contaminants persist in treated wastewater released into surface waterways. Pharmaceutical compounds are one of such emerging contaminants present in treated wastewater. In recent decades, there have been huge developments in the pharmaceutical field. These developments have improved the human quality of life. But the continuous use of pharmaceuticals and the disposal of expired medicines result in a great threat to the environment. Improper disposal of pharmaceuticals, human and animal excretions, pharmaceutical, industrial and hospital wastewater are the main channels of these drugs into the environment. These drugs will contaminate not only surface water but also groundwater. The majority of this contaminated water will reach the wastewater treatment plant. But the wastewater treatment plant is not specifically designed for the removal of pharmaceuticals, so these drugs are present in WWTP in higher concentrations. These pharmaceuticals are highly stable because of their chemical structure and the presence of active molecule groups. Their higher stability is one of the reasons for their presence in treated wastewater. And eventually these will find their way back to the human body.

The NSAIDs are one of such pharmaceutical groups present in treated wastewater since the usage of these types of drugs is higher because they are available without a prescription. Diclofenac (DIC), ibuprofen (IBU) and ketoprofen (KET) are the major ingredients in these drugs. Diclofenac is one of the most widely consumed NSAIDs so its detection in aquatic environments is also higher. It reduces pain and inflammation by blocking cyclooxygenase (COX) enzymes and prostaglandin production, which cause edema, pain and vasodilation. Ibuprofen is a propionic acid derivative drug with anti-inflammatory properties and can be used as an alternative to aspirin, and ketoprofen is an acidic painkiller and can be used to treat arthritis. The higher the usage of these drugs in daily life, the higher the possibility of their presence and detection in wastewater. Continuous inputs of these non-regulated emerging contaminants will cause potential risk to aquatic life, which can be calculated through an Environmental Risk Assessment. Even though the presence of these pharmaceuticals in WWTPs is a major concern, very few studies have been conducted in India. This chapter aims to summarize the available studies on the occurrence of NSAIDs in WWTPs situated in different regions of India and presents their removal by different treatment technologies.

2.2 DETECTION AND ANALYSIS OF NSAIDs

If NSAIDs are present in aquatic environments, they have a negative impact on aquatic life and human health. Usually, the concentrations of these drugs in aquatic environments will be in the range of μg/L or ηg/L. Even if they seem low, these amounts can cause a severe threat to aquatic life. Therefore, proper monitoring of the discharge of these drugs is essential, but detection of this lower concentration in a complex matrix such as wastewater is a great challenge. Advanced analytical techniques can overcome this problem to some extent. The important stages in the detection of NSAIDs from wastewater samples, along with the physico-chemical properties of a few NSAIDs, are summarized in this chapter.

2.2.1 Target Drugs and Their Mechanism of Action

Diclofenac (DIC); 2-[(2, 6-dichlorophenyl)amino] phenylacetic acid is the most consumed drug among the NSAIDs. This phenyl acetic acid compound is used to reduce inflammation and pain associated with different kinds of arthritis, such as rheumatoid arthritis and osteoarthritis. Since this

TABLE 2.1
Details of NSAIDs Discussed

Name of drug	Diclofenac	Ibuprofen	Ketoprofen
IUPAC name	2-[(2,6dicholorophenyl) amino] phenyl}acetic acid	2-[4-(2-methylpropyl)phenyl] propanoic acid	2-(3-Benzoylphenyl) propanoic acid
Chemical structure	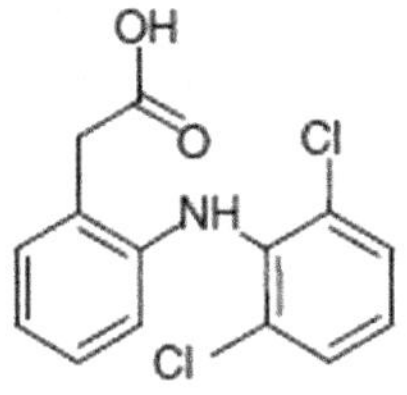	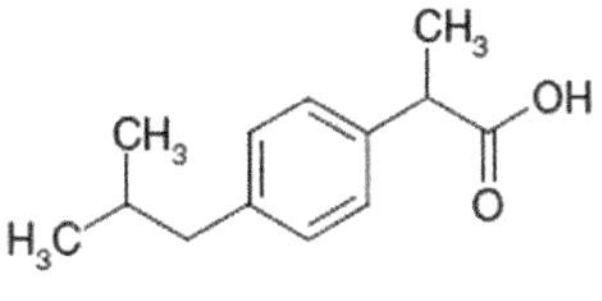	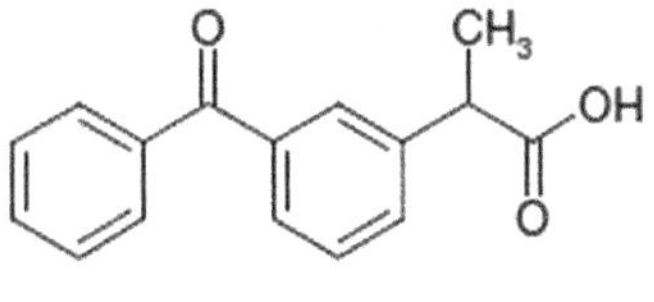
CAS No.	15307–86–5	15687–27–1	22071–15–4
Chemical formula	$C_{14}H_{11}Cl_2NO_2$	$C_{13}H_{18}O_2$	$C_{16}H_{14}O_3$
pKa dissociation constant (water, 20 C)	4.15	4.91	4.45
Log Kow	4.51	3.97	3.11

Source: Thalla and Vannarath (2020); Praveenkumarreddy et al. (2021).

belongs to an over-the-counter drug category, its presence is higher in aquatic environments. The European Union considers DIC as a priority molecule to be monitored in aquatic environments as per the Water Framework Directive (2013/39/EU).

Ibuprofen (IBU); (2-[4-(2-methylpropyl)phenyl]propanoic acid) is one of the important drugs included in the "Essential Drug List" compiled by the World Health Organization (WHO). It is mainly used to reduce pain and fever and has a vital role in the treatment of arthritis and rheumatic disorders. Usually the prescription volume for IBU is large and also it has a higher excretion rate compared to other NSAIDs. It can be used as an alternative to aspirin (Parolini 2020; Thalla and Vannarath 2020).

Ketoprofen (KET); (2-(3-benzoylphenyl) propanoic acid) is a NSAID used to treat arthritis and other painful conditions. This aryl propionic acid drug has both antipyretic and analgesic properties (Thalla and Vannarath 2020). Details of these target drugs are given in Table 2.1.

These NSAIDs reduce pain and inflammation by inhibiting the production of cyclooxygenase (COX) and prostaglandins. Prostaglandins act as physiological and pathological mediators in the therapeutic effects on inflammation, cancer, pain, pyrexia and neurological diseases. The mechanism is shown in Figure 2.1.

COX convert arachidonic acid (AA) to cyclic endoperoxides, which play an important role in pain, fever, inflammation, clotting, ovulation and bone metabolism. Prostaglandin (PG) is one endoperoxide. So when an NSAID inhibits COX, it will lead to the inhibition of PGs. These drugs reduce fever through the inhibition of PGE2 and an analgesic effect is attained by the inhibition of PGE2 and PGI2. They reduce edema by decreasing vasodilation through the inhibition of PGE2 and PGI2 (Terzi et al. 2018).

2.2.2 Sample Preparation Methods for Analyzing NSAIDs in Wastewater

Mostly, the concentration of pharmaceuticals present in treated wastewater will be very small. In order to quantify these trace amounts of pharmaceuticals, their metabolic and degradation products

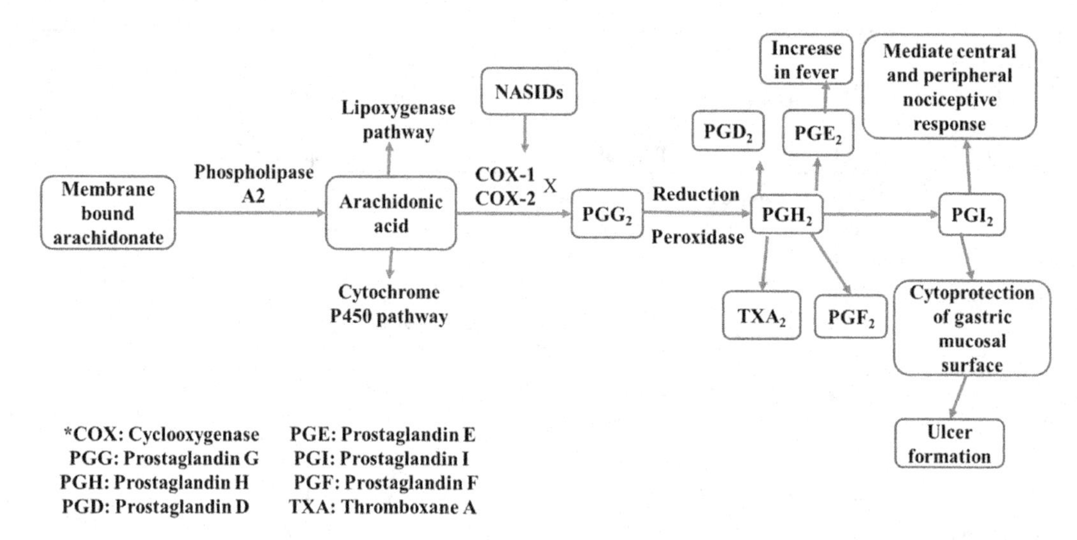

FIGURE 2.1 Action mechanism of NSAIDs.

Source: Izadi et al. (2020).

in aquatic environments, more accurate and sensitive analytical methods are required. Sample preparation is one important step in the analysis procedure, when it comes to a complex matrix such as wastewater. For the proper quantification of a small amount of drugs, a pre-concentration step is required. A clean-up step helps in reducing the matrix effect caused by interference, which results in the wrong measurement, by forcing the targeted molecule to leave uncharged when it enters the ionization chamber for analysis (Daniels et al. 2020).

Liquid-liquid extraction is one of the traditional methods for sample extraction, and it has been replaced by solid phase extraction (SPE). In this method, an SPE cartridge with packed sorbent will bind the target compounds from the sample due to the high affinity of sorbent material to the target compound. Then this target compound will be eluted using solvents with a higher affinity. A schematic representation is given in Figure 2.2.

Many commercially accessible SPE materials are available. When selecting a SPE sorbent, the physiochemical characteristics of both the target analyte(s) and the matrix chemicals must be taken into account. The popular Oasis HLB cartridges have a balance of hydrophilic and lipophilic characteristics, while others have high cation (Oasis MCX) or anion (Oasis MAX) exchange capabilities. Other commercial cartridges are specially designed to hold and release strong acids (Oasis WAX) or bases (Oasis WCX). Some SPE materials have C8 or C18 organic groups chemically bonded to silica (Patel et al. 2019).

Solid-phase extraction with Cleanert PEP cartridges has been used for simultaneous detection of NSAIDs (ketoprofen, ibuprofen and diclofenac) in wastewater with LC-qTof-MS in one study, which resulted in 75.5% recovery in a water sample (Zhao et al. 2014). Lacey et al. (2008) studied 20 pharmaceutical compounds (diclofenac and ibuprofen-NSAID) detected in wastewater using different cartridges, namely, Strata-X, Supelco C8, Supelco C18, Waters Oasis HLB, Varian Focus and MerckLiChrolut-EN, with Strata-X being the most efficient when combined with liquid chromatography-electrospray ionization mass spectrometry (LC-MS/MS). In a study by Afonso-Olivares et al. (2017), Oasis HLB was used for the quantification of ketoprofen, naproxen, ibuprofen and diclofenac in a WWTP situated in the south-east of the island of Gran Canaria. The extraction procedure was simplified to three steps, eliminating conditioning and the equilibration steps in their study due to the characteristics of the Oasis HLB cartridge. Oasis HLB has been used in many studies for NSAIDs' extraction. It has been combined with LC-MS/MS in research conducted at a municipal wastewater treatment plant in central Greece. They used two different cartridges, namely Oasis

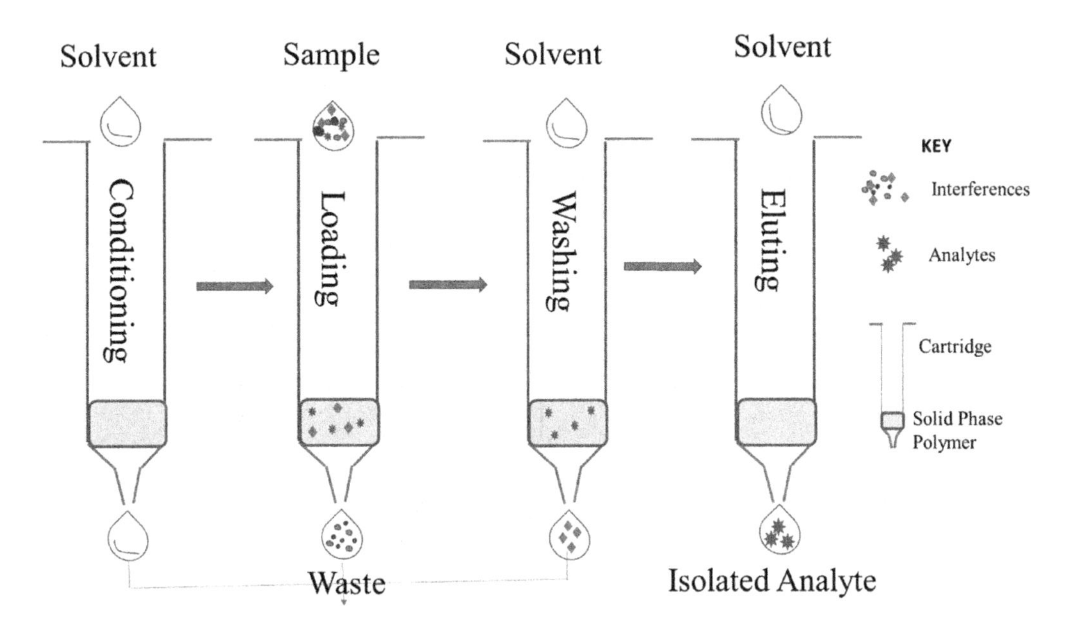

FIGURE 2.2 Schematic representation of SPE procedure.

Source: Daniels et al. (2020).

HLB and CNW HLB, with good results for both. Slightly higher recovery was obtained with Oasis HLB. This cartridge is efficient in the detection of NSAIDs after combining with another analytical method, such as GC-MS (Papageorgiou, Kosma, and Lambropoulou 2016). Praveenkumarreddy et al. (2021) extracted some NSAIDs from 100 ml of a wastewater sample using acetone to elute the analyte. Simultaneous extraction of acidic neutral and basic polar analytes in a wide range of pH values also made SPE a suitable sample preparation method (Daniels et al. 2020).

There are some disadvantages to the SPE method, such as the consumption of large amounts of solvents and reagents, it is labor-intensive and a multi-step operation. In order to overcome these problems, miniaturized extractions techniques have gained importance. Hollow fiber-protected liquid-phase microextraction (HF-LPME), dispersive liquid-liquid microextraction (DLLME), solid phase microextraction (SPME), and magnetic solid-phase extraction (MSPE) are widely recognized miniaturized extractions techniques (ibid.). In HF-LPME, through a liquid membrane, analytes of interest are extracted from an aqueous sample solution, and retained in the acceptor phase held in the lumen. The extracting (acceptor) phase of this technique is supported by a piece of porous polypropylene hollow fiber, offering stability while enabling the targeted chemicals to flow through.

García et al. (2009) improved the detection method of NSAIDs from river and wastewater effluent with the SPME method. Fiber conditioning, extraction of analytes into the SPME fiber from the aqueous sample, automated desorption and transference of analytes from the fiber to the LC system using the SPME interface and fiber cleaning before extraction of the next sample, are the steps involved in SPME. Conditioning of fiber will be done with a mobile phase. Then fiber will be inserted in a 3 ml sample whose pH is adjusted to 3 and the sample is stirred with a magnetic stirrer. For analysis, this fiber will be placed in a desorption chamber filled with mobile phase, analysis will take place with the help of this LC–SPME interface. It produced a satisfactory result. Another method is to inject the samples without any extraction in large volumes. Usually up to 5μl volume was used for LC-MS and GC-MS due to the limit of volume expansion into the liner (GC) and the matrix effect or peak quality for LC-MS. This has improved instrument sensitivity, lowered detection limits, along with reduced time for labor and reduced cost investment. The injection volume can be increased from 1–2 μL to 20μL and 100–200 μL for GC and LC respectively. Anumol et al. (2015)

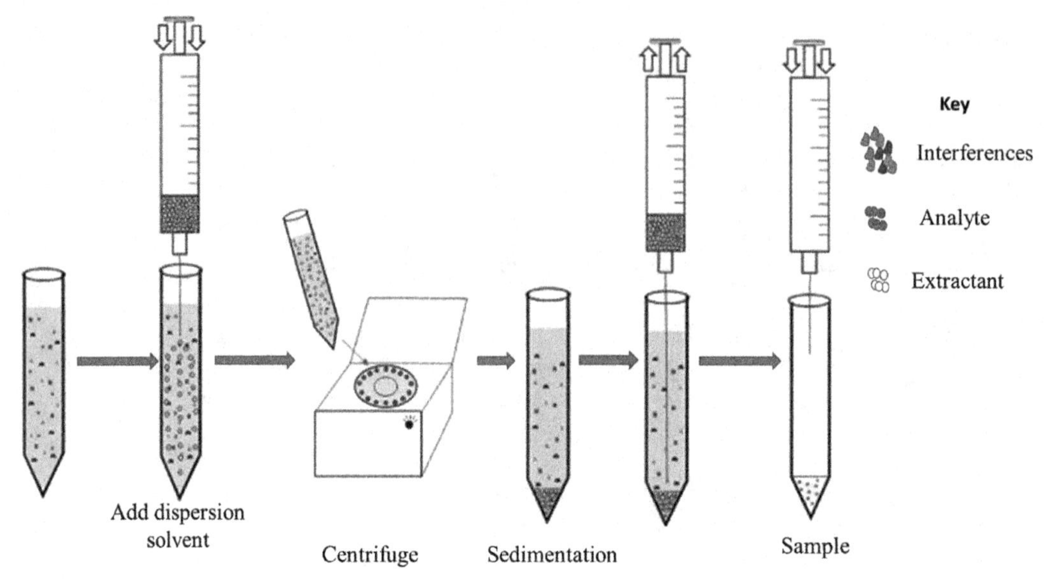

FIGURE 2.3 Schematic representation of DLLME.

Source: Daniels et al. (2020).

used this direct injection method to detect diclofenac in wastewater; 80 µl volume was optimized for injecting in that study. Along with HPLC-MS, 100% recovery was observed.

Ease of operation, rapidity and high recovery, as well as other variables that may be controlled to increase the extraction efficiency, make the dispersive liquid-liquid micro extraction (DLLME) technique extremely suitable for pharmaceutical analysis. In this, a mixture of extracting and dispersive solvent will be injected rapidly into a sample using a syringe. As a result, solvent droplets will be formed. Due to the high surface area contact between the sample phases and the extracting phase, mass transfer takes place between them (Figure 2.3). Typically, methanol, acetonitrile and acetone are applied as dispersing solvents and chlorinated organic compounds as extracting solvents.

Zgoła-Grześkowiak (2010) has developed a method for detecting NSAIDs from river water with the application of DLLME; 1 ml acetone and 80 µl chloroform were used as dispersing and extracting solvents respectively. The requirement of a few microliters of solvent to extract a wide variety of analytes from complex matrices is one of the main advantages of this technique.

Most of the NSAIDs are acidic compounds with a pKa value less than 5. These compounds are found largely in their ionized form at neutral pH and are retained poorly by lipophilic sorbents. For proper extraction, samples were acidified in all the above-mentioned methods. Methanol is the commonly used solvent to elute the compounds.

2.2.3 Sample Analysis Methods for NSAIDs in Wastewater

There are many techniques available for pharmaceuticals analysis, such as titrimetric, spectroscopic and chromatographic techniques. But chromatographic techniques, such as thin-layer chromatography (TLC), high performance thin-layer chromatography (HPTLC), high performance liquid chromatography (HPLC), and gas chromatography (GC), have replaced the titrimetric techniques due to their ability to detect small amounts of pharmaceuticals in complex matrices. Spectroscopic techniques are another group of analytical methods that quantify the reflection or transmission properties of a material as a function of the wavelength (Izadi et al. 2020).

Due to the low volatility of many polar pharmaceuticals, LC (liquid chromatography) is preferred to GC (gas chromatography). An additional derivatization step required before GC-MS analysis is the reason why. Derivatization was required due to the decomposition of compounds in a high temperature during GC analysis. To produce a more powerful analytical method, coupling of different techniques is necessary, so various detectors such as MS (mass spectrometry), TOF-MS (time-of-flight mass spectrometry), MS/MS (Mass Spec/Mass Spec), UV (ultraviolet), are coupled with chromatographic separations (Patel et al. 2019). But investigation done by Lacey et al. (2008) suggested that liquid chromatography-electrospray ionization mass spectrometry (LC–MS/MS) was more suitable and efficient than UV detection for simultaneous detection and identification of 20 pharmaceutical compounds from influent and effluent of WWTPs using solid phase extraction (SPE).

Most of the MS systems are based on triple-quadrupole via electrospray ionization (ESI) or atmospheric pressure chemicalionization (APCI) for environmental samples. However, time-of-flight (TOF) MS is recognized as a powerful analytical tool for the analysis of trace pharmaceuticals in environmental water due to accurate mass measurements and high resolution. Zhao et al. (2014) successfully detected DIC, KET, IBU in WWTP influent and effluent using liquid chromatography coupled with quadrupole time-of-flight mass spectrometry (LC-QTOF MS). Sample extracts were analysed by LC-QTOF MS for the simultaneous detection of ketoprofen, diclofenac and ibuprofen in negative ion mode after solid-phase extraction using Cleanert PEP cartridges. For separation and determination, a C18 reversed-phase column was employed. To improve the sample throughput and sensitivity, the pH of the extracts, the type of SPE columns and the MS conditions were optimized. Using these technologies various investigations have been conducted to detect NSAIDs in different water matrices.

2.3 OCCURRENCE OF NSAIDS IN WWTPs

2.3.1 Occurrence of NSAIDs in WWTPs: the Indian context

The presence of pharmaceuticals in wastewater treatment plants is causing more concern due to their higher concentrations in effluent rather than their concentration in surface water and groundwater in most cases. Potential natural processes such as biotransformation, photolysis, sorption, volatilization and dispersion might be the reason behind its lower concentration in surface water and groundwater. Moreover, the higher stability of pharmaceuticals, due to their chemical structure and presence of active molecules, makes them remain untreated in WWTP and also these plants are not specifically designed for the treatment of pharmaceuticals. The technologies and infrastructure implemented also play an important role in the removal of pharmaceuticals. A developing country like India has less efficient technologies and infrastructure compared to some other countries, which might be the reason behind the higher effluent concentrations of these drugs in WWTPs (Izadi et al. 2020; Praveenkumarreddy et al. 2021). Even though it is one of the countries with the greatest drug consumption and ranked 11th in over-the-counter drugs consumption in the world, very few studies are available on the presence of pharmaceuticals in wastewater in India (Subedi et al. 2017).

The presence and concentration of these drugs are highly influenced by the particular region, due to sales and usage patterns, the water consumption and water treatment methods employed there (Anumol et al. 2016). This review will summarize the reported studies by considering different regions of India. Based upon climatic, geographical and cultural features, the whole of India can be divided into six categories: the North Zone, the South Zone, the East Zone, the West Zone, the Central Zone and the North-east zone (Philip, Aravind, and Aravindakumar 2018). The reported studies in the period 2010–2021 show that more investigations have been performed in the South zone compared to the other zones. Obtained concentrations of NSAIDs, mainly diclofenac, ketoprofen and ibuprofen were presented in Table 2.1. These drugs have been detected in eight WWTPs in the South zone (Coimbatore, Udupi (1), Udupi (2), Mangalore, Manipal, Chennai (1), Chennai (2) and Chennai (3)). The treatment technologies adopted in each WWTP are given in Table 2.2.

TABLE 2.2
Summary of Target NSAIDs Mean and Range of Concentrations in WWTPs in the Indian Context

NSAIDs	WWTP	Region	Influent	Effluent	Sample collection date	References
Ibuprofen (µg/l)	Manipal WWTP	Southwestern India	22 (0.05–94)	0.14 (0.01–0.40)	25–31 March 2019	(Praveenkumarreddy et al. 2021)
			4.460	ND	July and August 2012	(Subedi et al. 2015)
	Udupi (1) WWTP	Southwestern India	6.08 (ND-23)	2.31 (0.08–6.7)	25–31 March 2019	(Praveenkumarreddy et al. 2021)
			1.2 (<LOQ-2.8)	0.98 (0.27–1.940)	19–25 May 2013	(Subedi et al. 2017)
	Udupi (2) WWTP	Southwestern India	0.834	0.145	July and August 2012	(Subedi et al. 2015)
	Mangalore WWTP	Southwestern India	5.1 (0.05–13)	1.41 (0.03–6.7)	25–31-March 2019	(Praveenkumarreddy et al. 2021)
			2109.875	22.71	May 2016	(Thalla and Vannarath 2020)
			43.51	0	September 2016	(Thalla and Vannarath 2020)
			1.4 (0.97–1.9)	0.63 (0.29–0.71)	19–25 May 2013	(Subedi et al. 2017)
	Coimbatore WWTP	South India	2.14	1.89	July and August 2012	(Subedi et al. 2015)
	Delhi (1) WWTP	North India	320	223	–	(Saxena et al. 2021)
	Delhi (2) WWTP	North India	252.8	275	–	(Saxena et al. 2021)
	Nagpur WWTP	Central India	ND	ND	–	(Saxena et al. 2021)
	Saidpur WWTP	North India	1.13	ND	July and August 2012	(Subedi et al. 2015)
	Beur WWTP	North India	0.686	0.204	July and August 2012	(Subedi et al. 2015)
	Ghaziabad WWTP	North India	2.38	NA	January 2010	(Singh et al. 2014)
	Lucknow WWTP	North India	1.43	NA	January 2010	(Singh et al. 2014)
Ketoprofen (µg/l)	Manipal WWTP	Southwestern India	41 (0.05–260)	0.56 (0.08–1.35)	25–31 March 2019	(Praveenkumarreddy et al. 2021)
			ND	0.00504	July and August 2012	(Subedi et al. 2015)
	Udupi (1) WWTP	Southwestern India	7.21 (0.04–37)	0.8 (ND-2.71)	25–31 March 2019	(Praveenkumarreddy et al. 2021)

	Udupi (2) WWTP	Southwestern India	0.00980	ND	July and August 2012	(Subedi et al. 2015)
	Mangalore WWTP	Southwestern India	2.8 (0.01–10.5)	0.57 (ND-2.70)	25–31 March 2019	(Praveenkumarreddy et al. 2021)
			2747.29	270.76	May 2016	(Thalla and Vannarath 2020)
			559.56	0	September-2016	(Thalla and Vannarath 2020)
	Coimbatore WWTP	South India	ND	ND	July and August 2012	(Subedi et al. 2015)
	Saidpur WWTP	North East India	.0396	0.0234	July and August 2012	(Subedi et al. 2015)
	Beur WWTP	North East India	0.0522	.0218	July and August 2012	(Subedi et al. 2015)
	Ghaziabad WWTP	North India	1.08	NA	January 2010	(Singh et al. 2014)
	Lucknow WWTP	North India	0.20	N/A	January 2010	(Singh et al. 2014)
Diclofenac µg/L	Manipal WWTP	Southwestern India	69 (ND-410)	2.44 (0.68–6.46)	25–3 March 2019	(Praveenkumarreddy et al. 2021)
	Udupi (1) WWTP	Southwestern India	29 (ND-82)	26 (0.5–126.5)	25–3 March 2019	(Praveenkumarreddy et al. 2021)
	Mangalore WWTP	Southwestern India	12 (ND-45)	13 (0.5–70.5)	25–3 March 2019	(Praveenkumarreddy et al. 2021)
			721.37	131.15	May 2016	(Thalla and Vannarath 2020)
			49.18	0	September 2016	(Thalla and Vannarath 2020)
	Chennai (1) WWTP	South India	3.1–5.3	2.3	Late summer 2013,winter 2013, early summer 2014	(Anumol et al. 2016)
	Chennai (2) WWTP	South India		1.7	Late summer 2013, winter 2013, early summer 2014	(Anumol et al. 2016)
	Chennai WWTP3	South India		1.1	Late summer 2013, winter 2013, early summer 2014	(Anumol et al. 2016)
	Ghaziabad WWTP	North India	3.12 (max: 25.68)	N/A	January 2010	(Singh et al. 2014)
	Lucknow WWTP	North India	0.36 Max: 1.43	N/A	January 2010	(Singh et al. 2014)
Aspirin	Manipal WWTP	Southwestern India	184 (ND-654)	0.44 (ND-1.5)	25–31 March 2019	(Praveenkumarreddy et al. 2021)
	Udupi (1) WWTP	Southwestern India	125 (ND-335)	0.49 (ND-1.2)	25–31 March 2019	(Praveenkumarreddy et al. 2021)

(continued)

TABLE 2.2 (Continued)
Summary of Target NSAIDs Mean and Range of Concentrations in WWTPs in the Indian Context

NSAIDs	WWTP	Region	Influent	Effluent	Sample collection date	References
	Mangalore WWTP	Southwestern India	157 (0.1–529)	0.69 (ND-4.4)	25–31 March 2019	(Praveenkumarreddy et al. 2021)
			2213.36	221.86	May 2016	(Thalla and Vannarath 2020)
			351.4	0	September 2016	(Thalla and Vannarath 2020)
Naproxen	Manipal WWTP	Southwestern India	217 (0.3–1370)	12.01 (0.02–70.1)	25–31 March 2019	(Praveenkumarreddy et al. 2021)
	Udupi (1) WWTP	Southwestern India	29 (ND-67)	14 (0.1–70.7)	25–31 March 2019	(Praveenkumarreddy et al. 2021)
	Mangalore WWTP	Southwestern India	11 (0.06–26)	3.59 (0.03–6.9)	25–31 March 2019	(Praveenkumarreddy et al. 2021)
			2132.48	173.075	May 2016	(Thalla and Vannarath 2020)
			115.385	0	September 2016	(Thalla and Vannarath 2020)
	Chennai (1) WWTP	South India	0.31–0.92	NA	Late summer 2013,winter 2013, early summer 2014	(Anumol et al. 2016)
	Chennai (2) WWTP	South India		NA	Late summer 2013,winter 2013, early summer 2014	(Anumol et al. 2016)
	Chennai (3) WWTP	South India		NA	Late summer 2013,winter 2013, early summer 2014	(Anumol et al. 2016)
	KSTP (1)	Southwestern India	0.134	ND	December 2011.	(Prabhasankar et al. 2016)
			0.01	0.148	July 2012	(Prabhasankar et al. 2016)
			0.547	0.013	September 2012	(Prabhasankar et al. 2016)
	KSTP (2)		0.118	ND	December 2011.	(Prabhasankar et al. 2016)
			.061	.047	July 2012	(Prabhasankar et al. 2016)
			0.219	.028	September 2012	(Prabhasankar et al. 2016)
	KSTP (3)		0.059	ND	December 2011.	(Prabhasankar et al. 2016)
			0.043	0.011	July 2012	(Prabhasankar et al. 2016)
			0.058	0.028	September 2012	(Prabhasankar et al. 2016)

	KSTP (4)		0.127	0.149	December 2011.	(Prabhasankar et al. 2016)
			0.014	0.016	July 2012	(Prabhasankar et al. 2016)
			0.002	0.010	September 2012	(Prabhasankar et al. 2016)
Acetaminophen	Udupi (1) WWTP	Southwestern India	9 (5.4–11)	0.69 (0.33–1.2)	19–25 May 2013	(Subedi et al. 2017)
	Mangalore WWTP	Southwestern India	4.5	0.34	19–25 May 2013	(Subedi et al. 2017)
	Nagpur (2) WWTP	Central India	13.25	1.5	April to June 2014	(Archana et al. 2016)
			30	11	May (summer season)	(Archana et al. 2017)
	WWTP (1)	Western India	86.8	-	November, April and June 2014–2015	(Mohapatra et al. 2016)
	WWTP (2)	Western India	7.1	-	November, April and June 2014–2015	(Mohapatra et al. 2016)

Recently Praveenkumarreddy et al. (2021) conducted a study in three wastewater treatment plants (Manipal, Udupi (1) and Mangalore) and one river situated in the South-western region of India. Using GC-MS technology they quantified some of the NSAIDS, namely aspirin, naproxen, diclofenac, ketoprofen and ibuprofen present in influent, clarifier and effluent of WWTPs. The authors found aspirin as the NSAID present in higher concentrations in influent among the considered drugs and the obtained order was aspirin > naproxen > diclofenac > ketoprofen >ibuprofen, while ketoprofen was the highest NSAID present in a similar study previously conducted by Thalla and Vannarath (2020) in Mangalore WWTP. The variation in consumption pattern and rate might be the reason behind this. Seasonal variation in the occurrence of these drugs was also considered in their study, which showed higher influent concentration in summer than in the monsoon season due to reduced consumption and dilution by storm water in the monsoon season.Activated sludge process is the treatment technology employed in most WWTPs (Table 2.3). Along with it, aerated lagoon and MBBR are also considered in some studies (Anumol et al. 2016; Saxena et al. 2021). Differences in

TABLE 2.3
Details of WWTPs and Treatment Technologies

WWTP	Treatment technology	Capacity (MLD)	Population served	State	References
Manipal WWTP	Activated sludge process	2	12,000	Karnataka	(Praveenkumarreddy et al. 2021)
Udupi (1) WWTP	Activated sludge process	12	150000 (domestic+ hospital WW)	Karnataka	(Subedi et al. 2017)
Udupi(2) WWTP	activated sludge process	2.0	10000	Karnataka	(Subedi et al. 2015)
Mangalore WWTP	USABR technology and activated sludge process	43.5	450000	Karnataka	(Thalla and Vannarath 2020)
Coimbatore WWTP	Activated sludge process	50	350000	Tamil Nadu	(Subedi et al. 2015)
Saidpur WWTP	Activated sludge process	45	350000	Bihar	(Subedi et al. 2015)
Beur WWTP	Activated sludge process	35	275000	Bihar	(Subedi et al. 2015)
GhaziabadWWTP	–	–	–	Northern India	(Singh et al. 2014)
Lucknow WWTP	–	–	–	Uttar Pradesh	(Singh et al. 2014)
Chennai 1 WWTP	Conventional activated sludge- Maturation pond	54	–	Tamil Nadu	(Anumol et al. 2016)
Chennai 2 WWTP	Activated sludge chlorination	40	–	Tamil Nadu	(Anumol et al. 2016)
Chennai 3 WWTP	Primary aerated lagoon secondary Clarification/ flocculation Final filtration with chlorine addition	1.5	15000	Tamil Nadu	(Anumol et al. 2016)
Delhi (1) WWTP	Activated sludge process	45	–		(Saxena et al. 2021)
Delhi (2) WWTP	Activated sludge process	636	–		(Saxena et al. 2021)
KSTP 2	Moving bed bioreactor	0.3	–	Maharashtra	(Saxena et al. 2021)
KSTP 3	Extended aeration activated sludge process	1.5	11000+ 2000 beds	Karnataka	(Prabhasankar et al. 2016)

TABLE 2.3 (Continued)
Details of WWTPs and Treatment Technologies

WWTP	Treatment technology	Capacity (MLD)	Population served	State	References
KSTP 4	Extended aeration activated sludge process	2	11000+ 2000 beds	Karnataka	(Prabhasankar et al. 2016)
Nagpur (2) WWTP	Extended aeration activated sludge process	2	9000	Karnataka	(Prabhasankar et al. 2016)
W WWTP 1	Extended aeration activated sludge process	.05	200 beds	Karnataka	(Prabhasankar et al. 2016)
WWTP 2	Activated sludge process	80	–	Maharashtra	(Archana et al. 2016)
	Facultative aerated lagoon	46	–	Western India	(Mohapatra et al. 2016)
	Activated sludge process	60	–	Western India	(Mohapatra et al. 2016)

the trend in concentration have been observed for some drugs in the same WWTP at different time periods. A similar observation was obtained for pharmaceuticals considered in studies conducted by Fick et al. (2009) and Larsson, de Pedro, and Paxéus (2007) at the PETL (Patancheru Enviro Tech Limited) WTP near Hyderabad, that received 1.5 MLD effluents from drug manufacturers in the vicinity in Patancheru. Concentration of all the drugs was found to be reduced in a study conducted after two years. This shows that a proper monitoring of these drugs in WWTPs is required to know the fluctuations in the occurrence with the time period.

As per the literature, around 47% of pharmaceutical manufacturing units in the country are located in the West zone, followed by the North (21%), the South (18%), the East (9%), the Central (4%) and the North-east (0.5%) zones (Philip, Aravind, and Aravindakumar 2018). Studies on the occurrence and fate of these pharmaceutical drugs in water matrices reported from the North, West, East and North-West zones were much fewer or limited in number. A recently reported study in northern India was done by Saxena et al. (2021). They investigated the presence of ibuprofen in three sewage treatment plants situated in urban and semi-urban areas of North India and observed target drug concentration below the detectable limit for sewage treatment plants (STPs) operating in semi-urban areas. This is due to the different usage pattern of the pharmaceutical in geo-spatially different urban and semi-urban areas as well as the difference in population served by STPs. Similarly, Singh et al. (2014) reported the maximum detection frequency for diclofenac in North India. The authors reported the presence of ibuprofen, ketoprofen and diclofenac in Lucknow and Ghaziabad WWTP with higher diclofenac concentration in the influent.

2.3.1.1 The South Zone

In the South zone, eight WWTPs (Coimbatore, Udupi (1), Udupi (2) Mangalore, Manipal, Chennai (1), Chennai (2) and Chennai (3)) were investigated for the presence of NSAIDs (mainly diclofenac, ketoprofen and ibuprofen) in reported studies. Among them ibuprofen was detected in five WWTPs (Coimbatore, Udupi (1), Udupi (2), Mangalore and Manipal). The influent concentration range of ibuprofen is 0.8–2109.875 µg/l in the South Indian region with a maximum concentration of 2109.875 µg/l observed in Mangalore WWTP in 2016 (in the summer season). There is a rise in influent concentration of ibuprofen by year (2012–2019), except in Mangalore WWTP. This rise might be due to the increased consumption of these drugs by communities in every year and highlights the importance of monitoring its presence in WWTPs. The lowest mean influent concentration for ibuprofen has been reported in Udupi (2) WWTP, which serves the smallest numbers of population compared to the other four WWTPs in the South Indian region. This shows the quality of the wastewater also influences the concentration. Even though all the treatment plants have implemented the activated sludge process as a secondary treatment, removal efficiency of ibuprofen varies among all the treatment plants due to the difference in operation and maintenance conditions. The decreased concentration value of effluent shows the fast biodegradation of ibuprofen. Preference for an aerobic biodegradation pathway (Vystavna et al. 2017) might be the reason behind this higher degradation. On comparing the removal efficiencies of ibuprofen in Mangalore WWTP and Udupi (1) WWTP in the same period under consideration, the WWTP with combined anaerobic and aerobic bio-degradation shows better removal properties.

In the case of diclofenac and ketoprofen, influent concentration ranges are 3–721.37 μg/l and 0.009–2747.29 μg/l, respectively. In most cases, diclofenac was present in higher concentrations in the influent than ibuprofen and ketoprofen. And higher effluent concentration of diclofenac shows less removal in the WWTP. Even negative removal efficiency was observed in some cases. Negative removal efficiency might be due to the deconjugation of metabolites present in the influent during the biological processes or desorption from the activated sludge tank of previously adsorbed drugs (Salgado et al. 2012). A decreased influent concentration was observed for all three target drugs (DIC, IBU, KET) in the monsoon season of the same year of 2016 (due to the reduced consumption of drug and dilution of contaminants by storm water) (Thalla and Vannarath 2020). In Chennai (3) WWTP, aerated lagoons had been implemented instead of grit chamber and primary clarifier, as the primary treatment, which showed higher removal of diclofenac. But in Chennai (1) and Chennai (2), significant removal of diclofenac was observed in the activated sludge process due to microbial degradation in AS. Chennai (3) WWTP had very little additional removal of these compounds in the secondary treatment because the biological degradation of these compounds had already taken place in the aerated lagoons. Chlorination was also efficient in the degradation of diclofenac (Anumol et al. 2016). Aspirin, naproxen and acetaminophen were other NSAIDs detected in South Indian WWTPs. Some of those WWTPs showed negative removal of naproxen.

2.3.1.2 The North Zone

Delhi (1) WWTP, Delhi (2) WWTP, Saidpur WWTP, Beur WWTP, Ghaziabad WWTP, Lucknow WWTP are situated in North India. Results from Delhi (1) WWTP, Delhi 2) WWTP (urban area) and Nagpur WWTP (semi-urban) clearly showed the difference in influent concentrations for ibuprofen. The concentration was higher in the urban area. Higher usage of these drugs in urban areas might be the reason for the difference.

2.3.1.3 The Central Zone, the East Zone, the West Zone, and the North-east Zone

Unfortunately, to the best of our knowledge, very few studies have been conducted in these regions. In future, more focus should be given to these regions to provide baseline data to form regulations. Investigations in Nagpur (2) WWTP showed the presence of acetaminophen in raw and treated wastewater, while in Nagpur (1) WWTP ibuprofen was not detected. Nagpur (1) WWTP implemented MBBR technology for treatment and was situated in a semi-urban area. Acetaminophen is the only NSAIDs studied in the western region of India. This showed influent concentrations of 86.8 and 7.1μg/l for two WWTPs with facultative aerated lagoon and activated sludge treatment techniques, respectively. To the best of the authors' knowledge, no other studies have been conducted for the presence of drug concentrations in WWTPs in other regions of India. So it is very important to give more attention to those regions to understand the current situation.

2.3.2 Environmental Risk Assessment

In India, only 31% of the total produced sewage reaches STPs. The remaining part is directly evacuated in the environment. Even most of the treated wastewater from STPs is also discharged into water bodies such as rivers (Narain and Sengupta 2016). Due to the lack of regulations for pharmaceuticals in treated wastewater, the main pathway for NSAIDs into the aquatic environment is enabled. In order to understand the potential ecotoxicological effects, an environmental risk assessment has to be performed.

Risk assessment of NSAIDs can be done by calculating the risk quotient (RQ) values. These are calculated based on the measured environmental concentration (MEC) and the predicted no effect concentration (PNEC) of resident organisms. PNEC values are derived from toxicological assays or ECOSAR software and are calculated from the lowest LC50 or EC50 for the aquatic organisms with an assessment factor (AF). EC50 is the concentration in wastewater that causes 50% inhibition of any particular activity, for example, the respiration rate. From these, the RQ is calculated as the ratio between the highest MEC and the lowest PNEC, where the MEC is the highest concentration found in wastewater. Predicted no-effect concentrations (PNECs) for aquatic organisms are derived by dividing either the EC50 by an assessment factor (AF) of 1,000 or by dividing the NOEC (no observed effect concentration) values by an AF of 10 (Shanmugam et al. 2014). PNEC can be calculated for acute toxicity, by using the lowest values of EC50 or LC50 divided by an assessment factor (AF) (usually 1000) .In the case of chronic toxicity, it is calculated by using NOEC divided by an assessment factor (AF). When one long-term NOEC is available for chronic toxicity, an AF of 100 is used, an AF of 50 is used when two long-term NOECs are available in two trophic levels, and an AF of 10 is used when long-term NOECs from all three trophic levels are available (Papageorgiou, Kosma and Lambropoulou 2016).

Acute toxicity refers to the consequences caused by either a single exposure or multiple exposures within a short time period and appears as lethal endpoints (e.g., mortality or immobilization). Chronic toxicity refers to the onset of negative consequences resulting from prolonged and repeated exposure to stressors, which appears as sub-lethal endpoints (e.g., growth inhibition, molecular or biochemical alterations, behavioral change) (Parolini 2020).

As mentioned earlier, the risk quotient (RQ) can be found using Equation (2.1):

$$\textit{Risk quotient} = \frac{\text{MEC}}{\text{PNEC}} \tag{2.1}$$

where

MEC = measured environmental concentration

PNEC = predicted no-effect concentration for a given compound

where RQ<0.1 indicates a low level risk, whereas RQ in between 0.1 and 1 is an indication of medium risk level. If RQ>1, this represents a higher risk (CHMP 2006). From MEC and PNEC values, it is possible to predict the concentration up to which low, medium and high level risk are observed.

2.4 REMOVAL OF NSAIDS IN WWTPs AND OTHER ADVANCED TECHNOLOGIES

2.4.1 Conventional WWTPs and Treatment Technologies

Municipal WWTPs are designed to remove substances, such as particulates, carbonaceous substances, nutrients and pathogens. The efficiency of WWTPs in removing these pollutants is usually high, but they are inefficient in the complete removal of pharmaceuticals. In order to obtain higher removal efficiency of these drugs, treatment processes have to be optimized by evaluating their fate and removal in each unit. Mostly WWTPs consist of primary, secondary and an optional tertiary or advanced treatment process. In a study based on globally available data on the occurrence and removal of pharmaceuticals from municipal WWTPs, it was found that approximately 60% of the investigated municipal WWTPs used only primary and biological secondary treatment processes with no tertiary or advanced treatment processes, whereas the remaining 40% did use tertiary or advanced treatment processes (Fawzi, Khasawneh, and Palaniandy 2021).

Primary and preliminary treatments are used for the removal of oil, grease, suspended solids and large floating organic matter. But the removal efficiency of this primary treatment is insignificant in the removal of NSAIDs. The difference in the treatment used can lead to different removal efficiencies, such as a WWTP using an aerated lagoon as a primary treatment was able to remove pharmaceuticals including diclofenac, and naproxen more than another two WWTPs using a grit chamber and a primary clarifier. In a developing country like India, the lagoon-based treatment is more convenient and in these lagoons, contaminants are subjected to oxidation, sedimentation, fermentation, photolysis, biodegradation and disinfection simultaneously (Mohapatra et al. 2016). But the overall removal efficiency of most of the compounds in a primary treatment process was less than 10% (Anumol et al. 2016). In another study, only 28% removal was observed for diclofenac through primary treatment and no reduction was observed for ibuprofen and naproxen (Luo et al. 2014), even though biodegradation, sorption and volatilization are the major removal mechanisms for NSAIDs in WWTPs (Papageorgiou, Kosma, and Lambropoulou 2016). Volatilization is less important compared to the other two mechanisms since water is usually covered in WWTPs. The main removal in the primary treatment process will be through a physical process such as adsorption onto larger organic matter which was removed by filtering and settling (Anumol et al. 2016). In contrast to these observed removal efficiencies, around 50% decrease in concentration of diclofenac was reported in the pre-treatment of sewage by grit removal and primary sedimentation in an STP situated in Sweden (Zorita, Mårtensson and Mathiasson 2009). Studies have reported that adsorption of diclofenac is higher on primary sludge than secondary sludge. Less than 5% sorption was observed on secondary whereas 5–15% on primary sludge. At neutral pH, the carboxylic acid moiety of DIC is negatively ionized, therefore the molecule repels negatively charged sludge. DIC becomes electrically neutral at an acidic pH, allowing it to adsorb to the sludge. The pH of the primary treatment procedure is often lower than that of the biological treatment unit (Vieno and Sillanpää 2014).

Generally, secondary and tertiary treatments show improved removal of NSAIDs better than primary treatment in WWTPs. Hydrophobicity, biodegradability and volatility of the drugs are some of the factors affecting the removal mechanisms. Removal of drugs by adsorption mainly depends on hydrophobicity. Generally, based on the octanol-water partition coefficient (KOW) value, a compound will show higher sorption potential if log KOW> 4 and similarly log KOW< 2.5 indicates low and 2.5 $<$log KOW$<$ 4 indicates medium sorption potential (Luo et al. 2014). Salgado et al. (2012) observed higher removal of diclofenac occurring through adsorption rather than by biodegradation and UV photolysis. Higher log KOW (4.51) of diclofenac might be the reason for its higher removal through adsorption. But diclofenac exhibited low (<25%) biodegradation due to the amount of N-H group and Cl atoms that slow down the development of bacteria in wastewater, whereas ibuprofen and ketoprofen were biodegraded to a much higher extent (>75%). Due to the higher biodegradation rate of ketoprofen and ibuprofen, less adsorption was observed even after having a high log KOW value. The authors found removal percentages of these mechanisms were 45% by biodegradation and 33% and 22% by adsorption to sludge and UV photolysis, respectively. Similarly, higher biodegradation was seen in aerated lagoon results in secondary treatment with very little additional removal (Anumol et al. 2016).

Luo et al. (2014) reported average removal efficiencies in conventional wastewater treatment as 91.4%, 75.5% and 51.7% for ibuprofen, naproxen and ketoprofen, respectively. Average 35.8% removal was observed for diclofenac. This low removal agreed with the values quoted in a review done by (Yamkelani et al. 2019). The reported removal efficiencies for diclofenac in studies conducted in France, Spain and Sweden were -20–50%, 14% and 17–69%, respectively. Whereas, in the same studies, the reported removal efficiency for other NSAIDs (ibuprofen, ketoprofen, and naproxen) was above 40% in most cases. Praveenkumarreddy et al. (2021) also observed a similar trend in removal. Some 58% and 0% median removal efficiencies were observed for diclofenac in Manipal and Udupi WWTPs of South India, while Thalla and Vannarath (2020) observed the highest removal for ibuprofen (98.92%) and the least removal for diclofenac (81.82%) in the summer season. Results obtained from studies conducted in WWTPs operating in five European countries obtain >90% removal for ibuprofen and 80% for naproxen followed by 39% diclofenac. Activated sludge process is the main treatment technology used in most of the WWTPs considered (Paxéus 2004). Even though ketoprofen and ibuprofen were structurally similar, ibuprofen shows more efficient elimination in most cases (Luo et al. 2014). Ibuprofen's primary biodegradation route, aliphatic chain oxidation, leads in the production of unique polar hydroxy (Ibu-OH) and carboxy (Ibu-COOH) metabolites, as well as carboxyhydratropic acid. All these metabolites have been detected in influent and effluent. The most common species in raw sewage are ibuprofen and Ibu-COOH while Ibu-OH is dominant in treated water (Paxéus 2004).

The activated sludge process (ASP) is a conventional treatment technology used to treat domestic wastewater. Microbial activity for nutrient removal and oxidation of carbon-rich material is the core part of this treatment. Mixing of wastewater with microbial culture in an aeration tank and recycling of sludge in each cycle is the method followed in ASP. Many microbes present in the sludge are capable of degrading pharmaceuticals. Studies have reported that diclofenac can be oxidized co-metabolically by *Pseudomonas putida* during active manganese oxidation. Most of such degradation studies gave good results in aerobic conditions. Fewer studies were found in anaerobic conditions. Biotransformation of naproxen and diclofenac was observed by the enzyme acetate kinase under anaerobic conditions (Kanaujiya et al. 2019). Disturbances in the activated sludge process reduced the removal rates of all acidic medicines, especially diclofenac and ibuprofen. Paxéus (2004) stated their respective removal efficiencies dropped <10% and <60% in a disturbed activated sludge process. In the case of ibuprofen, disturbances in the activated sludge process did not appear to influence Ibu-COOH removal (97%) but did result in a reduced removal of ibuprofen (43%) and virtually no removal of Ibu-OH (1%).

The coagulation flocculation process is another treatment usually implemented in WWTPs. Particulate debris, colloids and certain dissolved compounds can be removed via coagulation–flocculation. It can also remove some pharmaceuticals, such as diclofenac, in addition to these substances. During this treatment, the composition of wastewater can have a favorable or negative impact on micropollutant removal. The presence of dissolved humic acid results in a more effective elimination of diclofenac and ibuprofen. On the other hand, the presence of dissolved organic matter (DOM), particularly low molecular weight fractions, may inhibit micropollutant removal because DOM is removed more preferentially than drugs through coagulation, and negatively charged DOM may react with positively charged aluminum hydrolysis species, resulting in a reduction in the quantity of coagulant available to remove the compounds. Luo et al. (2014) quoted 12%, 21% and 31% removal of ibuprofen, diclofenac and naproxen, respectively, when alum is used as a coagulant. Diclofenac's removal was mainly via adsorption to floc, due to its high log KOW value. Upgrading the treatment methods and procedures for effluents generated by traditional WWTPs may minimize micro pollutant discharge into recipient water bodies and enhance the effluent quality for reuse. Tertiary treatment is such an option.

In the case of membrane processes like microfiltration (MF) and ultrafiltration (UF), micro pollutants are often inadequately removed during UF and MF because the membrane pore diameters are significantly bigger than micro pollutant molecular sizes. The nature of the membrane also influences the removal process. Ibuprofen retained 25% in the hydrophobic membrane but insignificant removal was observed in the hydrophilic membrane. This filtration process combines with the activated sludge biological treatment in a membrane bioreactor (MBR). MBRs remove micro pollutants by a combination of physical retention (adsorption), air stripping and biodegradation (Luo et al. 2014). Kim et al. (2014) reported naproxen and acetaminophen were removed by at least 99% through biological degradation in an MBR system. Compared to activated sludge, the MBR sludge has a lower particle size and hence a greater surface area for adsorption. Also, the content of inert matter is generally higher compared to activated sludge, owing to the prolonged sludge age applied in the MBR process, ibuprofen and acetaminophen showed high removal efficiency at lower SRT (sludge retention time) of 8 days, while removal of diclofenac was less at the same SRT (Kanaujiya et al. 2019). But studies compared AS at 7 days SRT, and MBR with 15 days and 65 days SRT, which results in 40% removal of diclofenac at AS and MBR (15 days) and 80% at MBR (65 days). This longer SRT and greater biomass content are responsible for the higher removal efficiency in MBR. Higher biomass leads to a lower food-to-microbes ratio, and the relative scarcity of biodegradable organic matter may force microbes to metabolize more recalcitrant compounds in the sewage. In agreement with this statement, another study showed DIC elimination ranged from 8% to 38% when SRT was 20–48 days, 59% when SRT was 62 days, and 53% when SRT was 322 days. An increase in HRT also improved the removal of ibuprofen and ketoprofen due to increased contact time between the micro-organisms and wastewater (Vieno and Sillanpää 2014). Adsorption is an important removal mechanism in the MBR process. A mass balance study on the removal of ketoprofen shows, out of a total 97% removed, 86% was removed by biodegradation and 11% by adsorption (Park et al. 2020).

The MBBR (moving bed biofilm reactor) system is another technology used in WWTPs. It is a biofilm-based reactor. Biofilms are formed on tiny surfaces in a reactor system. Biomass carriers (1–4 cm), such as kaldness, biosupport the materials, and these biosupport materials have been carefully developed. This increased the surface area available for biofilm development and it can be used in a variety of biological wastewater treatment applications systems (Kanaujiya et al. 2019). A comparison between activated sludge process and MBBR was done by Kompare et al. (2013) in a laboratory scale. The authors considered four non-steroidal anti-inflammatory drugs, ibuprofen, naproxen, ketoprofen and diclofenac. Average removal efficiencies obtained for these compounds were 86%, 74%, 78% and 48%, respectively, in ASP, while in MBBR an improved removal of 85%, 80% and 94% were observed for diclofenac, naproxen and ibuprofen. Four distinct doses (2, 4, 7, 10 mg/L) were chosen to feed the reactor in order to compare the degree of elimination in a study quoted by Fatehifar, Borghei and Ekhlasi (2018) and the final concentrations were measured using HPLC testing. Higher concentrations (up to 10 mg/L) resulted in greater elimination rates. Similar to other technology, HRT will affect the efficiency of removal of drugs found in wastewater. Micro-organisms eliminated a substantial amount of DIC in the first 5 hours, but there was no significant clearance in the next 5 hours in an experiment with HRTs for 5 h and 10 h. When HRT was 5 hours, IBU did not show considerable change. When comparing IBU removal rates at two different HRTs, it was discovered that a 10-hour retention period resulted in a considerably greater proportion of IBU removal because of more time required to react. The optimum HRTs for DIC and IBU were found to be 5 h and 10 h and removal efficiencies as 31–66% and 0–37.33%, respectively, in that study. But >90% removal was observed for IBU and aspirin in an MBBR process by Jiang et al. (2017). Diclofenac removal was 33.3% at the HRT of 24 h and an improved efficiency of 83.3% was shown at 18 h. Sorption is the main removal mechanism of diclofenac removal and it occurs at a faster rate (ibid.).

Any treatment process will have less efficiency compared to hybrid systems. Jiang et al. (ibid.) observed higher efficiency for a hybrid MBBR-MBR system. Since biodegradation plays an important role in the removal of NSAIDs, the attached growth pattern was beneficial for enriching slow-growing bacteria and creating a diversity of biocoenosis. An optimum HRT of 18 h was obtained among considered HRTs of 24, 18, 12 and 6 h.

Intermittently feeding in MBBR indicated higher efficiency than the usual operation of MBBR. Tang et al. (2017) conducted studies with intermittently fed MBBR using raw wastewater taken after primary clarification. More recalcitrant drugs like diclofenac were removed very rapidly in this study. The capability of biofilm developed on carrier in reactors with BOD-rich wastewater was the reason for this improved removal.

Biological nitrification and enitrification also have an influence on the removal process. Denitrification is the biological conversion of nitrate/nitrite to nitrogen gas, whereas nitrification is the biological conversion of ammonium to nitrite and nitrate. The nitrification process was discovered to be more efficient in the removal of ibuprofen and diclofenac and less removal was observed during denitrification or anoxic condition (Luo, Guo, Hao, Duc, Ibney, Zhang, et al. 2014; Vieno and Sillanpää 2014). But, in contrast, Zwiener and Frimmel (2004) reported higher removal of diclofenac (34-–38%) and less removal of ibuprofen (17–21%) during the anoxic condition than the oxic condition. The observed removal efficiencies were 64–70% and 1–4% for ibuprofen and diclofenac respectively during the oxic condition. Naproxen and ketoprofen also showed improved removal in aerobic conditions.

A disinfection treatment unit also showed some amount of removal for NSAIDs. The only NSAIDs that showed any substantial removal on chlorination at a WWTP using ASP were naproxen and diclofenac and they were removed due to their fast breakdown with chlorine. A WWTP with liquid chlorine was dosed in a batch manner into an equalization tank and allowed to sit for 30 minutes (Anumol et al. 2016).

2.4.2 Constructed Wetlands

Studies have proved that constructed wetlands (CW) can also be considered a treatment option for NSAIDs. Minimal energy input, easy operation and low cost maintenance make CW more attractive compared to advanced technologies like activated carbon, reverse osmosis, which has been shown to be effective in removing these drugs. A study conducted on CW built to replace a WWTP in Ukraine showed that diclofenac, ibuprofen and ketoprofen removal is effective in CW from hospital wastewater. Photo degradation, anaerobic reductive dehalogenation due to enhanced microbial activity in the rhizosphere, and sorption/sedimentation removal with the use of sand as a filter material with a high sorption capacity were the reasons behind the higher removal of diclofenac. Ibuprofen was the least removed among the considered drugs. An increased aeration condition has improved the removal of ibuprofen which shows a preference for an aerobic biodegradation pathway, and photo degradation was the key mechanism in the removal of ibuprofen. It is able to accumulate in plant tissues, so a higher density of macrophyte plantations can improve its removal efficiency (Vystavna et al. 2017).

Zhang et al. (2018) considered different design configurations of CW for the removal of acidic pharmaceuticals (ibuprofen, diclofenac and ketoprofen) from synthetic wastewater. Around 25.9–79% removal efficiency was obtained for ibuprofen in HF-CW with *Canna indica* in the summer season. Plant uptake and aerobic conditions have an important role in ibuprofen removal. But for diclofenac and ketoprofen, SF-CW have given the highest removal efficiency of 42.2–68.3% and 51.1–91.3%, respectively, as CWs provide favorable conditions for receiving solar radiation. Ketoprofen was a readily photodegradable compound and photolysis might be the reason behind the higher removal of diclofenac and ketoprofen in SF-CW.

Plant species plays an important role in the removal of NSAIDs. The study by Zhang et al. (ibid.) showed an increased 80% removal of diclofenac with *Scirpus validus*, whereas only 55% removal of diclofenac was observed with *Typha angustifolia* in CW (Yamkelani et al. 2019).

2.4.3 Adsorption

Among the treatment methods, adsorption is one of the simplest and lowest cost methods available for the removal of NSAIDs with high efficiency. Activated carbon (AC) is a common adsorbent with high adsorption capacity and a larger area. Of diclofenac, ibuprofen and ketoprofen, diclofenac showed higher capacity among studies conducted with AC. Ahmed (2017) gave the adsorption order as diclofenac > naproxen > ketoprofen > ibuprofen in his review. The high molecular weight, polarizability and hydrophobicity of DIC might be the reasons for its high adsorption.

Different materials can be used for the preparation of activated carbon. Abo El Naga et al. (2019) produced porous activated carbon from an agro-waste sugar cane bagasse for the removal of diclofenac from contaminated water. The adsorbent showed a fast adsorption rate and a high binding capacity toward DFC and it can be reused by washing the DFC-saturated adsorbent with acetone. The binding capacity obtained for the adsorbent was 315.0 mg/g which was higher than other carbon-based adsorbents derived from orange peel, peach stones, olive-waste cakes, potato peel waste, and graphene, etc. A high adsorption efficiency of 92.4% at pH 2 was obtained in his study. Manthiram Karthik and Philip (2021) investigated eight low-cost, commonly accessible materials' potential as adsorbents for the removal of pharmaceutically active substances like diclofenac and ibuprofen. Diclofenac showed higher removal than ibuprofen. Wood charcoal >natural zeolite >waste AAC blocks >LECA >blast furnace slag >natural pyrite >brickbats >sand was the order of preference of materials as adsorbents. The findings showed that the materials displayed quicker sorption in the beginning, but as time progressed, most of the active sites were saturated. Among the materials tested, wood charcoal reached equilibrium significantly more quickly than the others. Within two hours, almost 75% of the chosen chemicals were absorbed into the wood charcoal. At the end of 10 hours, DCF had a higher removal rate (70–80%) than IBU (65–68%). This is likely due to DCF's higher hydrophobicity (log KOW -4.51). Blast furnace slag, natural pyrite, brickbats, and sand, on the other hand, had slower sorption kinetics and reached equilibrium around 15 hours. This could be due to the fact that their surfaces have fewer active sorption sites.

2.4.4 Advanced Oxidation Processes

In advanced oxidization processes (AOPs), hydroxyl radicals (•OH) are produced in different ways and this highly reactive radical will react with pollutants leading their transformation to mineralized end products. The breakdown of a target molecule by the absorption of light that leads to its photolytic decomposition is called photolysis. Combining a light source with hydrogen peroxide (H_2O_2) leads to the formation of • OH. This can be enhanced by the presence of UV light. Studies have proved that ibuprofen can be eliminated completely by photolytic degradation with UV/TiO_2 (Kaur, Umar, and Kansal 2016). In a study conducted by Achilleos et al. (2014), only 4% photolytic degradation was obtained in the absence of TiO_2, which clearly shows the importance of the catalyst in the reaction. The rate of drug removal was affected by the drug type, the feed matrix, and the concentration of photocatalyst. Diclofenac was shown to disintegrate more quickly than ibuprofen and naproxen. In general, when the feed matrix varied, the efficiency of the drug removal changed in the following order: ultra pure water > tap water > secondary effluents > primary effluents (Darowna et al. 2014).

Ozonation is another efficient redox technology with a high degradation rate, non-selectivity and the disinfecting effect that exclusively reacts with double bonds as well as aromatic rings. Petrie et al. (2013) reviewed the removal of NSAIDs from municipal wastewater using the ozonation process and reported the removal efficiencies to be 46–62% for ibuprofen and 92–99%, 50–99%, 73–98% for diclofenac, naproxen and ketoprofen, respectively. Fenton is another AOP in which fenton's reagent oxidizes in the presence of ferrous salt and H_2O_2 in a simple redox reaction. Its efficiency increased in the presence of UV irradiations and this is called the photo-fenton process. Nadais et al. (2018) performed a study on the bio-electro-fenton process for the degradation of NSAIDs from water and wastewater. It is one of the AOPs which has received attention in the past few years. In this process the strong oxidant hydroxyl radical is formed using the electrons derived from bacterial oxidation of organic substrate. The treatment was analysed for ketoprofen, diclofenac, ibuprofen and naproxen in a microbial electrolysis cell based on the bio-electro-fenton process. After optimizing the parameters such as pH, Fe^{+2}, air-flow and applied voltage, 59–61% removal efficiency was obtained for ketoprofen, while diclofenac and ibuprofen were 87–97% and 80–86%, respectively. Since this treatment depends on hydroxyl radical, the organic matter present in the real wastewater matrix will compete with the NSAIDs, leading to lower removal efficiency. But it can be adopted as a promising and sustainable alternative for the treatment of wastewater containing low concentrations of NSAIDs.

2.5 CONCLUSION

NSAIDs are one group of pharmaceuticals considered to be emerging contaminants due to their presence in different water matrices such as wastewater and surface water. Even though their quantity ranges from µg/L–ηg/L, these can be a severe threat to aquatic life as a result of their direct disposal into water bodies. In India, very few studies have been conducted on the occurrence of NSAIDs in WWTPs. Those reported studies were reviewed in this chapter to understand the occurrence of NSAIDs in different regions of India. In contrast to the number of pharmaceutical manufacturing units' location, fewer investigations were done in the Northern region compared to the Southern region of India. In most cases, an increase in influent concentration of drugs was observed in every year, which highlights the importance of proper monitoring of NSAIDs in water matrices. An Environmental Risk Assessment can be performed to understand the potential eco-toxicological effects on aquatic organisms caused by these drugs. Due to the high risk caused by these drugs, proper removal of NSAIDs through different treatment technologies is essential. Even though WWTPs are inefficient in completely removing these drugs, different treatment units are capable of effective removal of certain pollutants. Removal mechanisms of NSAIDs were due to different treatment technologies, based on the physio-chemical properties of individual drugs, biodegradation, adsorption or other mechanisms can dominate. Factors like HRT and SRT are also important when considering the removal of these drugs.

REFERENCES

Abo El Naga, Ahmed O., Mohamed El Saied, Seham A. Shaban, and Fathy Y. El Kady. 2019. "Fast Removal of Diclofenac Sodium from Aqueous Solution Using Sugar Cane Bagasse-Derived Activated Carbon." *Journal of Molecular Liquids* 285: 9–19. https://doi.org/10.1016/j.molliq.2019.04.062.

Achilleos, A., E. Hapeshi, N.P. Xekoukoulotakis and D. Mantzavinos. 2014. "UV-A and Solar Photodegradation of Ibuprofen and Carbamazepine Catalyzed by TiO_2." *Separation Science and Technology* December: 37–41. https://doi.org/10.1080/01496395.2010.487463.

Afonso-Olivares, C., Z. Sosa-Ferrera, and J.J. Santana-Rodríguez. 2017. "Occurrence and Environmental Impact of Pharmaceutical Residues from Conventional and Natural Wastewater Treatment Plants in Gran Canaria (Spain)." *Science of the Total Environment* 599–600: 934–43. https://doi.org/10.1016/j.scitotenv.2017.05.058.

Ahmed, Muthanna J. 2017. "Adsorption of Non-Steroidal Anti-inflammatory Drugs from Aqueous Solution Using Activated Carbons: Review." *Journal of Environmental Management* 190: 274–82. https://doi.org/10.1016/j.jenvman.2016.12.073.

Anumol, Tarun, Arya Vijayanandan, Minkyu Park, Ligy Philip and Shane A. Snyder. 2016. "Occurrence and Fate of Emerging Trace Organic Chemicals in Wastewater Plants in Chennai, India." *Environment International* 92–3: 33–42. https://doi.org/10.1016/j.envint.2016.03.022.

Anumol, Tarun, Shimin Wu, Mauricius Marques Dos Santos, Kevin D. Daniels and Shane A. Snyder. 2015. "Rapid Direct Injection LC-MS/MS Method for Analysis of Prioritized Indicator Compounds in Wastewater Effluent." *Environmental Science: Water Research and Technology* 1 (5): 632–43. https://doi.org/10.1039/c5ew00080g.

Archana, G., Rita Dhodapkar and Anupama Kumar. 2016. "Offline Solid-Phase Extraction for Preconcentration of Pharmaceuticals and Personal Care Products in Environmental Water and Their Simultaneous

Determination Using the Reversed Phase High-Performance Liquid Chromatography Method." *Environmental Monitoring and Assessment* 188 (9): 512. https://doi.org/10.1007/s10661-016-5510-1.

Archana, G., Rita Dhodapkar and Anupama Kumar. 2017. "Ecotoxicological Risk Assessment and Seasonal Variation of Some Pharmaceuticals and Personal Care Products in the Sewage Treatment Plant and Surface Water Bodies (Lakes)." *Environmental Monitoring and Assessment* 189: 446. https://doi.org/10.1007/s10661-017-6148-3.

CHMP (Committee for Medicinal Products for Human Use) 2006. "Guideline on the Environmental Risk Assessment of Medicinal Products for Human Use." Doc. Ref.: EMEA/CHMP/SWP/4447/00 corr 1, London, 1 June.

Daniels, Kevin D., Minkyu Park, Zhenzhen Huang, Ai Jia, Guillermo S. Flores, Hian Kee Lee, Shane A. Snyder, et al. 2020. "A Review of Extraction Methods for the Analysis of Pharmaceuticals in Environmental Waters." *Critical Reviews in Environmental Science and Technology* 50 (21): 2271–99. https://doi.org/10.1080/10643389.2019.1705723.

Darowna, Dominika, Sara Grondzewska, Antoni W. Morawski and Sylwia Mozia. 2014. "Removal of Non-Steroidal Anti-Inflammatory Drugs from Primary and Secondary Effluents in a Photocatalytic Membrane Reactor." *Journal of Chemical Technology and Biotechnology* 89 (8): 1265–73. https://doi.org/10.1002/jctb.4386.

Fatehifar, Maryam, Seyed Mehdi Borghei and Ali Ekhlasi. 2018. "Application of Moving Bed Bio Fi Lm Reactor in the Removal of Pharmaceutical Compounds (Diclofenac and Ibuprofen)." *Journal of Environmental Chemical Engineering* 6 (4): 5530–5. https://doi.org/10.1016/j.jece.2018.08.029.

Fawzi, Omar, Suleiman Khasawneh and Puganeshwary Palaniandy. 2021. "Occurrence and Removal of Pharmaceuticals in Wastewater Treatment Plants." *Process Safety and Environmental Protection* 150: 532–56. https://doi.org/10.1016/j.psep.2021.04.045.

Fick, Jerker, Hanna Söderström, Richard H. Lindberg, Chau Phan, Mats Tysklind and D.G. Joakim Larsson. 2009. "Contamination of Surface, Ground, and Drinking Water from Pharmaceutical Production." *Environmental Toxicology and Chemistry* 28 (12): 2522–7. https://doi.org/10.1897/09-073.1.

García, M., D. Gil, F. Cañada, M.J. Culzoni, L. Vera-Candioti, G.G. Siano, H.C. Goicoechea and M. Martínez Galera. 2009. "Chemometric Tools Improving the Determination of Anti-Inflammatory and Antiepileptic Drugs in River and Wastewater by Solid-Phase Microextraction and Liquid Chromatography Diode Array Detection." *Journal of Chromatography A* 1216 (29): 5489–96. https://doi.org/10.1016/j.chroma.2009.05.073.

Izadi, Parnian, Parin Izadi, Rana Salem, Sifat Azad Papry, Sara Magdouli, Rama Pulicharla and Satinder Kaur Brar. 2020. "Non-Steroidal Anti-Inflammatory Drugs in the Environment: Where Were We and How Far We Have Come?" *Environmental Pollution* 267: 115370. https://doi.org/10.1016/j.envpol.2020.115370.

Jiang, Qi, Hao H. Ngo, Long D. Nghiem, Faisal I. Hai, William E. Price and Jian Zhang. 2017. "Bioresource Technology Short Communication Effect of Hydraulic Retention Time on the Performance of a Hybrid Moving Bed Biofilm Reactor-Membrane Bioreactor System for Micropollutants Removal from Municipal Wastewater." *Bioresource Technology* July. https://doi.org/10.1016/j.biortech.2017.09.114.

Kanaujiya, Dipak Kumar, Tanushree Paul, Arindam Sinharoy and Kannan Pakshirajan. 2019. "Biological Treatment Processes for the Removal of Organic Micropollutants from Wastewater: A Review." *Current Pollution Reports* 5: 112–28.

Kaur, Amandeep, Ahmad Umar and Sushil K. Kansal. 2016. "Heterogeneous Photocatalytic Studies of Analgesic and Non-Steroidal Anti-Inflammatory Drugs." *Applied Catalysis A, General* 510: 134–55. https://doi.org/10.1016/j.apcata.2015.11.008.

Kim, M., P. Guerra, A. Shah, M. Parsa, M. Alaee and S.A. Smyth. 2014. "Removal of Pharmaceuticals and Personal Care Products in a Membrane Bioreactor Wastewater Treatment Plant." *Water Science and Technology* 69 (11): 2221–9. https://doi.org/10.2166/wst.2014.145.

Kompare, Boris, Mojca Zupanc, Tina Kosjek, Martin Petkovšek, Ester Heath and Brane Širok. 2013. "Ultrasonics Sonochemistry Removal of Pharmaceuticals from Wastewater by Biological Processes, Hydrodynamic Cavitation and UV Treatment." *Ultrasonics Sonochemistry* 20: 1104–12. https://doi.org/10.1016/j.ultsonch.2012.12.003.

Lacey, C., G. McMahon, J. Bones, L. Barron, A. Morrissey and J.M. Tobin. 2008. "An LC-MS Method for the Determination of Pharmaceutical Compounds in Wastewater Treatment Plant Influent and Effluent Samples." *Talanta* 75 (4): 1089–97. https://doi.org/10.1016/j.talanta.2008.01.011.

Larsson, D.G. Joakim, Cecilia de Pedro and Nicklas Paxéus. 2007. "Effluent from Drug Manufactures Contains Extremely High Levels of Pharmaceuticals." *Journal of Hazardous Materials* 148 (3): 751–5. https://doi.org/10.1016/j.jhazmat.2007.07.008.

Luo, Yunlong, Wenshan Guo, Huu Hao, Long Duc, Faisal Ibney, Jian Zhang, Shuang Liang and Xiaochang C. Wang. 2014. "A Review on the Occurrence of Micropollutants in the Aquatic Environment and Their Fate and Removal during Wastewater Treatment." *Science of the Total Environment* 473–4: 619–41. https://doi.org/10.1016/j.scitotenv.2013.12.065.

Luo, Yunlong, Wenshan Guo, Huu Hao, Long Duc, Faisal Ibney, Jinguo Kang, Siqing Xia, Zhiqiang Zhang and William Evan. 2014. "Removal and Fate of Micropollutants in a Sponge-Based Moving Bed Bioreactor." *Bioresource Technology* 159: 311–19. https://doi.org/10.1016/j.biortech.2014.02.107.

Manthiram Karthik, R. and Ligy Philip. 2021. "Sorption of Pharmaceutical Compounds and Nutrients by Various Porous Low Cost Adsorbents." *Journal of Environmental Chemical Engineering* 9 (1). https://doi.org/10.1016/j.jece.2020.104916.

Mohapatra, Sanjeeb, Ching Hua Huang, Suparna Mukherji and Lokesh P. Padhye. 2016. "Occurrence and Fate of Pharmaceuticals in WWTPs in India and Comparison with a Similar Study in the United States." *Chemosphere* 159: 526–35. https://doi.org/10.1016/j.chemosphere.2016.06.047.

Nadais, Helena, Xiaohu Li, Nadine Alves, Cátia Couras, Henrik Rasmus Andersen, Irini Angelidaki and Yifeng Zhang. 2018. "Bio-Electro-Fenton Process for the Degradation of Non-Steroidal Anti-Inflammatory Drugs in Wastewater." *Chemical Engineering Journal* 338: 401–10. https://doi.org/10.1016/j.cej.2018.01.014.

Narain, S. and S. Sengupta. 2016. "Why Urban India Floods—Indian Cities Grow at the Cost of Their Wetlands." New Delhi: Centre for Science and Environment.

Papageorgiou, Myrsini, Christina Kosma and Dimitra Lambropoulou. 2016. "Seasonal Occurrence, Removal, Mass Loading and Environmental Risk Assessment of 55 Pharmaceuticals and Personal Care Products in a Municipal Wastewater Treatment Plant in Central Greece." *Science of the Total Environment* 543: 547–69. https://doi.org/10.1016/j.scitotenv.2015.11.047.

Park, Junwon, Changsoo Kim, Youngmin Hong, Wonseok Lee, Hyenmi Chung, Dong-hwan Jeong and Hyunook Kim. 2021. "Distribution and Removal of Pharmaceuticals in Liquid and Solid Phases in the Unit Processes of Sewage Treatment Plants." *International Journal of Environmental Research* 17 (3): 687.

Parolini, Marco. 2020. "Toxicity of the Non-Steroidal Anti-Inflammatory Drugs (NSAIDs) Acetylsalicylic Acid, Paracetamol, Diclofenac, Ibuprofen and Naproxen towards Freshwater Invertebrates: A Review." *Science of the Total Environment* 740: 140043. https://doi.org/10.1016/j.scitotenv.2020.140043.

Patel, Manvendra, Rahul Kumar, Kamal Kishor, Todd Mlsna, Charles U. Pittman and Dinesh Mohan. 2019. "Pharmaceuticals of Emerging Concern in Aquatic Systems: Chemistry, Occurrence, Effects, and Removal Methods." *Chemical Reviews* 119 (6): 3510–673. https://doi.org/10.1021/acs.chemrev.8b00299.

Paxéus, Nicklas. 2004. "Removal of Selected Non-Steroidal Anti-Inflammatory Drugs (NSAIDs), Gemfibrozil, Carbamazepine, β-Blockers, Trimethoprim and Triclosan in Conventional Wastewater Treatment Plants in Five EU Countries and Their Discharge to the Aquatic Environment." *Water Science and Technology* 50 (5): 253–60. https://doi.org/10.2166/wst.2004.0335.

Petrie, B. et al. 2013. "Fate of Drugs During Wastewater Treatment." *TrAC Trends in Analytical Chemistry*, 49: 145–59. https://doi.org/10.1016/j.trac.2013.05.007.

Philip, Jeeva M., Usha K. Aravind and Charuvila T. Aravindakumar. 2018. "Emerging Contaminants in Indian Environmental Matrices—A Review." *Chemosphere* 190: 307–26. https://doi.org/10.1016/j.chemosphere.2017.09.120.

Prabhasankar, Valiparambil Prabhakaranunni, Derrick Ian Joshua, Keshava Balakrishna, Iyanee Faroza Siddiqui, Sachi Taniyasu, Nobuyoshi Yamashita, Kurunthachalam Kannan, … Keerthi S. Guruge. 2016. "Removal Rates of Antibiotics in Four Sewage Treatment Plants in South India." *Environmental Science and Pollution Research* 23 (9): 8679–85. https://doi.org/10.1007/s11356-015-5968-3.

Praveenkumarreddy, Yerabham, Krishnamoorthi Vimalkumar, Babu Rajendran Ramaswamy, Virendra Kumar, Rakesh Kumar Singhal, Hirakendu Basu, Chikmagalur Mallappa Gopal, et al. 2021. "Assessment of Non-Steroidal Anti-Inflammatory Drugs from Selected Wastewater Treatment Plants of Southwestern India." *Emerging Contaminants* 7: 43–51. https://doi.org/10.1016/j.emcon.2021.01.001.

Salgado, R., R. Marques, J.P. Noronha and G. Carvalho. 2012. "Assessing the Removal of Pharmaceuticals and Personal Care Products in a Full-Scale Activated Sludge Plant." *Environmental Science and Pollution Research* 19: 1818–27. https://doi.org/10.1007/s11356-011-0693-z.

Saxena, Priyam, Isha Hiwrale, Sanchita Das, Varun Shukla, Lakshay Tyagi, Sukdeb Pal, Nishant Dafale and Rita Dhodapkar. 2021. "Profiling of Emerging Contaminants and Antibiotic Resistance in Sewage Treatment Plants: An Indian Perspective." *Journal of Hazardous Materials* 408: 124877. https://doi.org/10.1016/j.jhazmat.2020.124877.

Shanmugam, Govindaraj, Srimurali Sampath, Krishna Kumar Selvaraj, D.G. Joakim Larsson and Babu Rajendran Ramaswamy. 2014. "Non-Steroidal Anti-Inflammatory Drugs in Indian Rivers." *Environmental Science and Pollution Research* 21 (2): 921–31. https://doi.org/10.1007/s11356-013-1957-6.

Singh, Kunwar P., Premanjali Rai, Arun K. Singh, Priyanka Verma and Shikha Gupta. 2014. "Occurrence of Pharmaceuticals in Urban Wastewater of North Indian Cities and Risk Assessment." *Environmental Monitoring and Assessment* 186 (10): 6663–82. https://doi.org/10.1007/s10661-014-3881-8.

Subedi, Bikram, Keshava Balakrishna, Derrick Ian Joshua and Kurunthachalam Kannan. 2017. "Mass Loading and Removal of Pharmaceuticals and Personal Care Products Including Psychoactives, Antihypertensives, and Antibiotics in Two Sewage Treatment Plants in Southern India." *Chemosphere* 167: 429–37. https://doi.org/10.1016/j.chemosphere.2016.10.026.

Subedi, Bikram, Keshava Balakrishna, Ravindra K. Sinha, Nobuyoshi Yamashita, Vellingiri G. Balasubramanian and Kurunthachalam Kannan. 2015. "Mass Loading and Removal of Pharmaceuticals and Personal Care Products, Including Psychoactive and Illicit Drugs and Artificial Sweeteners, in Five Sewage Treatment Plants in India." *Journal of Environmental Chemical Engineering* 3 (Part 4A): 2882–91. https://doi.org/10.1016/j.jece.2015.09.031.

Tang, Kai, Gordon T.H. Ooi, Klaus Litty, Kim Sundmark, M.S. Kamilla, Christina Sund, Caroline Kragelund, Magnus Christenson and Henrik R. Andersen. 2017. "Removal of Pharmaceuticals in Conventionally Treated Wastewater by a Polishing Moving Bed Biofilm Reactor (MBBR) with Intermittent Feeding." *Bioresource Technology*. https://doi.org/10.1016/j.biortech.2017.03.159.

Terzi, Murat, Gamze Altun, Sedat Şen, Adem Kocaman, Arife Ahsen Kaplan, Kıymet Kübra Yurt and Süleyman Kaplan. 2018. "The Use of Non-Steroidal Anti-Inflammatory Drugs in Neurological Diseases." *Journal of Chemical Neuroanatomy* 87: 12–24. https://doi.org/10.1016/j.jchemneu.2017.03.003.

Thalla, Arun Kumar and Adhira Shree Vannarath. 2020. "Occurrence and environmental risks of nonsteroidal anti-inflammatory drugs in urban wastewater in the southwest monsoon region of India." *Environmental Monitoring and Assessment* 192 (3): 1–13. https://link.springer.com/content/pdf/10.1007/s10661-020-8161-1.pdf.

Vieno, Niina and Mika Sillanpää. 2014. "Fate of Diclofenac in Municipal Wastewater Treatment Plant—A Review." *Environment International* 69: 28–39. https://doi.org/10.1016/j.envint.2014.03.021.

Vystavna, Y., Z. Frkova, L. Marchand, Y. Vergeles and F. Stolberg. 2017. "Removal Efficiency of Pharmaceuticals in a Full Scale Constructed Wetland in East Ukraine." *Ecological Engineering* 108 (May): 50–8. https://doi.org/10.1016/j.ecoleng.2017.08.009.

Yamkelani, Nomchenge, Somandla Ncube, Precious Nokwethemba, Luke Chimuka and Lawrence Mzukisi. 2019. "Adsorbents and Removal Strategies of Non-Steroidal Anti-inflammatory Drugs from Contaminated Water Bodies." *Journal of Environmental Chemical Engineering* 7 (3): 103142. https://doi.org/10.1016/j.jece.2019.103142.

Zgoła-Grześkowiak, Agnieszka. 2010. "Application of DLLME to Isolation and Concentration of Non-Steroidal Anti-Inflammatory Drugs in Environmental Water Samples." *Chromatographia* 72 (7–8): 671–8. https://doi.org/10.1365/s10337-010-1702-y.

Zhang, Xiaomeng, Ruiying Jing, Xu Feng, Yunyu Dai, Ran Tao, Jan Vymazal, Nan Cai and Yang Yang. 2018. "Removal of Acidic Pharmaceuticals by Small-Scale Constructed Wetlands Using Different Design Configurations." *Science of the Total Environment* 639: 640–7. https://doi.org/10.1016/j.scitotenv.2018.05.198.

Zhao, Longshan, Ning Liang, Xiaowen Lun, Xin Chen and Xiaohong Hou. 2014. "LC-QTOF-MS Method for the Analysis of Residual Pharmaceuticals in Wastewater: Application to a Comparative Multiresidue Trace Analysis between Spring and Winter Water." *Analytical Methods* 6 (17): 6956–62. https://doi.org/10.1039/c4ay01200c.

Zorita, Saioa, Lennart Mårtensson and Lennart Mathiasson. 2009. "Occurrence and Removal of Pharmaceuticals in a Municipal Sewage Treatment System in the South of Sweden." *Science of the Total Environment* 407 (8): 2760–70. https://doi.org/10.1016/j.scitotenv.2008.12.030.

Zwiener, Christian and Fritz H. Frimmel. 2004. "LC-MS Analysis in the Aquatic Environment and in Water Treatment—A Critical Review: Part II: Applications for Emerging Contaminants and Related Pollutants, Microorganisms and Humic Acids." *Analytical and Bioanalytical Chemistry* 378 (4): 862–74. https://doi.org/10.1007/s00216-003-2412-1.

ABBREVIATIONS

AA	Arachidonic acid
AF	Assessment factor
APCI	Atmospheric pressure chemical ionization
ASP	Activated sludge process
COX	Cyclooxygenase
DIC	Diclofenac
DLLME	Dispersive liquid-liquid micro extraction
GC	Gas chromatography
HPLC	High performance liquid chromatography
HPTLC	High performance thin-layer chromatography
HRT	Hydraulic retention time
IBU	Ibuprofen
KET	Ketoprofen
KOW	Octanol-water partition coefficient
LC-MS	Liquid chromatography mass spectrometry
LC-QTOF MS	Liquid chromatography coupled to quadrupole time-of-flight mass spectrometry
MBBR	Moving bed bio-reactor
MBR	Membrane bioreactor
MEC	Measured environmental concentration
MF	Microfiltration
MWWTP	Mangalore wastewater treatment plant
ŋg/L	Nano grams per liter
NOEC	No observed effect concentration
PG	Prostaglandin
pKa	Dissociation constant
PNEC	Predicted no effect concentration
RQ	Risk quotient
SPE	Solid phase extraction
SRT	Sludge retention time
TLC	Thin-layer chromatography
TOF	Time-of-flight
UF	Ultrafiltration
µg/L	Micrograms per liter
WWTP	Wastewater treatment plant

3 Organochlorine and Organophosphate Pesticides and Emerging Pollutants in the Ganga River System

An Overview

Leena Singh[1,*] *and Aryaman Singh*[2]
[1]Department of Environmental Sciences Kirori Mal College, University of Delhi, Delhi, India
[2]NALSAR University of Law, Hyderabad, Telangana, India
*Corresponding author leenaplato@gmail.com

CONTENTS

3.1 Introduction 49
3.2 History of the Use of Pesticides 51
3.3 The Role of Pesticides in Water Quality 51
3.4 Review of Pesticides and their Presence in Water and Sediment of River System 53
3.5 Organochlorines and Organophosphates Pesticides in the Ganga River: Case Study of Silk City, Bhagalpur, Bihar, India 56
3.5.1 Lindane 57
3.5.2 Methyl Parathion 58
3.5.3 Endosulfan (α- and β-Endosulfan) 58
3.5.4 DDT (o,p'-DDT, orthopara-DDT and p,p'-DDT, parapara-DDT) 59
3.6 Regulatory Measures Relating to Emerging Contaminants 60
3.6.1 International Law 61
3.6.1.1 The Rotterdam Convention 61
3.6.1.2 The Stockholm Convention 61
3.7 Conclusion 62
References 62

3.1 INTRODUCTION

The increasing global population results in the growing demand for food which leads to a considerable increase in the production of agrochemicals, such as pesticides and fertilizers, causing continued contamination of air and water environments (Awasthi et al. 2018, 2020; Negi et al. 2019; Sirohi et al. 2021). Pesticide chemicals are one of the most important sources of pollution which often have disastrous consequences. They are classified in the category of emerging contaminants pollutants (ECPs), as they belong to synthetic chemicals that are not regularly monitored in the environment and are known to cause or suspected to cause adverse ecological and human health effects after entering the ecosystem or the environment. These chemicals are known as "contaminants of emerging concern"

DOI: 10.1201/9781003239956-4

(CECs). "Emerging contaminants are important because the risk they pose to human health and the environment is not yet fully understood" (Chowdhary and Raj 2020; Chowdhary et al. 2020). Trace amounts of these contaminants have been discovered in water throughout India's environmental matrices for a long time and pose a serious threat to biodiversity.

With the changes in farm practices and the growth of intensive agriculture, pesticide use has increased drastically over the last two decades. The residues cropping up in environmental matrices are a result of this dramatic increase in pesticide use for agricultural and non-agricultural purposes.

A pesticide is a mixture of substances, which is used to prevent and control the damage caused by a pest. A pesticide may be a chemical substance, biological agent (such as a bacteria, virus or fungi), an antimicrobial disinfectant, or any medium used against pests. Pests include insects, weeds, plant pathogens, nematodes, mollusks, fish, birds, mammals, and microbes that compete with humans for food, and can destroy property, or are a vector for disease or cause a nuisance. Yet there are also benefits as well as impacts in the use of pesticides which have potential toxicity to humans and other animals.

Pesticides are artificially synthesized substances used to fight and improve agricultural production (Shabbir et al. 2018; Negi et al. 2021). At present, a large number of pesticides of varying chemical natures are used throughout the world. The detection of these in environmental matrices such as soil, water and air is attributed to their widespread use.

Broadly, pesticides can be categorized into chemical pesticides and bio-pesticides. Chemical pesticides can be classified as organophosphate, organochlorine, carbamate, and pyrethroid. Bio-pesticides can be classified as microbial pesticides, plant-incorporated protectants (PIPS), and biochemical pesticides. Organochlorine and organophosphorus are the most important chemical pesticides. Organochlorine pesticides (OCPs) are persistent organic pollutants and are a concern around the world due to their chronic toxicity, persistence, and bioaccumulation (Willett et al. 1998). The threat posed by these compounds to ecosystems and human health is quite high. Many countries, especially developed nations, have banned or restricted the application of these chemicals; however, some developing countries, as a result of the pesticides low cost and versatility, continue to use them (Tanabe et al. 1994; Sarkar et al. 1997).

Organochlorines, with a reported half-life between 2–15 years, are highly persistent pesticides in the environment (The United States Environmental Protection Agency 1989; Augustjin-Beckers et al. 1994) and are immobile in most soils. Routes of loss and degradation include run-off, volatilization, photolysis, and biodegradation (aerobic and anaerobic) (Agency for Toxic Substances and Disease Registry 2005). Their extremely low solubility in water, particularly of DDT, means they tend to be retained in the soil and soil fractions, bound to a higher proportion of soil organic matter (WHO 1989). They have been detected in soil and underground water in many places, where they may be available to affect organisms (ibid.; US EPA 1989). This is probably due to their high persistence nature. Although they are immobile or slightly mobile, after several years, they may be able to eventually leach underground, especially in soils with little organic matter.

An organochlorine pesticide does not degrade and is concentrated through the food chain and produces a significant magnification of the original concentration at the end of the chain. It has been cited that the degradation of DDT in soils is 75–100% in 4–30 years. They stay in the environment for a long time so it is important that the pollution they cause is examined.

Commonly used organochlorines include DDT, lindane, endosulfan, aldrin, dieldrin, chlordane, heptachlor, etc. Organophosphates (OP) are a group of insecticides or nerve agents which act on the enzyme acetylcholinesterase (Chandra and Kumar 2015). The term is used often to describe virtually any organic compound containing phosphorus, especially when dealing with neurotoxins.

Organophosphate pesticides affect this enzyme in several ways and can cause poisoning even at low levels of exposure. Organophosphates degrade rapidly by hydrolysis on exposure to sunlight, air, and soil, but can be detected in food and drinking water in small amounts. Due to their easy degradation ability, they become an attractive alternative to the persistent organochlorines, such as DDT, aldrin, and dieldrin.

Although organophosphates degrade faster than organochlorines, they have greater acute toxicity, posing risk to people who may be exposed to large amounts. These pesticides are a very common cause of poisoning worldwide and are frequently and intentionally used in suicides in agricultural areas. Their toxicity is not limited to the acute phase, rather, chronic effects also have long been noted. Commonly used organophosphates include methyl parathion, parathion, malathion, diazinon, chlorpyrifos, dichloroos, phosmet, tetrachlorinphos, and azinphosmethyl.

The pathways of these pesticides into the rivers include run-off from non-point sources, discharge of industrial wastewater, wet or dry deposition, and other means. Therefore, residues of OCPs might ultimately pass on to people through the consumption of drinking water, fish, and agricultural food (Ochome et al. 1984; Kawano et al. 1988). The organochlorine pesticides show an affinity for particulate matter, and river and marine sediments are one of their main sinks. Thus, investigation of the sediment samples of the river can indicate the status of aquatic contamination.

Chlorinated pesticides like DDT and HCH are used extensively in India, both for agricultural and sanitary purposes. It is estimated that about 25,000 MT of chlorinated pesticides are used annually in India and DDT accounted for 40% of this group (Mathur 1993). However, most of these pesticides have now been banned in the country (UNEP 2002) but are still being used as pest control chemicals in agriculture and public health activities (malaria eradication, etc.) in developing countries.

The Ganga River system in India contains pesticides as the most potentially hazardous pollutants due to its socio-economic considerations. In the present study the qualitative and quantitative aspects of pesticide pollution are categorized as an emerging contaminant in the water and the sediments of the Ganga River system.

3.2 HISTORY OF THE USE OF PESTICIDES

Humans have been using pesticides since 2500 BC to protect their crops. The elemental sulfur dusting was the first pesticide used in Sumeria about 4500 years ago. In the fifteenth century, lead, mercury and arsenic, which were toxic chemicals compounds, were used on crops to kill pests. During the seventeenth and nineteenth centuries, nicotine sulfate was extracted from tobacco leaves and used as an insecticide, and pyrethrum, derived from chrysanthemums, and rotenone, derived from the roots of tropical vegetables, were used as pesticides. Paul Muller discovered that DDT was a very effective insecticide and it became the most widely used pesticide in the world in 1939. Graeme Murphy (2005) considers the 1940s and 1950s to be the start of the "pesticide era". Table 3.1 shows the chronology of pesticide development.

3.3 THE ROLE OF PESTICIDES IN WATER QUALITY

The aquatic environment receives pesticide residues through run-off, leaching, careless disposal of empty containers, washing of industrial equipment, etc. The primary movement of pesticides from agricultural land toward surface water is a result of surface run-off (Richards and Baker 1993). Factors such as the characteristics of the soil, agricultural practices, the weather, the topography, the chemical properties of the pesticides, and the environmental properties of the combination of pesticides and the weather impact the amount lost from fields in the form of surface run-off (Wagenet 1987; Leonard 1990). The combined effect of these factors on the temporal and spatial magnitude of pesticide concentrations and fluxes in large integrating river systems is largely unknown (Larson et al. 1995). The physical displacements and chemical reactions influence the persistence of the chemicals in the soil and have different environmental implications. The adequately

TABLE 3.1
Chronology of Pesticide Development

Period	Example	Source	Characteristics
1800–the 1920s	Early organics, nitro-phenols, chlorophenols, creosote, naphthalene, petroleum oils	Organic chemistry, by-products of coal gas production, etc.	Often lack specificity and were toxic to the user or non-target organisms
1945–1955	Chlorinated organics, DDT, HCCH, chlorinated cyclodienes	Organic synthesis	Persistent, good selectivity, good agricultural properties, good public health performance, resistance, harmful ecological effects
1945–1970	Cholinesterase inhibitors, organophosphorus compounds, carbamates	Organic synthesis, good use of structure-activity relationships	Lower persistence, some user toxicity, some environmental problems
1970–1985	Synthetic pyrethroids, avermectins, juvenile hormone mimics, biological pesticides	Refinement of structure-activity relationships, new target systems	Some lack of selectivity, resistance, costs, and variable persistence
1985–	Genetically engineered organisms	Transfer of genes for biological pesticides to other organisms and into beneficial plants and animals. Genetic alteration of plants to resist non-target effects of pesticides	Possible problems with mutations and escapes, disruption of microbiological ecology, a monopoly on products

Source: Stephenson and Solomon (1993), cited in FAO (n.d.).

soluble chemicals which are resistant to degradation may be transported in the water and reach the water bodies in significant amounts (Wauchope 1978; Wagenet 1987).

Throughout the world, several hundred pesticides of different chemical natures are currently used for agricultural purposes. They are found in various environmental matrices, such as soil, water, and air because of their widespread use. The interaction of pesticides with soils, surface waters, and groundwater is controlled by numerous physical, chemical, and biological processes, i.e., transformation; transfer, and transport. Transformation refers to a biological and chemical process that changes the structure of a pesticide or completely degrades it, i.e., microbial degradation (both aerobic and anaerobic). Transfer refers to how a pesticide is distributed between solids and liquids or between solids and gases, i.e., sorption. Transport is the movement from one environmental compartment to another, such as leaching of pesticides through the soil to groundwater, volatilization into the air, and run-off to surface water. Pesticides can enter river water through all of the above sources but the transport mechanism plays a distinctive role.

Organochlorine pesticides (OCPs) have low waters solubility and a strong affinity for particulate matter. As a result, sediment can serve as an ultimate sink for them. The combination of their physicochemical properties such as low aqueous solubility, moderate vapor pressure, an octanol-water partition coefficient, and persistence in the environment makes them capable of long-range transport. As a result of the extensive use of chlorinated pesticides in India, traces of the same have been detected in numerous sectors of the environment, which have shown a potential to biomagnify or accumulate in animal tissues, human blood, adipose tissue, and breast milk (Beg et al. 1989). The lipophilic nature of the pesticide and the cumulative accumulation of low concentrations of these in the body fat of mammals might pose potential hazards in the long run (Metcalf 1997).

3.4 REVIEW OF PESTICIDES AND THEIR PRESENCE IN WATER AND SEDIMENT OF RIVER SYSTEM

Pesticides are chemical substances that are employed in agriculture, forestry, and horticulture to combat animal and plant parasites. They have made a valuable contributions to world food production and hygiene, especially in the past 40 years, which has given rise to the many environmental problems related to the ample use of slowly degrading pesticides. The aquatic system is heavily polluted with several types of pesticides of a different chemical nature (Callahan et al. 1979; Edward et al. 1979; Bakre et al. 1990). India, being an agricultural country, has been using a huge number of pesticides to protect crops, so that makes pesticides and their related substances the most important pollutants. Many of the pesticides enter the aquatic environment as an intentional application (Nimmo 1985). The water of the Ganga river is affected by the toxicity of these chemicals, causing water and sediment pollution. Millions of people using the Ganga water routinely for drinking and outdoor bathing suffer enormously due to the pesticide pollution of the river.

Public awareness regarding the degradation of the environment has made the use of pesticides questionable and it is now necessary to re-evaluate their use in crop protection and public health. This review concerns the presence of organochlorine and organophosphate in the water and the sediment of the riverine system and its mutagenicity.

At the international level, there have been many studies on the degree of pollution in the river system with special emphasis on its pesticides profile. Callahan et al. (1979) studied 129 priority pollutants and their water-related environmental fate. Wang et al. (1980) reported that the water and sediment of Indian river lagoons contain 0.01 ppb of DDT. Albanis et al. (1986) studied organophosphorus and carbamates pesticide residue in the aquatic system of the Ioannia basin and Kalamas River in Greece. Sukamoto and Hayatsu (1990) studied the simple methods of monitoring the mutagenicity of river water. Fielding et al. (1991) reported the presence of pesticides in groundwater and drinking water, whereas Galassi et al. (1992) noted the toxic and chemical nature of pesticides in rivers and potable waters. Brambilla et al. (1993) reported on the fate of pesticides in river potable waters, whereas Albanis et al. (1994) reported on pesticide transportation in the estuaries of the Lowros and Anacthos River. Albanis et al. (1995) also evaluated the chemical parameters in the Aliakmon River in Greece. They also studied the multi-residue pesticide analysis in various water samples. Larson et al. (1995) studied the relations between pesticide use and riverine flux in the Mississippi River basin, whereas Angelidis et al. (1996) tested pesticide residues in the Evros River delta in Greece. Griffini et al. (1997) noted the presence of pesticides in the potable water of the Arno River. Gascon et al. (1998) monitored organonitrogen pesticide in the Ebro River and Gotz (1998) also reported on the pesticide concentration in the Louros River. Skoulikidis et al. (1998) studied the freshwater resources of rivers and lakes in Greece and their environmental state. Aguilar et al. (1991) monitored pesticides in river water using liquid chromatography and mass spectrometry. The methodology for the detection of organophosphorus and organochlorine pesticides was studied by Ballesteros and Parrado (2003), Totolin (2003), and Vagi et al. (2003). Albanis et al. (1998) monitored three rivers and a lake in the Epirus region for the detection of target pesticides, whereas the pesticides level in water and soil samples was reported by Ferencz et al. (2005) from Murus in Romania. Zehra et al. (2006) reported the concentration of organochlorine pesticides in the range of 7.68–269.4 ngl^{-1} and 23.11–316.5 $ng\ g^{-1}$ dry weight, respectively for 13 organochlorine pesticides studied in surface water and sediment from the Qiantang River in East China. Jung et al. (2006) have also studied organochlorine pesticides in Yannan lakes and rivers in China, whereas Erik et al. (2006) carried out a five-year survey of water-polluting pesticides in Hungary. The presence of OCP and OPPs were reported by Elisabeth et al. (2006) in the Queme River, which is one of the most important rivers in the Republic of Benin. Holvoet et al. (2007) conducted an intensive pesticide monitoring campaign in Belgium. Pasti et al. (2007) found 42 pesticides present from 11 points along the Ferrara River area, Italy. Various authors have studied the analysis, extraction, and removal of pesticides from water and sediments of water bodies (Ajmal et al. 1987; Ormad et al. 2007; Poolpak et al. 2007; Moral et al. 2008).

In India, the pesticide profile of river water and sediment has received little attention (Kusamran et al. 1994). A total of 114 cities discharge raw sewage into the Ganga River. But in many respects, pesticides pose a greater threat. Extensive farming, even on the river bed, uses a variety of pesticides. Most of these pesticides are organochlorine, which is known for high toxicity and bio-accumulation. Organochlorine pesticides such as hexa-chloro-cyclo hexane (HCH) and dichloro diphenyl trichloro ethane (DDT) account for two-thirds of the total consumption of pesticides in the country. Though DDT has been banned, it is commonly used to control the vectors of *Kala-azar* and malaria. To control the vector of *Kala-azar*, 2192.58 metric tons of DDT were sprayed in Bihar during 1995–1998. The catchment area of the Ganga River and its tributaries in Bihar are prone to this disease and DDT spraying is the only precautionary measure being taken by the state. As a result, a significant quantity of DDT residues ends up in the river through surface run-off.

A study by the Industrial Toxicological Research Centre in Lucknow (ITRC 1987) (Table 3.2) shows the presence of DDT in concentrations exceeding the US Environmental Protection Agency's criterion value of 0.001 ppb for freshwater aquatic life in the water of the Ganga. The concentration of DDT in the water of the Ganga was in the range of 0.07–143 ppb with a mean value of 13 ppb as proposed by the World Health Organization (WHO) for drinking water. The Ganga water is a major source of drinking water for the cities and villages located on its banks. The ITRC study also found as much as 0.5671 ppb high concentration of endosulfan, another organochlorine pesticide used in the agricultural sector, in the Ganga at Allahabad. Long-term exposure to endosulfan is supposed to affect the reproductive physiology in humans besides increasing the chances of breast cancer.

Nimmo (1985) worked on pesticide methodology and application, whereas Singh et al. (1987) worked on the analysis of pesticide residue in water. Agarwal et al. (1986) found DDT residue in the River Yamuna in Delhi. Sarkar and Sen Gupta (1988) reported DDT residues in sediments from the Bay of Bengal. Haldar et al. (1989) noted residues of endosulfan and DDT in Ganga water (Hooghly) from a selected stretch. Of 35 samples taken for study, DDT was reported in 12 samples in a concentration of 0.004–0.000006 ppm. The presence of endosulfan residues was very insignificant. Bakre et al. (1990) found organochlorine residues in water from the Mahala reservoir in Jaipur.

TABLE 3.2
Organochlorine Pesticides in the Ganga River System

	DDT(ppb)		HCH (ppb)	
Location	**Fish[1]**	**Water[2**]**	**Fish[1]**	**Water[2**]**
Rishikesh	NA	0.00382–0.0984	NA	0.00596–0.12428
Haridwar	3700	0.00234–0.11258	110	0.00414–0.15375
Kannauj	NA	0.00336–0.14948	NA	0.00802–0.15426
Kanpur	300	0.00846–0.17391	77	0.01488–0.35942
Allahabad	120	0.00242–0.13642	57	0.00684–0.2699
Varanasi	NA	0.00308–0.08404	NA	0.0091–0.15616
Patna	1300	0.00472–0.38488	110	0.01092–0.13098
Farakka	60	NA	28	NA
Uluberia	NA	0.0056–0.05742	NA	0.0094–0.1875

Notes: 1. See Kumar et al. (1999).
2. ITRC (1993).
NA: Data not available.
** Range of averages of annual minimum and maximum concentrations during 1986–1992.

ISGE (1990), under the Ganga project, has also reported the presence of pesticides between Kachla and Kannauj. Agnihotri et al. (1994) monitored different residues of organochlorine insecticides in the Ganges River water near Farrukhabad, with the presence of HCH and DDT in almost all the samples and residues of aldrin, endosulfan, and heptachlor were present in a huge number of samples. Rehana et al. (1995) studied the genotoxicity of the Ganges water at Narora in Uttar Pradesh and reported many pesticides such as DDT, BHC, aldrin, endrin, and dieldrin. The mutagenic activity of the Ganges water, especially with pesticide pollution in the river between Kachla to Kannauj in Uttar Pradesh, was also studied by them. The presence of organochlorines residues in the middle stream of the Ganga was reported in almost all the water samples collected along the bank of the river in the Varanasi region (Nayak et al. 1995). Rajendran Babu et al. (1997; 1999) reported the presence of pesticide residues in water from the River Kaveri. They also reported the presence of pesticide residue in surface sediment. Ghosh et al. (2000) noted the presence of DDT, HCH, and endosulfan residues in the cover stretch of the Ganga as 0.006, 0.003, and 0.0016 μgl^{-1} respectively, while in sediment as 0.047, 0.3000, and 0.009 μgkg^{-1} respectively. Pandit et al. (2001) studied the monitoring of organochlorine pesticide residues in the Indian marine environment. Rajendran et al. (2004) studied the ecotoxicological implications and their distribution of PCBs, HCHs, and DDTs in the Bay of Bengal. Sankararamakrishnan et al. (2004) conducted a survey and reported the presence of a higher concentration of organochlorine and organophosphorus pesticides in the surface and groundwater samples of the River Ganges in Kanpur, Uttar Pradesh. Malathion was found to be in much higher concentrations as per the water quality standards in the groundwater samples from industrial areas. Singh et al. (2004) studied the persistent organochlorine pesticide (OPPs) residues and their distribution in the Gomti River. Kole et al. (2005) studied the multi-residue analysis of pesticides for the evaluation of surface water quality of the Ganga in West Bengal. They state about 87.8% of 278 samples taken for the study were found to contain a combination of OCPs residues, such as HCH, DDT, and endosulfan in the range of 0.0004–18.78 μgl^{-1} and 62.5% were detected for OPPs such as dimethoate, malathion or methyl parathion in the range of 0.008–5.33 μgl^{-1}. Asma and Malik (2004) reported on the genotoxicity of the Yamuna River water at Okhla, Delhi, and reported the presence of DDT, BHC, dieldrin, endosulfan, aldrin, methyl parathion, and malathion at concentrations of 14, 25, 2.1, 114, 0.9, 0.6, 0.9, 1.7, and 1.9 μgl^{-1}, respectively.

A detailed investigation of the water quality of the Yamuna River with special reference to a pesticide profile from Delhi to Agra was conducted by the Central Pollution Control Board CBCB and the Ministry of the Environment and Forests in 1999–2005, and it reported the presence of various pesticides on several stretches from Delhi to Agra.

Recent studies have shown pesticides being abundantly used in the river basin and the agricultural fields along its bank (Singh et al. 2012; Chakarborty et al. 2016; PQRS 2017; Mondal et al. 2018).

In Bihar, not much emphasis has been given to studies of the pesticide profile of the River Ganges, except for a few reports on selected stretches of the river.

Anupma et al. (2002) have found the presence of HCHs, DDTs, aldrin, and endosulfan residues in Ganga water between Buxar and Rajmahal in Bihar. The concentrations of DDTs, HCHs, aldrin, and endosulfan in the river water were 0.019–1.663, 0.189–2.597, ND–0.8, and ND–0.862 μgl^{-1}, respectively. The concentration of these organochlorines exceeded the permissible limits recommended by US EPA. The study by Dr. Anupma and her team reveals a higher concentration of HCH in water than DDT, but it was less than the DDT found in fish of commercial importance, reflecting the greater bio-accumulative tendency of DDT in the Ganga River system.

Thus, from the above review, it is clear that pesticide pollution of the Ganga appears to be more critical than the point sources of the pollution, such as domestic and industrial discharge. In this chapter, four pesticides, both organochlorine and organophosphates, have been monitored for their presence, quantification, and ecotoxicity in river water and sediment: lindane, DDT, and methyl parathion as organochlorines, and endosulfan as organophosphates.

3.5 ORGANOCHLORINES AND ORGANOPHOSPHATES PESTICIDES IN THE GANGA RIVER: CASE STUDY OF SILK CITY, BHAGALPUR, BIHAR, INDIA

The River Ganga and its tributaries constitute the most extensive riverine system in India. In its journey from the Himalayas to the Bay of Bengal, the river system receives a huge burden of urban, agricultural, and industrial effluents which makes the water highly polluted and unpotable. A large number of dead bodies and the ashes of the dead are also regularly dropped in the Ganga throughout its stretch. The quantum of these pollutants has degraded the ecological health of the Ganga.

Bhagalpur is also known as Silk City and is the oldest divisional headquarter of eastern Bihar. The town is situated on the southern bank of the River Ganga at 85^{0}59' east longitude and 25^{0}15' north latitude. The elevation is about 172 feet from the mean sea level. The surface land is a plain and the river flows from west to east, separating north and south Bhagalpur. Several streams meet the River Ganga in the southern parts which have their origin in the hills of Santhal Parganas. Channan, the most important river of this region, is divided into about 15 smaller channels, of which only two can touch the River Ganga. One of these streams meets the Ganga at Champanagar, 3 km west of Bhagalpur, and it is one of the sampling sites in the present study. The Ganga River has moved from south to north in recent years, and presently the main course of the river is about 4–5 km north of the Nathnagar-Bhagalpur area. Bhagalpur is a flood-prone area in the middle Ganga plain. The flood situation is aggravated when the Ganga itself is in high spate in monsoon. As both the longitudinal, as well as the lateral slopes in the Bihar plains, are too low, the inundations spread over extensive areas.

The Jamania river channel from Nathnagar onwards flows along the Bhagalpur town parallel to the main Ganga and ultimately meets the River Ganga near Vikramshila Bridge, Barari, in Bhagalpur. In the monsoon season, the Jamania river channel swells due to the backflow of water from the main Ganga. Active braided channels meander, and oxbow lakes, which result from dynamic hydrological processes occurring within a low gradient alluvial plain, characterize the geomorphology of the Ganga in this area. The river channels are not deeply incised in this area, and exposed bank sediments are those of the modern aggrading flood plain systems. Detached flood plains with sand/mud deposits, 10–15 m thick, have developed over years in between the Jamania river channel and the main stem of the Ganga. Presently these detached flood plains are being extensively used by local people for agricultural purposes. Farmers are using synthetic fertilizers, pesticides, and insecticides extensively in agriculture. Toxic substances, such as fertilizers, pesticides, and insecticides, are drained from the flood plain agricultural fields on both the banks through run-off directly into the river and cause chemical pollution of the river.

In addition to that, at numerous points on the Ganga at Bhagalpur, garbage, excreta, and muck are being dumped into the river. The sewage of the town and other urban wastes are also being channelized to the river (about 5 mega gallons of sewage is received by the river per day from Bhagalpur). To this is added the burden of other human activities like bathing, washing of clothes, excessive navigation, brick making and immersion of dead bodies, etc. This causes considerable contamination of the river water and upsets the ecological balance of the river.

However, very few studies are available on the presence and their concentration of various pesticide residues in the surface water and sediments along the Gangetic plains of Bhagalpur, Bihar, and its impact on river water quality, the extent of pollution-tolerant forms, and the regenerative capacity of the river. In this case study, an attempt has been made to identify and quantify the pesticide present at this stretch by various researchers in the middle Ganga plains. The study reveals that there is a variety of organochlorines and organophosphate pesticide residue and various other toxic trace metals in the water as well as in the sediment of the Ganga River, which have future health risks to people and organisms consuming the water for drinking sources from the Ganga River basin. However, very scanty data was available on pesticide concentration in water and sediments of the river in these stretches of the middle Ganga plains.

There is a need for systematic monitoring and strict law enforcement to develop a plan of action to manage the environmental threats arising due to these contaminants of concern (CECs) and to improve the environmental protection of this area. More studies are needed to know the ecotoxicity and bio-accumulation of these pesticides in the food chain and their impacts on the ecosystem and human health.

3.5.1 Lindane

Lindane is an organochlorine insecticide. It is also known as BHC and HCH. The solubility of lindane in water is 10 mgl^{-1} with a half-life of 18 hours. BIS (1992) has recommended the absence of pesticides in drinking water with a maximum permissible limit of 0.001 mgl^{-1}.

The WHO classifies lindane as "moderately hazardous". Lindane is released into the environment during and after agricultural application through volatilization into the atmosphere, where it has long-range transport potential and can be deposited by rainfall. Lindane in the soil can leach to the surface and even groundwater and can bio-accumulate in the food chain (US EPA 2006). It is long-lived in soil, generally by adsorption and that depends on the organic matter, whereas the increasing solubility of lindane results in an increase in mobility. The adsorption of lindane is also dependent on the pH. Under alkaline conditions, it is higher compared to normal conditions. Its adsorption also depends on the type of soil and it is greater in sandy soils. Its accumulation is also influenced by the increase in temperature.

The US EPA has determined that lindane does not contaminate drinking water above the Agency's level of concern (ibid.). However, exposure to lindane in large amounts not only affects the nervous system but also produces a range of symptoms like headache and dizziness to convulsions and more rarely death (Agency for Toxic Substances and Disease Registry 2005). Exposure to β-HCH, an isomer of lindane, can alter thyroid hormone levels and could affect brain development (Alvarez et al. 2008).

Lindane has been reported by various researchers in both the water and the sediments of the River Ganga at various stretches in various concentrations. Singh et al. (2011) have reported lindane in the River Ganga at various stretches of Bhagalpur in the concentration from ND to 0.074 μgl^{-1}in river water and 18.97–392.60 ng g^{-1} in river sediments, which is quite high compared to its concentration in river water. Sankaramakrishnan et al. (2004) also reported lindane as not being detected in the river water of Ganges at Kanpur except for a few stretches where the concentration was 0.190–1.671 μgl^{-1}. Rehana et al. (1995) reported the presence of lindane in the water of the River Ganga at Narora, in Uttar Pradesh. Anupma et al. (2002) monitored the presence of HCH in the range of 0.019–1.663 μgl^{-1}in the River Ganga between Buxar in Bihar, and Rajmahal in Jharkhand. Ghosh et al. (2000) also reported the presence of HCH in the lower stretch of the River Ganga and found the concentration was 0.003μgl^{-1}. Agnihotri et al. (1994) noted a higher concentration of HCH in the river water of Ganga near Farrukhabad. The CPCB (1999–2005) reported a higher concentration of T.BHC (lindane) in the water of Yamuna, where a maximum of 5517.79 ngl^{-1} was observed downstream at Agra in the 2005 monsoon. In the water of the Gomti River, the total BHC concentration was reported to be relatively higher, i.e. 0.02–4846.0 ngl^{-1} (Singh et al. 2004). Nayak et al. (1995) found the highest concentration of lindane in the range of BDL–99517.0 ngl^{-}in the Ganga River.

There is scanty data on the investigation and distribution of organochlorines and organophosphorus pesticide residues in river sediments worldwide. Some of the works which show their presence in river sediments are (Ghosh et al. 2000), where the presence of HCH had been reported in the lower stretch of the Ganga in the concentration of 0.300 μg kg^{-1} = 0.3 ng g^{-1}. Agnihotri et al. (1994) found the presence of HCH in the soils of the Indo-Gangetic plain near Farrukhabad in northern India. Singh et al. (2004) reported a higher concentration of BHC in the river sediment of the Gomti in the range of 0.1–1650 ng g^{-1}, which showed that contamination of BHC is widespread in the Gomti River. The CPCB (1999–2005) had reported the presence of BHC in the range of

ND–801.19 ng g^{-1} in the sediment of the River Yamuna from Delhi to Agra. Rajendran (2004) found higher concentrations of HCH isomers in the Bay of Bengal. Some international authors also noted the presence of lindane (BHC and HCH) in the river sediment of the Qiantang River, East China (Zhou et al. 2006), and the Mae Klong river of Central Thailand (Poolpak et al. 2007).

3.5.2 Methyl Parathion

Methyl parathion, an organophosphate, has a half-life of 10 days to 2 months. It has a solubility of 24 mgl^{-1}. This pesticide is extensively used in agriculture, residential landscaping, and pest control. Being an organophosphate, it can degrade quickly by hydrolysis on exposure to sunlight, air, and soil but it has greater acute toxicity for humans and aquatic organisms. In mammals, methyl parathion is a nerve toxin and damages the central nervous system. Acute poisoning takes the form of bouts of sweating, increased production of saliva, diarrhea, bronchitis, heart attack, and may even result in a coma. In aquatic organisms, it is highly toxic to fish which cause deformation. It inhibits the growth of algae.

Methyl parathion has been reported by researchers either in low concentration or as absent in Ganga water. Singh et al. (2012) reported the absence of methyl parathion in all the stretches of Bhagalpur in river water and sediments. Sankaramakrishnan et al. (2004) also reported the non-presence of methyl parathion in the water of the River Ganga at Kanpur in all the stretches monitored. A lower concentration of methyl parathion was reported by Rehana et al. (1995), i.e. 0.41μgl^{-1} in the River Ganga at Narora in Uttar Pradesh. The CPCB (1999–2005) have observed methyl parathion in the river water only for two consecutive years (1999, 2000) in seven years of study of the River Yamuna. Aleem and Malik (2004) reported 1.7 ngl^{-1} of methyl parathion in the water of the River Yamuna at Okhla in Delhi.

There are few reports of the presence of methyl parathion in the sediment of river bodies because, in soil and aquatic systems, methyl parathion degrades to methyl-paroxon, by metabolic conversion, by hydrolysis to p-nitrophenol and dimethyl-o-thiophosphoric acid, or nitro group reduction to methyl amino-parathion (Eichelberger and Lichtenberg 1971; Sharmila et al. 1988).

3.5.3 Endosulfan (α- and β-Endosulfan)

Endosulfan is an organochlorine which is a very toxic pesticide, as per the US EPA toxicity list. The solubility of α-endosulfan in water is 0.33 mgl^{-1}which has 50 days half-life in soil and 5 weeks in water but the β-isomer has a longer half-life i.e., 150 days under neutral conditions. According to (BIS 1992), it is recommended not to use pesticides in drinking water. Due to its very high toxicity, as recommended by the US EPA, it is considered a restricted use pesticide (RUP). So its critical limit is neither assigned by the US EPA nor by the WHO for drinking water. However. Australian Standards recommend (0.05 μgl^{-1} = 50 ngl^{-1}) of endosulfan as a guideline value (GV) for drinking water.

Endosulfan is considered a very toxic pesticide, responsible for many fatal pesticide poisoning incidents. Endosulfan is a xenoestrogen, which is an endocrine disruptor and can cause reproductive and developmental damage in animals and humans. It is a neurotoxin in insects as well as mammals. Endosulfan is highly toxic to aquatic organisms and has a bio-accumulating effect, especially in fish.

Endosulfan rapidly degrades mainly into water-soluble compounds but remains persistent in the soil environment. The breakdown of endosulfan is faster in water (5 weeks) in neutral conditions but in acidic conditions or basic conditions, it breaks down into endosulfan sulfate and endosulfan diol, having a similar structure and having an estimated half-life of 9 months to 6 years. According to the US EPA, ,"endosulfan is a very persistent chemical which may stay in the environment for lengthy periods, particularly in acid media." This might also be a reason for the presence of endosulfan in the river water of the Ganga.

Endosulfan, with its isomers α- and β-endosulfan, has been reported by many researchers at high concentrations in various stretches of the River Ganga. Singh et al. (2011) reported a higher concentration of α-endosulfan in the river water as compared to its isomer β-endosulfan. Its value ranged from BDL–739 ngl^{-1}in river water and β-endosulfan ranged from ND–157.30 ngl^{-1}in river water. In river sediments, the value for α-endosulfan ranged between 35.58–50.47 ng g^{-1} and, for β-endosulfan, the value ranged between 34.40–303.09 ng g^{-1}, which was higher compared to river water. Anupma et al. (2002) reported the presence of endosulfan in the range of ND–0.862 μgl^{-1} in the River Ganges between Buxar in Bihar and Rajmahal in Jharkhand. Agnihotri et al. (1994) noted the presence of a higher concentration of α-endosulfan over another isomer of endosulfan in the River Ganges water near Farrukhabad. Haldar et al. (1989) found the presence of both the isomers of endosulfan, i.e. α- and β-endosulfan in the river water of Ganga at Hooghly. Kole et al. (2005) reported endosulfan in the range of 0.004–18.78 μgl^{-1} in the river water of Ganga in West Bengal. Singh et al. (2004) found the presence of endosulfan in the range of 0.2–1372.0 ngl^{-1} in the River Ganges. Sankaramakrishnan et al. (2004) reported the absence of endosulfan in the river water of the Ganges at Kanpur. Aleem and Malik (2004) observed 114 ngl^{-1} of endosulfan in the Yamuna river water at Okhla in Delhi. The CPCB (1999–2005) reported the maximum concentration of endosulfan to be 4591.08 ngl^{-1} downstream of the River Yamuna at Mathura.

In sediments, endosulfan with its isomers were reported by (Ghosh et al. 2000), who found 0.009 $\mu g\ kg^{-1}$ (= 0.009 ng g^{-1}) of endosulfan in the slower stretch of the River Ganga. Agnihotri et al. (1996) reported the presence of α-endosulfan in higher concentration compared to β-endosulfan in the soils of the Indo-Gangetic plain of Farrukhabad in northern India. Singh et al. (2005) reported the concentration of endosulfan to be in the range of BDL–72.6 ng g^{-1} in the sediment of the River Gomti. The CPCB (1999–2005) reported the presence of endosulfan in the River Yamuna from Delhi to Agra where the concentration ranged between BDL–612.18 ng g^{-1}. Some international authors have also found endosulfan in the sediment of the Queme River in the Republic of Benin (Elisabeth et al. 2006) and the sediment of the Mae Klong River of Central Thailand (Poolpak et al. 2007).

3.5.4 DDT (o,p′-DDT, orthopara-DDT and p,p′-DDT, parapara-DDT)

DDT is also an organochlorine insecticide which has a similar structure to the pesticides dicofol and methoxychlor. It is almost insoluble in water but has a greater affinity to organic solvents, fats, and oils. DDT is a persistent organic pollutant with a half-life of 2–15 years and is immobile in most soils. Its half-life is 28 days in river water. According to (BIS 1992), the absence of pesticides is desirable for drinking water. The WHO (2004) had given a guideline value (the value at which the evidence of a hazard is present, but the available information on health effects is limited) as 1 mgl^{-1}. Under the Stockholm Convention, DDT has been banned for agricultural use worldwide, but it is a part of many organochlorine insecticides which are still used worldwide. So, this might be the reason for the presence of DDT along with its metabolites in river water and sediments. It is also reported that about 31 banned or restricted pesticides in other countries are still in use in India and about 350,000 tonnes since 1985 and 7000 tonnes in 2001–2002 of DDT have been used (www.cseindia.org/2002).

DDT and its metabolic products magnify through the food chain and are stored mainly in the body fat. DDT has been classified under "moderately toxic" by the US National Toxicological Program and "moderately hazardous" by the WHO. DDT is a toxicant across a certain range of phyla. DDT was a major reason for the decline of the bald eagle in the 1950s and 1960s (www.fws.gov/endangeredi/b/masaboh.html). DDT and its breakdown products are toxic to embryos and can disrupt calcium absorption, thereby impairing eggshell quality (US EPA). DDT is considered to be very toxic to aquatic life, especially fish, and is moderately toxic to amphibians where it can bio-accumulate, leading to long-term exposure to high concentrations. In human beings, a higher concentration of DDT leads to neuropsychological and psychiatric symptoms. DDT and its metabolites are reported to be a probable human carcinogen. It is also regarded as an endocrine disruptor and affects the reproductive system.

DDT with its isomers orthopara (o, p') and parapara (p, p') DDT has been found by many researchers in the water and the sediments of the River Ganges at various stretches. Singh et al. (2012) reported T- DDT with its isomers in the range of ND–489 ngl^{-1}, which is very low in river water, and of the two isomers of DDT, orthopara-DDT (o,p'-DDT) was present in higher concentrations compared to parapara-DDT (p,p'-DDT). In river sediments, it ranged from 29.19–4203.4 ng g^{-1}, which was quite high compared to river water. Of the two isomers of DDT, o,p'-DDT was found in higher concentrations compared to p,p'-DDT. Rehana et al. (1995) found 1.36 μgl^{-1} of DDT in the River Ganges at Narora in Uttar Pradesh. Rehana et al. (1995) also reported the presence of DDT in the Ganges water between Kachla and Kannauj in Uttar Pradesh in the range of 3.33, 5.19–5.33 μgl^{-1} Nayak et al. (1995) reported 135.0–66516.0 ngl^{-1} of DDT in the middle stream of the River Ganga but Sankaramakrishnanet al. (2004) reported the absence of DDT in the water of the Ganges at Kanpur. Kole et al. (2005) found 0.008–5.53μgl^{-1} of DDT in the water of the Ganga in West Bengal.

In river sediments, DDT with its isomers were reported by Agnihotri et al. (1996), who found DDT with p,p'-DDT dominating o,p'-DDT in the soils of the Indo-Gangetic plain near Farrukhabad in northern India. Ghosh et al. (2000) found DDT in the concentration of 0.047 μg kg^{-1} (= 0.047 ng g^{-1}) in the lower stretch of the River Ganga. Singh et al. (2004) found T-DDT in the concentration ranging between 0.2–509.0 ng g^{-1}. The CPCB (1999–2005) reported the presence of T-DDT in the range of BDL–349.95 ng g^{-1}. The presence of DDT in the sediments has been reported by Elisabeth et al. (2006) in the Queme River in the Republic of Benin. Zhou et al. (2006) found DDT in the range of 9.41–69.66 ng g^{-1} and (Poolpak et al. 2007) found the range of 0.08–1.83 μg g^{-1} (= 80–1830 ng g^{-1}) in the river sediments.

3.6 REGULATORY MEASURES RELATING TO EMERGING CONTAMINANTS

As this chapter has shown, many studies have reported the presence of pesticides in the various rivers around the globe. It has been identified by many regulatory bodies around the world, such as the REACH regulation registers of the European Union, the Contaminant Candidate List of the United States of America (CCL), and the POPs (Persistent Organic Pollutants) of the Stockholm Convention, pesticides form a major proportion of these lists.

The OECD (Organisation for Economic Co-operation and Development) has defined contaminants of emerging concern (CECs) as contaminants of a large array which have only lately appeared in water, or those for which concern has recently spiked, since the levels at which they have been detected in water matrices are unexpectedly high. Moreover, an accurate assessment of the risk posed by these pesticides to human and environmental health is yet to be completed.

Since these contaminants are relatively new and unorthodox when compared to traditional pollutants, CECs have become a regulatory headache. While CECs have been identified as including a large number of classes, including pesticides, individual legislation aimed at correcting the levels of pesticides in water are relatively absent at the national, transnational, and international levels.

The US government through the EPA has defined contaminants of emerging concern as those contaminants which occur or are likely to occur in public water sources at a level high enough to warrant a risk to human life. However, routine testing for these chemicals understood as ECs has only recently been initiated. At regular intervals, the EPA publishes the list of contaminants found in water (in higher-than-expected categories) that may harm human health. This list is known as the Contaminant Candidate List (CCL), and these contaminants are analyzed and tested, but it is left to the administrator to judge whether they need to be regulated or not. In the third such list published by the EPA (CCL 3), many pesticides were identified. These included hydroxycarbofuran, acetaldehyde, acetochlor ethane sulfonic acid (ESA), acetochlor oxanilic acid (OA), ethylene glycol, bromochloro methane, and TPTH among others. In January 2016, it was decided not to regulate only 4 of the 116 chemicals in the CCL 3, and almost all pesticides were included in advisories to be regulated.

The latest list published in February of 2021 (CCL 4) made the final determination to regulate two contaminants, perfluorooctanesulfonic acid (PFOS) and perfluorooctanoic acid (PFOA). While EPA health advisories are non-enforceable and non-regulatory, they have prompted several states to regulate and monitor the level of ECs, to ensure access to safe drinking water. EPA advisories for PFOA (per fluoro octatonic acid) and PFOS have been adopted by many states as the regulatory standard for drinking water (US EPA 2021). Along similar lines, effective regulations for other compounds such as pesticides need to be formulated in parts per billion/parts per million ranges as well.

While not decided on a large scale as of yet, organophosphates are deemed to be one of the leading candidates to be included in the next CCL (Yang et al. 2019).

While it has not been recommended for regulation right now, one can hope that with stronger evidence of the effects of these chemicals, the EPA will be forced to take action. Therefore, there is a periodic revision of results as and when more scientific data become available.

Regulatory measures focused on mitigating both the use and effects of the substances may generally be classified under two headings based on jurisdiction: international and domestic.

While international agreements such as the Stockholm and Rotterdam Conventions are tasked with the regulation of more toxic pollutants since they are more likely to have been recognized by many countries as toxic, domestic laws, as shown above, also play a very important part in mitigating the use and effects of these pollutants.

While international regulations must be the preferred form of embargoing contaminants, it is essential to recognize that international law, in such cases, takes root in domestic laws. A chemical is more likely to be put on the watch list of the Stockholm Convention and the Food and Agricultural Organization (FAO) if many domestic jurisdictions have red-flagged its use. In most cases, international and domestic law go hand in hand, while, in cases of conflict, international law must prevail (El-Ashry 1972).

3.6.1 International Law

3.6.1.1 The Rotterdam Convention

The Rotterdam Convention was created with the aim of identifying and moving toward a cooperative responsibility to restrict and mitigate the use and trade of such chemicals which were deemed to harm the environment. Relying on the Prior Informed Consent doctrine, while heavily reliant on the consent of the state, it is almost a loophole-free doctrine since parties are supposed to obtain consent before exporting or importing any such chemicals.

Aldrin, carbofuran, DDT, dieldrin, endosulfan, HCH, and lindane are among the 53 chemicals to have been put on the PIC list as per the Rotterdam Convention. This list, however, is nonexhaustive and by consent of the member parties may be updated every 6 months. Of these 53 chemicals listed, 35 of them are pesticides.

These lists, recommended for listing and candidate chemicals respectively are composed of some of those chemicals which would be identified as emerging contaminants both in a legal and environmental sense. These include chemicals such as carbosulfan and acetochlor.

3.6.1.2 The Stockholm Convention

The Stockholm Convention, drafted in 2004, remains the largest legal document, legally binding on the highest number of states, which deals with the regulation of many emerging contaminants, especially those which have been classified as persistent organic pollutants (POPs), which stay in the environment for long periods.

The convention comes from the combined realization of parties that no state acting alone can successfully counter the menace posed by these compounds. Unilateral actions would bring no benefit in the fight against emerging contaminants since their consequences are not region-specific, but through cross-border carriers, such as rivers, pose an international risk. Moreover, with almost 184 parties and 150 ratifications, the convention is a strong binding legal instrument in the international field.

Annex A and Annex B of the Stockholm Convention include the pesticides to have been classified as POPs. These include pesticides to be discontinued for use (Annex B), such as DDT, and severely restricted for use (Annex A), such as aldrin, alpha hexachlorocyclohexane (Alpha-HCH), beta hexachlorocyclohexane (Beta-HCH), chlordane, chlordecone, dieldrin, endrin, heptachlor, lindane, mirex, pentachlorophenol and its salts and esters (PCP), technical endosulfan and its related isomers, and toxaphene.

These ECs are not regulated or tested for on a routine basis, at either international or domestic levels. Thus, their ecotoxicity is not very well understood. Hence there is a dearth of regulations regarding the same.

As has been pointed out throughout the chapter, most international regulatory authorities flag pesticides as an important heading under their respective categories of emerging contaminants (CECs for the US EPA, POPs for the Rotterdam Convention) and many such restricted chemicals are indeed pesticides. However, most of these restrictions are either only advisory, and even more are generally focused, rather than pesticide-focused. Pesticides as emerging contaminants need to be treated and checked right at the source, as has also been discussed in this chapter, they form one of the biggest risks to human life if left unchecked. It is of the utmost importance that newer and more stringent legislations are formulated to identify these pesticides even in traces of quantities to restrict and regulate the use of pesticides and their trade and phase them out in favor of less toxic alternatives. Only when such targeted actions become commonplace at the national and international levels, the menace of pesticides as ECs be checked.

3.7 CONCLUSION

The overview of this chapter shows that most of our river systems are contaminated by pesticides. Lindane, endosulfan, DDT, methyl parathion are very commonly found in the river systems. Their concentration levels, in water as well as in sediments of the rivers, are found to be above the standards prescribed by international or national regulatory authorities. This poses a risk to the river ecosystems. The presence of pesticides at significant levels, in these river systems, indicates widespread usage of regulated pesticides and has led to a negative impact on river health and its biodiversity. Even though pesticides like DDT are banned in most countries, the presence of the same in rivers shows that such regulations are only loosely enforced.

Pesticides, having been classified as one of the headings under CECs, have been put on the regulatory watch list by many international bodies, such as the Rotterdam Convention, Domestic bodies such as the US EPA have also come forward with advisories to monitor and limit the levels of pesticides to ensure safe drinking water. However, these regulations are only advisory. Strict laws regulating the levels of pesticides in water, and substituting or phasing them out are required at both international and domestic levels to check and rectify the harms posed by these pesticides.

REFERENCES

Agarwal, H.C., Mittal, P.K., Menon, K.B. and Pillai, M.K.K. "DDT residues in the river Jamuna in Delhi India." *Water, Air, & Soil Pollution*, 28 (1986): 89–104.

Agency for Toxic Substances and Disease Registry. *Toxicology Profile for Alpha-, Beta, Gamma, and Delta-Hexachlorocyclohexane*, Washington, DC: U.S. Department of Health and Human Services, August (2005).

Agnihotri, N.P., Gajbhiye, V.T., Kumar, M., Mohapatra. S.P., and Agnihotri, N.P. "Organochlorine insecticide residue in Ganga river, water near Farrukhabad." *Environmental Monitoring and Assessment*, 30(2) (1994): 105–112.

Agnihotri, N.P., Kulshrestha, G., Gajbhiye, V.T., Mohapatra, S.P., and Singh, S.B Organochlorine insecticide residues in agricultural soils of the Indo-Gangetic plain. *Environment Monitoring and Assessment,* 40(3) (1996): 279–288.

Aguliar, C., Ferrer, I., Borull, F., Marce, R.M., and Barcelo, D. Monitoring of pesticides in river water based on sample previously stored in polymeric of cartridges followed by on line solid–phase extraction–liquid chromatography–diode array detection and conformation by atmospheric pressure chemical ionization mass spectrometry. *Analytica Chimica Acta*, 386 (1991): 237–248.

Albanis, T.A., Danis, T.G., and Kourgia, M. "Transportation of pesticides in—estuaries of Anxious. Lodias and Aliakmon rivers (Greece)." *The Science of the Total Environment*, 156 (1994): 11–22.

Albanis, T.A., Danis, T.G., Voutsa, D., and Kouimtzis, T. "Evaluation of chemical parameters of Aliakmon river Northern Greece, Part III." *Journal of Environmental Science and Health*, 30 (1995): 1945–1956.

Albanis, T.A., and Hela, D.G. "Pesticide's concentration in Louros river and their fluxes into the marine environment." *International Journal of Environmental Analytical Chemistry*, 70 (1998): 105–120.

Albanis, T.A., Promnis, P.J., and Sdokos, A.T. "Organophosphorous and carbamates pesticide residues in an aquatic system of Ioannia basin and Kalamas River (Greece)." *Chemosphere*, 15(8) (1986): 1023–1034.

Aleem, A., and Malik, A. "Genotoxicity of the Yamuna River water at Okhla (Delhi) India." Department of Agricultural Microbiology, Faculty of Agricultural Sciences, Aligarh Muslim University (2004). Available at: www.ScienceDirect.com.

Alvarez, P. M., Rabas-Fito N., and Torrent, M. "Thyroid disruption at birth due to prenatal exposure to beta-hexachlorocyclohexane." *Environment International*, 34 (2008): 737–740.

Angelidis, M.O., and Albanis, T.A. "Pesticides residue and heavy metals in the Evros river delta, N.E. Greece." *Toxicological Environmental Chemistry*, 53 (1996): 35–44.

Anupma, K., Sinha, R. K.; Gopal, K.; and Lata, S. "Concentration of organochlorines in Ganges River dolphins from Patna, Bihar." *Journal of Environmental Biology*. 23(3) (2002): 279–281.

Augustijn-Beckers, P.W.M., Hornsby, A.G., and Wauchope, R.D. "SCS/ ARS/ CES pesticides properties database for environmental decision making." *Additional Properties Preview of Environmental Contamination and Toxicology*, 137 (1994): 1–82.

Awasthi, M.K., Wong, J.W.C., Kumar, S., Awasthi, S.K., Wang, Q., Wang, M., Ren, X., … Zhang, Z. "Biodegradation of food waste using microbial cultures producing thermostable α-amylase and cellulase under different pH and temperature." *Bioresource Technology*, 248 (2018): 160–170. https://doi.org/10.1016/j.biortech.2017.06.160.

Awasthi, S.K., Sarsaiya, S., Awasthi, M.K., Liu, T., Zhao, J.C., Kumar, S., and Zhang, Z.Q. "Changes in global trends in food waste composting: Research challenges and opportunities." *Bioresource Technology*, 299 (2020): 122555. https://doi.org/10.1016/j.biortech.2019.122555.

Bakre, P.P., Misra, V., and Bhatnagar, P. "Organochlorine residues in water from the Mahala water reservoir. Jaipur, India." *Environmental Pollution*, 63 (1990): 275–281.

Ballesteros, E., and Parrado, M. J. "Continuous solid-phase extraction and gas chromatographic determination of organophosphorus pesticides in natural drinking water." *Journal of Chromatography, A*, 1029 (2003): 297–273.

Beg, M.V., Saxena, R.P., Kidwai, R.M., Agarwal, S.N., Siddiqui, F., Sinha, R., Bhattacharjee, B.D., and Ray, P.N. *Toxicology Map of India*, vol. I: *Pesticides*. Lucknow: Industrial Toxicology Research Centre (1989)..

Brambilla, A., Rindone, B., Polesello, S., Galasii, S., and Balestrine, R. "The fate of triazine pesticides in river Po water." *Science of Total Environment*, 132, (1993): 339–348.

Bureau of Indian Standards. *Specifications for Drinking Water,* (IS: 105000). New Delhi: Government of India (1992).

Callahan, M.A., Shmak, M., Gbel, N., et al. *Water-Related Environment Fate of 129 Priority Pollutants.* Washington, DC: EPA (1979).

Chandra, R., and Kumar, V. "Biotransformation and biodegradation of organophosphates and organohalides." In Chandra, R. (Ed.) *Environmental Waste Management.* Boca Raton, FL: CRC Press (2015). DOI:10.1201/b19243-17.

Chowdhary, P., Hare, V., Singh, A.K., Pandit, S., and Chaturvedi, P. "Emerging environmental contaminants: sources, consequences, and future challenges." In Chowdhary, P., and Raj, A. (Eds.) *Contaminants and Clean Technologies*. Boca Raton, FL: CRC Press (2020).

Chowdhary, P., Raj, A., Verma, D., and Akhter, Y. (Eds.). *Contaminants and Clean Technologies*. Boca Raton, FL: CRC Press (2020).

CPCB (Central Pollution Control Board Ministry of Environment and Forests). Water quality status of Yamuna River. Assessment and Development of River Basin Series: ADSORBS/ 41/ 2006–07 New Delhi: Government of India (1999–2005).

Edward, Z., Janina, L., Stanislaw, K., and Maria, W. "Cadmium and lead in human body fluids, air and drinking water in the area of Cracow." *Water, Air, & Soil Pollution*, 45 (1979): 219–234.

Eichelberger, J.W., and Lichtenberg, J.J. "Persistence of pesticides in river water." *Environmental Science and Technology*, 5(6) (1971): 541–544.

EI-Ashry, M.T. "Foreword to Jon Martin Trolldalen, International Environmental Conflict Resolution (1992)." In *Stockholm Declaration on the Human Environment: Report of the United Nations Conference on the Human Environment*, U.N. Doc. A/Conf.48/14 Corr. 1 (1972), (June 16, 1972). reprinted in 11 I.L.M. 1416, 1420.

Elisabeth, Y., Pozou, A., Michel, B., Carnelis, A.M., et al. "Organochlorine and organophosphorus pesticides residues in the Queme River Catchment in the Republic of Benin." *Environment International*, 32 (2006): 616–623.

Erik, M., Ernst, A., Hegedus, G., Darvas, B., and Andras, S. "Monitoring water-polluting pesticides in Hungary." *Microchemical Journal*, 85 (2006): 88–97.

FAO (Food Agricultural Organization) "Chronology of pesticide development." (n.d.b). Available at: www.fao.org.

Ferencz, L., Jelinek, I.L., and Pasztor, J. "Pesticides level in water and soil samples from Mures County (Romania)." *Revista de Medicinasi Farmacie*, 51(4) (2005): 241–246.

Fielding, M., Barcelo, D., Helweg, P., Glassi, S., et al. "Pesticides in ground and drinking water: EU Water Pollution Research, Report." Brussels: EU, 27 (1991).

Galassi, S., Guzzela, L., Mingazzni, M., Vigano, L., Capri, S., and Sora, S. "Toxicological and chemical characterization of organic micropollutants in river Po water." *Water Research*, 26 (1992): 19–17.

Gascon, J., Salak, J.S., Qubina, A., and Barcelo, D. "Monitoring of organochlorine pesticides in Ebro River." *Preliminary Loading Estimates Analyst*, 123 (1998): 941–945.

Ghosh, S., Das, A.K., and Vass, K.K. "DDT, HCH, and endosulfan residues in the lower stretches of river Ganga." *Central Inland Capture Fisheries Research Institute*, 27(4) (2000): 161–164.

Gotz, R., Bauer, O.H., Friesel, P., and Roch, K. "Organic trace compounds in the water of river Elbe near Hamburg." *Part II. Chemosphere*, 36 (1998): 2108–2118.

Griffini, O., Bao, M.L., Barbieri, C., Burrini, D., and Pantani, F. "Occurrence of pesticides in Arno river and in potable water." *Bulletin of Environmental Contamination and Toxicology*, 59 (1997): 202–209.

Haldar, P., Raha, P., Bhattacharya, P., Choudhary, A., and Aditya. N. "Studies on the residue of DDT and endosulfan occurring in Ganga water, India." *Journal of Environmental Health*, 3(1) (1989): 156–161.

Holvoet, K., Seuntjens, P., Mannaerls, R., DeSchepper, V., and Vanrolleghem, P.A., Risk Assessment Project Group. "The dynamic water sediment system: Results from an intensive pesticide monitoring campaign." *Water Science Technology*, 55(3) (2007): 177–182.

ISGE. "Integrated Study of Ganga Ecosystem between Kachla to Kannauj. Ganga project report." New Delhi: Dept. of Environment, Ministry of Environment and Forest, Government of India (1990).

ITRC (Industrial Toxicological Research Centre) "Analysis of pesticide residue in water. ITRC manual No. 1." Lucknow: Industrial Toxicology Research Centre (1987).

ITRC (Industrial Toxicological Research Centre) "Sixth Annual Progress Report (July 1991–June 1992), Measurements on Ganga water quality: Heavy metal and pesticides." Lucknow: ITRC (1992).

Jung, Y., Wenjing, Z., Yunfen, S., Weisong, F., and Xinhue, W. "Monitoring of organochlorine pesticides using PFU systems in Yuhnan lakes and river China." *Chemosphere*, 66 (2006): 219–255.

Kawano, M.T., Jhoue, T., Wada, H., Hidaka, H., and Tatsukawa, R. "Bioconcentration and residue patterns of Chlordane compounds in marine animals: Invertebrates, fish, mammals, and seabirds." *Environmental Science and Technology*, 22 (1988): 792–797.

Kole, R.K., Alam, S., Mukherjee, P., Sardar, D., Kole, S., and Saha, T. "Multiresidue analysis of pesticides for evaluation of surface water quality of the River Ganga in West Bengal." Paper presented at International Congress: Crop Science and Technology, Glasgow (2005).

Kumar, K.S., Kannan, K., Sinha, R.K., Tanabe, S., and Giesy, J.P. "Bioaccumulation profiles of polychlorinated biphenyl congeners and organochlorine pesticides in Ganges river dolphins." *Environmental Toxicology and Chemistry*, 8(7) (1999): 1511–1520.

Kusamran, W.R., Wakabayashi, K., Oguri, A., Tepsuwan, A., Nagao, M., and Sugimura, T. "Mutagenicities of Bangkok and Tokyo River waters." *Mutation Research*, 325 (1994): 99–104.

Larson, S.J., Capel, P.D., Croolshy, D.A., Dangg, S.D., and Sandstrom, M.V. "Relation between pesticides use and riverine flux in the Mississippi river basins." *Chemosphere*, 31 (1995): 3305–3321.

Leonard, R.A. "Movement of pesticides into surface waters." In Cheng, H. (Ed.). *Pesticides in the Soil Environment: Processes, Impacts and Modeling*. Madison, WI: Soil Science Society of America (1990). pp. 303–349.

Mathur, S.C. "Pesticides industry in India." *Chemical Engineering World*, 19 (1993): 7–15.

Metcalf, R.L. "Pesticides in the aquatic environment." In Khan, M.A.Q. (Ed.) *Pesticides in the Environment*. New York: Plenum Press (1997).

Mondal, R., Mukherjee, A., Biswas, S., and Kole, R.K. "GC-MS/MS determination and ecological risk assessment of pesticides in aquatic system: A case study in Hooghly River basin in West Bengal, India." *Chemosphere*, 206 (2018): 217–230.

Moral, A., Sicilia, M.D., Rubio, S., and Pérez-Bendito, D. "Multifunctional sorbents for the extraction of pesticide multi residue from natural water." *Analytica Chima Acta*, 608 (2008): 61–72.

Murphy, G. *Resistance Management-Pesticide Rotation*. Ontario: Ministry of Agriculture, Food and Rural Affairs (2005).

Nayak, A.K., Raha, P., and Das, A.K. "Organochlorine pesticides residue middle stream of Ganga River. India." *Bulletin of Environment Toxicology*, 55(1), (1995): 68–75.

Negi, S., Mandpe, M., Hussaina, A., and Kumar, S. "Collegial effect of maggots larvae and garbage enzyme in co-composting of food waste with wheat straw and biomass waste." *Journal of Cleaner Production*, 258 (2019): 120854. https://doi.org/10.1016/j.jclepro.2020.120854.

Negi, S., Rani, A., Hu, A.H., and Kumar, S. "Pesticide pollution: management and challenges." In Mehmood, M.A., Hakeem, K.R., Bhat, R.A., and Dar, G.H. (Eds.) *Pesticide Contamination in Freshwater and Soil Environs*. New York: Routledge (2021), pp. 69–88.

Nimmo, D.R. "Pesticides.". In Rand, G., et al. (Eds.) *Fundamentals of Aquatic Toxicology: Methods and Applications*. Washington, DC: United States Department of Energy's Office of Scientific and Technological Information (1985).

Oehme, M., and Mano, S. "The long-range transport of organic pollutants to the Arctic." *Fresenius Zeitschrift für Analytische Chemie*, 319 (1984): 141–146.

Ormad, M.P., Claves, N.A., and Matesanz, J.M. "Pesticide's removal in the process of drinking water production." *Chemosphere*, 71 (2007): 97–106.

Pandit, G.G., Rao, A.M.M., Jha, S.K., Krishnamoorthy, T.M., Kale, S.P., Raghu K., et al. "Monitoring of organochlorine pesticide residues in the Indian marine environment." *Chemosphere*, 44 (2001): 302–305.

Pasti, L., Nava, E., Marco, M., Bignami, S., and Dondi, F. "GC/MS analysis of pesticides in the Ferrara area (Italy) surface water." *Annali di Chemica.*, 97(5–6) (2007): 359–372.

Poolpak, T., Pokethityoo, P., Krauatrachue, K.M., Arjarasirikoon, U., and ThanWani W.N. "Residue analysis of organochlorine pesticide in the Mae Klong River of Central Thailand." *Journal of Hazardous Material*, 156 (2007): 230–239.

PQRS (Physician Quality Reporting System) "Report." Washington, DC: Centers for Medicare and Medicaid Services (2017).

Rajendran, R.B., Imagewa, T., Tao, H., and Ramesh, R. "Distribution of PCBs, HCHs and DDTs and their ecotoxicological implication in Bay of Bengal, India." *Environmental International*, 31 (2004): 503–512.

Rajendran Babu, R., and Subramaniam, A.N. "Pesticide residues in water from River Kaveri, South India." *Chemical Ecology*, 13 (1997): 223–236.

Rajendran Babu, R., Venugopalan, V.K., and Ramesh R. "Pesticide residues in air from coastal environment, South India." *Chemosphere*, 39 (1999): 1699–1706.

Rehana, Z., Malik, A., and Ahmad, M. "Genotoxicity of the Ganga water at Narora (UP)." *Life Science Mutation Research*, 367 (1995): 187–193.

Richards, R.P., and Baker, D.B. "Pesticide concentration pattern in agricultural drainage networks in the Lake Erie basin." *Environmental Toxicology and Chemistry*, 12 (1993): 13–26.

Sankaramakrishna, N., Sharma, A.K., and Singh, R. "Organochlorine and organophosphorous pesticide residue in groundwater and surface water of Kanpur, Uttar Pradesh, India." *Environmental International*, 31 (2004): 113–120.

Sarkar, A. , Nagaranjan, R. , Chaphadkar, S. , Pal, S. , and Singhbal, S.Y.S. "Contamination of organochlorine pesticides in sediments from the Arabian Sea along the west coast of India." *Water Research*, 31 (1997): 195–200.

Sarkar, A., and Sen Gupta, R. "DDT residues in sediments from the Bay of Bengal." *Bulletin of Environmental Contamination and Toxicology*, 41 (1988): 664–669.

Shabbir, Md., Singh, M., Maiti, S., Kumar, S., and Saha, S.K. "Removal enactment of organo-phosphorous pesticide using bacteria isolated from domestic sewage." *Bioresource Technology*, 263 (2018): 280–288. https://doi.org/10.1016/j.biortech.2018.04.122.

Sharmila, M., Ramanand, K., Adhya, T.K., and Sethunathan, N. "Temperature and persistence of methyl parathion in a flooded soil." *Soil Biology and Biochemistry*, 20(3) (1988): 399–401.

Singh, K.P., Jamkoo, R., and Ray, P.K. "Analysis of Pesticides Residue in Water: ITRC Manual No. I." Lucknow: Industrial Toxicology Research Center (1987).

Singh, K.P., Malik, A., Mohan, D., and Takroo, R. "Distribution of persistent organochlorine pesticide residues in Gomti river, India." *Chemosphere*, 74: (2005). 146–154.

Singh, K.P., Mohan, D., Sinha, S., and Dalwani, R. "Impact assessment of treated/untreated wastewaters toxicants discharged by sewage treatment plants on health, agricultural and environmental quality in the wastewater disposal area." *Chemosphere*, 55 (2004): 227–255.

Singh, L., Choudhary, K.S., and Singh, K.P. "Organochlorine and organophosphorous pesticides residues in water of River Ganga at Bhagalpur, Bihar, India." *International Journal of Research in Chemistry and Environment*, 1(1) (2011): 77–84.

Singh, L., Choudhary, K.S., and Singh, K.P. "Pesticide concentration in water and sediment of River Ganga at selected sites in middle Ganga plain." *International Journal of Environmental Science*, 3(1) (2012): 260–274.

Sirohi, R., Tarafdar, A., Gaur, V.K., Singh, S., Sindhu, R., Rajasekharan, R., Madhavan, A., … Pandey, A. "Technologies for disinfection of food grains: Advances and way forward." *Food Research International*, 145 (2021): 110396. https://doi.org/10.1016/j.foodres.2021.110396.

Skoulikidis, N.T., Bertahas, J., and Kounsouris, T. "The environmental state of the freshwater resource in Greece." *Environmental Ecology*, 36 (1998): 1–17.

Sukamato, H. and Hayatsu, H. "A simple method for monitoring mutagenicity of river water. Mutagen in Yodo river system, Kyoto-Osaka." *Bulletin of Environmental Contamination and Toxicology*, 44 (1990): 521–528.

Tanabe, S., Iwata, H., and Tatsukawa, R. "Global contamination by persistent organochlorine and their ecotoxicological impact on marine mammals." *Science of the Total Environment*, 154 (1994): 163–177.

Totolin, M. "The extraction and clean-up of organochlorine pesticides and PCBs from water using solid-phase extraction and gas chromatography analysis." *Environmental Engineering and Management Journal*, 2(1) (2003): 25–28.

UNEP. *Global Reports on Regionally Based Assessment of Persistent Toxic Substances.* Geneva: UNEP Chemicals (2002).

US EPA (United States Environmental Protection Agency). "Environmental Fate and Effects Division, Pesticides Environmental Fate, Online Summary: DDT (p,p')." Washington, DC: EPA (1989).

US EPA (United States Environmental Protection Agency). "Guidance for 2006 assessment, listing and reporting requirement pursuant of the CLEAN WATER ACT 2006." (2006). Available at: www.epa.gov

US EPA (United States Environmental Protection Agency). "Drinking-Water Health Advisories for PFOA and PFOS," US EPA. (2021). Available at: https://epa.gov/ground-water-and-drinking-water/drinking-water-health-advisories-pfoa-and-pfos.

US EPA (United States Environmental Protection Agency). "Toxicity profiles, Ecological Risk Assessment 1 Region 5 superfund/I." Washington, DC: US EPA (n.d.).

Vagi, M.C., Pelsas, A.S., Kostopoudous, M.N., and Lekkhas, T.D. "Monitoring of pesticides residue in the surface water of Greece." Paper presented at Conference, Laboratory of Water and Air Quality. Department of Environmental Studies (June 4–6, 2003) (Report).

Wagenet, R.J. "Process influencing pesticides loss with water under conservation tillage." In Logan, T.J., Davinsion, J.M., Baker, J.L., and Overcash, M.R. (Eds.) *Effect of Conservation Tillage on Ground Water Quality: Nitrates and Pesticides*. Chelsea, MI: Lewis (1987), pp. 189–204.

Wang, W.-C., and Evans, R.L. "Nutrients and quality in impounded water." *Journal of American Water Works, Association*, 62(8) (1970): 510–514.

Wauchope, R.D. "The pesticides content of surface water draining from agricultural fields—A review." *Journal of Environmental Quality*, 7 (1978): 459–472.

WHO (World Health Organization). *Environmental Health Criteria 83, DDT and Its Derivatives—Environmental Effects*. Geneva; WHO (1989).

WHO (World Health Organization). *Guidelines Values for Chemicals of Health Significance in Drinking Water. Guidance Manual for Drinking Water Quality Monitoring and Assessment,* Delhi: NEERI, Nagpur and NICD: A p-12. (2004).

Willett, K.L., Ulrich, E.M., and Hites, R.A. "Differential toxicity and environmental fates of haxachlorocyclohexhane isomers." *Environmental Science and Technology*, 32 (1998): 2197–2207.

Yang, J., Zhao, Y., et al. "A review of a class of emerging contaminants: The classification, distribution, intensity of consumption, synthesis routes, environmental effects and expectation of pollution abatement to organophosphate flame retardants (OPFRs)." *International Journal of Molecular Science*, 20(12) (2019): 2874.

Zehra, R., Zhou, L., Yank, K., and Chen, Y. "Distribution of organochlorine pesticides in surface water and sediments from Qiantang River, East China." *Journal of Hazardous Material, A*, 137 (2006): 68–75.

Zhou, L., Yank, K., Chen, Y. "Distribution of organochlorine pesticides in surface water and sediments from Qiantang River, East China." *Journal of Hazardous Material, A*, 137 (2006): 68–75.

4 Sources, Spread, and Surveillance of Antimicrobial Resistance

A Global Concern

Sneha Suresh,[1] *Preethy Chandran,*[2,*] *and Joseph Kadanthottu Sebastian*[3]
[1,2]School of Environmental Studies, Cochin University of Science and Technology, Kalamassery, Kochi, Kerala, India
[3]Department of Life Sciences, CHRIST, Bangalore, Karnataka, India
[*]Corresponding author Email: preethychandran@cusat.ac.in

CONTENTS

4.1 Introduction 69
4.2 Sources and Spread of AMR 70
 4.2.1 Antibiotics as an Emerging Contaminant 70
 4.2.2 Sources of Antibiotic Resistance 71
  4.2.2.1 Agriculture as the Sources 73
  4.2.2.2 Sewage as the Sources 73
  4.2.2.3 The Environment as the Source 73
4.3 Spread of Antimicrobial Resistance 73
4.4 Removal Strategies 74
4.5 Why AMR Is a Threat to Health 76
4.6 Antimicrobial Resistance Surveillance 77
4.7 Conclusion 79
References 79

4.1 INTRODUCTION

Antibiotics were a revolutionary discovery in the health sector which also highly influenced socio-economic progress. A wide spectrum of antibiotics has been discovered as a solution to several diseases (Arya et al. 2019). This reflects a progressive development in the health sector. Antibiotics belong to different classes, according to their mechanism of action. They contain compounds of both natural and artificial origin. It has recently been discovered that the unpredictable threat of diseases may be caused by the antibiotics themselves. Improper disposal of unused and expired drugs facilitates the chance of contamination in aquatic systems. Antibiotics, analgesics, and psychiatric drugs were found predominantly in the water environment of the South Asian regions (Chaturvedi et al. 2020; Chaturvedi, Chowdhary et al. 2021; Chaturvedi, Shukla et al. 2021; Khan et al. 2020). The introduction of antibiotic residues, even in very low concentrations, can influence the environmental microflora. These may develop resistance in pathogens and commensal bacteria. From

DOI: 10.1201/9781003239956-5

another perspective, this gives rise to a complex threat of diseases because the same antibiotics fail to provide a cure.

A resistant microbe is described as a microbe that is not killed by an antimicrobial agent after a standard course of treatment (WHO 1998). The use of antimicrobials in any dose for a certain period forces the microbes to adapt or die and it is the surviving microbes that produce drug-resistant genes. These may be transferred to other strains within the same genus and species to other unrelated species (Wise et al. 1998).

Bacterial resistance can be understood as natural and acquired, based on the origin of the resistance (ibid.; Chaturvedi et al. 2020). Natural resistance to an antibiotic is present in most bacteria as part of evolution. Such antibiotic evolution dates back billion of years. Those genes are located in chromosomes and are unable to transfer to other bacteria (Hawkey 1998). Acquired resistance is the resistance that is gained in the course of the lifecycle and is specific. This occurs when a bacterium is previously exposed to antibiotics and gradually has developed resistance (Tomasz et al. 1995). Mutation and horizontal gene transfer are the main modes of mechanism involved in acquired resistance. This resistance is transferable to other bacteria through different kinds of resistance elements.

Recent studies show that antimicrobial resistance (AMR) is one of the major threats challenging the modern world (Dafale et al. 2020). Its complex nature makes it difficult to understand, even using advanced technologies in the health sector. The intense spread of AMR worldwide, episodes of new disease outbreaks, and the re-emergence of old diseases are all connected to each other. AMR outbreaks result in high mortality, such as respiratory tract infections, which cause 3.5 million deaths in children annually. Worldwide distribution of penicillin-resistant *pneumococci* is another example of an outbreak that was first reported in Australia and Papua New Guinea .Typhoid is also endemic, which causes 16 million cases each year with approximately 700,000 deaths (Komolafe 2003). The World Health Organization (WHO) has declared AMR to be one of the top 10 global public health threats facing humanity and has established a surveillance network and guidelines to tackle the AMR problem worldwide.

The presence of multiple antibiotic-resistant bacteria in different environment matrices was revealed in recent studies. These are termed superbugs which also involve pathogens. Methicillin-resistant *Staphylococcus aureus* (MRSA), vancomycin-resistant *enterococci* (VRE), drug-resistant *Escherichia coli*, *Pseudomonas aeruginosa*, *Stenotrophomonas maltophilia*, and *Burkholderia cepacian*, etc., are some of multi-resistance pathogens studied (WHO 2021).

This chapter focuses on understanding AMR spread and its mitigation. It comprises sources and pathways of antibiotics, ARB (antibiotic-resistant bacteria), and ARGs (antibiotic-resistant genes). The efficiency of both conventional and advanced removal strategies for AMR is discussed. Threats of AMR in the health care sector are highlighted and the factors that facilitate AMR spread are explained. The current surveillance systems on AMR are also discussed to emphasize the relevance of surveillance on AMR.

4.2 SOURCES AND SPREAD OF AMR

4.2.1 Antibiotics as an Emerging Contaminant

The era of antibiotics started with the discovery of penicillin in the 1940s (Table 4.1). Then the world witnessed a progressive development in the field of disease treatment. Scientists initially extracted antibiotic compounds from natural sources and then the trend of artificial production was started. It found its application in different sectors such as growth promoters in livestock farming, in agriculture to control bacterial diseases, and in bee-keeping, other than as chemotherapeutic agents.

Indiscriminate use of antibiotics leads to its continuous input into the environment. Their level of persistence in various environmental matrices and high impact at lower concentrations make

TABLE 4.1
Timeline of Antibiotics: Year of Production, Name of Antibiotic, and Class

Antibiotic drug discovery: timeline		
Period	**Antibiotic**	**Class**
1908	Salvarasan	Arsenical
1932	Prolonsil	Sulfonamide
1940–50	Gramicidine	Peptide
	Pencillin	β-lactam
	neomycin	Aminoglycoside
	Streptomycin	Aminoglycoside
	Cephalosporin	β-lactam
1950–60	Chloramphenicol	Phenylpropanoid
	chlortetracycline	Tetracycloine
	Polymyxin	Lipopeptide
	Erythromycin	Macrolide
	Vanomycin	Glycopeptide
	Virginiamycin	Streptogramin
1960	Rifamycin	Ansamycin
1962	Nalidixic acid	Quinolone
2000	Linezolid	Oxazolidinone
2003	Daplomycin	Lipopeptide

Source: Wright (2010).

antibiotics a concern as an emerging pollutant. Impact studies should be conducted along with the introduction of antibiotics (or any such newly introduced compounds) as there is a huge chance of their input into the environment. Such studies have only recently been conducted and thus reflect the flaws in the implementation of regulation techniques (Chaturvedi, Chowdhary et al. 2021; Chaturvedi, Shukla et al. 2021; Chowdhary et al. 2020a).

Quantification of antibiotic residues in a different environment is important in order to know the future impacts. Unlike conventional pollutants, practical hindrances are there in emerging contaminants. The first factor is its lower concentration in the environmental matrix. Second, antibiotic residues do not occur as their original parent forms in the environment. They can be transformed into different metabolites by the action of microorganisms and by physical or chemical reactions. Third, the difference in the regulation of concentration limits or tolerance levels of antibiotics in the environment. The concentration varies between the higher μg/L range in hospital effluents, and the lower μg/L range in municipal wastewater. The concentration is extremely different in the various aquatic environments. The amount of usage of different classes of antibiotics has increased at a higher rate. Over 8000 tons of antibiotics are used to promote animal growth in China every year, without any regulations (Ben et al. 2009). Lack of environmental regulations and standards is a crucial hindrance to be overcome. Antibiotics are not regulated through the current European environmental water quality standards due to the gap area in widespread environmental contamination and intrinsic hazard (Carvalho et al. 2016).

4.2.2 Sources of Antibiotic Resistance

Understanding the source of antibiotic contaminants in the environment matrix under investigation is essential as a general prediction is not applicable. It may change from place to place. Important

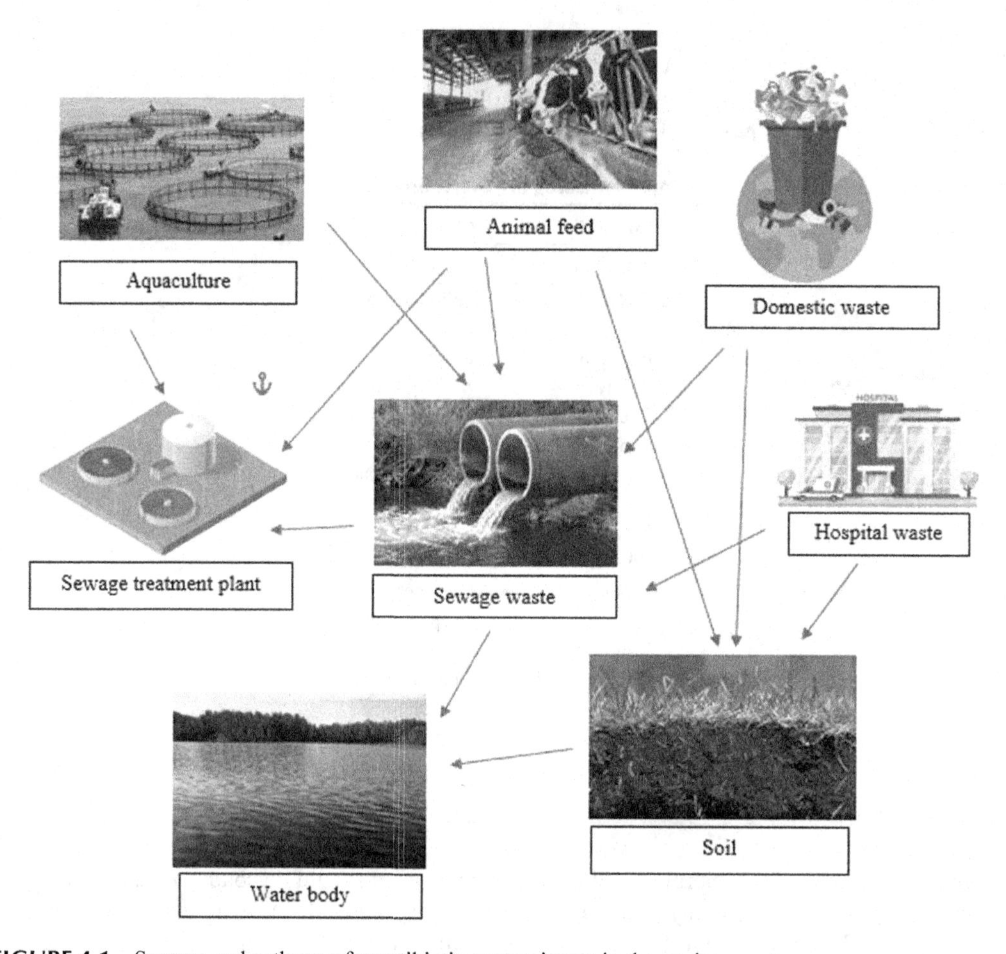

FIGURE 4.1 Sources and pathways for antibiotic contaminants in the environment.

sources, in general, include hospitals, private households, sewage, etc. (Figure 4.1). Improper disposal of unused pharmaceuticals is a major source of antibiotics residues in domestic wastes. Hospital wastes are of important concern as they contain both contaminants and dangerous pathogens. Aquaculture and antibiotic product units can directly contaminate surface water. Thus, the sewage treatment plant is important as it is the key point where surface water meets the wastewater. Several studies have shown that some antimicrobial compounds, especially polar ones, are not completely removed during wastewater treatment processes.

Antibiotics are also used as a supplement in animal feed as growth promoters. Through excretion, they reach the soil environment and contaminate water sources (Levy 1992). The pharmaceutical industry is also important to mention as a major source. The presence of antibiotic residues in a particular environment may be of remote origin, thus tracing all the possible sources is crucial. The general prediction may not be applied and investigation is needed to trace the sources and pathway in each case study. This has to be precisely carried out as future studies will depend on these investigations.Antibiotic compounds are possibly degraded in the environment by different factors. Thus, knowing their pathway is also important in order to evaluate their fate and impacts. Most of the studies focus on parent compounds as the degraded standards are not commercially available. As degraded compounds also have a negative impact, new strategies and methodology are needed to trace degraded compounds (Hernandez et al. 2007).

4.2.2.1 Agriculture as the Sources

Manure applied to improve crop growth, in turn, can be the source of antibiotic-resistant bacteria. Manure can be considered a "hot spot" of bacteria carrying ARGs. When soils are treated with manure, bacteria-carrying ARG are introduced into the soil. In soil, ARG is likely to be horizontally transferred to soil bacteria, a process that is enhanced by manure. Irrigation water is also regarded as one of the important bacterial contamination sources in vegetable growth.

4.2.2.2 Sewage as the Sources

Urban wastewater treatment plants (WTPs) are one of the most important sources of antibiotic resistance from humans to the environment. The sewage entering the WTPs combines the excreta and antibiotic residues. It carries pathogens, ARBs, resistance genes, and associated mobile genetic elements. As sewage carries pathogens and resistance genetic elements, WTPs become the hotspot of AMR pathogens. This exposure facilitates the development of antibiotic resistance by the pathogens.

Many studies have reported the presence of antimicrobial-resistant bacteria, including *enterococci*, and the number of pathogens in samples of crude inflow, treated effluent, and sludge collected in municipal sewage treatment plants. Data from studies indicate that the use of antimicrobials has created a large pool of resistance genes in sewage. This challenges the efficiency of conventional sewage treatment processes to address this emerging contaminant.

4.2.2.3 The Environment as the Source

The sources that lead the AMR bacteria in the environment can be of human or animal origin or directly from industries. Either accumulation of antibiotic residues in the environment develops resistance in bacteria or resistance bacteria are directly exposed to the environment. Genetic elements that have embedded resistance circulate in the environment and alter the microbial system. The water environment is one of the matrices that is highly exposed to such entry from human-animal sources and industries.

The extent of the threat of environmental sources is that these resistant genes are able to be transferred to other species of bacteria, including pathogens. Understanding the extent of the environmental resistome and its mobilization is of paramount importance in the present scenario.

The most prevalent Gram-negative pathogens, such as *Escherichia coli*, *Salmonella enterica*, and *Klebsiella pneumoniae*, cause a variety of diseases in humans and animals. A strong correlation between antibiotic use in the treatment of these diseases and antibiotic-resistance development has been observed over the past half-century. For example, after introducing the β-lactam class of antibiotics, new inactivating enzymes were identified in several studies. The inactivating enzyme for the β-lactam class of antibiotics is the β-lactamases. In studies, up to 1000 resistance-related β-lactamases were identified (Davies and Davies 2010).

Most of the antibiotics are not completely metabolized and released in unmetabolized forms which enter the food chain and affect various ecological niches through bio-accumulation. The persistence period of antibiotics ranges from 1 to 3466 days in the environment. ARBs are the most widespread zoonotic pathogens commonly found in the environment. The bacteria have a shorter generation time and are liable to strong selection pressure from host immunity as well as antimicrobial drugs (Dafale et al. 2020).

4.3 SPREAD OF ANTIMICROBIAL RESISTANCE

The key factor that makes AMR a challenging threat is its potential to spread. This involves complex phenomena and processes of transfer. This makes it impossible to trace the mechanism in a one-way approach. AMR spread is not restricted to any one of the environmental systems but its presence

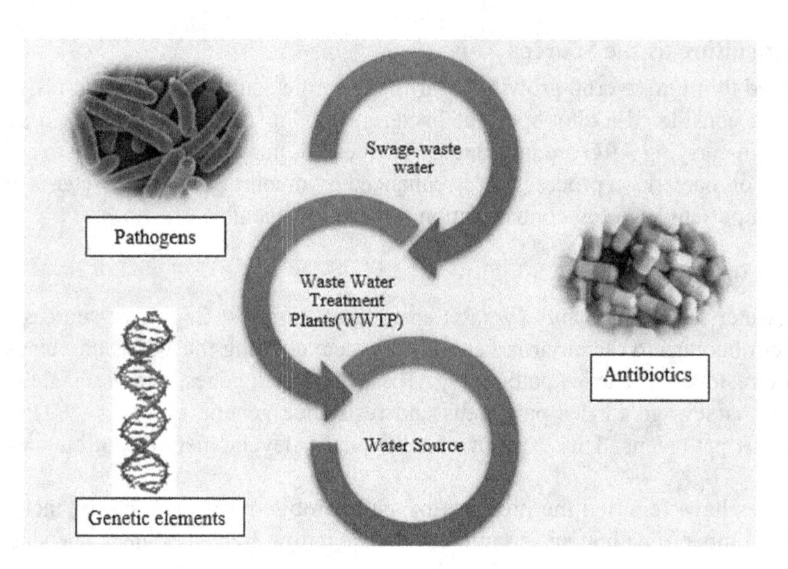

FIGURE 4.2 Wastewater treatment plants (WWTPs) as the transition point of AMR.

extends to all spheres of the environment. How the acquisition and transmission of resistance genes occur, what are the factors that make its potential spread, are essential questions to be answered. Genetic studies have revealed some of the extent of the understanding, but they still do not go far enough..

Wastewater treatment plants (WWTPs) have become the major point of AMR transmission. These WWTPs are the transition point from wastewater to natural water sources (Figure 4.2). Sewage, wastewater from hospitals, pharmaceuticals, etc. reach natural water sources containing antibiotic residues, AMR bacteria, and resistance genes. Commensal microorganisms and pathogens will be exposed to antibiotic residues. These persistent antibiotics enable the microflora to acquire resistance. The microorganism that gain resistance can impart the attribute to other strains via the horizontal gene transfer mechanisms (HGT). Resistant genes can be directly transmitted between strains of the same or different genus. Mechanisms of transmission of genes vary in different groups of microorganisms. This is again a challenging element as a general conclusion cannot apply unless the mechanism of each microorganism is studied separately.

Genetic elements such as transposons, integrons, Insertion Sequence (IS) elements, and the "new" Insertion Sequence Common Region (ISCR) elements are studied as elements responsible for the transfer and the increase in antibiotic resistance (Lekunberri et al. 2017).

A new type of genetic element, the common region (CR or ISCR), has been identified as being closely associated with the spread of many antibiotic resistance genes. They belong to two groups: (1) those that form complex class 1 integrons (termed ISCR1); and (2) those that are associated with non-class 1 integrons (ISCR2–12) (Lekunberri et al. 2017; Walsh 2006). In each case study, genetic elements that play a major role in AMR spread may be different. Hence the genetic study became essential in AMR investigations.

4.4 REMOVAL STRATEGIES

Emerging contaminants always challenge the lack of advanced wastewater treatment systems. Antibiotic residues and other emerging contaminants have advantages over conventional WWTPs. Conventional-type WWTPs have limitations on the removal of antibiotic residues from media. There are great discrepancies in the removal efficiencies of ECs in WWTPs between countries. This could be due to the limitations of previously used sampling methods (Hernandez et al. 2007; Chowdhary and Raj 2020; Chaturvedi, Chowdhary et al. 2021 Chaturvedi, Shukla et al. 2021). Lower concentration, difficulties in detection, and complex structures make the ECs monitoring complicated.

Many studies have attempted to address the treatment of ARBs and related genetic elements from wastewater. Degradation of antibiotics and inactivation of antibiotic resistance genes (ARGs) in cephalosporin C fermentation (CEPF) residues was such a study (Chu et al. 2020). The antibiotic removal percentages of different treatment methods were 86%, 80%, and 72–87% by gamma, ozonation, and thermal treatment, respectively. ARGs removal by gamma, ozonation, and thermal treatments ranged from 27–74%.

Hybrid carbon membranes could effectively remove 98.9% of Tetra Cycline Hydrochloride (TCH) from water by vacuum filtration. Hybrid carbon membranes constitute thick graphene oxide (GO) and activated carbon (AC) with a thickness of 15 μm (Liu et al. 2017).

Ionizing radiation applied to the antibiotic fermentation residues resulted in a reduction in the abundance of antibiotics resistance genes by 89–98% at 10 kGy irradiation. Over 99% of total bacteria were removed and antibiotic-resistant bacteria were less than the detection limit after 10 kGy irradiation (Chu et al. 2020).

Hiller et al. (2019) reviewed different advanced and conventional treatment strategies that have been studied to address the removal of antibiotic resistance bacteria and genetic elements. The evaluation of the fate of antibiotic resistance genes during six different full-scale municipal wastewater treatment processes reveals the aerated tanks are the point where horizontal gene transfer occurs the most. Anaerobic or anoxic tanks can reduce the occurrence of antibiotic resistance. The pan-genome of *E. coli* in WWTPs and hospitals showed no difference between influent and effluent of WWTPs. Thus conventional WWTPs can reduce the number of bacteria in the environment, but they fail to reduce the pathogenic potential for antibiotic resistance. The wastewater pan-genome was larger than a clinical pan-genome of similar size due to possible horizontal gene transfer in the wastewater. Available studies reported limited AMR removal efficiencies, varying between inefficient up to 2 log units' removal by sand filtration (Table 4.2).

TABLE 4.2
Advanced and Conventional Treatment Strategies for the Removal of Antibiotic Resistance Bacteria and Genetic Elements

Treatment process	ARB removal (log10 CFU/100 mL)	ARG removal (log10 genes/100 mL)
Conventional wastewater treatment		
Activated sludge	1.1–1.8	1–4.5
Trickling filter	0.4–0.9	
Submerged aerated filter	2.2–2.5	
Membrane bioreactor (MF)	2.7–5.4	2.5–7
Membrane bioreactor (UF)		2.6–7
Physical separation processes		
Sand filtration	0.2–0.5	0.1–2.5
GAC filtration	0.5–0.6	
Microfiltration		1
Ultrafiltration	0.9	1.7–7
Nanofiltration	0.9	5.9
Oxydation and disinfectant processes		
UV irradiation	0.2–4	
Ozonation	1.4–1.8	0.5
Chlorination	0.5–3	0.5

Source: Hiller et al. (2019).

The disinfection efficiency of chlorine depends on chlorine exposure, which is often calculated simply as dosage or reactor effluent concentration multiplied by contact time. As inactivation rates for chlorination of numerous bacteria strains are well documented in the literature, the same mechanisms for the inactivation of ARB and not-resistant bacteria can be expected.Membranes are a very efficient way to remove ARGs which depend on pore size and the molecular weight cut-off of membranes. The authors suggested DNA colloidal interactions as an efficient removal mechanism in membrane treatment. A systematic study of the impact of other parameters in WWTPs such as water quality, fouling, flux, etc. has yet to be performed.

4.5 WHY AMR IS A THREAT TO HEALTH

The health sector is now facing the challenges of the emergence of new diseases and the re-emergence of old diseases. This implies that disease-causing microorganisms are gaining an extra-ordinary ability, so that they can create this recent scenario that previously was unknown in history. This history should be read along with the development of antibiotics. There is a huge gap in the development of antibiotics and that of treatment strategies for antibiotics. This is the core issue of the modern world, in more detail, the lack of an efficient waste management system corresponding to the development of new facilities.

The unnecessary prescription and overuse of drugs are attributed to the increased concentration of these contaminants in the environment. The unnecessary prescription of drugs has increased in developed countries, and in many developing countries over-usage of drugs is an issue. Minimum laboratory diagnostic facilities and improper management strategies are the major challenges in developing countries. The over-usage of drugs and over-treatment enable a competitive advantage for antibiotic-resistant bacteria.

Misuse by unskilled practitioners and the general public is another major issue. Usage of poor-quality antibiotics and improper disposal are common in developing countries. Expired drugs receive new labels, are dumped without a label, or donated, these are common practices (Komolafe et al. 2003). These are highly challenging issues that increase the threat of resistant microbes in the environment. International travel is another mode of transmission of antibiotic-resistant microbes. A person who is exposed to specific resistant microbes in one country, is able to carry and spread them to another country.

This issue has a stronger impact when common pathogens become untreatable. Several incidents can be traced from different regions of the globe. A woman in Nevada died from *Klebsiella pneumoniae* which was found to be untreatable to all available classes of antibiotics.

AMR in hospitals and communities is a major threat to public health (Moellering 2007). Prescribing antibiotics is becoming a common part of disease treatment. The important concern is whether the community follows the scientific usage or not. Mostly the consumption, dosage, and course of treatment of antibiotics may not be in a scientific manner. An incomplete course of treatment may lead to the worst situation of resistance. Microbial flora in the human body can become familiarized to a spectrum of antibiotics. The fecal flora and microbiota upper respiratory are such prominent groups in the human body that later become the major source of resistance genes.

The introduction of antibiotic residues in the environment is the other challenging aspect. Even very low concentrations can create resistance in microbes when they persist for a long time in the environment. Most of the antibiotics residue were developed ages ago, but no efficient system has been developed to inhibit their introduction into the environment. That is, their introduction to the environment is not a recently occurring phenomenon, and they have had more than enough time to acquire resistance. Recent assessments of antibiotic resistance in environmental microbes show it is a worldwide issue and has become a challenging threat to the health sector at the global level. A spectrum of pathogens is already reported as being resistant to the existing antibiotics. Most of the commensal microbes are also recognized as resistant to antibiotics.

Years of interaction have provided these microflorae an adequate period to recognize most of the commercially available antibiotics and adapt themselves for resistance. Genetic and evolution studies have an important role in revealing the mechanism of acquiring resistance, transmission, etc. Studies show resistant genes evolved million years ago, that is, years before the discovery of antibiotics. The presence of antibiotics is the triggering element that activates the genes to express phenotypically. That is, the scientific clue to resist the antibiotics is itself embedded within these early prokaryotes.

The WHO has recognized AMR as one of the major threats to the modern world. No sector of the environment is safe from AMR threats as its spread is unpredictable. As they travel, the transport of food products is also contributing to the risks of the spread of resistant genes. Chugh (2008) has reviewed the scenario of AMR in various pathogens at the global level.

4.6 ANTIMICROBIAL RESISTANCE SURVEILLANCE

Episodes of new diseases and the emergence of old diseases are strongly connected to AMR development in bacteria. Most of the AMR assessments occurred after the outbreak. However, such pre-analysis studies of AMR are important to predict such disease outbreaks. The intense spread of new diseases at an alarming rate points to the need for surveillance of AMR.

Public water resources, rivers, other drinking water sources, etc., need to be analyzed for the presence of antibiotic residues, AMR microbes, and related genes. Special attention needs to be directed to the efficiency of wastewater treatment plants. Conventional treatment systems are no longer effective in removing the emerging contaminants. Human and animal fecal samples are important in the surveillance of AMR to understand the direct spread. The COVID-19 pandemic is considered the demo which explains how surveillance can be carried out globally. It is an example that worldwide surveillance is practically possible even in a continuous manner for a short period. The case can be considered an answer to the questions on the economic feasibility of surveillance versus epidemics outbreaks. Despite the necessity, only a few AMR surveillance studies have been reported. This huge gap has to be filled through the integration of both the research community and government bodies to prevent the dangerous consequences of AMR in the future.

An epidemic survey commonly provides all the basic information on different kinds of epidemics. The information is incomplete when the question is about the treatment's success. AMR surveillance provides the status of antimicrobial resistance gained by the pathogens, so that it is possible to predict the effectiveness of the current treatment system and the need for advanced antibiotics (Chowdhary and Bharagava 2019; Chowdhary, Hare et al. 2020).

The importance of an epidemic survey is also required for AMR surveillance. The data on affected persons, sources, the causing pathogens are inadequate to prevent the death rate in epidemic outbreaks. The success of the antibiotic treatment is not possible unless the antibiotic resistance of the pathogen is analyzed. And this is different from region to region, based on the exposure to antibiotic residues in the particular environment. Thus, it is essential to understand the AMR status of pathogens in epidemic outbreaks. As most of the AMR studies occur only after such outbreaks, immediate treatment will become impossible. Most of the antibiotics are produced years before and are long-lasting (see Table 4.1). And, it technically makes the treatment fail, as, during this long period, pathogens gain the resistance to the antibiotics. New generation antibiotics also may reflect a similar threat to what the modern world is facing now: the problem of the less efficient waste treatment system and more increased pollution than ever before.

Recent studies depict AMR as a serious issue and focus on its assessment. Antibiotic susceptibility tests using a disk diffusion method are considered the standard culture-based method. Molecular techniques using PCR facilities provide a culture-independent approach. The conventional method is time-consuming and addresses only a small fraction of the bacteria population. Advanced molecular methods provide more detailed data than conventional systems. Both have their advantages and disadvantages. Most studies focus on the selected pathogens or epidemic-causing pathogens. The WHO has identified seven bacteria with high levels of antibiotic resistance, including Gram-positive and Gram-Negative bacteria (Table 4.3). Extended-spectrum β-lactamase and carbapenemase-producing *Enterobacteriaceae* and methicillin-resistant *Staphylococcus aureus* (MRSA) are the most dangerous among them, due to mortality risks.

Indeed, emerging pathogens may also cause unpredictable consequences by acquiring resistance. The conventional method is not effective in addressing the issue of emerging pathogens.

TABLE 4.3
Bacteria with High Levels of Antibiotic Resistance

Gram-negative bacteria	Gram-positive bacteria
Neisseria gonorrhoeae	*Streptococcus pneumoniae*
Escherichia coli	*Staphylococcus aureus*
Klebsiella pneumoniae	
Shigella spp	
Non-typhoidal salmonella	

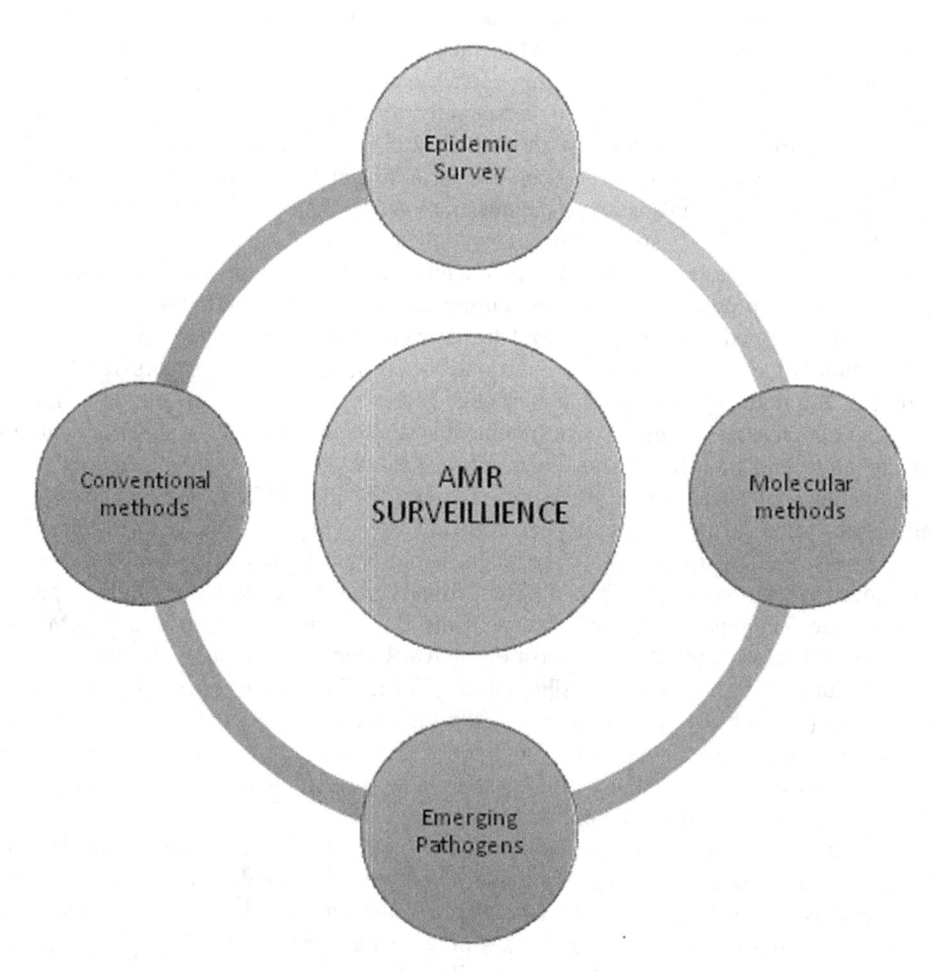

FIGURE 4.3 AMR surveillance and related tools.

TABLE 4.4
GLASS-Focused Surveillance and Special Survey Activities

Sl. No	Surveillance and special survey activities
1.	GLASS Emerging Antimicrobial Resistance Reporting (EAR) for the timely detection, reporting, risk assessment, and monitoring of emerging resistance
2.	Surveillance of invasive fungal bloodstream infections caused by Candida spp.
3.	One Health2 surveillance through the WHO integrated, multi-sector surveillance based on the extended-spectrum beta-lactamase (ESBL) *E. coli* Tricycle project. The project was pilot-tested in six low-income (LICs) and lower-middle-income (LMICs) countries and implemented in a further nine countries in 2020.
4.	The Enhanced Gonococcal Antimicrobial Surveillance Programme (EGASP) has been pilot-tested in the Philippines and Thailand. Activities initiated in Cambodia in 2020 provided an unprecedented set of clinical data related to resistant *N. gonorrhoeae* and have allowed the finalization of the EGASP protocol.
5.	Point prevalence surveys (PPS) on antimicrobial use (AMU) in hospitalized patients in 34 countries
6.	Planned assessment of attributable mortality of AMR bloodstream infections (BSIs)

Metagenomics is a highly promising tool to identify the unknown bacteria from an environmental sample directly. Thus, an integrated approach will be more effective to give real results.

The Global Antimicrobial Resistance Surveillance System (GLASS) is the first system that enables harmonized global reporting on AMR, established in 2015, in which currently 109 countries and territories worldwide have enrolled (Figure 4.3). The fourth GLASS report summarizes the 2019 data on AMC surveillance from 15 countries and AMR data on 310,6602 laboratory-confirmed infections reported by 24,803 surveillance sites in 70 countries (Table 4.4). The European Antimicrobial Resistance Surveillance Network (EARS-Net) and the Central Asian and European Surveillance of Antimicrobial Resistance (CAESAR) network are examples of large regional AMR surveillance networks.

4.7 CONCLUSION

Environmental waste management is the area which has to be developed enough to meet the current problem. In practice, most countries use basic conventional treatment systems even though the emerging contamination issues are rising at an alarming rate. The removal efficiency of conventional treatment systems and the impact of relevant parameters are yet to be studied systematically. The introduction of a sufficient number of new efficient antibiotics is the immediate action to be considered. Proper replacement of current antibiotics in a period is the way to increase the efficiency of treatment against pathogens. AMR surveillance is inevitable in this current scenario of new and old emerging diseases. The extension of the current surveillance systems and the development of more regional surveillance systems are necessary. The practical success of AMR surveillance depends on the integration and smooth coordination between the scientific society and governments.

REFERENCES

Arya, S.R., Chavan, D., and Kumar, S. (2019). Recent trends in bio-processing of antibiotic residues and its resistant genes in solid waste. *Biological Processing of Solid Waste*. Boca Raton, FL: CRC Press. https://crcpress.com/Biological-Processing-of-Solid-Waste/Li-Awasthi-Zhang-Kumar/p/book/9781138106420

Ben, W., Qiang, Z., Pan, X., and Chen, M. (2009). Removal of veterinary antibiotics from sequencing batch reactor (SBR) pretreated swine wastewater by Fenton's reagent. *Water Research*, 43(17), 4392–4402.

Carvalho, I. T. and Santos, L. (2016). Antibiotics in the aquatic environments: A review of the European scenario. *Environment International*, 94, 736–757.

Chaturvedi, P., Chowdhary, P., Singh A., Pandey, A., and Gupta, P. (2021). Dissemination of antibiotic resistance genes, mobile genetic elements, and efflux genes in anthropogenically impacted riverine environments. *Chemosphere*, 273, 129693. https://doi.org/10.1016/j.chemosphere.2021.129693.

Chaturvedi, P., Shukla, P., Giri, B.S., Chowdhary, P., Pandey, A., and Gupta, P. (2021). Prevalence and hazardous impact of pharmaceutical and personal care products and antibiotics in the environment: A review on emerging contaminants. *Environmental Research*, 194, 110664. https://doi.org/10.1016/j.envres.2020.110664.

Chaturvedi, P., Singh, A., Chowdhary, P., Pandey, A., and Gupta, P. (2020). Occurrence of emerging sulfonamide resistance (*sul1* and *sul2*) associated with mobile integrons-integrase (*intI1* and *intI2*) in riverine systems. *Science of the Total Environment*. https://doi.org/10.1016/j.scitotenv.2020.142217.

Chowdhary, P. and Bharagava, A. (Eds.) (2019). *Emerging and Eco-Friendly Approaches for Waste Management*. Singapore: Springer.

Chowdhary, P., Hare, V., Singh, A.K., Pandit, S., and Chaturvedi, P. (2020). Emerging environmental contaminants: Sources, consequences, and future challenges. In Chowdhary, P., Raj, A., Verma, D., and Akhter, Y. (Eds.) *Contaminants and Clean Technologies*. Boca Raton, FL: CRC Press.

Chowdhary, P., Raj, A., Verma, D., and Akhter, Y. (Eds.) (2020a). *Contaminants and Clean Technologies*. Boca Raton, FL: CRC Press.

Chowdhary, P., Raj, A., Verma, D., and Akhter, Y. (Eds.) (2020b). *Microorganisms for Sustainable Environment and Health*. Oxford: Elsevier.

Chu, L., Chen, D., Wang, J., Yang, Z., Yang, Q., and Shen, Y. (2020b). Degradation of antibiotics and inactivation of antibiotic resistance genes (ARGs) in Cephalosporin C fermentation residues using ionizing radiation, ozonation and thermal treatment. *Journal of Hazardous Materials*, 382, 121058.

Chugh, T.D. (2008). Emerging and re-emerging bacterial diseases in India. *Journal of Biosciences*, 33(4), 549–555.

Dafale, N.A., Srivastava, S., and Purohit, H.J. (2020). Zoonosis: An emerging link to antibiotic resistance under "One Health Approach." *Indian Journal of Microbiology*, 60(2), 139–152.

Davies, J. and Davies, D. (2010). Origins and evolution of antibiotic resistance. *Microbiology and Molecular Biology Reviews*, 74(3), 417–433.

Hawkey, P.M. (1998). The origins and molecular basis of antibiotic resistance. *BMJ*, 317(7159), 657–660.

Hernández, F., Sancho, J.V., Ibáñez, M., and Guerrero, C. (2007). Antibiotic residue determination in environmental waters by LC-MS. *TrAC Trends in Analytical Chemistry*, 26(6), 466–485.

Hiller, C.X., Hübner, U., Fajnorova, S., Schwartz, T., and Drewes, J.E. (2019). Antibiotic microbial resistance (AMR) removal efficiencies by conventional and advanced wastewater treatment processes: A review. *Science of the Total Environment*, 685, 596–608.

Khan, H.K., Rehman, M.Y.A., and Malik, R.N. (2020). Fate and toxicity of pharmaceuticals in water environment: An insight on their occurrence in South Asia. *Journal of Environmental Management*, 271, 111030.

Komolafe, O.O. (2003). Antibiotic resistance in bacteria: An emerging public health problem. *Malawi Medical Journal*, 15(2), 63–67.

Komolafe, O.O., James, J., Kalongolera, L., and Makoka, M. (2003). Bacteriology of burns at the Queen Elizabeth Central Hospital, Blantyre, Malawi. *Burns*, 29(3), 235–238.

Lekunberri, I., Subirats, J., Borrego, C.M., and Balcázar, J.L. (2017). Exploring the contribution of bacteriophages to antibiotic resistance. *Environmental Pollution*, 220, 981–984.

Levy, S.B. (1992). *The Antibiotic Paradox: How Miracle Drugs Are Destroying the Miracle*, New York: Plenum Press.

Liu, M.K., Liu, Y.Y., Bao, D.D., Zhu, G., Yang, G.H., Geng, J.F., and Li, H.T. (2017). Effective removal of tetracycline antibiotics from water using hybrid carbon membranes. *Scientific Reports*, 7, 43717.

Moellering Jr, R.C., Graybill, J.R., McGowan Jr, J.E., and Corey, L. (2007). Antimicrobial resistance prevention initiative—an update: Proceedings of an expert panel on resistance. *American Journal of Infection Control*, 35(9), S1–S23.

Tomasz, A. and Munoz, R. (1995). β-Lactam antibiotic resistance in Gram-positive bacterial pathogens of the upper respiratory tract: A brief overview of mechanisms. *Microbial Drug Resistance*, 1(2), 103–109.

Walsh, T.R. (2006). Combinatorial genetic evolution of multiresistance. *Current Opinion in Microbiology*, 9(5), 476–482.

Wise, R., Hart, T., Cars, O., Streulens, M., Helmuth, R., Huovinen, P., and Sprenger, M. (1998). Antimicrobial resistance is a major threat to public health, *BMJ*, 317: 609–610.

WHO (World Health Organization) (1998). Antimicrobial resistance, Fact Sheet No. 194. Geneva: WHO.

WHO (World Health Organization) (2021). Global Antimicrobial Resistance and Use Surveillance System (GLASS) Report: 2021. Geneva: WHO.

Wright, G.D. (2010). Antibiotic resistance in the environment: A link to the clinic? *Current Opinion in Microbiology*, 13(5), 589–594.

5 Black Trail in Blue Oceans

Cristina Carapeto
Universidade Aberta, Portugal
Corresponding author Email: cristina.carapeto@uab.pt

CONTENTS

5.1 Introduction 83
5.2 Major Sea Pollutants 84
5.3 Black Trail and Sewage 90
5.4 Conclusion 95
References 95

5.1 INTRODUCTION

Most humans feel an extraordinary attraction to the sea. We just have to look at coastal seas and see that beaches are a major place for entertainment during summer time. Even in bad weather, most people find it very uplifting to walk close to the water. To swim and dive just for the pleasure of being in the water are also very attractive for a good number of people, let alone those who are dedicated to the study of marine life.

In the open oceans, people enjoy spending their time in small boats, yachts or on cruises just to be surrounded by water with blue skies above and birds cruising the air.

In the sea, the sound around us is simplified. The sound of water is far simpler than the sound of voices, the sound of music or the sound of a city. Our visual input is also simplified and we look at the horizon with no walls or buildings to disturb the perspective. All this gives us a feeling of peace and relaxation. This may be the appeal of nature to humankind or it may represent a stronger bond to the place where life began. Whatever the reason, the truth is that, since ancient times, oceans have always been a mystery to fear and a mystery to discover.

Our relationship with the oceans has always been a complex one. Oceans supply fresh water, oxygen, act to moderate climate, influence weather, affect our health and are an enormous source of food and a "road" to elsewhere. As the world's population and the standards of living grow, we all need to understand the impact of human species on the oceans. Furthermore, we need to understand how important it is to protect the oceans from human activities since the use of the oceans' resources has always been exploitative. Today, the main key dimensions of human-ocean interactions are: (1) the use of oceans' bio-resources; (2) the use of oceans' mineral resources; (3) scientific research; (4) transportation and harbour development; (5) recreation and tourism; and (6) military purposes.

The use of oceans' bio-resources is one of the key dimensions of human-ocean interactions, including fishing, seaweeds and other resources for consumption. Many nations dedicate huge efforts to benefit from the oceans' mineral resources, such as oil and natural gas, but phosphorite and manganese nodules are also important.

Scientists have studied the oceans to understand how different life forms have adapted and evolved in the same environment and how we may protect them since they are Earth's patrimony. They also try to understand the geologic and geochemical processes involved in the oceans' evolution and the interactions between the oceans, the atmosphere and climate change.

The ocean has always functioned as a means of transportation not only for goods but also for people, and harbours have a significant role in this process since they are places of arrival and

DOI: 10.1201/9781003239956-6

departure. With the advent of the tourism industry, oceans became not only an important means of transportation of goods and people but they also became the "ground" of floating cities, i.e. ocean-going luxury cruise liners. Although recreation and tourism are in general considered a "non-invasive" dimension of human-ocean interaction, the exponential growth in the cruise industry observed in the last few years makes the "non-invasive" description quite debatable.

Finally, the military involvement with the oceans is less visible to the public but still no less harmful. Since ancient times a nation's ability to rule the seas has been seen as a measure of its military strength, which does not make this activity acceptable, especially when modern military vessels use nuclear power instead of galleons and classic pirates' techniques.

Although marine tourism is now an important industry that improves the standards of living of many people around the world via job creation, it would be irresponsible to forget that this industry imposes a high toll on the environment. Marine tourism is a type of tourism that includes recreational activities with a focus on the marine environment (Gedik and Mugan-Ertugral 2019). In 2014, the European Commission estimated that coastal and marine tourism employed approximately 3.2 million people and represented over 1/3 of the maritime European economy (EC 2014). Though these numbers are impressive, the stress that marine tourism places on the natural environment and people's heritage and culture is also extensive. To mention just a few of these problems: damage to coral reefs, the garbage problem, the damage to marine ecosystems by the construction of harbours, the accidental harm to marine animals or their hunting for sport, the biological effects of environmental pollutants (biocides, ballast water, fuel oil), and many others.

5.2 MAJOR SEA POLLUTANTS

Ocean dumping has been practised for centuries, during which time the nature of marine debris has changed dramatically. Decades ago, much of human waste was made of organic, degradable materials. Now, synthetic elements are abundant in solid waste. Coastal seas and oceans are the final deposit place for a variety of these materials, such as plastic beverage bottles, packing straps, tarpaulins, and synthetic fishing lines.

Marine litter may be defined as the persistent substantial matter of human origin while marine debris is associated with municipal improper waste disposal, industrial activity in the coastal area or in rivers from the catchment area, fishing, tourism, military and many other sources (Novac et al. 2020). Given the high plastic consumption patterns, this type of litter has become the major waste affecting marine environments. Most of this waste comes from the river flows (Vlachogianni et al. 2018). Macroplastics as marine debris are of dimensions that marine organisms can see and avoid but under the action of marine environment they tend to be broken up into smaller parts, becoming microplastics which then are decomposed into even smaller pieces, the so-called nanoplastics. It is this last category that can easily be swallowed by marine animals and later passed into the human food chain (Schulz et al. 2013; Hahladakis 2020).

Sheavly and Register (2007) reviewed the available data on marine debris found worldwide and indicated that the dominant types and sources of debris are from what we consume (including food wrappers, beverage containers, cigarettes and related smoking materials), what we use in transporting ourselves by sea, and what we harvest from the sea (fishing gear). Already in 1991, the United Nations Joint Group of Experts on the Scientific Aspects of Marine Pollution (GESAMP) determined that land-based sources accounted for up to 80% of the world's marine pollution (GESAMP 1991). In 2020, GESAMP recognized that plastics and other floating debris were not the only pollutants affecting the oceans, but also that chemical pollutants, or new entities, remained poorly described. This happens due to the complexity of synthetic chemicals, uncertainties about appropriate thresholds or precautionary levels, the substantial number of new chemicals generated and released into the environment each year, and a lack of understanding of how these chemicals interact with each other once they are released (GESAMP 2020).

TABLE 5.1
Classification of Plastic Litter According to Their Dimensions

Authors	Macroplastics	Mesoplastics	Microplastics	Nanoplastics
Gregory (1996)				< 0.5 nm
Thompson et al. (2004)			Around 0.2 mm	
Arthur et al. (2008)			< 5 mm	
Imhof et al. (2012)	> 20 mm	5 mm–20 mm	< 5 mm	
González et al. (2016)	> 25 mm	5 mm–25 mm	< 5 mm	
Gago et al. (2016)			< 5 mm	
Hartman et al. (2019)	> 1 cm	1 mm–10 mm	1 μm–1000 μm	
Rodríguez et al. (2020)		> 5mm	1 mm–5 mm	

TABLE 5.2
Fractions of Crude Oil Refinery

	Boiling temperature (approximately) ºC	Molecular size of the fractions
Oil gases	30	C_3–C_4
Light gasoline, benzene*	30–140	C_4–C_6
Naphtha	120–175	C_7–C_{10}
Kerosene	165–200	C_{10}–C_{14}
Diesel	175–365	C_{15}–C_{20}
Fuel and waste	350	C_{20} +

Source: Carapeto (1999).

Note: *Benzine = benzene with impurities.

Table 5.1 lists the classification of plastics into macro-, meso-, micro- and nanoplastics, according to different researchers, and it can be seen that there are no uniform measures when referring to these bits of debris. This lack of agreement makes the task of specifying marine pollution even more difficult.

Oil pollution continues to be an important source of pollution in the marine environment. The first widespread use for petroleum products was in lamps, where mineral oil began to replace vegetable or whale oil in the mid-nineteenth century. It was shipped in barrels carried in the hold, like whale oil or any other liquid cargo, but even so was liable to spillage. Steamers did not engage, at first, in the petroleum trade for fear of fire, so the first true tankers were sailing vessels fitted out with lower hopper tanks. Much of the early pollution of European coasts was by fuel from the bilges and bunker-tanks of marine vessels equipped with oil-fired engines. Crude oil was mostly processed in petroleum-producing regions and tankers carried the refined products via commercial routes (Nelson-Smith 1970). Petroleum is a naturally occurring complex mixture made up predominantly of organic carbon and hydrogen compounds. It also frequently contains significant amounts of nitrogen, sulphur and oxygen together with smaller quantities of nickel, vanadium and other elements.

When oil pollution is discussed, it includes both the pollution caused by crude oil and its refined products. The basic refining technique is based on the knowledge that the various components of raw oil have boiling points that are characteristic of them and, therefore, can be separated by distillation. As, in general, hydrocarbons have higher boiling points as their molecular size increases, distillation allows the mixture to be separated into fractions of similar molecular weight (fractional distillation). It can be said that the process is simple. The crude oil is heated in a sequence of temperatures that gradually increase and the different fractions that separate are collected. Table 5.2 indicates the different fractions that can be collected during the distillation of oil, at different temperatures.

Light gasoline is the basis for the manufacture of gasoline used in motor vehicles. With the increasing use of cars, the need for gasoline has increased considerably compared to other products extracted from oil and has exceeded the amount that could be obtained by the simple process of fractional distillation. For this reason, refineries started using catalytic processes, without abandoning distillation, to obtain a higher molecular fraction yield between C_3 and C_4. The use of higher molecular weight hydrocarbons makes it possible to obtain the desired fractions and supply the market with the required quantities.

Other products obtained from distillation are:

- Naphtha, the third fraction obtained from distillation, is used in the petrochemical industries.
- Kerosene has several uses as a fuel, namely in lighting.
- Diesel is used in several engines suitable for the use of this fuel.
- Fuel, a liquid fraction of relatively heavy hydrocarbons, feeds ships and factories and the higher molecular weight fractions are used for other applications, for example, for the production of tar.

With regard to water pollution, it should be noted that the various products of oil refining have different degrees of toxicity to organisms. Thus, a spill of 300 tonnes of gasoline into a body of water will have different effects on aquatic life, compared to a spill of crude oil or fuel, of the same magnitude. The toxicity of hydrocarbons is directly related to their resistance to degradation and to their tendency and ease of incorporation into the various compartments of the ecosystem. Ultimately, all components of crude oil are degradable by bacteria, albeit at different rates. The smallest compounds, straight or branched, are the ones that degrade the fastest, while the cyclic compounds are the slowest. High molecular weight compounds tend to form balls that, owing to their small surface in relation to the volume, undergo an extremely slow degradation.

Whenever there is an oil spill in aquatic ecosystems, a series of processes begins that modify its composition (Figure 5.1).

The toxic effects of oil can be divided into two categories:

1. effects associated with the lining of organisms;
2. effects due to the toxic properties of hydrocarbons.

The former are more directly related to high molecular weight hydrocarbons, which are less soluble in water, less likely to evaporate and which confer many of the viscosity properties to oil, and can cover the bodies of both waterfowl and intertidal organisms. The second category of toxic effects relates to aromatic hydrocarbons (the most toxic), cycloalkanes and alkanes, which, if ingested by organisms, will be incorporated into lipids and other tissues, disrupting the normal functioning of the biological activities.

Fuel oil (also known as heavy oil, marine fuel, bunker, furnace oil, or gasoil) is a fraction obtained from petroleum distillation. It includes distillates and residues (the lighter fractions and the heavier fractions respectively). Broadly speaking, fuel oil is any liquid petroleum product that is burned in a furnace or boiler for the generation of heat or used in an engine for the generation of power, except oils having a flash point of approximately 40 °C (104 °F) and oils burned in cotton or wool-wick burners.

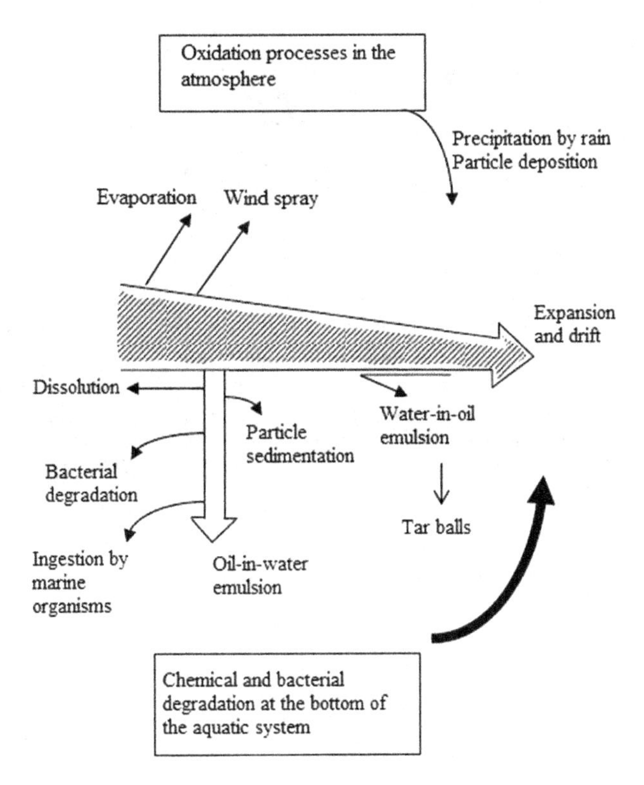

FIGURE 5.1 Main physical, chemical and biological processes that command the fate of oil pollution.

Heavy fuel oil (HFO) is a category of fuel oils of a tar-like consistency. HFO is predominantly used as a fuel source for the propulsion of marine vessels due to its relatively low cost compared to cleaner fuel sources such as distillates. Most cruise ships burn HFO, which is the dirtiest fossil fuel available. Most of these ships also do not have any diesel particulate filters or selective catalytic converters to clean the exhaust (while these technologies are standard for road vehicles like trucks).

In 2017, the luxury cruise brands owned by Carnival Corporation & PLC emitted in European seas alone 10 times more disease-causing sulphur oxide than did all of Europe's 260+ million passenger vehicles (T&E 2019). In the study done by T&E in 2019, conclusions pointed out that even a relatively small number of cruise ships (Figure 5.2) emit vast amounts of air pollution and that high emissions are due to insufficient stringency of the marine fuel quality and engine emissions standards.

Even assuming that ships did fully comply with the existing SO_X and NO_X standards in place and comparing ship emissions with car emissions (with the ship emissions being divided by those of the passenger cars), the final results found by T&E (2019) are likely to be on the conservative side. This means that they underestimated the comparative extent of air pollution from cruise ships *versus* cars (Figure 5.3), concluding that even a relatively small number of cruise ships emits vast amounts of air pollution (even though the scope of their analysis was limited to continental Europe and surrounding islands only).

In 2019, McCarthy compared SOx emissions from cruise ships and cars in European port cities in 2017 (Figure 5.4) providing an overview of the European cities suffering the worst environmental impact from the cruise industry in 2017 (McCarthy 2019). Ground-level NO_2 and $PM_{2.5}$ concentrations represent a serious public health concern in Europe and in many other places.

FIGURE 5.2 Cruise ship.

Source: The Green Optimistic.com. In Fiala (2018).

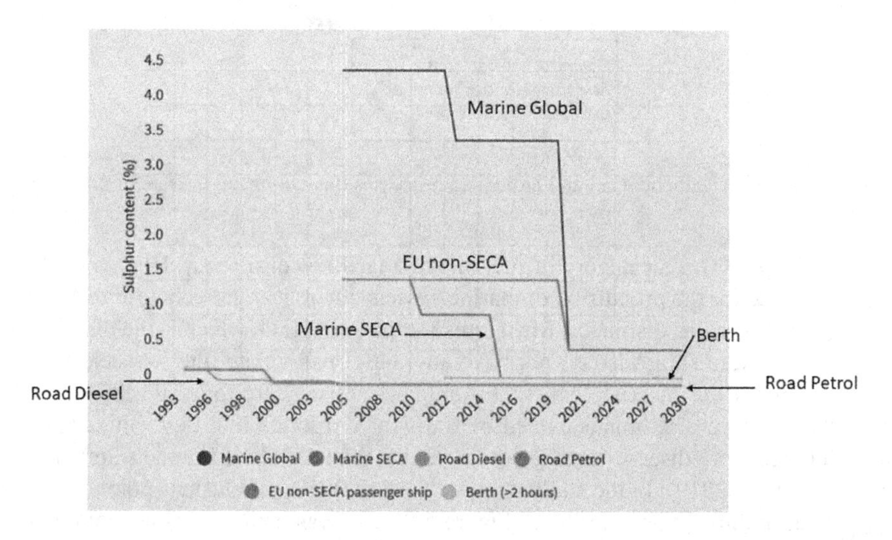

FIGURE 5.3 EU sulphur standards for marine and road fuels.

Source: Transport & Environment (2019).

The unique situation created by the SARS-CoV-2 virus pandemic (COVID-19) led to a sudden decrease in economic activities (OECD 2020) and mitigation measures as the lockdown and the implementation of distance working caused a decrease in traffic in all the cities around the world. As a direct consequence, the concentrations of nitrogen dioxide (NO_2) dropped in many cities although the same tendency was not followed by fine particulate matter ($PM_{2.5}$) which showed a more limited and less consistent decrease (Beloconi et al. 2021). Messaoud and Smiti (2020) have illustrated the same results with images from Sentinel-5P satellite before quarantine and during quarantine in both France and Italy (Figures 5.5 and 5.6).

Although this decrease in NO_2 in city centres may be attributed to the halt of motor vehicles, the fact is that in many coastal cities the same effect was enhanced because tourist cruise liners were prohibited. What perhaps many people do not realize is that each time a tourist cruise liner arrives at

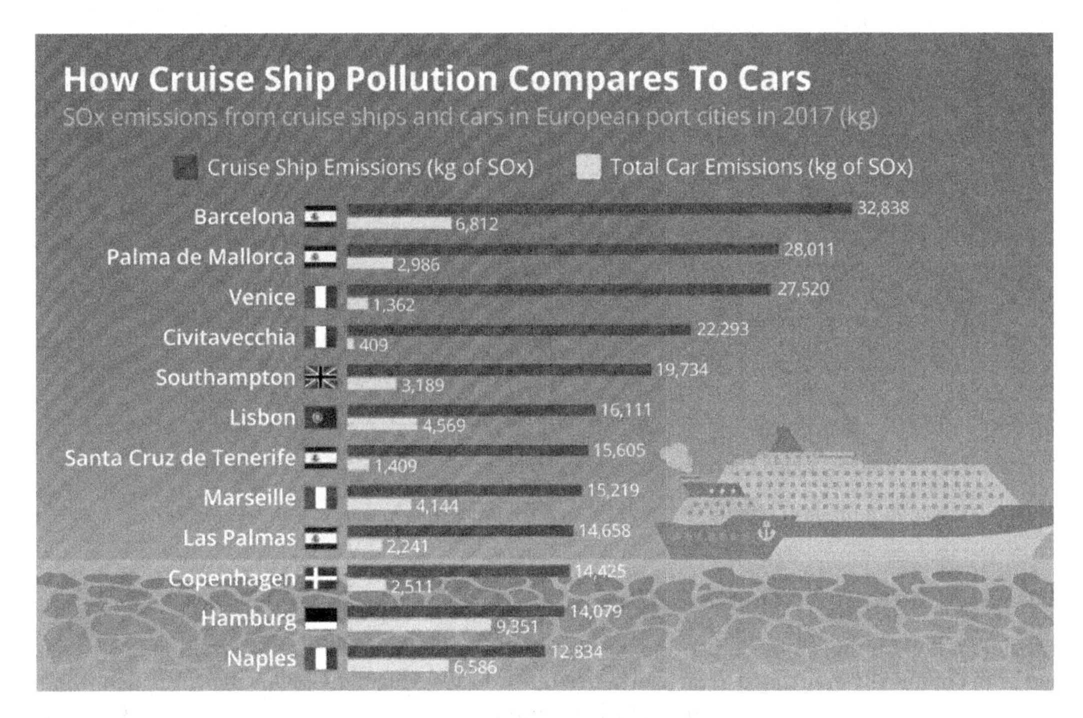

FIGURE 5.4 Cruise ship pollution compared to car pollution in European cities in 2017.

Source: How cruise ship pollution compares to cars. Available at: https://statista.com/chart/18351/emissions-from-cruise-ships-and-cars-in-european-port-cities/

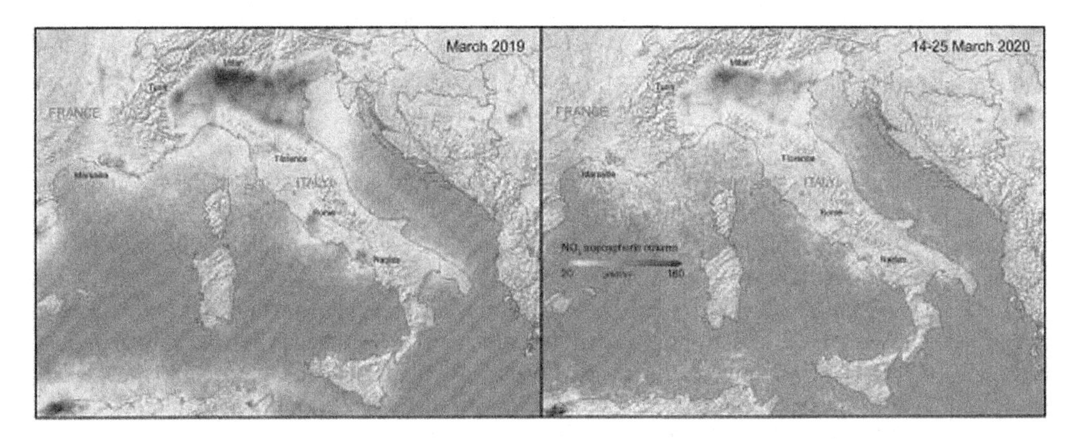

FIGURE 5.5 Nitrogen dioxide concentrations over Italy.

Source: Messaoud and Smiti (2020).

a given port, it is similar to accepting a new city in that same space. Tourist cruise liners are floating cities that have to keep running 24 hours a day to supply their customers with electricity, food, running water, laundry, swimming pools, entertainment and everything else they need. This means that the ship engines keep on working to provide energy for all these needs because the majority of ports do not have the capacity to supply enough electricity.

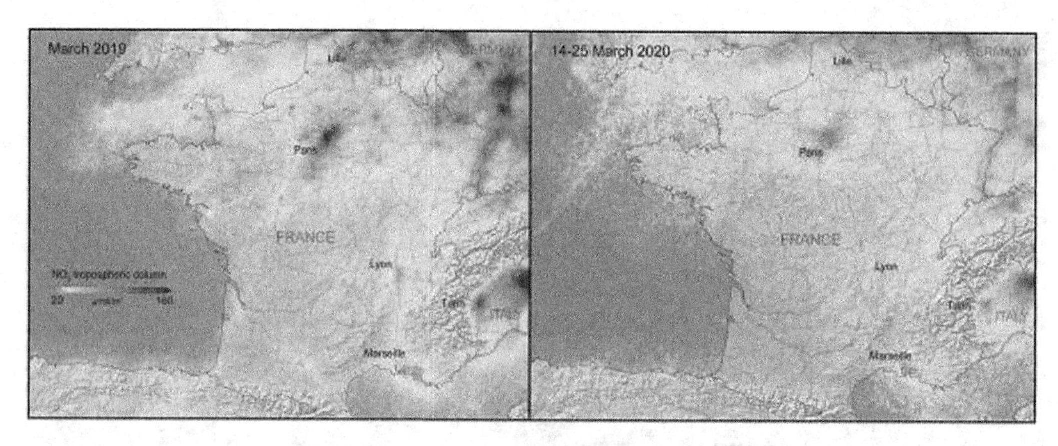

FIGURE 5.6 Nitrogen dioxide concentrations over France.

Source: Messaoud and Smiti (2020).

5.3 BLACK TRAIL AND SEWAGE

The heavy fuel oil that ships burn is the dirtiest of all the fuels used for transportation, after coal had been completely eradicated as an alternative in the 1950s. Heavy fuel oil is the waste that is left over from the production of gasoline and diesel in refineries. A thick, black, and strong-smelling liquid that has no other use than serving as fuel for ships (or for paving roads). HFO is the bottom-of-the-barrel leftovers from the oil refining process. When spilled in a body of water, it is nearly impossible to completely clean up and burning it emits more climate-warming black carbon (BC) than other fuels. Ships use this to power their engines and in the process emit a wide variety of pollutants that have important health and climate change impacts. The problem with cruise ships is not just the quantity of what is emitted, but also the type of their emissions. Ship emissions have been responsible for around 400,000 premature deaths worldwide owing to the development of lung cancer and cardiovascular disease, and around 14 million childhood asthma cases per year (Sofiev et al. 2018).

Fine particulate matter ($PM_{2.5}$), sulphur oxides (SOx), and nitrogen oxides (NOx) emerge from ships' smokestacks and together they damage human health and ecosystems, leading to premature mortality and morbidity (Corbett et al. 2007; EPA 2009; Winebrake et al. 2009). SOx emissions from sulphate (SO_4) aerosols increase human health risks and contribute to the acidification of terrestrial and aquatic ecosystems.

Cruise ships are real floating cities that have to be up and running even while they are docked in ports. Therefore, they have to burn fuel oil that, despite using less sulphur when the ship is in port, means they still have a huge impact in terms of ultra-fine particles, which are those that most easily penetrate human bodies and that carry a whole series of other adjacent pollutants into the human body and end up having serious health consequences.

The International Maritime Organization (IMO) has already proposed new global standards to limit sulphur (S) in fuel oil, decreasing the current limit of 3.5% S to 0.5% S (by mass) after January 1, 2020. These standards, if applied, will reduce sulphate aerosols and will provide health benefits to exposed populations. With the implementation of the low-sulphur fuel standards in 2020, it is expected to avoid premature deaths of 266,300 people per year (in the range of 138,500–395,700), which means a reduction of around 34%, and childhood asthma morbidity due to shipping will decline by 54% (Sofiev et al. 2018).

In the case of Europe, the European Commission estimates that every year 50,000 people die prematurely due to atmospheric pollution from ships.

The recent fourth IMO Greenhouse Gas Study on greenhouse gases emitted by ships globally, completed in the summer of 2020 (IMO 2021) reveals that emissions from shipping continue to grow. In six years, between 2012 and 2018, they increased 10%, up to 1076 million tons. It is mainly carbon dioxide. In the specific case of black carbon, the increase was 12%.

In addition to being a pollutant, black carbon is identified as an element to be taken into account in climate change, due to its short-term impact, especially in regions such as the Arctic, where it is 37 times more powerful than carbon dioxide, owing to a phenomenon called albedo: the reflection of sunlight from the white surface of snow and ice. Black carbon (which is a kind of soot) can darken the surface of the snow, causing radiation from the sun to be absorbed, generating heat and thereby aggravating the melting. While shipping in the high Arctic is currently limited by sea ice, observations in the Arctic boundary layer suggest that shipping around the periphery of the Arctic Ocean is an important source of black carbon (Xie et al. 2007). One particular type of Arctic ship activity has already increased substantially during the last two decades, namely sightseeing cruises. Major touristic destinations in the Polar Regions are the Antarctic Peninsula, Greenland and Svalbard (Spitsbergen), and concern is rising that cruise ship emissions are affecting the pristine polar atmosphere and fragile ecosystems (Eckhardt et al. 2013). Comer et al. (2020) showed that emissions of these ultrafine particles grew by 85% in the Arctic between 2015 and 2019.

The prediction is that if drastic measures are not taken in the way ships are built and operated, there could be 30% more greenhouse gas emissions in this sector by 2050 compared to 2008 levels. This will be in stark contrast to what has been promoted by the United Nations and the European Union: to reduce emissions to zero by the middle of the century to be able to meet the objectives set out in the Paris Agreement (limiting the increase in the planet's temperature to 1.5 degrees Celsius above the temperature of pre-industrial levels).

Figure 5.7 gives an overview of fuel use by fuel type for each flag state operating in the Arctic in 2019, and Figure 5.8 shows the relative share of each fuel used by each flag state.

However, the problem with cruise ships doesn't end with the problem of air pollution. The fact is that the maritime environment also suffers with this continued transatlantic travelling. Besides the problem of plastics already mentioned as well as the accidental oil spillages, one also has to take into account the noise pollution and light pollution that greatly disturb marine life. Noise produced by cruise ships' engines, propellers, generators and bearings can cause marine species such as whales to accidentally collide with the vessels or abandon their natural habitat. This affects life on the open

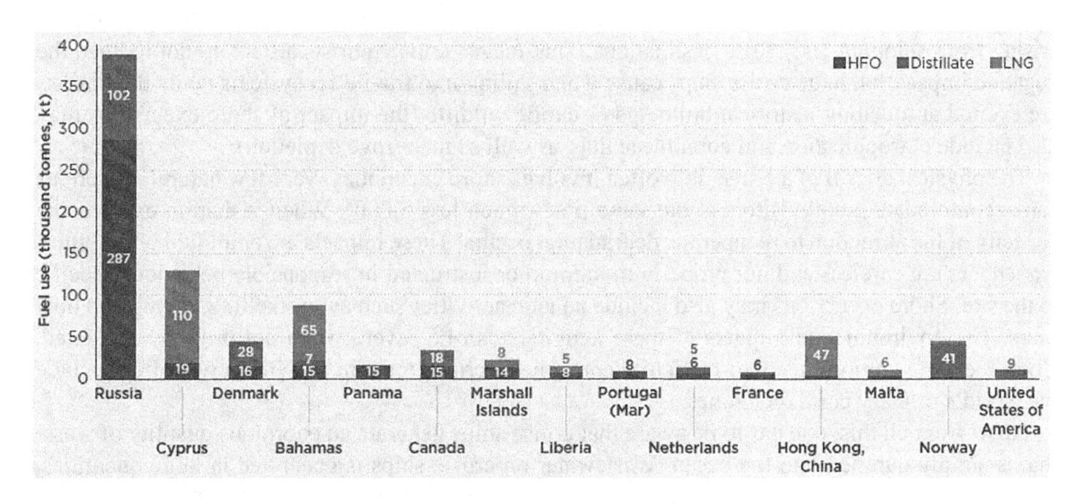

FIGURE 5.7 Top 15 countries using fuel in the Arctic in 2019.

Source: Comer et al. (2020).

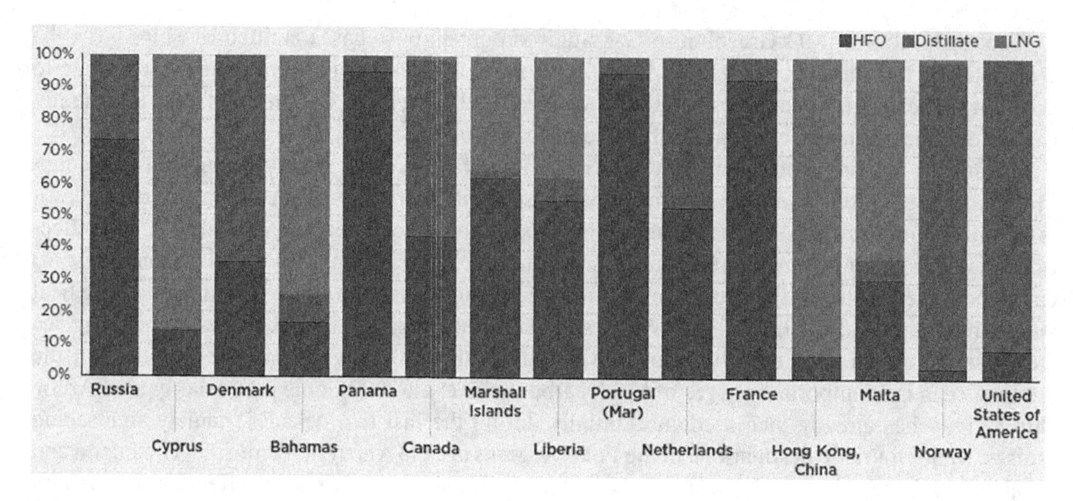

FIGURE 5.8 Top 15 countries: share of fuel use by fuel type in 2019.

Source: Comer et al. (2020).

sea as well as near ports and, above all, in fragile ecosystems. Noise pollution also occurs when ships arrive and as large numbers of passengers disembark. In addition to affecting marine life, they also have an effect on the local inhabitants of the port cities. If the primary impacts are disruption to aquatic systems, pollution and environmental degradation, the impact on the destination port city and pressure on the local culture because of inland visiting time of a large number of passengers is another key factor to be considered. If one pays attention, it is easy to see that the images that cruise lines offer to the tourism market do not correspond to an authentically sustainable version of tourism, genuinely respectful towards the environment and natural ecosystems. Although today the concept of sustainability seems to occupy an important role in the cruise industry, and environmental issues appear to be fundamental for the cruise companies, the truth is that this type of tourism is far from being environmentally friendly or sustainable.

Cruise passengers are interested in their tourist experience and cruise companies will do anything necessary to meet their expectations. Unfortunately, cruise passengers are fascinated by visiting attractive destinations, but they do not seem interested in decreasing the sustainability problems arising from shipping and cruise destinations. That means cruise tourists are likely not to heed the negative impact that huge cruise ships cause when sailing into fragile ecosystems while the tourists are excited at touching historical buildings or exotic wildlife (the impact of shore excursions may also include eutrophication and coral bleaching, as well as mangrove depletion).

The physical arrival of a cruise ship often involves shore excursions. Very few natural attractions can accommodate 2,000 visitors at the same time (much less 5,000). When visitation exceeds the capacity of the attraction to recuperate, degradation occurs. These impacts are amplified when cruise passengers are careless and not properly monitored or instructed in responsible behaviour specific to the site. Shore excursions may also include aquatic activities such as snorkelling, diving and boat tours. The environmental impacts of these activities can be severe when not properly regulated. Coral reefs are environments of particular concern for cruise tourism and itineraries often include the world's primary coral reef zones.

Apart from all this, one has to be aware that cruise ships generate an enormous quantity of waste that is simply dumped into the ocean. Wastewater on cruise ships is generated in large quantities, which means that the quality of this wastewater is of the foremost importance. Wastewater on ships can be divided into sanitary and bilge wastewater. Sanitary wastewater can still be divided into two categories: black water and grey water.

The former refers to faecal wastewater coming from toilets and medical facility sinks. Grey water comes from cabin sinks and showers, laundering, galley sinks, air conditioning condensate, and salon sinks.

Black water generally hosts many pathogens of concern to human health, namely *E. coli*, and when these waters reach swimming areas and shellfish beds, there is a potential risk to human health and to the environment by increasing the rate of waterborne diseases. In addition, nutrients in sewage, such as nitrogen and phosphorus, promote excessive algal growth, which consumes oxygen in the water and can lead to fish kills and destruction of other aquatic life. Cruise ships generate, on average, 34 litres/day/person of sewage, and a medium cruise ship (3,000 passengers and crew) can generate an estimated 60,000–120,000 litres per day of sewage (CRS 2010).

Grey water contains a variety of pollutant substances like bleach, strong acids from some cleaning products, giving water a low pH, or strong alkalis (including many detergents, phosphates, whiteners, and foaming agents giving water a high pH), oil and grease, suspended solids and organic particles (Peric 2016). It may also include faecal coliform bacteria (CRS 2010).

Both sanitary wastewater and grey water have a negative impact on the environment. For the former, the treatment system installed on cruise ships can make a huge difference as long as regulations are followed by all companies around the world.

The International Maritime Organization (IMO), a body of the United Nations, sets international maritime vessel safety and marine pollution standards. Cruise ships flagged under countries that are signatories to MARPOL (the International Convention for the Prevention of Pollution from Ships) are subject to its requirements, regardless of where they sail, and member nations are responsible for vessels registered under their respective nationalities. Six Annexes of the Convention (I. prevention of pollution by oil; II. pollution by noxious liquid substances carried in bulk; III. pollution by harmful substances; IV. pollution of the sea by sewage; V. different types of garbage, including plastics; VI. limits on sulphur oxide, nitrogen oxide, and other emissions) cover the various sources of pollution from ships but they are not enough by themselves to protect the marine environment unless all sovereign states ratify and implement the convention. However, critics have pointed out deficiencies with this regulatory structure because the marine sanitation devices (MSD) regulations only cover discharges of bacterial contaminants and suspended solids, forgetting many more pollutants, such as chemicals, pesticides, heavy metals, oil, and grease that may be released by cruise ships as well as land-based sources (ibid.).

On a ship, oil often leaks from engines or engine maintenance works and mixes with the water in the bilge (which is in the lowest part of the hull of the ship). Bilge water also may contain solid wastes and pollutants containing high amounts of oxygen-demanding material, oil, and other chemicals, as well as soaps, detergents, and degreasers used to clean the engine room. These chemicals can kill marine organisms if they are discharged. Amounts vary, depending on the size of the ship, but large vessels often have additional waste streams that contain sludge or waste oil and oily water mixtures that can inadvertently get into the bilge. A typical large cruise ship will generate an average of 8434 litres of oily bilge water for each 24 hours of operation (Sweeting and Wyne 2006) (Figure 5.9) and, according to the Environmental Protection Agency, bilge water is the most common source of oil pollution from cruise ships (EPA 2008).

Nevertheless, apart from wastewater, cruises also discard into the ocean a vast quantity of solid wastes. The Cruise Ship Discharge Assessment Report from 2008 specifies that the solid waste which is generated on board cruise ships, includes food waste, plastics, glass, paper, wood, cardboard, incinerator ash, and metal cans (EPA 2008). Naturally, the quantity of this type of waste produced on board varies in accordance with the size of the ship, number of passengers and crew members, as well as with consumption. It is estimated that an average cruise ship of 3,000 passengers and crew generates about 50 tons of solid waste in one week (Pallisa et al. 2017).

As noted by Sliškovic et al. (2018), as the tourism industry continues to fight to grow, seeking new economic niches to explore, and putting profit as the most important part of the equation, one

FIGURE 5.9 Cruise liner dumping untreated sewage and wastewater into the ocean.

Source: Ecobrasil; www.ecobrasil.eco.br.

must be aware that the negative effects and consequences that arise from tourism should not be neglected. To achieve smart growth of this industry, all stakeholders must be involved. The waste generated on board cruise ships in the near future will become a problem, especially if we think that most ports do not have the capacity to receive and treat the disposal of 3,000–5,000 people each time a cruise liner arrives (ibid.). Furthermore, as there are no proper management policies, even if the waste is properly separated and treated on board, the same procedure may not be followed at the port, owing to the lack of facilities.

Although cruise ships have significantly improved their waste management procedures, this has to be in parallel with the destination authorities improving their monitoring of cruise ship activities and environmental quality levels.

With globalization, the mass cruise tourism industry grew as another part of that same globalization. However, the negative effects of its disproportionate growth are now evident as they disturb marine ecosystems, increase air pollution, exacerbate socio-territorial exclusions and damage precarious regional livelihoods (Renaud 2020). Since the new pandemic situation, the tourism scenario has completed changed with frontiers closed and tourist cruises forbidden. In this dramatic situation, with people being infected and dying, economies struggling to survive while scientists do their best to find a solution for humanity, maybe it is time to put into perspective our priorities as a global planet and recognize that we have to move towards a more sustainable tourism. In what concerns the mass cruise tourism industry, destinations have to consider if the economic benefits linked to cruises are worth the social and environmental impacts and they might plan to forbid mega-ships that exceed a certain number of passengers from docking. On the other hand, any destinations accepting mega-ships in their ports should be equipped with facilities that are essential (e.g. waste treatment, energy supply so that ships do not have to keep on burning fuel) and should have a progressive taxation system according to the size of the ship to make it possible to compensate for some environmental impacts.

One must realize that in many parts of the world mass cruise tourism is not vital to local economies. It is more vital to the tourism industry itself, which means that it is a question of favouring a transformation of cruise tourism towards more increased sustainability.

5.4 CONCLUSION

As the environmental crisis worsens and humanity faces the inevitable climate problems, it is urgent we rethink our choices. If we can forgive ourselves for the fact that for too many years (since the Industrial Revolution) we have not been aware that the way we were building our society was leading us towards global disaster, today that excuse is no longer acceptable. Today we have the scientific knowledge to understand the consequences of our actions, we have the technology to, if not reverse some already irremediable errors, at least stop the progress of others. We just have to choose between a thriving economy in the short term or a more modest but sustainable one. It is necessary to consider if crossing the oceans on large ships for the sheer pleasure of being a tourist compensates for the destruction of marine life and the pressure put on other cultures that are visited. It is also necessary to think whether embarking on a trip to the Moon as a tourist and then signing conventions to limit carbon dioxide emissions really makes sense for an intelligent species that we claim to be. Whatever our choices, planet Earth will not disappear and a new Big Bang will not be necessary for a new origin of life. Perhaps not even *Homo sapiens* will be extinct and the natural biogeochemical cycles will continue as the laws of physics command them. However, it is worth pondering about how happy (or unhappy) will be those individuals of our species who will be lucky (or unlucky) enough to save themselves from extinction.

REFERENCES

Arthur, C., Baker, J. and Bamford, H. (eds) (2009). Proceedings of the International Research. Workshop on the Occurrence, Effects and Fate of Microplastic Marine Debris. September 9–11, 2008. NOAA Technical Memorandum NOS-OR&R-30.

Beloconi, A., Probst-Hensch, N.M., and Vounatsou, P. (2021). Spatio-temporal modelling of changes in air pollution exposure associated to the COVID-19 lockdown measures across Europe. *Science of the Total Environment*, 787: 147607.

Carapeto, C. (1999). *Poluição das Águas*. Universidade Aberta (ed.), Lisbon, Portugal: Universidade Aberta.

Comer, B., Osipova, L., Georgeff, E., and Mao, X. (2020). The International Maritime Organization's proposed Arctic heavy fuel oil ban: Likely impacts and opportunities for improvement. White Paper 2020. The International Council for Clean Transportation.

Corbett, J.J., Winebrake, J.J., Green, E.H., Kasibhatla, P., Eyring, V., and Lauer, A. (2007). Mortality from ship emissions: A global assessment. *Environmental Science & Technology*, 41(24): 8512–8518.

CRS (Congressional Research Service) (2010). Cruise ship pollution: background, laws and regulations, and key issues. CRS report for Congress. Congressional Research Service. RL32450.

Eckhardt, S., Hermansen, O., Grythe, H., Fiebig, M., Stebel, K., Cassiani, M., Baecklund, A., and Stohl, A. (2013). The influence of cruise ship emissions on air pollution in Svalbard—a harbinger of a more polluted Arctic? *Atmospheric Chemistry and Physics*, 13: 8401–8409.

EPA (Environmental Protection Agency) (2009). *Regulatory Impact Analysis: Control of Emissions of Air Pollution from Category 3 Marine Diesel Engines.* Washington, DC: U.S. Environmental Protection Agency.

EPA (Environmental Protection Agency) (2008). *Cruise Ship Discharge Assessment Report.* Washington, DC: U.S. Environmental Protection Agency, Oceans and Coastal Protection Division, Office of Wetlands, Oceans, and Watersheds, Office of Water.

European Commission. (2014). Coastal and maritime tourism. Available at: https://ec.europa.eu/oceans-and-fisheries/index_en

Fiala, M. (2018). Maritime industry aims for zero emission future. Available at: https://greenoptimistic.com/maritime-industry-aims-for-zero-emission-future-20181009/

Gago, J., Galgani, F., Maes, T. and Thompson, R.C. (2016). Microplastics in seawater: Recommendations from the Marine Strategy Framework Directive implementation process. *Frontiers in Marine Science*, 3: Article 2019.

Gedik, S. and Mugan-Ertugral, S. (2019). The effects of marine tourism on water pollution. *Fresenius Environmental Bulletin*, 28(2): 863–866.

GESAMP (Group of Experts on the Scientific Aspects of Marine Pollution) (1991). *The State of the Marine Environment*. London: Blackwell Scientific Publications, 146.

GESAMP (Group of Experts on the Scientific Aspects of Marine Pollution) (2020). Reports to GESAMP. In Kershaw, P.J., Carney Almroth, B., Villarrubia-Gómez, P., Koelmans, A.A., and Gouin, T. (IMO/FAO/UNESCO-IOC/UNIDO/WMO/IAEA/UN/ UNEP/UNDP/ISA Joint Group of Experts on the Scientific Aspects of Marine Environmental Protection) (eds). *Proceedings of the GESAMP International Workshop on Assessing the Risks Associated with Plastics and Microplastics in the Marine Environment*, No. 103, 68. GESAMP.

González, D., Hanke,G., Tweehuysen, G., Bellert, B., Holzhauer, M., Palatinus, A., Hohenblum, P., and Oosterbaan, L. (2016). Riverine litter monitoring—options and recommendations. MSFD GES TG Marine Litter Thematic Report, JRC Technical Report; EUR 28307; DOI:10.2788/461233.

Gregory, M.R. (1996). Plastic 'scrubbers' in hand cleansers: a further (and minor) source for marine pollution identified. *Marine Pollution Bulletin*, 32(12): 867–871.

Hahladakis, J.N. (2020). Delineating the global plastic marine litter challenge: Clarifying the misconceptions. *Environmental Monitoring and Assessment*, 192(5): 267.

Hartmann, N.B., Huffer, T., Thompson, R.C., Hassellov, M., Verschoor, A., Daugaard, A.E., Rist, S., … Wagner, M. (2019). Are we speaking the same language? Recommendations for a definition and categorization framework for plastic debris. *Environmental Science and Technology*, 53, 1039–1047.

Imhof, H.K., Laforsch, C., Schmid, J., Niessner, R., and Ivleva, N.P. (2012). A novel, highly efficient method for the separation and quantification of plastic particles in sediments of aquatic environments. *Limnology and Oceanography: Methods*, 10(July): 524–537. https://doi.org/10.4319/lom.2012.10.524

IMO (International Maritime Organization) (2021). *Fourth IMO GHG Study 2020 Executive Summary*. London. International Maritime Organization. Available at: www.imo.org

McCarthy, N. (2019). How cruise ship pollution compares to cars. *Cruise Ship Pollution*. Available at: https://statista.com/chart/18351/emissions-from-cruise-ships-and-cars-in-european-port-cities.

Messaoud, T.A., and Smiti, A. (2020). The COVID-19 pandemic: What about air pollution? Paper presented at 4th International Conference on Advanced Systems and Emergent Technologies (IC_ASET). DOI: 10.1109/IC_ASET49463.2020.9318299.

Nelson-Smith, A. (1970). The problem of oil pollution of the sea. *Advances in Marine Biol*ogy 8: 215–306.

Novac, V, Moraru, L., Gasparotti, C. and Eugen Rusu, E. (2020). Black sea marine litter pollution related to naval operations. E3S Web of Conferences 180, 04018. https://doi.org/10.1051/e3sconf/202018004018

OECD (Organisation for Economic Co-operation and Development) (2020). *OECD Economic Outlook*, vol. 2020, Issue 2. Paris: OECD Publishing. https://doi.org/10.1787/39a88ab1-en.

Pallisa, A.A., Papachristou, A.A., and Platias, C. (2017). Environmental policies and practices in cruise ports: Waste reception facilities in the Med. *Spoudai: Journal of Economics and Business*, 26: 54–70.

Peric, T. (2016). Wastewater pollution from cruise ships in coastal sea area of the Republic of Croatia. *Scientific Journal of Maritime Research*, 30: 160–164.

Renaud, L. (2020). Reconsidering global mobility—distancing from mass cruise tourism in the aftermath of COVID-19. *Tourism Geographies*, 22(3): 679–689, DOI: 10.1080/14616688.2020.1762116.

Rodríguez, C., Fossatti, M., Carrizo, D., Sánchez-García, L., Teixeira de Mello, F., Weinstein, F., and Lozoya, J.P. (2020). Mesoplastics and large microplastics along a use gradient on the Uruguay Atlantic coast: Types, sources, fates, and chemical loads. *Science of the Total Environment*, 721: 137734.

Schulz, M., Neumann, D., Fleet, D.M., and Matthies, M. (2013). A multi-criteria evaluation system for marine litter pollution based on statistical analyses of OSPAR beach litter monitoring time series. *Marine Environmental Research*, 92: 61–70.

Sheavly, S.B., and Register, K.M. (2007). Marine debris & plastics: Environmental concerns, sources, impacts and solutions. *Journal of Polymers and the Environment*, 15: 301–305.

Sliškovic, M., Boljat, H.U., Jelaska, I., and Mrcelic, G.J. (2018). Review of generated waste from cruisers: Dubrovnik, Split, and Zadar Port case studies. *Resources*. 7(4): 72 (Special Issue Renewable Resources, Clean Resources, Future Resources). https://doi.org/10.3390/resources7040072.

Sofiev, M., Winebrake, J.J., Johansson, L., Carr, E.W., Prank, M., Soares, J., Viral, J., … Corbett, J.J. (2018). Cleaner fuels for ships provide public health benefits with climate tradeoffs. *Nature Communications*, 9: 406, DOI: 10.1038/s41467-017-02774-9.

Sweeting, J.E.N., and Wyne, S.L. (2006). A shifting tide: Environmental challenges and cruise industry responses. In R.K. Dowling (ed.) *Cruise Ship Tourism*, Wallingford: CABI Publishing.

Thompson, R.C., Olsen, Y., Mitchell, R.P., Davis, A., Rowland, S.J., John, A.W.G., McGonigle, D., and Russell, A.E. (2004). Lost at sea: Where is all the plastic? *Science (New York, N.Y.)*, 304(5672): 838. https://doi.org/10.1126/science.1094559

T&E (Transport & Environment) (2019). *One Corporation to Pollute Them All; Luxury Cruise Air Emissions in Europe*. Brussels: European Federation for Transport and Environment AISBL

Vlachogianni, T., Fortibuoni, T., Ronchi, F, Zeri, C., Mazziotti, C., Tutman, P., Varezić, D.B., Palatinus, A., … Scoullos, M. (2018). Marine litter on the beaches of the Adriatic and Ionian Seas: An assessment of their abundance, composition and sources. *Marine Pollution Bulletin*, 131(Part A): 745–756.

Winebrake, J.J., Corbett, J.J., Green, E.H., Lauer, A., and Eyring, V. (2009). Mitigating the health impacts of pollution from oceangoing shipping: An assessment of low-sulfur fuel mandates. *Environmental Science & Technology*, 43: 4776–4782.

Xie, Z., Blum, J.D., Utsunomiya, S., Ewing, R.C., Wang, X., and Sun, L. (2007). Summertime carbonaceous aerosols collected in the marine boundary layer of the Arctic Ocean, *Journal of Geophysical Research*, 112: D02306, DOI:10.1029/2006JD007247.

6 Health Hazards of Food Allergens and Related Safety Measures

Sonia Morya,[*,1,2] *Nishi Singh,*[3] *and Chinaza Godswill Awuchi*[4]

[1]School of Agriculture, Lovely Professional University, Phagwara, India
[2]Warner School of Dairy Technology, Sam Higginbottom University of Agriculture, Technology and Sciences, Allahabad, India
[3]National Dairy Research Institute, Karnal, India
[4]School of Natural and Applied Sciences, Kampala International University, Kampala, Uganda
*Corresponding author Email: sonia.morya8911@gmail.com

CONTENTS

6.1 Introduction 99
6.2 Food Hypersensitivity 100
 6.2.1 Food Allergy 100
 6.2.2 Epidemiology 101
6.3 Food Allergens and Common Features 101
6.4 Factors Influencing Food Allergenicity 103
 6.4.1 Cross-Reactivity 103
 6.4.2 Types of Food Allergies 103
 6.4.3 Immediate or IgE-mediated Food Allergy 104
6.5 Food Allergy Management 105
 6.5.1 Immunotherapy 105
 6.5.2 Diagnosis 105
 6.5.3 Differential Diagnosis 106
6.6 Food Allergy Prevention 107
 6.6.1 Introduction of Allergenic Foods 107
 6.6.2 Avoidance of Allergens 107
 6.6.3 Nutritional Facts Labels 107
 6.6.4 Hypoallergenic Food Production 108
6.7 Challenges and Future Prospects 108
6.8 Conclusion 109
Acknowledgments 109
References 109

6.1 INTRODUCTION

Humans need foods that provide energy for activity, growth, repair, and to support all functions of the body. Food consumption also promotes a sense of fulfillment, culinary pleasure and serves as a source of shared entertainment (Awasthi et al. 2020; Sirohi et al. 2021). However, for a small proportion of

DOI: 10.1201/9781003239956-7

the population, consumption of specific types of foods, even in minute quantities, can lead to adverse allergic reactions (Sicherer and Sampson 2006; Rona et al. 2007; Sirohi et al. 2020). Food allergies (FA) are observed as a serious public health concern that affects both children and adults worldwide. They account for serious morbidity and though deaths are rare (Umasunthar et al. 2013; Chowdhary et al. 2021), still no remedial treatments or efficient prevention methods exist. A global survey, conducted by the joint efforts of the World Allergy Organization (WAO) and the Worldwide Universities Network (WUN) in 2012, reports that a total of 10% of the food allergy prevalence occurs in children (aged below five years) in Europe, and 7–10% in the USA, Asia, and Africa altogether. Scientifically more than 170 types of foods are identified as causing food allergies, whereas nine foods (and their derived products) are known to be major allergens, comprising more than 90% of all reported food allergic reactions. These prime allergens include milk, soybeans, peanuts, tree nuts, eggs, crustaceans and mollusks, seafood, mustard, sesame, and gluten-containing cereals.

Today the increase in cases related to FA is an intriguing phenomenon, mainly attributed to lifestyle changes, such as a change in living conditions, alimentary habits, and intense use of antibiotics, resulting in immune deviation and thus evolution of the allergy in present times (Von Mutius 2004). FA are not only known to negatively influence the quality of life for patients and families but also pose a huge financial burden (Gupta et al. 2013; Walkner et al. 2015). Genetic factors which also serve as risk factors associated with the development of FA include ethnicity (Strachan 1989; Vierk et al. 2007; Branum and Lukacs 2009; Liu et al. 2010), sex (Osborne et al. 2010; Du Toit et al. 2015; Peters et al. 2015), atopic family history (Hourihane et al. 1996; Lichtenstein and Svartengren 1997), atopic dermatitis, and other inherited polymorphisms. FA do not appear as a result of germline genetic alterations alone but, instead, one or numerous environmental exposures, or absence thereof, can provoke epigenetic modifications that disrupt the deficient immunologic state of tolerance to foods. However, the composite interplay linking genetic and environmental factors is responsible for giving rise to the FA. According to the hygiene hypothesis, lack of initial childhood exposure to infectious agents, symbiotic microorganisms (for example, probiotics or gut flora), and parasites increase the vulnerability to allergic diseases by compromising the development of the innate immune system (Strachan 1989). Various foods can provoke allergic reactions and, thus, the threshold dosage required to induce these reactions varies for different individuals. In some patients, even small quantities of offending substances from food can trigger severe allergic reactions, whereas others can withstand allergen quantities which are several times higher in magnitude (Taylor et al. 2004; Perkin et al. 2016). Similarly, the intensity of symptoms of these allergies varies markedly among patients, and sometimes food conditions, such as production, processing, and other pattern effects can modify the molecular constitution of allergens in food; hence, their immunogenic properties and recognition. Today, food allergens management all through the food value chain and diagnosis of its related diseases remain a stern challenge to both the food industry and health care experts (Alvarez and Boye 2012). In this chapter, we will present an outline of the different types of food hypersensitivities, the classification of food allergies, according to the mechanism of allergic responses, a summary of major food allergens and their source, and strategies for food allergens management and prevention.

6.2 FOOD HYPERSENSITIVITY

Abnormal reaction or any adverse clinical response associated with the ingestion of food is broadly known as food hypersensitivity. However, based on the path-physiological mechanism of reactions, these can be classified into food allergy and food intolerance.

6.2.1 Food Allergy

Food allergy is an adverse immunological reaction resulting from the ingestion, inhalation or atopic contact of the offending food in the sensitized hosts. A series of these immunological reactions can be arbitrated by IgE antibodies or by other immune cells such as T-cells. These types of reactions often recur every time the food is consumed and are dose-independent. Commonly, any food (especially those containing offending proteins) can manifest several allergic responses in sensitive individuals. Allergy-provoking proteins may be enzymes or inhibitors, binding or structural proteins with various biological functions (Chapman et al. 2000; Sicherer and Sampson 2018).

6.2.2 Epidemiology

FA are globally widespread and have become a severe public health crisis. Even though accurate epidemiological facts and figures are still lacking, it is obvious that the occurrence of food allergies has increased significantly in the past two decades, especially in Western nations, where a rough hike of 10% has been recorded among preschool children (Osborne et al. 2011; Comberiati et al. 2019). It is projected that more than 220 million people suffer from food allergies worldwide (Jones and Burks 2017; Dunlop and Keet 2018; Sicherer and Sampson 2018). Incidences of food allergies are mainly reported in children (1–3 years) and infants (6–8%). This rate further dips moderately and affects nearly 4% of adults (Venter et al. 2008; Schussler et al. 2017; De Martinis et al. 2020b). Food allergy is one of the most frequent causes of anaphylaxis in children (Bock et al. 2001), and has been reported to severely affect children suffering from eosinophilic esophagitis (Spergel et al. 2009).

In developed countries, foods such as wheat, cow's milk, peanuts, fish and shellfish, eggs, soybeans, and walnuts are largely responsible for causing frequent allergies in children (Benedè et al. 2016). On the other hand, the recent exponential increase in both adult and geriatric populations (particularly in Western nations), with ecological and lifestyle shifts, has greatly altered the health physics of FA. Clinical symptoms caused due to allergic foods in the elderly exhibit irregular features and immunopathogenesis patterns, thereby contributing to diagnostic complications (De Martinis et al. 2020a; De Martinis et al. 2020b).

The rise in food allergies is the outcome of the effect of environmental components and lifestyles with an inherited predisposition (De Martinis et al. 2017; Genuneit et al. 2017). The maximum occurrence of food allergies has been recorded in Australia, increasing to up to 10% of the total number of infants (Caraballo et al. 2016; Tang and Mullins 2017). In 2011, the WAO projected that 220–250 million individuals may suffer from FA globally. About 11–26 million Europeans suffer from food allergy, adding another huge global health burden (Mills et al. 2007). Incidences of growing frequencies of FA rates parallel to those of Western nations have been documented in India, China, and other developing nations, including Africa. The financial growth of nations and the extension of the globalization phenomenon are directly linked to a potential rise in FA prevalence (De Martinis et al. 2020a).

Several studies have concluded that there exists a definitive relationship between the individual's epigenetics, the environment, and their way of life; which in turn has a huge impact on the susceptibility and pathology related to the development of FA. Researchers have stated that Asian origin children, who were born in Australia, were two times more susceptible to nuts allergy, compared to Asian children who relocated to Australia at an age of 5 years who were more protected (Panjari et al. 2016). This further confirms both the genetics and ecological impacts on the development of allergic diseases in individuals (De Martinis et al. 2020a).

6.3 FOOD ALLERGENS AND COMMON FEATURES

According to the WAO, food allergens are naturally-occurring proteins in foods or its derivatives that are the cause of unusual immune responses. Virtually all foods containing proteins as a constituent possess the capacity to induce allergic reactions in an individual, who happens to be sensitized to proteins.

Food allergies have gained the attention of medical research communities worldwide but their incidence around the globe is alleged to be growing, by 8% in children and 2% in adults. Concern

TABLE 6.1
Major Food Allergens From Offending Food Sources Which Induce Food Allergies in Children and Adults

Food source	Offending proteins	Food allergens	References
Peanuts	Vicilin, conglutin, glycinin	*Ara h*1 to *Ara h*9 *Ara h*10 and *Ara h*11	Krause et al. 2009; Palmer et al. 2005; Rabjohn et al. 1999
Soya bean	β-conglycinin, 11S globulin, profiling	*Gly m*1 and *Gly m*2, *Gly m*3 *Gly m*4, *Gly m*5, *Gly m*6	Holzhauser et al. 2009
Eggs	Ovomucoid, ovalbumin, ovotransferrin, and lysozyme	*Gal d*1 to *Gal d*4, *Gal d*6	Amo et al. 2010
Wheat	Profilin, nonspecific lipid transfer protein LTP1, agglutinin isolectin 1, omega-5 gliadin, thioredoxin, glutenin	*Tri a*12, *Tri a*14, *Tri a*18, *Tri a*19, *Tri a*25, *Tri a*26	Palacin et al. 2007; Matsuo et al. 2005
Cow's milk	α-lactalbumin, β-lactoglobulin, serum albumin, immunoglobulin, and caseins	*Bos d*4 to *Bos d*8	Natale et al. 2004
Mustard seeds	2S albumin 11S globulin, nonspecific LTP, profiling	*Sin a*1, *Sin a*2, *Sin a*3, *Sin a*4	Sirvent et al. 2009
Sesame seeds	2S albumin, vicilin-type globulin, oleoresins, 11S albumins	*Ses i*1, *Ses i*2, *Ses i*3, *Ses i*4, *Ses i*5, *Ses i*6, *Ses i*7	Beyer et al. 2007
Almond	Nonspecific LTP, profilin, acidic ribosomal protein, amandin	*Pru du*63, *Pru du*4, *Pru du*5, *Pru du*6	Tawde et al. 2006
Walnut	2S albumin, 7S vicilin-like globulin, LTP, 11S legumin-like globulin	*Jug r*1, *Jug r*2, *Jug r*3, *Jug r*4	Wallowitz et al. 2006
Cashew	7S vicilin-like protein, 11S globulin, 2S globulin	*Ana o*1, *Ana o*2, *Ana o*3	Robotham et al. 2005
Hazelnut	Profilin, LTP and 11S globulin like seed storage protein, vicilin-like 7S, 2S albumin	*Cor a* 1.04, *Cor a* 1, *Cor a* 2, *Cor a* 8, *Cor a* 9, *Cor a* 11, *Cor a* 12, *Cor a* 13, *Cor a* 14	Zuidmeer et al. 2009; Akkerdaas et al. 2006
Brazil nut	2S seed storage albumin, 11S globulin legumin-like protein	*Ber e*1, *Ber e*2	Beyer 2008
Shrimp	Tropomyosin	*Pen i*1	Shanti et al. 1993

lists are based on how severe the patient's reactions are on exposure to definite allergens and the regional or global incidences of food allergies. Guidelines concerning which food to regard as contaminants may vary worldwide, even though the present FAO/WHO Codex General Standard for Labelling of Pre-packaged Foods consists of a defined note of eight foods or food substances and/or their derivatives. A summary of food allergens reported to be the most common among children and adult populations is listed in Table 6.1. In many western and Asian nations, nine different foods or their groups have been listed as the basis of 90% of these allergic reactions: peanuts, soy, milk, egg, tree nuts, fish, seafood, sesame, and cereals. However, depending upon the conventional conditions that co-exist with food allergies, allergens can be of two types.

1. *Class 1 Traditional food allergens*: These are small water-soluble glycoproteins typically ranging from 10–70 kDa in size. They are also highly resistant to enzymatic action, heat, and low pH conditions and encourage allergic sensitization reactions, mainly via the gastrointestinal tract, and are accountable for systemic reactions (Sampson 1999).
2. *Class 2 Food allergens/oral allergy syndrome (OAS) allergens*: These are highly analogous with proteins present in pollens, such as apple, kiwis, tomatoes, celery, and green peppers.

They are also heat-labile and easily disposed of during digestion by the action of the gastric juices and enzymes. OAS is due to sensitization to labile proteins that are typically borne via the respiratory tract. Further, IgE antibodies to pollens identify similar epitopes on the food proteins of plant origin (Egger et al. 2006).

6.4 FACTORS INFLUENCING FOOD ALLERGENICITY

The frequency of sensitization to particular food allergens differs based on the age and type of the considered population. Several animal models have been developed to examine both the molecular and cellular events leading to sensitization and anaphylaxis occurring in humans (Galand et al. 2016; Hussain et al. 2018). It has been found that when the food allergen is orally administered to an animal, it induces tolerance but can also result in sensitization and severe allergic reactions. These reactions in developed animal models are affected by many factors (presented in Table 6.2) which might surpass the epithelial barrier. These models also imply that sensitization can happen through several other sites, such as the oral cavity, the respiratory tract or even the skin, as opposed to the gastrointestinal tract, where the oral tolerance is usually a response that avoids the allergen (Noti et al. 2014; Barcik et al. 2016; Bogh et al. 2016; Smit et al. 2016; Chowdhary et al. 2020).

6.4.1 Cross-Reactivity

Cross-reactivity is mostly defined by the structure of proteins. Two proteins are termed as cross-reactive only if they share similar structural characteristics. Phyletically associated mammals express identical milk proteins with amino acid sequence homology, further developing medical cross-reactivity among cow, goat and sheep milk (D'Auria et al. 2005; Jarvinen and Chatchatee 2009). Clinical cross-reactivity involving peanut and several legumes are extremely rare. Oral allergy syndrome (OAS) is a type of contact allergy occurring due to atopic contact with raw fruits and vegetables. OAS involves around 50% of pollen-allergic adults and exhibits the most common food allergies occurring in adults. It results from the cross-reactivity between the allergenic proteins in pollens and plant foods (Valenta and Kraft 1996; Sicherer 2001).

6.4.2 Types of Food Allergies

The pathogenesis of allergy to food antigens begins with a sensitization phase which may occur in the oral cavity, the gastrointestinal tract, the skin, and rarely in the respiratory tract. Upon ingestion,

TABLE 6.2
Different Factors Influencing the Food Allergenicity in Animal Models

Food-borne factors	Intrinsic factors
• Allergen type and exposure dose • Factors damaging epithelial-barrier (detergents, toxins, alcohol, or any other unfamiliar ingredients) • Exposure pathway • Type of microbial adjuvant(s) in food allergen, their dose, etc. • Food model effects, e.g. glycosylated sugars, lipids, globular or fibrous proteins, etc. • Thermal processing temperature	• Age and immunity status • Sex and ethnicity • Chronic medical conditions (rhinitis, atopy, immuno-deficiency, dermatitis, etc.) • Barrier defects • Use of antacids • Gut micro-biome

general food proteins are broken down by the action of gastric juices and the digestive enzymes present within the stomach and intestine. The residual undigested food proteins/peptides are then recognized as a foreign intruder and the body begins to build an immune-defensive reaction (Reboldi et al. 2016). Any further consumption of allergen-based food can trigger allergic responses that may exhibit any of the following three forms (Gell and Coomb's classification) depending upon the immunological and type of effector mechanism responsible for cell and tissue injury involved: type I, IgE mediated or immediate; type III, IgM/IgG immune complex-mediated; and type IV, T-cell mediated or delayed-type hypersensitivity (Coombs and Gell 1975; Lee and Burks 2006; Sicherer and Sampson 2009).

6.4.3 Immediate or IgE-mediated Food Allergy

This is a common/classic type of allergy which happens shortly after eating, i.e. an allergy occurs within a few minutes to a few hours after ingestion of the offending food by the host and hence, is called immediate-type reaction. Populations with decreased tolerance to the allergen, especially in industrialized nations, are more likely to suffer from such a type of hypersensitivity. Around 1–2% of the global population suffers from IgE-mediated food allergies (Sicherer and Sampson 2018), wherein infants and young children (with 5–8% infants $\leq$ 3 years) represent the most susceptible group (Sicherer et al. 2010). Recent studies have reported that nearly half of the patients diagnosed with type 1 food allergies have experienced at least one grave anaphylactic shock, mainly in their early years of childhood and adolescence.

Immediate allergy can further be classified into two phases: sensitization and the effector phase. After the initial exposure to the ingested food allergen via the gastro-intestinal tract, a primary sensitization appears which then activates the series of allergic responses. These food allergens are otherwise extremely stable proteins that exhibit resistance to gastric juices, digestive enzymes, and heat denaturation. IgE is a primary antibody in humans with atopic disorders and an emblem of all allergic sensitization occurring in the body. These antibodies have also been reported to mediate immediate-phase allergies by inducing mast cells and basophiles degranulation. However, IgE antibodies then circulate in the blood and bind to the surface of mast cells and basophiles. Mast cells and basophiles enclose granules, which are filled with effective inflammatory mediators, such as cytokines, histamine, proteases, prostaglandin, and other chemotactic factors. However, after sensitization and re-exposure to the allergen, to which the patient has formerly produced IgE antibodies, the allergen binds with the IgE antibody present on the surface of basophiles and mast cells. This sets in the effector phase, which activates a complex chain of reactions that results in the release of histamine and other chemo-mediators (prostaglandin D_2, leukotriene D_4, and kinins, heparin, etc.), resulting in the degranulation of the basophiles and mast cells. Eosinophils and neutrophils may perhaps also be involved in generating the whole spectrum of immediate hypersensitivity reaction, in addition to basophiles and mast cells. Chemical mediators like histamines are the substances responsible for giving rise to the symptoms of the allergy and evoking severe clinical illustration. The release of many of these chemo-mediators occurs quickly, i.e. within 5 minutes after the reaction between the ingested allergen and the IgE antibody. The acuteness of these allergies depends on the sensitivity of the person and the ingested amount of the allergen. The release of histamine from mast cells typically finishes within 30 minutes. But the release of other chemo-mediators is slower and, hence, their effects are more severe and long-lasting.

These reactions may take place in two phases. The initial stage or first symptoms withdraw on their own or at times with medication and they tend to reappear in 4–6 hours. Symptoms may vary from gentle to localized conditions and are witnessed as gastrointestinal discomforts, such as vomiting, nausea, diarrhea, and enteric cramping, or at times involve the dermis, leading to dermatitis, urticaria, pruritis eczema or hives, angioedema, or mild itching. It can also engage the respiratory pathway, in which the patient may experience bronchial asthma, rhinitis, or laryngeal edema.

Anaphylactic shock defines the most substantial and disastrous expression of immediate hypersensitivity reactions as it dilates the blood vessels leading to constriction of airways; in reaction to allergens in food. Practically, every organ in humans can be influenced, even though the reactions concerning the cutaneous, respiratory, cardiovascular, gastrointestinal and neurological pathways are quite common.

6.5 FOOD ALLERGY MANAGEMENT

Management of food allergies mostly involves not consuming the allergenic food and also make plans in case exposure occurs (Sicherer and Sampson 2014). The plans may include wearing medical alert jewelry and giving adrenaline (epinephrine). Some food allergies in children usually resolve with age, such as milk, soy, and eggs allergies, while some, including shellfish and nuts allergies, often do not (ibid.). Total avoidance of known allergenic foods is the best way to treat food allergies. Allergens can be introduced into the body through the consumption of any food that contains the allergen, or by touching the mouth, nose, or eyes after touching a surface that is exposed to the allergen. For individuals with extreme sensitivity, prevention includes avoiding inhaling or touching the allergenic food. Complete avoidance is challenging since declaring the presence of traces of allergens in food is not compulsory.

6.5.1 Immunotherapy

Immunotherapy has been used for decades to mitigate food allergies. If systemic reaction, such as anaphylaxis, occurs as a result of accidental ingestion of food, epinephrine (a catecholamine hormone and neurotransmitter) can be used. For severe reactions, a second dose of the catecholamine hormone (and neurotransmitter) may be required. The affected individual has to be taken to an emergency room for more treatment. Steroids and antihistamines are other treatment measures. Auto-injectors of epinephrine are portable devices that dispense single-dose epinephrine for the treatment of anaphylaxis. Adrenaline (epinephrine) is the first-line treatment for anaphylaxis (severe reaction to an allergen). If timely administered, epinephrine can effectively reverse the allergic effects. Epinephrine ameliorates airway obstruction and swelling, and enhances the circulation of blood; there is an increase in heart rate and tightened blood vessels, which improves circulation to the organs of the body. Epinephrine can be made available through a prescription with an auto-injector. Antihistamines, on the other hand, can ameliorate some mild symptoms of allergic reactions; however, antihistamines are not effective for treating all anaphylaxis symptoms. Antihistamines work by blocking histamine actions, which dilate the blood vessels, making them leaky to the plasma protein. Additionally, histamine induces itching by its actions on the terminals of the sensory nerves. Diphenhydramine is an antihistamine mostly administered for treating food allergies. Steroids, including glucocorticoid steroids, can be used to relax the cells of the immune system which face attacks by chemicals discharged by allergic reactions. However, this treatment as a nasal spray is not used for the treatment of anaphylaxis; it relieves symptoms only in areas that come into contact with the steroid. One other reason the steroids are not to be used for the treatment of anaphylaxis is their delay in relieving inflammations. Steroids can be administered through the oral route or by injection, through which all the body parts can be treated and reached; however, it requires a long time for its effect to take place or be felt.

6.5.2 Diagnosis

Diagnosing food allergy mostly relies on medical history, oral food challenge, food-specific IgE antibodies blood tests, a skin prick test, or an elimination diet (ibid.). In the case of the skin-prick test, a small board made of protruding needles can be used. The food allergens are normally placed directly on the skin or the board, then the board is directly placed on the skin, which punctures the skin, allowing the allergens to enter the body. If the appearance of a hive occurs, the individual is regarded as being positive for the allergy. The skin prick test can only work for the IgE antibodies. The allergic reactions due to other antibodies are not detected by the skin prick test. A skin prick test can be done easily and the results are made available in just a couple of minutes. Many allergists make use of different testing devices, such as a "bifurcated needle", which resembles a two-prong fork, or a "multitest", which resembles a tiny board made with many pins sticking out from it. These tests make use of a small level of the expected allergen placed into the testing device (or onto the skin); this device is positioned onto the skin to break through or to prick the skin's top layer, causing the small level of the allergen to enter beneath the skin. Anywhere the individual is allergic, a hive will form. In general, this test yields negative or positive results. It is more suitable to discover quickly whether an individual is allergic, or not, to a specific food, as it detects IgE. Skin testing does not predict whether or not a reaction will occur or which type of reaction may arise if someone ingests a specific allergen. It can, however, confirm an allergy in light of a patient's history of reactions to a particular food. Allergies not mediated by IgE are not detected using this method. Another test, known as patch testing, determines if a specific substance results in a skin allergic inflammation. The patch test determines delayed food reactions (Rokaite et al. 2014). Blood testing can also be used to test for allergies, although this has the similar disadvantage of only detecting IgE allergens; blood testing does not detect all likely allergens. Radioallergosorbent testing (RAST) can be employed to detect IgE antibodies in certain allergens. The predictive values obtained from the specific RAST type are compared to the RAST score. If the predictive values are lower than the RAST score, there is a high possibility that the individual is allergic. This test has the advantage of testing several allergens at once. CAP-RAST has higher specificity compared to RAST and can show the level of IgE in each allergen.

Food challenges can be used to test allergens other than the ones resulting from IgE allergens. The allergens are usually taken as a pill, through direct ingestion of the allergen. The individual is monitored for symptoms and signs. Food challenges have a problem in that they have to be done in a hospital with careful monitoring, because of the likelihood of anaphylaxis. Food challenges, such as placebo-controlled, double-blind food challenges, are golden standards for diagnosing food allergies, such as the reactions not mediated by IgE, but are not always performed (Turnbull et al. 2015). Blind food challenges are more about packaging suspected allergens in capsules, administering them to patients, as well as monitoring the patients for symptoms or signs of allergic reactions. All recommended methods for the diagnosis of food allergies have to be done by a well-trained allergist, who will assess the history of the patient and the reactions or symptoms observed after ingestion of the food. If there is a feeling that the reactions or symptoms are typical of a food allergy, the allergist will conduct allergy tests. Other tools for diagnosing non-IgE or eosinophilic mediated reactions include biopsy, colonoscopy, and endoscopy.

6.5.3 Differential Diagnosis

Differential diagnosis can be used to assess food allergies. Differential diagnoses commonly used include:

1. Hereditary angioedema (C1 Esterase inhibitor deficiencies) is a very rare disease, which causes angioedema attacks, but can exclusively present with occasional diarrhea and abdominal pain; it may be mistaken for angioedema triggered by allergy.
2. Celiac disease results from lasting gluten intolerance, and is not just simple intolerance nor an allergy, but a lasting, multiple-organ autoimmune ailment that mostly affects the small intestine (Tommasini et al. 2011). Gluten is present in wheat oats, barley, rye, etc.
3. Irritable bowel syndrome (IBS).

4. Lactose intolerance, which mostly develops gradually with time (often in years), but could severely occur in young people in some instances. It is due to the deficiency of lactase (an enzyme), and not an immune reaction.

6.6 FOOD ALLERGY PREVENTION

6.6.1 Introduction of Allergenic Foods

Potential exposure to allergens in early life could be protective (Sicherer and Sampson 2014; Ierodiakonou et al. 2016). Breastfeeding for at least 4 months could prevent wheezing, cow milk allergy, and atopic dermatitis in early childhood (Greer et al. 2008), especially, early exposure to peanuts, eggs, etc. decreases the risks of allergy to these (Ierodiakonou et al. 2016). Studies recommend giving peanuts at 4 to 6 months with the inclusion of precautionary measures for infants at high risk (Chan et al. 2018). The previous guidelines that advise the delay of the introduction of peanuts are now believed to contribute to the recent increase in peanut allergies (Du Toit et al. 2015).

6.6.2 Avoidance of Allergens

Strict diets can be maintained to avoid allergic reactions. Determining the level of allergenic foods that elicit a reaction is challenging, so that total avoidance of allergenic foods can be considered. In certain instances, hypersensitive reactions are triggered through allergen exposures via alcohol, cosmetics, blood transfusions, participation in sports, kissing, inhalation, and skin contact (Du Toit et al. 2015; Chan et al. 2018).

6.6.3 Nutritional Facts Labels

Nutritional facts labels play an important role in helping individuals avoid foods they are allergic to. As a response to the risks that some foods pose to people who have food allergies, many countries have instituted labeling policies requiring food products to make consumers aware if their products have major allergens or allergens byproducts along with the intentionally added ingredients. The priority allergens differ according to country. Table 6.3 shows the priority allergenic foods in most countries.

In many countries, labeling is used to prevent the occurrence of food allergies. In many countries, such as the USA, where the Food Allergen Labeling and Consumer Protection Act of 2004 (FALCPA) makes it compulsory for companies to specify on food labels if packaged food products contain any prioritized food allergens intentionally added (wheat, soy, tree nuts, fish, shellfish, eggs, peanuts, cow's milk, etc.), laws mandate that intentionally added ingredients likely to cause allergic reactions are clearly stated on the food label (USFDA 2004, 2020). These ingredients have their origin in 1999 from the WHO Codex Alimentarius Commission (USFDA 2004, 2017; Allen et al. 2014). Many countries in Africa and Asia also require allergenic foods or ingredients to be specified on the food label. In addition to these major allergenic foods (wheat, soy, tree nuts, fish, shellfish, eggs, peanuts, cow's milk, etc.), the EU requires an additional listing of sulfites, sesame, lupin, mustard, celery, and molluscs as food allergens (USFDA 2015). The United States FDA issued a request in 2018 to consider labeling information for sesame to protect those allergic to sesame (USFDA 2018). In November 2020, it was agreed that food producers should voluntarily state that when pulverized seeds of sesame are used as a formerly unspecified flavor/spice, the label has to be termed "flavor (sesame)" or "spice (sesame)" (USFDA 2020). The European Union Food Information for Consumers Regulation 1169/2011 makes it a requirement for food producers to provide information about allergy on foods sold without being packaged, for instance, in sandwich bars, bakeries, deli counters, and outlets of catering.

TABLE 6.3
Allergenic Foods That Are Priority in Labeling Laws in Selected Countries

Foods	United Kingdom	European Union	Australia	United States	Canada	References
Peanuts	Prioritized	Prioritized	Prioritized	Prioritized	Prioritized	European Union 2011; USFDA 2021
Tree nuts	Prioritized	Prioritized	Prioritized	Prioritized	Prioritized	
Milk	Prioritized	Prioritized	Prioritized	Prioritized	Prioritized	
Eggs	Prioritized	Prioritized	Prioritized	Prioritized	Prioritized	
Fish	Prioritized	Prioritized	Prioritized	Prioritized	Prioritized	
Shellfish	Mollusc and Crustaceans	Mollusc and Crustaceans	Prioritized	Crustaceans	Mollusc and Crustaceans	
Soy	Prioritized	Prioritized	Prioritized	Prioritized	Prioritized	
Wheat	Included under gluten	Included under gluten	Prioritized	Prioritized	Includes triticale	
Sesame seeds	Prioritized	Prioritized	Prioritized	Voluntary	Prioritized	
Mustard	Prioritized	Prioritized	Not prioritized	Not prioritized	Prioritized	
Sulphites	Prioritized	Prioritized, >10 mg/kg	Not prioritized	Not prioritized	Prioritized	
Gluten	Prioritized	Prioritized	Not prioritized	Not prioritized	Prioritized	
Celery	Prioritized	Prioritized	Not prioritized	Not prioritized	Not prioritized	
Lupin	Prioritized	Prioritized	Prioritized	Not prioritized	Not prioritized	

6.6.4 Hypoallergenic Food Production

Elimination of food allergens using food processing is an effective means of preventing allergens. Though most antigenic determinants and allergens can tolerate several food processing techniques and methods to some extent, studies have demonstrated that specific food processing processes can have effects on allergenicity (Fu et al. 2019; Chinaza et al. 2020a; Awuchi, Hannington et al. 2020; Awuchi, Victory and Amagwula 2020; Awuchi, Victory et al. 2020). In food processing, fermentation, enzymatic, chemical, and physical methods can alter the allergens' activity. Protein is an essential food macronutrient and is the major category of allergen in our foods. The methods used in food processing can result in changes in the physicochemical properties and spatial structures of the macromolecules of protein and consequently destroy the protein allergenic epitopes (Fu et al. 2019). Four families of protein and subfamilies account for most allergens in foods such as legumes, including prolamins and cupins (seeds storage proteins), profilins, as well as pathogenesis-related proteins. Most of the approaches used in producing hypoallergenic crops include: (1) genetic transformation, which can silence the natural genes encoding protein allergens, and (2) germplasm lines, which involve screening for reduced content or absence of specific protein allergens (John et al. 2009). These strategies have been reported as recording huge success in producing peanuts and soybeans cultivars with decreased protein allergens (ibid.). Heat treatment, irradiation, ultrasound, ultrahigh-pressure treatment, pulsed ultraviolet technology, chemical modification, fermentation, glycosylation, etc. have also been shown to eliminate several allergens in foods (Fu et al. 2019; Chinaza et al. 2020a; Awuchi, Hannington et al. 2020; Awuchi, Victory and Amagwula 2020; Awuchi, Victory et al. 2020).

6.7 CHALLENGES AND FUTURE PROSPECTS

The prevalence of food allergies is both under- and over-reported. Many affected individuals, especially in developing countries, do not report the allergy; they often prefer to avoid the food in question.

Perceptions of self-diagnosed food allergies outweigh the true rates of food allergies because many individuals mistake non-allergic intolerance as food allergies and non-allergic symptoms as allergic responses. In addition, health experts who treat allergic reactions do not always report all the cases (Hadley 2006). Several desensitization techniques are being studied (Nowak-Węgrzyn and Sampson 2011). Research areas include sublingual immunotherapy (SLIT), oral tolerance induction (OTI) for oral immunotherapy (also called specific OTI (SOTI)), and omalizumab (an anti-IgE antibody).

Studies are being done on the effects of increasing polyunsaturated fatty acids intake during lactation, in pregnancy, in early childhood, and through infant formula on the consequent risks of food allergies' development during childhood and infancy. Some fatty acids have been associated with allergies. Consuming omega-3, long-chain FAs when pregnant seems to decrease the risk of medically diagnosed allergy mediated by IgE, food allergy, and eczema in the first year of life (ibid.: 98; Gunaratne et al. 2015; Best et al. 2016), however, the effects do not last more than the first year. Results of studies involving breastfeeding mothers taking diets high in polyunsaturated fatty acids were reportedly inconclusive (Waidyatillake et al. 2018). Supplementing the diets of infants with oils rich with PUFAs did not affect the risk of food allergy, asthma or eczema, into childhood or as infants (Schindler et al. 2016). Studies have been done on prebiotics, probiotics, and a combination of both (known as synbiotics) for preventing or treating child and infant allergies. There seems to be some eczema treatment benefits (Osborn and Sinn 2013; Cuello-Garcia et al. 2015; Chang et al. 2016), rhinoconjunctivitis, but not wheezing, or asthma (Zuccotti et al. 2015). The evidence was inconsistent on avoiding food allergy and as a consequence, the approach is not yet recommended (Osborn and Sinn 2013; Cuello-Garcia et al. 2015). Further studies are required to improve the current treatment measures or probe novel techniques for treating food allergies.

6.8 CONCLUSION

Nowadays, allergies to food are a worldwide problem which is increasing daily. The World Allergy Organization (WAO) has not supplied data on the occurrence of food allergies. In the current situation, the only way to prevent this disease is to avoid the allergy-causing food. Avoidance of allergy-causing food also leads to malnutrition, so avoidance cannot be a solution. Food can be diagnosed properly and allergens must be removed from the food. Food malnutrition can be compensated for by using prebiotics and probiotics in foods. Many different research studies must be conducted in this field to resolve the problem of food allergy without leading to malnutrition.

ACKNOWLEDGMENTS

The authors wish to acknowledge all the authors whose work is cited in this chapter.

REFERENCES

Akkerdaas, J. H., F. Schocker, S. Vieths, S. Versteeg, L. Zuidmeer, S.L. Hefle, et al. 2006. Cloning of oleosin, a putative new hazelnut allergen, using a hazelnut cDNA library. *Molecular Nutrition and Food Research* 50(1): 18–23.

Allen, K.J., P.J. Turner, R. Pawankar, S. Taylor, S. Sicherer, G. Lack, et al. 2014. Precautionary labelling of foods for allergen content: Are we ready for a global framework? *World Allergy Organization Journal* 7: 1–14. https://doi.org/10.1186/1939-4551-7-10

Alvarez, P., and J. Boye. 2012. Food production and processing considerations of allergenic food ingredients: A review. *Journal of Allergy* 1: 222–230.

Amo, A., R. Rodríguez-Pérez, J. Blanco, J. Villota, S. Juste, I. Moneo, et al. 2010. Gal d 6 is the second allergen characterized from egg yolk. *Journal of Agricultural and Food Chemistry* 58: 7453–7457.

Awasthi, S.K., S. Sarsaiya, M.K. Awasthi, T. Liu, J.C. Zhao, S. Kumar, and Z.Q. Zhang. 2020. Changes in global trends in food waste composting: Research challenges and opportunities. *Bioresource Technology*. 299: 122555. https://doi.org/10.1016/j.biortech.2019.122555

Awuchi, C.G., T. Hannington, S.I. Victory, and I.O. Amagwula. 2020. Food additives and food preservatives for domestic and industrial food applications. *Journal of Animal Health* 2: 1–16.

Awuchi, C.G., S.I. Victory, and I.O. Amagwula. 2020. Nutritional diseases and nutrient toxicities: A systematic review of the diets and nutrition for prevention and treatment. *International Journal of Advanced Academic Research* 6: 1–46.

Awuchi, C.G., S.I. Victory, I.O. Amagwula, and C.K. Echeta. 2020. Health benefits of micronutrients (vitamins and minerals) and their associated deficiency diseases: A systematic review. *International Journal of Food Sciences* 3: 1–32.

Barcik, W., E. Untersmayr, I. OPali-Sch€oll, L. O'Mahony, and R. Frei. 2016. Influence of microbiome and diet on immune responses in food allergy models. *Drug Discovery Today: Disease Models* 17: 71–80.

Benedè, S., A.B. Blázquez, D. Chiang, L. Tordesillas, and M.C. Berin. 2016. The rise of food allergy: Environmental factors and emerging treatments. *EBioMedicine* 7: 27–34.

Best, K.P., M. Gold, D. Kennedy, J. Martin, and M. Makrides. 2016. Omega-3 long-chain PUFA intake during pregnancy and allergic disease outcomes in the offspring: A systematic review and meta-analysis of observational studies and randomized controlled trials. *The American Journal of Clinical Nutrition* 103: 128–143. https://doi.org/10.3945/ajcn.115.111104

Beyer, K., 2008. Identification of a new Brazil nut allergen—Ber e 2. *Journal of Allergy and Clinical Immunology* 121(2): S247.

Beyer, K., G. Grishina, L. Bardina, and H.A. Sampson. 2007. Identification of 2 new sesame seed allergens: Sesi 6 and Sesi 7. *Journal of Allergy and Clinical Immunology* 119(6): 1554–1556.

Bock, S.A., A. Munoz-Furlong, and H.A. Sampson. 2001. Fatalities due to anaphylactic reactions to foods. *The Journal of Allergy and Clinical Immunology* 107(1): 191–193.

Bogh, K.L., J. van Bilsen, R. Glogowski, I. Lopez-Exposito, G. Bouchaud, C. Blanchard, et al. 2016. Current challenges facing the assessment of the allergenic capacity of food allergens in animal models. *Clinical and Translational Allergy* 6: 21.

Branum, A.M., and S.L. Lukacs. 2009. Food allergy among children in the United States. *Pediatrics* 124: 1549–1555.

Caraballo, L., J. Zakzuk, B.W. Lee, N. Acevedo, J.Y. Soh, M. Sánchez-Borges, et al. 2016. Particularities of allergy in the Tropics. *The World Allergy Organization Journal* 9(1): 1–44.

Chan, E.S., E.M. Abrams, K.J. Hildebrand, and W. Watson. 2018. Early introduction of foods to prevent food allergy. *Allergy, Asthma & Clinical Immunology* 14: 1–9. https://doi.org/10.1186/s13223-018-0286-1.

Chang, Y.S., M.K. Trivedi, A. Jha, Y.F. Lin, L. Dimaano, and M.T. Garcia-Romero. 2016. Synbiotics for prevention and treatment of atopic dermatitis: A meta-analysis of randomized clinical trials. *JAMA Pediatrics* 170: 236–242. https://doi.org/10.1001/jamapediatrics.2015.3943

Chapman, M.D., A.M. Smith, L.D. Vailes, L.K. Arruda, V. Dhanraj, and A. Pomes. 2000. Recombinant allergens for diagnosis and therapy of allergic disease. *The Journal of Allergy and Clinical Immunology* 106: 409–418.

Chinaza, G.A., K.E. Chinelo, and S.I. Victory. 2020a. Diabetes and the nutrition and diets for its prevention and treatment: A systematic review and dietetic perspective. *Health Sciences Research* 6: 5–19.

Chowdhary, P., A. Gupta, E. Gnansounou, A. Pandey, and P. Chaturvedi. 2021. Current trends and possibilities for exploitation of grape pomace as a potential source for value addition, *Environmental Pollution* 278: 116796.

Chowdhary, P., A. Raj, D. Verma, and Y. Akhter (Eds). 2020. *Microorganisms for Sustainable Environment and Health.* Oxford: Elsevier.

Comberiati, P., G. Costagliola, S. D'Elios, and D. Peroni. 2019. Prevention of food allergy: The significance of early introduction. *Medicina* 55: 323. https://doi.org/10.3390/medicina55070323

Coombs, R.R.A., and P.G.H. Gell. 1975. Classification of allergic reactions responsible for clinical hypersensitivity and disease. In Gell, P.G.H., Coombs, R.R.A., and Lachman, P.J. (Eds) *Clinical Aspect of Immunology*, 3rd ed. Oxford.: Blackwell Scientific Publications, pp. 575–596.

Cuello-Garcia, C.A., J.L. Brożek, A. Fiocchi, R. Pawankar, J.J. Yepes-Nuñez, L. Terracciano, et al. 2015. Probiotics for the prevention of allergy: A systematic review and meta-analysis of randomized controlled trials. *Journal of Allergy and Clinical Immunology* 136: 952–961. https://doi.org/10.1016/j.jaci.2015.04.031

D'Auria, E., C. Agostoni, M. Giovannini, E. Riva, R. Zetterstrom, R. Fortin, et al. 2005. Proteomic evaluation of milk from different mammalian species as a substitute for breast milk. *Acta Paediatrics* 94: 1708–1713.

De Martinis, M., M.M. Sirufo, and L. Ginaldi. 2017. Allergy and aging: An old/new emerging health issue. *Aging and Disease* 8(2): 162–175.

De Martinis, M., M.M. Sirufo, and L. Ginaldi. 2020a. Osteoporosis: Current and emerging therapies targeted to immunological checkpoints. *Current Medicinal Chemistry* 27(37): 6356–6372.

De Martinis, M., M.M. Sirufo, M. Suppa, D. Di Silvestre, and L. Ginaldi. 2020b. Sex and gender aspects for patient stratification in allergy prevention and treatment. *International Journal of Molecular Sciences* 21(4): 1535.

Dunlop, J.H., and C.A. Keet. 2018. Epidemiology of food allergy. *Immunology and Allergy Clinics* 38: 13–25.

Du Toit, G., G. Roberts, P.H. Sayre, H.T. Bahnson, S. Radulovic, A.F. Santos, et al. 2015. Randomized trial of peanut consumption in infants at risk for peanut allergy. *New England Journal of Medicine* 372: 803–813. https://doi.org/10.1056/NEJMoa1414850.

Egger, M., S. Mutschlechner, N. Wopfner, G. Gadermaier, P. Briza, and F. Ferreira. 2006. Pollen-food syndromes associated with weed pollinosis: An update from the molecular point of view. *Allergy* 61: 461–476.

European Union. 2011. Regulation (EU) No 1169/2011 of the European Parliament and of the Council of 25 October 2011 on the Provision of Food Information to Consumers, Amending Regulations (EC) No 1924/2006 and (EC) No 1925/2006 of the European Parliament and of the Council, and Repealing Commission Directive 87/250/EEC, Council Directive 90/496/EEC, Commission Directive 1999/10/EC, Directive 2000/13/EC of the European Parliament and of the Council, Commission Directives 2002/67/EC and 2008/5/EC and Commission Regulation (EC) No 608/2004.

Fu, L., B.J. Cherayil, H. Shi, Y. Wang, and Y. Zhu. 2019. Food processing to eliminate food allergens and development of hypoallergenic foods. *Food Allergy* 123–146. https://doi.org/10.1007/978-981-13-6928-5_6

Galand, C., J.M. Leyva-Castillo, J. Yoon, A. Han, M.S. Lee, A.N. McKenzie, et al. 2016. IL-33 promotes food anaphylaxis in epicutaneously sensitized mice by targeting mast cells. *Journal of Allergy and Clinical Immunology* 138: 1356–1366.

Genuneit, J., A.M. Seibold, C.J. Apfelbacher, G.N. Konstantinou, J.J. Koplin, S. La Grutta, et al. 2017. Overview of systematic reviews in allergy epidemiology (OSRAE) of the EAACI interest group on epidemiology. *Allergy* 72: 849–856.

Greer, F.R., S.H. Sicherer, and A.W. Burks. 2008. Effects of early nutritional interventions on the development of atopic disease in infants and children: The role of maternal dietary restriction, breastfeeding, timing of introduction of complementary foods, and hydrolyzed formulas. *Pediatrics* 121: 183–191. https://doi.org/10.1542/peds.2007-3022.

Gunaratne, A.W., M. Makrides, and C.T. Collins. 2015. Maternal prenatal and/or postnatal n-3 long chain polyunsaturated fatty acids (LCPUFA) supplementation for preventing allergies in early childhood. *Cochrane Database of Systematic Reviews* 7. https://doi.org/10.1002/14651858.CD010085.pub2

Gupta, R., D. Holdford, L. Bilaver, A, Dyer, J.L. Holl, and D. Meltzer. 2013. The economic impact of childhood food allergy in the United States. *JAMA Pediatrics* 167: 1026–1031.

Hadley, C. 2006. Food allergies on the rise? Determining the prevalence of food allergies, and how quickly it is increasing, is the first step in tackling the problem. *European Molecular Biology Organization Reports* 7: 1080–1083. https://doi.org/10.1038/sj.embor.7400846

Holzhauser, T., O. Wackermann, B.K. Ballmer-Weber, C. Bindslev-Jensen, J. Scibilia, L. Perono-Garoffo, et al. 2009. Soybean (Glycine max) allergy in Europe: Gly m 5 (β-conglycinin) and Gly m 6 (glycinin) are potential diagnostic markers for severe allergic reactions to soy. *Journal of Allergy and Clinical Immunology* 123(2): 452–458.

Hourihane, J.O., T.P. Dean, and J.O. Warner. 1996. Peanut allergy in relation to heredity, maternal diet, and other atopic diseases: Results of a questionnaire survey, skin prick testing, and food challenges. *British Medical Journal* 313: 518–521.

Hussain, M., L. Borcard, K.P. Walsh, M. Pena Rodriguez, C. Mueller, B.S. Kim, et al. 2018. Basophil-derived IL-4 promotes epicutaneous antigen sensitization concomitant with the development of food allergy. *Journal of Allergy and Clinical Immunology* 141: 223–234.

Ierodiakonou, D., V. Garcia-Larsen, A. Logan, A. Groome, S. Cunha, J. Chivinge, et al. 2016. Timing of allergenic food introduction to the infant diet and risk of allergic or autoimmune disease: A systematic review and meta-analysis. *The Journal of the American Medical Association* 316: 1181–1192. https://doi.org/10.1001/jama.2016.12623

Jarvinen, K.M., and P. Chatchatee. 2009. Mammalian milk allergy: Clinical suspicion, cross-reactivities and diagnosis. *Current Opinion in Allergy and Clinical Immunology* 9: 251–258.

John, J.R., K.W. Arthur, M.W. Sandra, and B.A. Wesley. 2009. Hypoallergenic legume crops and food allergy: Factors affecting feasibility and risk. *Journal of Agricultural and Food Chemistry* 58: 20–27.

Jones, S.M., and A.W. Burks. 2017. Food allergy. *New England Journal of Medicine* 377: 1168–1176.

Krause, S., G. Reese, S. Randow, D. Zennaro, D. Quaratino, P. Palazzo, et al. 2009. Lipid transfer protein (Ara h 9) as a new peanut allergen relevant for Mediterranean allergic population. *Journal of Allergy and Clinical Immunology* 124(4): 771–778.

Lee, L.A., and A.W. Burks. 2006. Food allergies: Prevalence, molecular characterization, and treatment prevention strategies. *Annual Review of Nutrition* 26: 539–565.

Lichtenstein, P., and M. Svartengren. 1997. Genes, environments, and sex: Factors of importance in atopic diseases in 7–9-year-old Swedish twins. *Allergy* 52: 1079–1086.

Liu, A.H., R. Jaramillo, S.H. Sicherer, R.A. Wood, S.A. Bock, A.W. Burks, et al. 2010. National prevalence and risk factors for food allergy and relationship to asthma: Results from the National Health and Nutrition Examination Survey 2005–2006. *The Journal of Allergy and Clinical Immunology* 126: 798–806.

Matsuo, H., K. Kohno, H. Niihara, and E. Morita. 2005. Specific IgE determination to epitope peptides of ω-5 gliadin and high molecular weight gluten in subunit is a useful tool for diagnosis of wheat dependent exercise-induced anaphylaxis. *Journal of Immunology* 175(12): 8116–8122.

Mills, E.N., A.R. Mackie, P. Burney, K. Beyer, L. Frewer, C. Madsen, et al. 2007. The prevalence, cost and basis of food allergy across Europe. *Allergy* 62: 717–722.

Natale, M., C. Bisson, G. Monti, A. Peltran, L. PeronoGaroffo, S. Valentini, et al. 2004. Cow's milk allergens identification by two-dimensional immunoblotting and mass spectrometry. *Molecular Nutrition and Food Research* 48(5): 363–369.

Noti, M., B.S. Kim, M.C. Siracusa, G.D. Rak, M. Kubo, A.E. Moghaddam, et al. 2014. Exposure to food allergens through inflamed skin promotes intestinal food allergy through the thymic stromal lymphopoietin-basophil axis. *Journal of Allergy and Clinical Immunology* 133: 1390–1399.

Nowak-Węgrzyn, A., and H.A. Sampson. 2011. Future therapies for food allergies. *The Journal of Allergy and Clinical Immunology* 127: 558–573. https://doi.org/10.1016/j.jaci.2010.12.1098

Osborn, D.A., and J.K. Sinn. 2013. Prebiotics in infants for prevention of allergy. *The Cochrane Database of Systematic Reviews* 3. https://doi.org/10.1002/14651858.CD006474.pub3

Osborne, N.J., J.J. Koplin, P.E. Martin, L.C. Gurrin, A.J. Lowe, M.C. Matheson, et al. 2011. HealthNuts investigators. Prevalence of challenge-proven IgE-mediated food allergy using population-based sampling and predetermined challenge criteria in infants. *The Journal of Allergy and Clinical Immunology* 127: 668–676.

Osborne, N.J., J.J. Koplin, P.E. Martin, L.C. Gurrin, L. Thiele, M.L. Tang, et al. 2010. The HealthNuts population-based study of paediatric food allergy: Validity, safety and acceptability. *Clinical and Experimental Allergy* 40: 1516–1522.

Palacin, A., S. Quirce, A. Armentia, M. Fernández-Nieto, L.F. Pacios, T. Asensio, et al. 2007. Wheat lipid transfer protein is a major allergen associated with baker's asthma. *Journal of Allergy and Clinical Immunology* 120(5): 1132–1138.

Palmer, G.W., D.A. Dibbern, A.W. Burks, G.A. Bannon, S.A. Bock, H.S. Porterfield, et al. 2005. Comparative potency of Ara h 1 and Ara h 2 in immunochemical and functional assays of allergenicity. *Clinical Immunology* 115(3): 302–312.

Panjari, M., J.J. Koplin, S.C. Dharmage, R.L. Peters, L.C. Gurrin, S.M. Sawyer, et al. 2016. Nut allergy prevalence and differences between Asian-born children and Australian-born children of Asian descent: A state-wide survey of children at primary school entry in Victoria, Australia. *Clinical and Experimental Allergy* 46: 602–609.

Perkin, M.R., K. Logan, A. Tseng, B. Raji, S. Ayis, J. Peacock, et al. 2016. Randomized trial of introduction of allergenic foods in breastfed infants. *New England Journal of Medicine* 374: 1733–1743.

Peters, R.L., K.J. Allen, S.C. Dharmage, C.J. Lodge, J.J. Koplin, A.L. Ponsonby, et al. 2015. Differential factors associated with challenge-proven food allergy phenotypes in a population cohort of infants: A latent class analysis. *Clinical and Experimental Allergy* 45: 953–963.

Rabjohn, P., E.M. Helm, J.S. Stanley, C.M. West, H.A. Sampson, A.W. Burks, et al. 1999. Molecular cloning and epitope analysis of the peanut allergen Ara h 3. *Journal of Clinical Investigation* 103(4): 535–542.

Reboldi, A., T.I. Arnon, L.B. Rodda, A. Atakilit, D. Sheppard, and J.G. Cyster. 2016. IgA production requires B cell interaction with subepithelial dendritic cells in Peyer's patches. *Science* 352: 6287.

Robotham, J.M., F. Wang, V. Seamon, S.S. Teuber, S.K. Sathe, H.A. Sampson, et al. 2005. Ana o 3, an important cashew nut (Anacardiumoccidentale L.) allergen of the 2S albumin family. *Journal of Allergy and Clinical Immunology* 115(6): 1284–1290.

Rokaite, R., L. Labanauskas, and L. Vaideliene. 2014. Role of the skin patch test in diagnosing food allergy in children with atopic dermatitis. *Medicina (Kaunas)* 40: 1081–1087.

Rona, R.J., T. Keil, C. Summers, D. Gislason, L. Zuidmeer, E. Sodergren, et al. 2007. The prevalence of food allergy: A meta-analysis. *The Journal of Allergy and Clinical Immunology* 120: 638–646.

Sampson, H.A. 1999. Food allergy. Part 1: Immunopathogenesis and clinical disorders. *The Journal of Allergy and Clinical Immunology* 103(5 Pt 1): 717–728.

Schindler, T., J.K. Sinn, and D.A. Osborn. 2016. Polyunsaturated fatty acid supplementation in infancy for the prevention of allergy. *The Cochrane Database of Systematic Reviews* 10: CD010112. https://doi.org/10.1002/14651858.CD010112.pub2

Schussler, E., J. Sobel, J. Hsu, P. Yu, D. Meaney-Delman, L.C. Grammer, et al. 2017. Workgroup report by the joint task force involving American Academy of Allergy, Asthma & Immunology (AAAAI); Food Allergy, Anaphylaxis, Dermatology and Drug Allergy (FADDA) (adverse reactions to foods committee and adverse reactions to drugs, biologicals, and latex committee); and the Centers for Disease Control and Prevention Botulism Clinical Treatment Guidelines Work group-Allergic Reactions to Botulinum Antitoxin: A systematic review. *Clinical Infectious Diseases* 66: S65–S72.

Shanti, K.N., B.M. Martin, S. Nagpal, D.D. Metcalfe, and P.V.S. Rao. 1993. Identification of tropomyosin as the major shrimp allergen and characterization of its IgE-binding epitopes. *Journal of Immunology* 151: 5354–5363.

Sicherer, S.H. 2001. Clinical implications of cross-reactive food allergens. *Journal of Allergy and Clinical Immunology* 108: 881–890.

Sicherer, S.H., and H.A. Sampson. 2006. Food allergy. *The Journal of Allergy and Clinical Immunology* 117(suppl. Mini-Primer): S470–S475.

Sicherer, S.H., and H.A. Sampson. 2009. Food allergy: Recent advances in pathophysiology and treatment. *Annual Review of Medicine* 60: 261–277.

Sicherer, S.H., and H.A. Sampson. 2014. Food allergy: Epidemiology, pathogenesis, diagnosis, and treatment. *The Journal of Allergy and Clinical Immunology* 133: 291–307. https://doi.org/10.1016/j.jaci.2013.11.020.

Sicherer, S.H., and H.A. Sampson. 2018. Food allergy: A review and update on epidemiology, pathogenesis, diagnosis, prevention, and management. *The Journal of Allergy and Clinical Immunology* 141: 41–58.

Sicherer, S.H., R.A. Wood, D. Stablein, A.W. Burks, A.H. Liu, S.M. Jones, et al. 2010. Immunologic features of infants with milk or egg allergy enrolled in an observational study (Consortium of Food Allergy Research) of food allergy. *The Journal of Allergy and Clinical Immunology* 125: 1077–1083. https://doi.org/10.1016/j.jaci.2010.02.038

Sirohi, R., A. Tarafdar, V.K. Gaur, S. Singh, R. Sindhu, R. Rajasekharan, A. Madhavan, … A. Pandey. 2021. Technologies for disinfection of food grains: Advances and way forward. *Food Research International* 145: 110396. https://doi.org/10.1016/j.foodres.2021.110396

Sirvent, S., O. Palomares, A. Vereda, M. Villalba, J. Cuesta- Herranz, and R. Rodríguez. 2009. NsLTP and profilin are allergens in mustard seeds: Cloning, sequencing and recombinant production of Sin a 3 and Sin a 4. *Clinical and Experimental Allergy* 39(12): 1929–1936.

Smit, J.J., M. Noti, and L. O'Mahony. 2016. The use of animal models to discover immunological mechanisms underpinning sensitization to food allergens. *Drug Discovery Today: Disease Models* 17: 63–69.

Spergel, J.M., T.F. Brown-Whitehorn, J.L. Beausoleil, J. Franciosi, M. Shuker, R. Verma, et al. 2009. 14 years of eosinophilic esophagitis: Clinical features and prognosis. *Journal of Pediatric Gastroenterology and Nutrition* 48(1): 30–36.

Strachan, D.P. 1989. Hay fever, hygiene, and household size. *British Medical Journal* 299: 1259–1260.

Tang, M.L.K., and R.J. Mullins. 2017. Food allergy: Is prevalence increasing? *Internal Medicine Journal* 47: 256–261.

Tawde, P., Y.P. Venkatesh, F. Wang, S.S. Teuber, S.K. Sathe, and K.H. Roux. 2006. Cloning and characterization of profilin (Pru du 4), a cross-reactive almond (Prunus dulcis) allergen. *Journal of Allergy and Clinical Immunology* 118(4): 915–922.

Taylor, S.L., S.L. Hefle, C. Bindslev-Jensen, F.M. Atkins, C. Andre, C.A.F.M. Bruijnzeel-koomen, et al. 2004. A consensus protocol for the determination of the threshold doses for allergenic foods: How much is too much? *Clinical and Experimental Allergy* 34 : 689–695.

Tommasini, A., T. Not, and A. Ventura. 2011. Ages of celiac disease: From changing environment to improved diagnostics. *World Journal of Gastroenterology* 17:3665–3671. https://doi.org/10.3748/wjg.v17.i32.3665

Turnbull, J.L., H.N. Adams, and D.A. Gorard. 2015. Review article: The diagnosis and management of food allergy and food intolerances. *Alimentary Pharmacology & Therapeutics* 41: 3–25. https://doi.org/10.1111/apt.12984

Umasunthar, T., J. Leonardi-Bee, M. Hodes, P.J. Turner, C. Gore, P. Habibi, et al. 2013. Incidence of fatal food anaphylaxis in people with food allergy: A systematic review and meta-analysis. *Clinical and Experimental Allergy* 43: 1333–1341.

USFDA (U.S. Food and Drug Administration). 2004. Food Allergen Labeling and Consumer Protection Act of 2004. Washington, DC: USFDA.

USFDA (U.S. Food and Drug Administration). 2015. Food Allergen Labelling and Information Requirements under the EU Food Information for Consumers Regulation No. 1169/2011: Technical Guidance. Washington, DC: USFDA.

USFDA (U.S. Food and Drug Administration). 2017. Have food allergies? Read the label. Washington, DC: USFDA.

USFDA (U.S. Food and Drug Administration). 2018. Statement from FDA Commissioner Scott Gottlieb, M.D., on the FDA's new consideration of labeling for sesame allergies. Washington, DC: USFDA.

USFDA (U.S. Food and Drug Administration). 2020. FDA issues draft guidance for industry on voluntary disclosure of sesame when added as flavoring or spice. Washington, DC: USFDA.

USFDA (U.S. Food and Drug Administration). 2021. Guidance for Industry: Questions and Answers Regarding Food Allergens (Edition 4). Center for Food Safety and Applied Nutrition. Available at: https://fda.gov/regulatory-information/search-fda-guidance-documents/guidance-industry-questions-and-answers-regarding-food-allergens-edition-4

Valenta, R., and D. Kraft. 1996. Type 1 allergic reactions to plant-derived food: a consequence of primary sensitization to pollen allergens. *Journal of Allergy and Clinical Immunology* 97: 893–895.

Venter, C., B. Pereira, K. Voigt, J. Grundy, C.B. Clayton, B. Higgins, et al. 2008. Prevalence and cumulative incidence of food hypersensitivity in the first 3 years of life. *Allergy* 63: 354–359.

Vierk, K.A., K.M. Koehler, S.B. Fein, and D.A. Street. 2007. Prevalence of self-reported food allergy in American adults and use of food labels. *The Journal of Allergy and Clinical Immunology* 119: 1504–1510.

Von Mutius, E. 2004. Influences in allergy: Epidemiology and the environment. *The Journal of Allergy and Clinical Immunology* 113(3): 373–379.

Waidyatillake, N.T., S.C. Dharmage, K.J. Allen, C.J. Lodge, J.A. Simpson, G. Bowatte, et al. 2018. Association of breast milk fatty acids with allergic disease outcomes: A systematic review. *Allergy* 73: 295–312. https://doi.org/10.1111/all.13300

Walkner, M., C. Warren, and R.S. Gupta. 2015. Quality of life in food allergy patients and their families. *Pediatric Clinics of North America* 62: 1453–1461.

Wallowitz, M., W.R. Peterson, S. Uratsu, S.S. Comstock, A.M. Dandekar, and S.S. Teuber. 2006. Jug r 4, a legumin group food allergen from walnut (Juglans regia Cv. Chandler). *Journal of Agricultural and Food Chemistry* 54(21): 8369–8375.

Zuccotti, G., F. Meneghin, A. Aceti, G. Barone, M.L. Callegari, A. Di Mauro, et al. 2015. Probiotics for prevention of atopic diseases in infants: Systematic review and meta-analysis. *Allergy* 70: 1356–1371. https://doi.org/10.1111/all.12700

Zuidmeer, L., C. Garino, J. Marsh, A. Lovegrove, M. Morati, S. Versteeg, et al. 2009. Isolation, cloning and confirmation as an allergen of the 2S albumin from hazelnut. *Allergy* 64(90): 235.

Part II

Remediation and Management of Pollutants

7 The Role of Microbes in Environmental Contaminants' Management

Pravin Khaire,[1,*] *Vajramma Boggala,*[2] *Akshay Mamidi,*[3] *and Tanaji Narute*[1]
[1]Department of Plant Pathology and Agricultural Microbiology, PGI, MPKV, Rahuri, India
[2]Division of Plant Pathology, Indian Agricultural Research Institute, New Delhi, India
[3]Department of Genetics and Plant Breeding, PJTSAU, Rajendranagar, T.S., India
*Corresponding author Email: pravinkhaire26893@gmail.com

CONTENTS

7.1 Introduction 118
7.2 The Contribution of Microbes to the Clean-Up of Environmental Pollutants 119
7.2.1 Different Strategies for Environmental Contaminants Management 120
7.2.1.1 Metabolic Engineering (ME) 120
7.2.1.2 Optimizing Recombinant DNA (rDNA) 123
7.2.1.3 Plasmids 123
7.2.1.4 Expression Systems 124
7.2.1.5 Post-Transcriptional Processing 124
7.2.1.6 Transposons 124
7.2.1.7 Family Shuffling and Genome Rearranging 124
7.2.1.8 Genome Rearranging (GR) 125
7.3 Environmental Contaminants (ECs) 125
7.4 Role of Microbes in Some Important Water Pollutants Management 126
7.4.1 The Food Processing Industries 126
7.4.2 The Pharmaceutical Industry 127
7.4.3 The Textile Industry (TI) 127
7.4.4 The Petrochemical Industries (PCI) 129
7.4.5 The Explosives Industry 130
7.4.6 The Distillery Industry (DI) 130
7.4.7 Heavy Metals (HM) Remediation 132
7.4.8 Microbial Management Strategies for Pesticides 132
7.4.9 Microbial Remediation of Hydrocarbons (HCs) 136
7.4.10 Plastic Biodegradation 138
7.4.10.1 Microbial Adherence 138
7.4.10.2 Fragmentation 140

DOI: 10.1201/9781003239956-9

7.4.10.3 Assimilation....140
7.4.10.4 Mineralization....141
7.5 Conclusion....141
References....141

7.1 INTRODUCTION

The global population is continuously rising at an alarming level and creating a demand for the rapid expansion of industries, food security, healthcare and transport to cater to the requirements of the increasing populations and ultimately putting high pressure on the available resources. The population explosion, mining, metallurgical smelting, electroplating, industrial growth, agrochemicals, and fertilizers have polluted the Earth's environmental matrices, such as air, soil, and water, by releasing various harsh metal channels into the environment in the form of agro-industrial waste, inland raw sewage (Zhang et al., 2012). But without proper management of the environment, it is very difficult to live and sustain the quality of life with all these new developments, which are unfavorable to the environment. Because the standard of living is inextricably tied to the reliability of the natural atmosphere, global attention has turned to ways to sustain and protect the environment from contaminants. Solving this problem is possible by using microbial remediation strategies. Microorganisms are ubiquitous in nature because of their versatility in nutrition and habitat. Microorganisms' nutritious potential has been used to biodegrade contaminants in the land, water, and the wind (Abatenh et al., 2017). Aboriginal microorganisms found in contaminated environments are playing a special role in resolving the majority of the problems associated with environmental remediation (Azubuike et al., 2016). In nature, different microorganisms, such as fungi, bacteria, and archaebacteria, have the ability to dissolve bacteria, remove toxins, and sometimes even accumulate hazardous particles, restoring the initial local habitat and preventing additional pollution (Demnerova et al., 2005; Tang et al., 2007; Strong et al., 2008).

Genetically modified organisms (GMOs) have recently gained popularity due to their improved ability to apply categorical residues, such as hydrocarbons and agrochemicals from contaminated environments. A genetically modified organism is a microorganism whose genetic make-up has been altered using gene-editing strategies, namely, recombinant DNA (rDNA) technology. Distinct biotechnological aspects, such as metabolic engineering (ME) as well as rDNA techniques, are used to develop an effective eco-remediation tactic. ME of microbes pathways may provide a superior method for increasing microbes' ability to rectify large amounts of waste. rDNA advances, such as the development of new reporter genes, the reconfiguration of gene regulatory channels, and strategies for reporting GMOs, can serve as new bio-remediation system applications. Amelioration of degradative features is mainly dependent on the ability to investigate the biological diversity and metabolic diversity of microorganisms (Fulekar, 2009), and genes that encode for biodegradative enzymes are almost always found in microbes' chromosomal and extrachromosomal DNA.

GEMs (genetically engineered microorganisms) have shown promise for bio-remediation applications in soil, surface water, and induced sludge ecosystems, with improved degradative abilities for a large number of different pollutants. Previously, a variety of options for improving degradative output through GE approaches emerged. Rate-limiting moves in earlier studied metabolic pathways, for example, can be genetically altered to increase the degradation efficiency, or entirely new metabolic pathways can be incorporated into bacteria strains for the degradation of previously recalcitrant compounds. GEMs mainly involve three activities: (1) alteration of enzyme high affinity and specificity; (2) pathway construction and regulation; and (3) bioprocess progression, supervising, and regulation. Extremely important bacterial genes are brought on in a particular chromosome, but the genes-encoding enzymes needed for the catabolism of certain unusual substrates might well be brought on plasmids. Plasmids have been linked to catabolism. As a result, GEMs can be used effectively for biodegradation activities, which reflects a science horizon with massive consequences in future.

7.2 THE CONTRIBUTION OF MICROBES TO THE CLEAN-UP OF ENVIRONMENTAL POLLUTANTS

Microorganisms are nature's classic recyclers and work as an essential element in nutritional chains, which are a vital piece of life's biological stability. Microbial strategies entail the clean-up of toxic pollutants using fungi, bacteria, algae and yeast. In the presence of toxic molecules on any dumping sites, microbes can withstand a wide range of temperatures, working below zero and also at high temperatures. Microbes' ability to adapt and their biological system have made them well suited for the treatment process (Prescot et al., 2002). A microbe consortium brought microbial processes in various climates by using organic and inorganic chemicals (carbon) as the primary energy source for the microbes' activity. Microbial strategies therefore represent a promising, new environmental alternative to conventional strategies, such as physical and chemical methods. Microorganisms' remediation strategies are mainly classified into two groups: (1) aerobic; and (2) anaerobic microorganisms. *Acinetobacter*, *Arthrobacter*, *Corynebacterium*, *Pseudomonas*, *Sphingomonas*, *Nitrosomonas*, *Nocardia*, *Achromobacter*, *Flavobacterium*, *Rhodococcus*, *Mycobacterium*, *Alcaligenes*, *Xanthobacter*, *Bacillus* and other aerobic bacteria are able to degrade complex molecules (Singh et al., 2014). The use of such aerobic bacteria to clean up contaminants, such as agrochemicals, hydrocarbons, alkanes, and polyaromatic compounds, is gaining popularity (Figure 7.1). They can also deal with chlorinated aromatic compounds, polychlorinated biphenyls, and dechlorination and conversion of the solvents trichloroethylene and chloroform. Anaerobic bacteria, such as *Dechloromonas aromatica*, which oxidizes benzene anaerobically, are not as commonly used as aerobic bacteria.

The use of microbial strategies is not limited to one field; they can be found in the remediation of heavy metals, dyes, hydrocarbons, petroleum and its products present in the environmental matrices. A thin layer of oil caused by spills from container ships into the marine ecosystem has become a

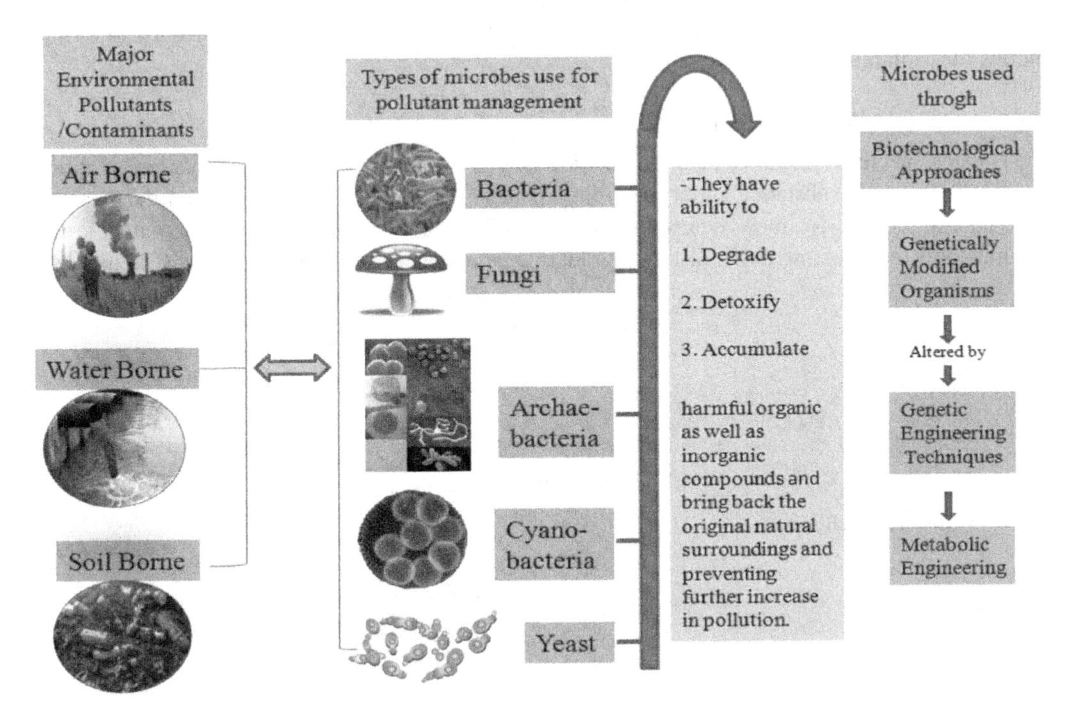

FIGURE 7.1 An overview of the major environmental contaminants and the microbes used for remediation.

Source: P.B. Khaire.

common occurrence. Many aquatic microbes can feed on oil, and most of them actually create powerful and effective surface-active molecules which can liquefy oil in water and facilitate its separation. The microbial emulsifier, unlike chemical surfactants, is non-toxic and biodegradable. *Pseudomonads*, various *corynebacteria*, *mycobacteria*, and yeast are among the microorganisms able to degrade petroleum. Several algae and bacteria generate lure that entice heavy substances that are eliminated from the food web in large amounts. *Geobacter metallireducens* is a recent example, in addition to the huge number of bacteria which can mitigate metals. It eliminates uranium, a radioactive huge drain, from mining wastewater and polluted groundwaters. Microbial cells that have died can still be beneficial in strategies for bio-remediation (BR). These findings imply that further investigation of microbial diversity will almost certainly result in the scientific breakthrough of several microbes with special properties beneficial in environmental remediation and clean-up.

7.2.1 Different Strategies for Environmental Contaminants Management

Recovery and recycling approaches are based on waste clean-up and shipping for therapies. The fundamental strategies are in situ and ex situ applications. The term "in situ" refers to the fact that it deals with the degradation of contaminants in the saturated soils and groundwater. Here chemotaxis plays an important role because microbial organisms with chemotactic abilities can move in to contaminated areas. Hence there is a need to enhance the chemotactic abilities of microbes for effective in-situ strategy. Figure 7.2 presents different in situ strategies.

Ex situ BR procedures necessitate the excavation of polluted soils or the force of groundwater to enhance biodegradation, and it is less cost-effective than the in situ tactic. It is divided into two types: slurry and solid phase structures. Figure 7.3 presents the ex situ strategies.

In situ methods have many advantages, including the fact that they do not entail excavation of the polluted soils (resulting in less site disruption and dust) and that parallel therapies of groundwater and soil are feasible, making them cost-effective. It also has drawbacks, as it is time-consuming and there can be problems applying the additives. Figure 7.4 is an overview of both in situ and ex situ options for contamination management.

7.2.1.1 Metabolic Engineering (ME)

The isolated species or strain can degrade a wide range of pollutants. It is the strain's innate ability to degrade the toxic substance under ideal circumstances. Even so, if we use the similar strain in scale-up experiments for mass deterioration, it is difficult because its economic output is lower. The metabolic engineering (ME) of bio-degradative pathways in microorganisms may be the most

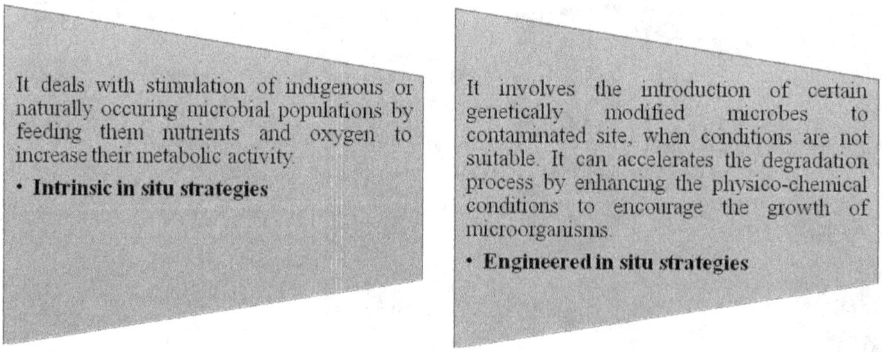

FIGURE 7.2 Different in situ strategies.

Source: B. Vajramma.

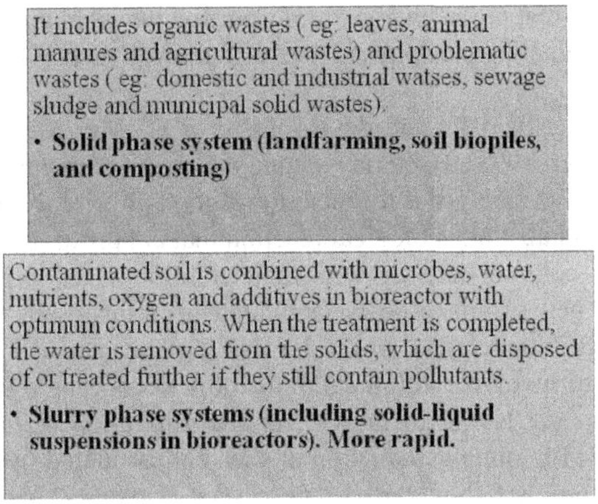

It includes organic wastes (eg: leaves, animal manures and agricultural wastes) and problematic wastes (eg: domestic and industrial watses, sewage sludge and municipal solid wastes).
• Solid phase system (landfarming, soil biopiles, and composting)
Contaminated soil is combined with microbes, water, nutrients, oxygen and additives in bioreactor with optimum conditions. When the treatment is completed, the water is removed from the solids, which are disposed of or treated further if they still contain pollutants.
• Slurry phase systems (including solid-liquid suspensions in bioreactors). More rapid.

FIGURE 7.3 Different ex situ strategies.

Source: B. Vajramma.

Types
In situ
Natural attenuation/ Intrinsic
Enhanced/ Engineered
Bioslurping
Bioventing
Biosparging
Biostimulation
Bioaugmentati on
Ex situ
Permeable reactive barrier
Slurry phase
Bioreactor
Solid phase
Biopiling
Biofilter
Windrow composting
Land farming

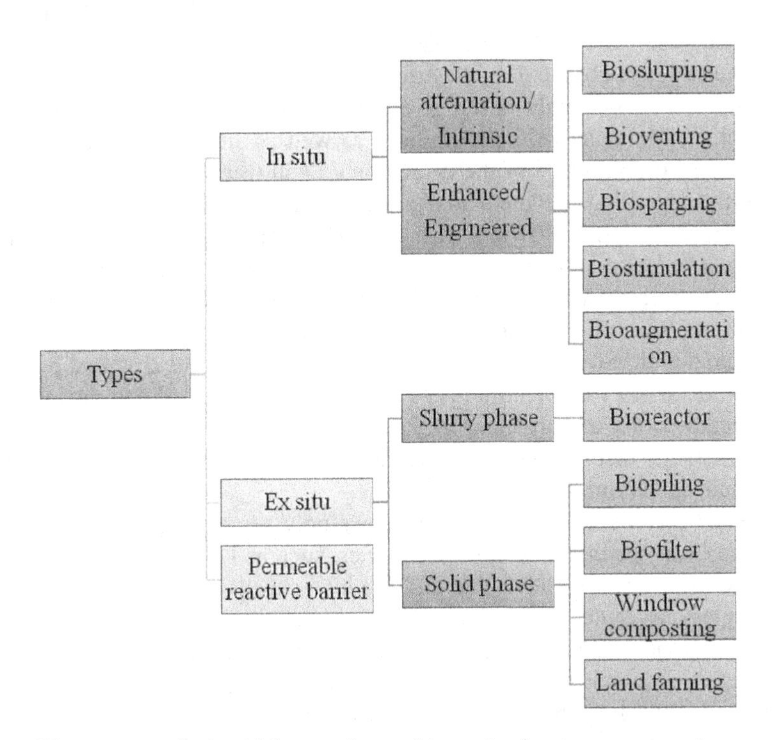

FIGURE 7.4 Different types of microbial strategies used to contaminants management.

Source: Aapted by B. Vajramma from Sharma (2021); Sasikumar et al. (2003).

effective method of increasing the host microorganism's ability to rectify large-scale waste (Chen et al., 1999). The following considerations must be kept in mind when attempting to build the metabolic activity:

1. Recognition of genes that encode the enzymes' function in the catabolic pathway (CP) that secrete the enzymes.

2. Recognition of genes that are in charge of transporting the compound.
3. Recognition of the signaling pathways that control the key metabolic pathways (MP).

Vast amounts of data are critical for the development of GEMs, according to the CP and the co-MP (Stapleton and Sayler, 1998). Bacteria residing in any experimental parameters are revealed to signals that must be analyzed in order for them to respond physiologically. For pollutant clean-up, it is important to formulate a physiological reaction that adapts to the specific regulatory catabolic operons that allow cells to keep up with their conditions for growth (Sato et al., 1997). The horizontal move of genetic make-up is facilitated by transposons and extra-chromosomal DNA. The morphological, biochemical, physiological, genetic, and CP of microbes, as well as the regulatory gene system involved with them, must now be thoroughly understood. There have been countless studies on the layout of the CP for the mitigation of toxic components. Using *Rhodococcus erythropolis* for sulphur reduction at sites contaminated by fossil fuels is a one-of-a-kind example. The bacteria's CP is constructed in such a manner that the bacteria can cut up the sulphur without splitting the ring while also maintaining the fuel content (Izumi et al., 1994). Numerous findings of genetic engineering in *Pseudomonas* are also available, for example, gene editing and incorporation of dsz bunch of genes in *Pseudomonas* strains that can decontaminate the substance more effectively than *Rhodococcus* sp. (Gallardo et al., 1997). While the similar strain with the dsz bunch is used and the EGSOX pathway is transmuted, the ability to eliminate sulphur rises fourfold when matched to the *Rhodococcus* strain. Knowledge of the structure, features, and method of control of new biocatalysts can give early chances for transmuting MPs by changing the genetic structure. Mutagenesis, both arbitrary as well as site-specific, can be used to tailor enzymes in MPs. The modification of the cytochrome P450 (Stevenson et al., 1998) to clean the receptor sites of the degrading enzyme haloalkane dehydrogenase is a typical case of site-directed mutagenesis (Holloway et al., 1998). A further case of enzyme engineering (EE) can be found in *Xanthobacter autotrophicus*. In metabolism, *Xanthobacter autotrophicus* can use dichloroethane as its sole source of energy. The haloalkane dehydrogenase enzyme can convert the chlorine atom to a hydroxyl group. Several coalescences of enzyme linkers and enzyme device clusters were devised by recognizing the structural composition of the enzyme. The eventual resulting variants have been proven to be more efficient in the structure's replacement of chlorine. On the contrary, no mutant was ready to use a more complex compound as a medium, such as trichloroethylene (TCE). This is due to the fact that a rational structure may occasionally perform badly in the degradation of involute contaminants.

The site-directed method is restricted as it allows only some patterns to transmute at a specific point in time. Non-specific or illogical strategies, such as DNA shuffling, arbitrary priming, and expanded extension, can be advisable and produce excellent outcomes for the degradation of more challenging molecules (Stemmer, 1994; Kuchner and Arnold, 1997; Harayama, 1998; Shao et al., 1998). Of these, DNA rearranging to help develop genes appears to be the most promising strategy because elongated sequences can be investigated for numerous locations of alteration (Crameri et al., 1998). Such an approach is also used to boost the genetically identical areas for enzymes in a degradation pathway, as well as incorporating such reconfigured enzymes into host microorganisms. This method, for example, was used to reconfigure the arsenic resistance operon. In the lack of any physical alteration, the different versions demonstrated a 40-fold rise in pollutants' resistance as well as a 12-fold boost in the arc gene product. Such outcomes are also assumed in particular change-led reconfigurations, but organic changes in process parameters allow microorganisms to adapt to those circumstances, creating organisms that are more reliable for a mass task. The imprecise tactic allows the rapid and detailed analysis of sequence diversity in the eco-system, which has resulted in the creation of new hybrid enzymes or pathways with the required characteristics.

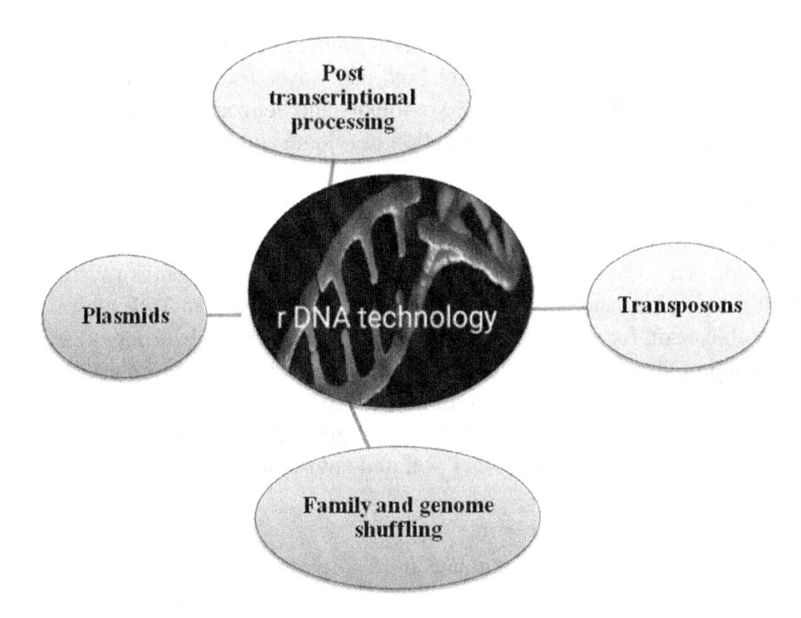

FIGURE 7.5 Important methods in rDNA technology.

Source: Adapted by B. Vajramma from Sanghvi et al. (2020).

7.2.1.2 Optimizing Recombinant DNA (rDNA)

For the mitigation of recalcitrant compounds, innovative methodologies for modifying the genetic material of microbes are being devised. Development of newer expression vectors for gene affirmation, reconfiguration of gene regulating networks (GRN) to manage tenacity and strategies for reporting GMOs are among the tactics used. The respective issues must be taken into account when designing and implementing molecular tools:

1. Meticulous planning of the artificial and power source pathways that drive the growth of cells.
2. The length and width of the genetic vectors (GVs) and the host organism's systems.
3. Sustainable management of the extra-chromosomal GVs, used to test for genetic changes (Sanghvi et al., 2020).

Figure 7.5 is an overview of the processes used in the rDNA technology.

7.2.1.3 Plasmids

Particular gene sets in any microbe are related to particular contaminant degradation pathways. This can be observed at both the chromosomal and extra-chromosomal levels. Plasmids are the most common type of extra-chromosomal genetic information used in cloning and appearance processes. Plasmids contain a huge percentage of genes that are implicated in the full or partial deterioration of the contaminant. Only some bunches of genes have been recognized for use in the chemical categorical degradative route. Plasmids are a quickly available and accessible structure for the discovery of genes in pathways, due to their character traits such as an easy molecular structure and screening tests (Sayler et al., 1990). Numerous plasmids are associated with the surface metabolism which exists in nature. Among the most frequently used plasmids for degradation is one that has been marked in the degradation of toluene and naphthalene. Such plasmids carry complete genetic operons responsible for the change of toluene and xylene (TOL) to intermediate metabolites. The top and bottom salicylate oxidation pathways are both found in the naphthalene 7 plasmid's genetic operon. Benzoate is the prevalent activator for both pathways. Toluene is degraded by converting xylene to toluene, which will then be degraded to toluate and benzoate. The meta cleavage pathway is used in the lower pathway again for deterioration of toluate and benzoate.

7.2.1.4 Expression Systems

7.2.1.4.1 Inducible Promoters

Inducible expression is required in most natural and engineered microorganism systems. Lac expression systems offer versatility and complete elucidation. Producers from TOL plasmid, such as Pm, are readily accessible again for genes that code for toluene and naphthalene degradation (de Lorenzo et al., 1993). Throughout transcription, the XylS regulatory protein activates the CP for toluene degradation. The promoter gene Pu is powered up in the upper pathway through toluene degradation. Throughout transcription, the regulatory proteins XylR and NahR are stimulated. Soon, the expression system is able to react to metal and other polluted environmental sites.

7.2.1.5 Post-Transcriptional Processing

Post-transcriptional change is used to regulate the expression of genes associated with enzyme segments (Carrier and Keasling, 1997a). This centralized process removes the need for a large number of strength-providing promoters. This tactic, also known as mRNA stability, involves inserting external DNA into the 5' non-coding region of a particular gene of interest (Carrier and Keasling, 1997a, 1997b). To ensure the service's dependability, a DNA material is intended for the creation of hairpin structures in mRNA at the 5' ends. The composition of AG in the subordinate structure was really the reason for incorporating the hairpin structure. Structures with a large AG initiation have high protein levels and outstanding mRNA reliability, which is essential for the regulation of gene expression. Such methodologies, which use a single promoter as a control point, will increase the probability of producing creative enzymes at various levels and in different pathways (Sanghvi et al., 2020).

7.2.1.6 Transposons

Transposons have a unique DNA sequence that allows them to move to any random spot on a chromosome. Transposons induce vague changes, but they were said to have been the ideal workable system for GMOs because they produce more reliable clones than plasmids (Davison et al., 1987; Ramos et al., 1994; Perez-Martin and de Lorenzo, 1996). The exact mechanism of transposon-mediated mitigation is still unknown. There is also a need for understanding the intensity of transposable element flow, the characteristics of the goods that result from it, and the phenotypic consequences. These transposable elements have been reported to be used in heavy metal polluted sites. For example, Huang et al. (2010) proposed a transposon-dependent system for mercury-polluted sites. They found the transposon on the chromosome of *Bacillus megaterium* MB1. The transposon was responsible for encoding wide-ranging mercury resistance.

7.2.1.7 Family Shuffling and Genome Rearranging

7.2.1.7.1 Family Rearranging (FR)

FR is the rearranging of DNA to a set of similar genes and the mixture of such genes that improves focused progression (Crameri et al., 1998). The best evidence is the biphenyl dioxygenase (bphA) family rearranging in the strain of *Bacillus* and *Pseudomonas* (Kumamaru et al., 1998). This altered enzyme substantially enhanced the degradation of recalcitrant pollutants, such as polychlorinated biphenyls (PCBs), toluenes, and many others. This method was also used to degrade PCBs by changing important segments of a specific bphA in a gene by *Rhodococcus* and *Burkholderia* sp. (Barriault et al., 2002).

7.2.1.8 Genome Rearranging (GR)

GR is the mixing of chromosomes throughout many bacteria to enhance the behavioral integrity of the entire organism. GR creates the genetically altered strain providing superior phenotypic traits. In sequence for the genomes to mix well, the mutant is detached after subsequent cycles of protoplast fusion. GR is beneficial in modifying multitrait phenotypes that are tough to alter actively. These approaches can be used particularly for those phenotypes where all modifications are unusual, requiring a complicated function to be developed that keeps significant growth. These methods were used to bioremediate pentachlorophenol by mutants created through a long sequence of protoplast fusions. It was revealed that these issues resulted in a 10-fold rise in pentachlorophenol degradation, as well as a massive increase in tolerance (Dai and Copley, 2004). This procedure has been used as a quantification way of assessing the health of mutants produced after repetitive rounds of protoplast fusion, and it may be one of the quality standards in perfect circumstances where the bacteria was used for the phase.

7.3 ENVIRONMENTAL CONTAMINANTS (ECs)

ECs have increased in recent decades as a result of rising human activity, population growth, urbanization, forest destruction, industrial growth, indiscriminate agricultural practices, and inappropriate use of energy supplies. Nuclear waste, heavy metals, chemical fertilizers, fungicides, pesticides, herbicides, hydrocarbons, and greenhouse gases are examples of toxic contaminants causing environmental and public health concerns. The majority of ECs can be divided into three categories: air, soil, and water contaminants.

1. *Air contaminants*: The burning of fuels in automobiles is one of the most significant contributors to air pollution, which releases carbon monoxide (CO), So_2 and No_2 into the atmosphere and creates smog (a CO-based dense layer of cloud matter) and acid rains, respectively. Air pollutants cause asthma, bronchitis and cancer. Presently burning paddy stubble mainly in the Punjab and Haryana causes severe air pollution (Smog+Fog) in Delhi. Burning paddy stubble indirectly causes environmental pollution by emitting pollutants such as greenhouse gases, NO_2, SO_2, HCl and also traces of furans and dioxins (Jenkins et al., 2003). It is also a source of aerosol particles, such as PM10 etc., affecting the air quality by creating fog (Engling et al., 2009; Chang et al., 2013; Hayashi et al., 2014). As a result of the aforesaid concerns, there is an urgent need to adopt eco-friendly paddy stubble management immediately. Thus, new strategies such as microbial decomposition of stubble may serve the purpose. Among all the microbial agents, fungi is the most basic form, as a result, it colonizes solid things quickly and its lignocellulolytic activity is considered one of the associated technological strategies to enhance the lignocellulose degradation potential in stubbles (Viji and Neelanarayanan, 2015) (Table 7.1). Thus, beneficial fungal species works as ideal microbes in the bio-degradation of the organic wastes that are lignocellulosic.

TABLE 7.1
Several Fungal Species Having Lignocellulolytic Activity

Sr. No.	Fungal species	References
1	*Fusarium* sp., *Aspergillus terreus*, *Paecilomyces fusisporous*, *Micromonospora Coriolus versicolor*	(Phutela and Sahni, 2012; Goyal and Sindhu, 2011)
2	*R. oryzae*, *A. oryzae* and *A. fumigatus* (Mixed culture for paddy straw decomposition)	(Viji and Neelanarayanan, 2015)

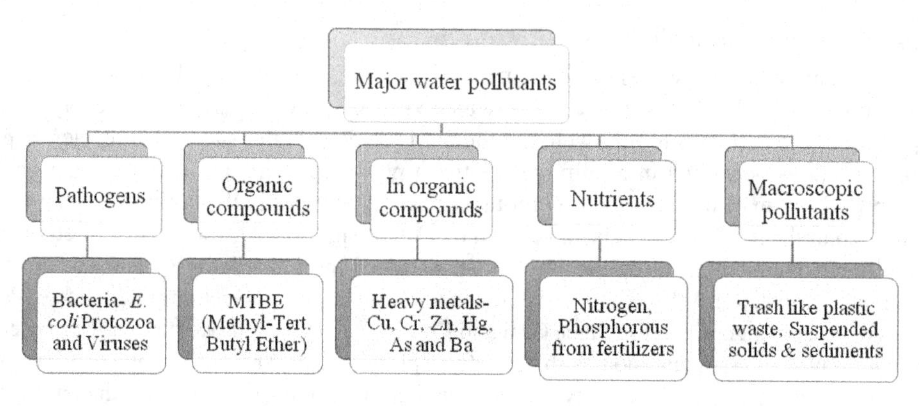

FIGURE 7.6 Major water pollutants.

Source: B. Vajramma.

2. *Soil contaminants*: Soil is being degraded by different source pollutant: heavy metals, pesticides, municipal garbage (contains discarded materials from homes and industry contains paper, plastic and organic matter) (Burken and Schnoor 2003). Sources of heavy metals come from atmospheric deposition, sewage, irrigation, industry and use of pesticides, fertilizers and improper disposal of chemicals and toxins. (Zang et al., 2011). Soil is mainly polluted by indiscriminate use of fertilizers, agro chemicals, fungicides, pesticides, and herbicides, and antibiotics from hospitals are major sources of soil contaminants and cause biodiversity loss and inhibit microbial activity and nutrient cycles (Su et al., 2014).
3. *Water contaminants*: Among the vital resources on Earth, water is a valuable habitat to a wide range of living organisms. Pollution levels in the aquatic habitats have led to a loss of clean water activity. The main cause of water pollution is the release of untreated wastewater from industries into water bodies, canals, rivers, lakes, ponds and oceans. Run-off from agricultural lands, marine pollutants or oil spills are other causes (Figure 7.6). Excess phosphorus results in the eutrophication in water bodies and finally increased biological oxygen demand (BOD). Heavy metal pollution in water can occur due to industrial accidents and leaching from waste disposal. Biomagnification of water pollutants leads to serious health hazards and almost 98% of pesticides are classified as acutely toxic to aquatic life, especially fish (de la Cruz et al., 2014).

7.4 ROLE OF MICROBES IN SOME IMPORTANT WATER POLLUTANTS MANAGEMENT

Water pollutants mainly come from industries: food, pharmaceutical, textiles, petrochemical, explosives and distillery processes, etc.

7.4.1 The Food Processing Industries

The main water-using food industries consist of dairy products, meat, poultry, bacon processing, vegetable oil, sugar refining and brewing. Wastewater released from the food industries has many contaminants that deplete oxygen, affecting life in natural water bodies (Joshi, 2012; Spina et al., 2012) and the major quality parameters of effluents are odor, pH, temperature, total dissolved solids (TDS), total suspended solids (TSS), biological oxygen demand (BOD) and chemical oxygen demand (COD) (Garg and Chaudhary, 2017), which jeopardize human and animal health (Emmanuel et al., 2013). Over the decades, the indiscriminate release of untreated waste from the food industries has been the subject of disputes between industries and the government. In most cases, the standard pollutant management strategies are not effective in removing the pollutant completely. Thus, before going for the wastewater treatment, a detailed study of each pollutant released from different food industries is needed. Additionally, a thorough full knowledge of industrial manufacturing operations and the pollutant's cumulative effect on the environmental ecosystem is crucial. For wastewater treatment, chemical, physical, or biological practices can be used, or a integrated treatment of this (Emmanuel et al., 2013). Fungi, bacteria and algae can play a potential role in wastewater treatment, which uses nutrients, for example, phosphorus, nitrogen, etc. from industrial organic waste matter. Most common microbes in an activated sludge include *Burkholderiales*, *Rhodocyclales*, *Sphingobacteriales*, *Anaerolineales*, *Rhizobiales*, *Planctomycetales*, *Clostridiales*, and *Xanthomonadales*.

7.4.2 The Pharmaceutical Industry

Today, India stands fourth in volume and 13th in value terms in the pharmaceutical industry. Pharmaceutical effluents mainly contain different chemicals, for example, antibiotics, steroids (plant and animal), hormones, anti-inflammatory drugs, anti-depressants and analgesics, which are hazardous to both terrestrial and aquatic ecosystems. The different antibiotics and hormones in pharmaceutical effluents include ibuprofen (IBU), carbamazepine, diclofenac, 4-aminophenol, naproxen (NAP), gemfibrozil (GEM), fluoxetine (FLU), triclocarban (TCC), diphenhydramine (DPA), sulfamethazine (SMI), 17b-trihydroxyestra-1, sex hormones, ethinylestradiol (EE), 3, 5, 14(10)-tetrene, estriol (E3), etc., (Suresh and Abraham, 2018).

Advanced physiochemical methods, such as oxidation processes (oxidation, ozonation, perozonation, photocatalysis), ultrasonic irradiation and fenton reactions increase the amount of biodegradability of pharma-industrial wastage. These methods were successful in removing the majority of suspended materials and colloidal organic substances, but they were unsuccessful in removing refractory compounds. Only microbial processes can mineralize these stable recalcitrant organic compounds, which, when combined with other promising advanced physiochemical methods, provide better solutions for pharmaceutical waste treatment and aid in the sustainable use of sources.

Microbes have traditionally been used in waste treatment processes, such as sequence batch reactors, anaerobic sludge reactors, anaerobic film reactors, activated sludge, anaerobic filters, membrane and batch reactors. The properties of wastewater are important in biological wastewater treatment. Some of the variables that need to be improved to increase the effectiveness of mineralization in biodegradation are organic load, temperature, the presence of recalcitrant chemicals and toxins, the form of the organisms, dissolved oxygen, retention time, and pH. With these advanced technologies, humans can produce inexpensive and waste-free goods, and this has good potential for the future. Table 7.2 presents different microbes able to degrade pharmaceutical wastewater.

7.4.3 The Textile Industry (TI)

The TI is one of the most water-intensive industries as it needs large quantities of water especially for dying and rinsing operations (Ntuli et al., 2009). As a result, massive amounts of wastewater with varying compositions are produced, which comprise a range of byproducts, auxiliary chemicals, and residual dyes, causing the water to turn brightly colored, alkaline, and therefore raise the BOD, the amount of suspended particles, and the temperature, all of which are causing public concern these days (U.S. Environmental Protection Agency, 1997; Mohan et al., 2007). Chemical, physical, and biological approaches can be used to treat toxic elements from the TI. Microbial wastewater treatment employs a wide range of bacteria, algae, yeasts, and fungi, etc. that can eat the effluent components and grow in number (biomass), making them capable of accumulating and degrading various contaminants; and this is the most cost-effective approach available (Table 7.3). This results

TABLE 7.2
Different Microbes Able to Degrade Pharmaceutical Wastewater

Strain	Process	References
Fungi		
Aspergillus sp., *Penicillium* sp., *Bjerkandera adusta,* MUT 2295.	COD reduction	(Angayarkanni et al., 2003; Mohammad et al., 2006; Spina et al., 2012).
Bacteria		
Klebsiella sp. *Streptomonas* sp., *Aeromonas* sp., *Enterobacter* sp.	Phenol and complex organic compounds, Resorcinol	(Ghosh et al., 2004)
Comamonas, Arthrobacter, Rhodococcus, Ralstonia.		(Kavitha and Beebi, 2003)
Bacillus etthermoglucosidasius *A7,*	Phenol/cresol	(Duffner et al., 2000)
P. putida, MTCC 1194, P. fluorescence	Phenol and catechol	Kumar et al., 2005; Agarry and Solomon, 2008)
Rhodobactor spheroids	COD removal	(Madukasi et al., 2010)

TABLE 7.3
Microorganisms Involved in Textile Industry Contaminants Management

Microrganism	Type	References
Fungi	*Toxicant*	
Umbelopsis isabellina, Rhizopus, oryzae Aspergillus foetidus and *Penicillium geastrivous* etc.	Decolorization of dyes or adsorption of dyes.	(Yang et al., 2003).
Pleurotus ostreatus and *Irpex lacteus* more efficient than *Dactylospora haliotrepha* and *Aspergillus ustus*	Degradation of dyes	(Marimuthu et al., 2013).
Fusarium sp., *Phanerocheate chrysosporium, Ralstonia* sp., *Corious versicolor.*	Phenol biodegradation	(Nair et al., 2008)
Bacteria		
E. coli, Arthrobacter sp., *Brevebacterium casei, Bacillus* sp., *Pseudomonas aeruginosa, Acinetobacter* sp.	Chromium degradation	(Srivastava et al., 2007; Das and Mishra, 2010).
M. luteus, L. denitrificans and *N. atlantica*	Azo dyes from textile industries	(Hassan et al., 2013)
Bacillus spp. ETL-2012, *B. pumilus* HKG212, *P. aeruginosa*	Textile dye (Remazol Black B), Sulfonated di-azo dye, Reactive red HE8B, RNB dye	(Maulin et al., 2013; Yogesh and Akshaya, 2016; Das et al., 2015)
B.cereus, Exiguobacterium aurantiacums, Acinetobacter baumanii and *E. indicum*	Azo dye effluents	(Kumar et al., 2016)
Klebsiella oxytoca, B. firmus, Staphylococcus aureus and *B. macerans*	Textile effluents and vat dyes	(Adebajo et al., 2017)
Microorganisms involved in remediation of dyes	*Compound*	
B. subtilis strain NAP1, NAP2, NAP4	Oil-based paints	(Phulpoto et al., 2016)
Myrothecium roridum IM 6482	Industrial dyes	(Jasinska et al., 2012; Jasinska et al., 2013; Jasinska et al., 2015)
Pycnoporus sanguineous, Phanerochaete chrysosporium and *Trametes trogii*	Industrial dyes	(Yan et al., 2014)
Penicillium ochrochloron	Industrial dyes	(Shedbalkar and Jadhav, 2011)

in the breakdown of complex, harmful molecules into simple, less toxic ones, lowering BOD and COD. The most popular procedures for the wastewater treatment in the TIs are decolorization, degradation, and adsorption by microbial biomass (Fu and Viraraghavan, 2001). The activated sludge technique is extensively used for the biological treatment. In this procedure, wastewater is homogenized in a tank after chilling and screening to produce an identical effluent with various characteristics. After that, acid (H_2SO_4 or HC_l) is used to generate an acidic medium (pH 6.5–8.5) for bacterial action. The wastewater is then directed to a biological oxidation tank. Here microorganisms break down the effluent into simple compounds. For optimal microbial activity, proper dissolved oxygen (DO), temperature, pH and food must be maintained.

7.4.4 The Petrochemical Industries (PCI)

The PCI have now permeated every aspect of human existence, from food and nutrition security to infrastructure and facilities; healthcare to social and physical infrastructure; home goods to construction; automobiles to electronics and communication; agricultural sectors to horticulture, and so on. PC products have heavily polluted all facets of life and diminished natural environments, especially in the PC production zone, resulting in soil, water, and air pollution. The PCI contaminants contain a broad range of less biodegradable organic chemicals at high concentrations in the environment, which include aromatic and aliphatic hydrocarbon (HC) phenols, formaldehyde, cyanide, and octanols (Yeruva et al., 2015). Petroleum and synthetic rubber, detergent, natural gas, plastic and the manufacturing industries create the majority of them. Physicochemical treatment technologies now in use, such as photocatalytic oxidation, flocculation-coagulation, and membrane extraction, have some limitations in terms of operational and investment costs, as well as waste creation (Madaeni and Eslamifard, 2010; Verma et al., 2010). Bioremediation, or the use of microbes in wastewater treatment, is the only way to accomplish environmental friendliness, efficacy, safety, simplicity and long-term goals of cost (Table 7.4). They have a reputation for degrading and cleaning polluted places (Thangaraj et al., 2007) and are known as chemoorganotrophs because they use petrochemicals as carbon sources and electron donors to produce ATP. Some microorganisms make

TABLE 7.4
Different Microorganisms Suitable for PC's Wastewater Management

Microbial species	PC's contaminants	References
Sphingobacterium	PAHs, Phenolic compounds and oils (catalytic cracking, hydro cracking, distillation unit, spent caustic)	(Khongkhaem et al., 2011; Chaudhary and Kim, 2016)
Pseudomonas	Phenolic compounds	(Qu et al., 2015;
Rudaea, *Brevundimonas*, *Comamonas*	PAHs, COD (lube oil, ballast water)	Gomez-Acata et al., 2016)
Flavobacterium sp., *Alcaligenes odorans*, *Caulobacter* sp. *Burkholderia cepacia* and *Acinetobacter calcoaceticus*	n-alkanes fraction of hydrocarbons	(Cao and Wang, 2004; Mohanty and Mukherji, 2008; Thangaraj et al., 2007)
Ralstonia and *Janibacter terrae*,	Dibenzofuran, Dibenzothiophene and Phenol	(Margesin et al., 2005)
Rhodococcus rhodochrous and *Sphingomonas paucimobilis*.	Fluorene biphenyls, benzyl sulfide aniline, 4-chlorophenol atrazine, salicylate.	(Thangaraj et al., 2007)
Flavobacterium, *Acidovorax* and *Cytophaga*	Alkanes, aromatic, polycyclic hydrocarbons	(Shokrollahzadeh et al., 2008)
Bacillus sphaericus and *Pseudomonas putida*	Ammonia and Odor H_2S	(Moller et al., 1996)

use of the hydrophobic properties of hydrocarbons to break them down by producing biosurfactants (Das and Mukherjee, 2007). Microbes, particularly bacteria, have developed specific methods for the complete breakdown of water-insoluble hydrocarbons, such as adhesion and the formation of extracellular polysaccharides as emulsifying agents to extend contact time (Hisatsuka et al., 1971). Petrochemicals can be degraded both anaerobically and aerobically. A standard aerobic technique underlying enzyme oxygenases, which contribute molecular oxygen to the lowered material, and alcohols, is being developed originally from aliphatic HCs that are then worked on by dehydrogenases and are oxidized to carboxylic acid, which is then oxidized. Monooxygenase and dioxygenase enzymes hydroxylate the rings of aromatic and polycyclic aromatic HCs. Biodegradation of oil can be accomplished in anaerobic environments using sulfate-reducing bacteria (Holba et al., 1996). According to studies, methano-genic and sulfo-genic bacteria are using the oxidative approach to break down a number of different HCs, such as benzene, toluene, and alkanes in nitric-free and strict anaerobic environments (Ward and Brock, 1978; Grabic-Galic and Vogel, 1987).

7.4.5 The Explosives Industry

Manufacturing explosives used by military defense systems has increased in the modern era of military strength. Octahydro-1,3,5,7-tetranitro-1,3,5,7-tetrazocine, also called HMX, NC (nitrocellulose), hexahydro1,3,5-trinitro-1,3,5-triazine, popularly called RDX, and 2,4,6-trinitrotoluene (TNT) are examples of common energetic explosives. Water and water sources and soil from the nearby area of these operations become barren due to the creation of pollutant substances during manufacturing, packing, transportation, and burning. Explosives and their by-products are resistant to degradation in the environment, are hazardous and mutagenic in nature, and are kept on the USEPA's priority list. TNT, a common World War II explosive, has been demonstrated to cause muscle soreness, aplastic anemia, cyanosis, cataracts, and other side effects (Yang et al., 2008; Kalderis et al., 2008). Microbes like fungi or bacteria work as a key component in the bioremediation of various environmental matrices contaminated with explosives (Table 7.5).

Microbial profiling of explosives-contaminated ground water at a specific location revealed a total of 1605 operational taxonomic units (OTU), of which 96 were bacterial taxa, with *Rhodococcus* (prevalent), *Actinobacteria* and *Proteobacteria* dominating this contaminated groundwater (Wang et al., 2016). Soil-native bacteria have the capability to degrade explosives such as TNT and Pentaerythritol Tetranitrate (PETN) at a slow rate and efficiency (Amin et al., 2017). According to Mercimek et al. (2015), microbial degradation of TNT into its different metabolites of 2,4-dinitrotoluene (2,4-DNT) and 4-aminodinitrotoluene (4-ADNT) by *Pseudomonas aeruginosa* is dependent on concentration, time and growth of cell biomass in the medium. Abiotic factors have a negligible impact on explosives decomposition but to some extent contribute to TNT photolysis and RDX volatilization (Sisco et al., 2015). Neither the salinity of the water or UV radiation have any effect on TNT and RDX degradation in saline water, while UV radiation did have a minor effect on TNT degradation. Soil amendments, such as biosurfactants (mono rhamnolipid) enhance cell membrane permeability and the bioavailability of explosives, resulting in increased biodegradation. Explosives decomposition is also reliant on the type of explosives, the environment (presence or absence of oxygen) and the presence of metabolizing enzymes in the bacterial cell, such as Xen A and Xen B reductases (flavoprotein oxidoreductase family, involved in the aerobic and anaerobic degradation). Under low oxygen conditions, the transition of RDX (no direct reduction of RDX nitro groups) through methylenedintramine to formaldehyde was seen in *Pseudomonas* species (Fuller et al., 2009).

7.4.6 The Distillery Industry (DI)

By using agro produce, such as sugar beet molasses, wheat, cassava, barley, corn, sugar cane juice, rice and molasses, the DI generates waste in the form of spent wash (Wilkie et al., 2000;

TABLE 7.5
Use of Different Beneficial Microbes for Explosives Degradation

Microorganism	Explosive	References
Bacteria		
Clostridium Sp., *Pseudomonas* sp. MS-4	CL-20	(Panikov et al., 2007) (Crocker et al., 2006)
Acinetobacter noscomialis, Sulfate Reducing and Methanogenic Consortia	TNT	(Sangwan et al., 2015) (Boopathy et al., 1998)
Clostridium Sp.	RDX, HMX	(Crocker et al., 2006)
Bacillus sphaericus JS905	4-nitrophenol	(Kadiyala and Spain, 1998)
Rhodococcus sp.	2,4-dinitrophenol	(Takeo et al., 2003)
Desulfovibrio sp. Strain B, *Pseudomonas* sp. JX165	TNT, DNT, Nitrobenzene	(Boopathy et al., 1993)
Acenitobacter noscomialis	TNT (Aqueous phase) into diaminonitrotoluenes, hydroxylaminodinitrotoluene, and aminodinitrotoluene.	(Sangwan et al., 2015)
Rhodoccus sp. (Aerobic conditions)	RDX→Ring cleavage→NH_4+CO_2 (Denitration)	(Fournier et al., 2002)
Gordonia and *Williamsia* sp. (Aerobic conditions)	RDX→nitrite and formaldehyde	(Thompson et al., 2005)
Fungi		
Trichoderma, *Alternaria*, *Aspergillus* etc	TNT	(Bennett et al., 1995)
Phanerochaete chrysosporium in aqueous phase	CL-20 (Hexanitrohexaazaisowurtzitane).	(Karakaya et al., 2009)

Kawa-Rygielska et al., 2007; Chowdhary et al., 2018; 2020). At various phases of the manufacturing process, 10–15 liters of wastewater are created for every liter of alcohol produced. If it is not treated appropriately, spent wash creates a serious environmental risk. It is has a low pH, having a high temperature, dark color and with a significant proportion of inorganic and organic matter, depending on the raw material being used in the industry. The dark color of spent wash inhibits light penetration into the water bodies, negatively impacting flora and animals. Once dumped in soil, it reduces soil fertility and alkalinity while raising the levels of heavy metals, making it dangerous to many microorganisms. Phenolic chemicals, proteins, caramels from overheated sugars, and furfurals formed by acid hydrolysis are among the additional compounds that contribute to the color of the spent wash (Kort, 1979). Distillery effluent should therefore be treated before being released into the environment. The current physicochemical treatments are based on the chemical composition of spent wash and have significant limitations in terms of cost, chemicals, sludge production, and byproduct formation (Boopathy and Senthilkumar, 2014).

Researchers all over the world are currently interested in microbial wastewater treatment. It is both cost-effective and environmentally beneficial. Melanoidin breakdown is catalyzed by microbial enzymes, such as manganese peroxidases, sugar oxidases laccases and lignin peroxidases (Couto et al., 2005; Freitas et al., 2009). The degradation mechanism involves the adsorption of compound onto cells, then is incorporated inside and decomposed by intracellular enzymes involving active oxygen and sugar molecules in the reaction mixture. The productive oxygen species formed by enzymatic oxidation in the existence of sugar, such as sucrose glucose, sorbose, maltose, lactose and xylose, is mainly hydrogen peroxide (Watanabe et al., 1982; Pant and Adholeya, 2007). Melanoidin is decolored and removed through enzymatic degradation by microbial species, which then use the

TABLE 7.6
Microorganisms Used to Decolorize Spent Wash and to Degrade Melanoidin

Fungal species	References
Coriolus hirsutus; *C. versicolor*; *Geotrichum candidum*; *P. chrysosporium*; *Flavodon flavus*; *Aspergillus niger*; *Rhizoctonia* sp. D-90; *Fusarium* sp. *Trametes versicolor*	(Dehorter and Blondeau 1993; Miyata et al., 1998; Lee et al., 2000; Thakkar et al., 2006)
Bacterial species	
Lactobacillus hilgardii WNS; *Bacillus* sp; *Alicaligens* sp.; *Pseudomonas fluorescenes*; *Acetobacter acetii*	(Ohmomo et al., 1988; Kambe et al., 1999; Dahiya et al., 2001; Sirianuntapiboon et al., 2004. Bharagava et al., 2009)
Algae/Cyanobacteria	
Oscillatoria boryna; *Lyngbya*; *Synechocystis*; *Chlorella* sp; *Nostoc muscorum*	(Kalavathi et al., 2001; Valderrama et al., 2002)

pigment as a nitrogen and carbon source, flocculation by extracellular materials, and adsorption on alive and dead biomass (Chandra et al., 2009; Pant and Adholeya, 2010).

Melanoidins from distillery effluent have been reported to be decolorized or degraded by the most commonly investigated fungi, bacteria, algae, and cyanobacteria (Table 7.6). In comparison to bacteria and fungus, microalgae are particularly useful in bioremediation because they use contaminants as feeder and can grow in harsh environments. Microalgae can also be harvested for the value products they produce, such as ethanol, methane, livestock feed, and biofertilizer. White rot fungus is a commonly used microorganism with an extracellular, nonspecific complex enzyme system that can break down melanoidins, polyaromatic and lignin chemicals even when nutrients are limited (Benito et al., 1997).

7.4.7 Heavy Metals (HM) Remediation

HMs, which include metalloids or metallic elements, are creating serious concern as persistent, nondegradable bioaccumulation in nature (Pushkar et al., 2015). Major HMs from different sources include Hg, Cr, Cd, Pb, Ar, Ni, Cu, and Zn etc. (Shiomi and Naofumi, 2015) (Figure 7.7). The majority of HM contaminants that accumulate over time come from natural biological processes and human activities, such as forest fires, weathering of minerals, volcanic activities, erosion and particles released from vegetation. The discharge of untreated effluents from the tanning, plastics, automobile, mining and electroplating industries into ecosystems is creating health hazards to the public. Thus, there is increasing need for novel, effective, microbial strategies over conventional methods to treat heavy metals.

Microbes can absorb HMs at their cellular binding sites, alter their oxidation state by immobilizing them and form a complex with extracellular polymers, then undergo mineralization to form metabolic intermediates (used as primary substrates for cell growth), allowing the HMs to be easily removed (Dixit et al., 2015) (Table 7.7).

7.4.8 Microbial Management Strategies for Pesticides

Any substance that restricts the spread of obnoxious attackers such as pathogens, insects, rodents, weeds or any damaging lifeforms is known as a pesticide, which include insecticides, nematicides, acaricides, rodenticides, fungicides, bactericide, herbicides, disinfectants, growth regulators, etc. Pesticide use soared after World War II by the introduction of pesticides like DDT (Dichloro diphenyl trichloroethane), benzene hexachloride (BHC), organochlorins (OCs) and others. Even

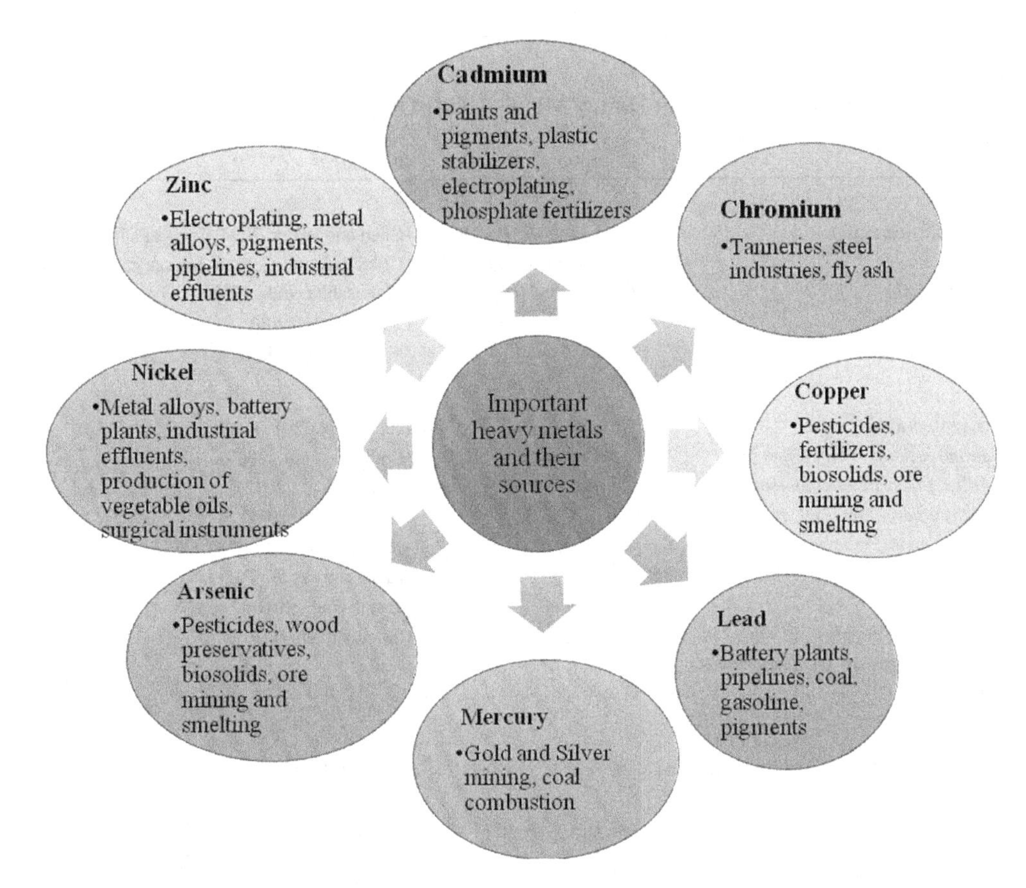

FIGURE 7.7 Major heavy metals and their sources

Source: B. Vajramma.

though they are effective against numerous pests, prolonged over-use of pesticides leads to its accumulation in environmental matrices, i.e. soil and water bodies (rivers, canals, pods, groundwater) via leaching and run-off from applied surfaces, and ultimately ends in the development of resistant organisms (Damalas, 2009). Additionally, over-use of pesticides also causes biomagnification in higher organisms through the food chain (Bernardes et al., 2015). Figure 7.8 classifies pesticides into their different sectors.

Among all the inhabiting soil microbes, some microbiotas have the potential to break down and bio-transform some residue of pesticides by taking the necessary rate of carbon and other growth-promoting substances from those pesticides. This microbial ability acts as key to treating the polluted areas with pesticides (Qiu et al., 2007; Aktar et al., 2009). Usually, microbiota thrive in contaminated sites by their pesticide-degrading enzymes, other parameters (ecological, physiological, biochemical, and molecular) also serve as major factors in degradation and the biotransformation of pesticides (Iranzo et al., 2001; Karigar and Rao, 2011; Ortiz-Hernandez et al., 2013). The first reports of microbial potential to degrade organophosphates was in 1973, when Flavobacterium species were reported (Singh and Walker, 2006). Moreover, numerous pesticides-degrading microorganisms consisting of actinomycetes, fungi, bacteria and algae were reported (Parte et al., 2017; Kumar et al., 2018). Organophosphates (OPs) have broad-spectrum use in controlling insects and nematodes, which acts as strong nerve poisons in the target organism with cholinesterase inhibitor activity, which disturbs the nerve impulse conduction throughout the synapse (Table 7.8). Rapid convulsions (spasms) in the voluntary muscles finally lead to paralysis and death

TABLE 7.7
Different Microorganisms Used in Heavy Metals (HM) Degradation

Microorganisms	Compound	References
Yeast		
Saccharomyces cerevisiae	Heavy metals, Pb, Hg, Cr, Ni, Cu, and Ni	(Chen and Wang, 2007; Talos et al., 2009; Machado et al., 2010; Infante et al., 2014)
Candida tropicalis	Cd, Cr, Cu, Ni, Zn	(Mattuschka et al., 1993)
C. utilis	Cd	(Kujan et al., 2006)
Pichia guilliermondii	Cu	(Mattuschka et al., 1993)
Fungi		
Agaricus bisporus	Cd, Zn	(Nagy et al., 2017)
Ganoderma lucidum, *Penicillium* spp.	Ar	(Loukidou et al., 2003)
Aspergillus, *Mucor*, *Penicillium* and *Rhizopus*	Cd, Cu, Fe	(Fulekar et al., 2012)
Aspergillus niger, *A. foetidus*, and *Penicillium simplicissimum*	Mn, Co, Ni, V, W, Zn, Mo, Fe	(Anahid et al., 2011)
A. fumigatus	Pb	(Ramasamy et al., 2011)
A. versicolor	Cr, Ni, Cu	(Tastan et al., 2010)
A. versicolor, *A. fumigatus*, *Terichoderma* sp., *Paecilomyces* sp., *Cladosporium* sp., *Microsporum* sp.	Cd	(Soleimani et al., 2015)
Bacteria		
Bacillus safensis (JX126862) strain (PB-5 and RSA-4)	Cd	(Priyalakshmi et al., 2014)
Pseudomonas aeruginosa, *Aeromonas* sp.	Cu, U, Cr, Ni	(Sinha et al., 2011)
Aerococcus sp., *Rhodopseudomonas palustris*	Cr, Pb, Cd	(Sinha and Paul, 2014; Sinha and Biswas, 2014)
Cunninghamella elegans	Heavy metals	(Tigini et al., 2010)
P. fluorescens and P. aeruginosa	Fe 2+, Zn2+, Pb2+, Mn2+ and Cu2	(Paranthaman and Karthikeyan, 2015)
Lysinibacillus sphaericus CBAM5	Co, Cu, Cr and Pb	(Pena-Montenegro et al., 2015)
Microbacterium profundi strain Shh49T	Fe	(Wu et al., 2015)
Streptomyces longwoodensis	Pb	(Friss and Myers-Keit, 1986)
Important microorganisms used for oil degradation		
Fusarium sp.	Oil	(Hidayat and Tachibana, 2012)
Alcaligenes odorans, *Bacillus subtilis*, *Corynebacterium propinquum*, *Pseudomonas aeruginosa*	Oil	(Singh et al., 2013)
B. cereus	Diesel oil	(Maliji et al., 2013)
Aspergillus niger, *Candida glabrata*, *C. krusei* and *Saccharomyces cerevisiae*	Crude oil	(Burghal et al., 2016)
B. brevis, *P. aeruginosa* KH6, *B. licheniformis* and *B. sphaericus*	Crude oil	(El-Borai et al., 2016)
P. aeruginosa, *P. putida*, *Arthobacter* sp., and *Bacillus* sp	Diesel oil	(Sukumar and Nirmala, 2016)
P. cepacia, *B. cereus*, *B. coagulans*, *Citrobacter koseri* and *Serratia ficaria*	Diesel oil, crude oil	(Kehinde and Isaac, 2016)

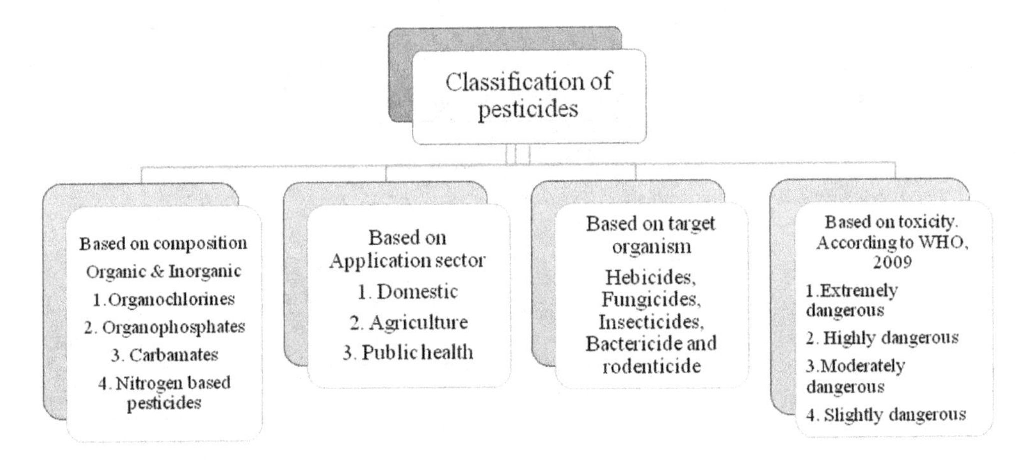

FIGURE 7.8 Classification of pesticides.

Source: B. Vajramma.

TABLE 7.8
Microorganisms Involved in Organophosphorus Chemicals Degradation

Microorganisms	References
Bacteria	
Flavobacterium sp.	(Singh and Walker, 2006)
Pseudomonas (most efficient)	(Abo-Amer, 2012)
Brevibacterium sp.	(Jariyal et al., 2018; Sidhu et al., 2019)
Microbacterium esteraromaticum	(Ou et al., 1994; Megharaj et al., 2003; Caceres et al., 2009; Cabrera et al., 2010)
Rhodococcus sp.	(Singh and Walker, 2006),
Bacillus sp.	(Madhuri and Rangaswamy, 2009; Tang and You, 2012; Aziz et al., 2014)
Serratia marcescens	(Abo-Amer, 2011; Cycon et al., 2013)
Sphingomonas paucimobilis	(Karpouzas et al., 2005)
Fungi	
white-rot fungi, *Trichoderma* sp. *Aspergillus* sp. (prominent pesticide degraders)	(Huang et al., 2018)
Phanerochaete chrysosporium	(Eissa et al., 2006)
Cladosporium cladosporioides	(Gao et al., 2012)
A. fumigatus	(Pandey et al., 2014)
A. niger	(Liu et al., 2001a; Pandey et al., 2014)
Penicillium raistrickii, and *A. sydowii*	(Alvarenga et al., 2014)
Marine Algae (Chlorellaceae)	
Chlorella, *Scenedesmus*, and *Stichococcus*	(Megharaj et al., 1987; Caceres et al., 2009)
Cynobacteria	
Anabaena	(Caceres et al., 2008; Ibrahim and Essa, 2010)
Oscillatoria	(Salman and Abdul-Adel, 2015)
Nostoc	(Megharaj et al., 1987; Caceres et al., 2008; Ibrahim et al., 2014)

(*continued*)

TABLE 7.8 (Continued)
Microorganisms Involved in Organophosphorus Chemicals Degradation

Microorganisms	References	
Potential of microorganisms for different pesticides degradation	*Compound*	
Bacteria		
Pseudomonas putida, *Acinetobacter* sp.	Ridomyl MZ 68 MG, Fitoraz	(Monica et al., 2016; Hussaini et al., 2013)
Arthrobacter sp.	Malation	
Bacillus sp, *Staphylococcus* sp.	Endosulfan	(Mohamed et al., 2011)
Enterobacter sp.	Chlorpyriphos	(Niti et al., 2013)
Pseudomonas sp., *Photobacterium* sp.	Methyl parathion	(Ravi et al., 2015)
Fungi		
Coriolus versicolor, *Stereum hirsutum*, *Auricularia auricula*, *Avatha discolor*, *Hypholoma fasciculare*, *Agrocybe semiorbicularis*, *Pleurotus ostreatus*, *Flammulina velupites*, and *Dichomitus squalens*	Triazine, phenylurea, dicarboximid, chlorinated organophosphorus compounds	(Hussaini et al., 2013)
White-rot fungi	Terbuthylazine, lindane, metalaxyl, DDT, gammahexachlorocyclohexane (g-HCH), dieldrin, diuron, chlordane mirex, aldrin, heptachlor atrazine, etc.	(Watanabe et al., 2001)

in rare cases (Yadav and Devi, 2017). Cadusafos comes under the group OPs, is a well-known nematicide, extensively used for crop protection and its persistent usage is lethal to mammals as it contaminates surface and groundwater.

7.4.9 Microbial Remediation of Hydrocarbons (HCs)

Of all the environmental contaminants of HCs from petroleum products (gasoline, diesel), lubricants, surfactants, and pesticides are a serious problem in the soil and water ecosystem. Among them, petroleum HCs have higher toxicity and are of major concern in the world as accidents (spillage) when exploring for oil, its extraction, transportation, and storage leaks into the marine and soil environments, finally contaminate groundwater via seepage and the leaching process.

Generally, most of the contaminants come under cyclic HCs, which are either aromatic or alicyclic in nature. Among them, polycyclic aromatic hydrocarbons (PAHCs) occupy the majority of HCs, are more toxic and persist in soil over a long period of time (Pawlak et al., 2008). Some of the sources of environmental release of PAHs include incomplete combustion of materials, for example, oil, coal, wood, and gasoline in marine ecosystems accidentally. Some 16 carcinogenic PAHs were identified by the US Environmental Protection Agency (USEPA) which are hazardous to human health, and to minimize their usage and to protect the environment, USEPA passed legislative restrictive measures on their release (Liu et al., 2001b). There are different biological, chemical and physical strategies that can be used to remediate environments contaminated by HCs (Table 7. 9), of which, bio-strategies using microbes are important, efficient degradation tools (Logeshwaran et al., 2018). Of all the microorganisms involved in bioremediation, bacteria are the most effective in the removal of low-molecular weight PAHs from contaminated environments. However, fungi are also more effective in PAH degradation than bacteria in a few instances; (1) in high molecular weight

TABLE 7.9
Different Microorganisms Having Hydrocarbon Degradation Potential

Microorganism	Hydrocarbon	References
Bacteria		
P. alcaligenes P. mendocina and *P. putida P. veronii*, *Achromobacter*, *Flavobacterium*, *Acinetobacter*	Petrol and diesel polycyclic aromatic hydrocarbons toluene	(Safi et al., 2015; Sani et al., 2015)
P. putida	Monocyclic aromatic hydrocarbons, e.g. benzene and xylene.	(Sarang et al., 2013)
Pseudomonas sp.	High mol.wt hydrocarbons like pristane (C19) and benzopyrene (C20)	(GuermoucheM'rassi et al., 2015).
Pseudomonas sp. LGM2	Degrade balkanes (linear and branched chain) as well as PAHs (low and high molecular weight)	
Alcaligenes odorans, *Bacillus subtilis*, *Corynebacterium propinquum*, *P. aeruginosa*	Phenol	(Singh et al., 2013)
Achromobacter, *Rhodococcus*, *Oerskovia paurometabola*, *Pantoea*, *Sejongia*, *Microbacterium*, and *Arthrobacter* sp.	PAHs (naphthalene, Phenanthrene(PHE) and pyrene)	(Haleyur et al., 2018).
Phingopyxis, *Rhodobacter*, and *Hyphomonas* sp.	Naphthalene and phenanthrene (PHE),	(Mahjoubi et al., 2018).
Acinetobacter sp., *Pseudomonas* sp., *Ralstonia* sp. and *Microbacterium* sp.	Aromatic hydrocarbons	(Simarro et al., 2013)
Microbial consortia		
Mycobacterium strains	Pyrene degradation	
Novosphingobium and *Ochrobactrum*	Pyrene intermediates degradation	(Wanapaisan et al., 2018).
Bacillus sp.	Improved the pyrene bioavailability with its biosurfactant producing ability	(Mangrove sediment-derived consortia from Thailand)
Consortia of bacterial species like *B. pumilus* KS2 + *B. cereus* R2 (Isolated from crude oil extraction fields in Assam.)	Hydrocarbons	(Patowary et al., 2016).
Stenotrophomonas sp. and *Pseudomonas* sp. (Endophytic bacterial strains)	Within 7 days it degrades more PHE (> 90%)	(Zhu et al., 2016).
Fungi		
Penicillium chrysogenum	Monocyclic aromatic hydro carbons (MAHs), benzene, xylene, toluene and phenol compounds	(Pedro et al., 2014; Abdulsalam et al., 2013)
Gloeophyllum trabeum, *Tyromyces palustris*, *Trametes versicolor*	Hydrocarbons (HCs)	(Karigar and Rao, 2011)
Phanerochaete chrysosporium	Biphenyl and triphenylmethane	(Erika et al., 2013)
A. niger, *A. fumigatus*, *F. solani* and *P. funiculosum*	Hydrocarbon	(Al-Jawhari, 2014)
Coprinellus radians	PAHs, methylnaphthalenes, and dibenzofurans	(Aranda et al., 2010)
Gleophyllum striatum	Anthracene, pyrene, Dibenzothiophene	(Yadav et al., 2011)

(*continued*)

TABLE 7.9 (Continued)
Different Microorganisms Having Hydrocarbon Degradation Potential

Microorganism	Hydrocarbon	References
white-rot fungi, *Bjerkandera adusta*, *Trametes versicolor*, *P. chrysosporium*, and *Pleurotus* sp.,	Hydrocarbon	(dos Santos Bazanella et al., 2013).
Pestalotiopsis palmarum	Carbon utilization from crude oil	(Naranjo Briceno et al., 2019).
Yeast		
Candida viswanathii	PHE, benzopyrene	(Hesham et al., 2012)
Green algae *Cyanobacteria*, *Bacillus licheniformis* and diatoms	Naphtalene	(Sivkumar et al., 2012; Lin et al., 2010)

PAHs' degradation, whereas bacteria have the capability to degrade smaller molecular weight PAHs; (2) fungi can degrade HCs very well in waterless and anoxygenic conditions, both these conditions are favorable to PAHs' accumulation in the environment; and (3) the fungi are better at penetrating the soil and reaching xenobiotics as they have an easily penetrable mycelium. Many fungi have been studied extensively for their high degradative ability in dealing with particular PAHs, due to their diversely adopted terrestrial habitats, highly evolved mechanisms, their enzymes, such as catalases, laccases, peroxidases and extracellular enzymes involved in mineralization, to degrade lignin, which produces Co_2 (Singh, 2006; Norton, 2012). Among all fungal genera, white-rot fungi with high ligninolytic activity are considered the main PAHs decomposers (Deshmukh et al., 2016). Some of the fungal enzymes, for example, cytochrome p450 monooxygenase and ligninolytic enzymes, have a high capacity to degrade PAHs (Cresnar and Petric, 2011). Damare et al. (2012) reported that a few groups of fungi which were present in marine environments are considered the best candidates in hydrocarbon degradation as they produce novel enzymes, secondary metabolites, polyunsaturated fatty acids, polysaccharides (PUFA) and biosurfactants.

7.4.10 Plastic Biodegradation

Plastics are polymers with a nondegradable nature that results in accumulation in the environment, however, a few microorganisms (fungi, bacteria and yeasts) have opened the era of green remediation with their ability to degrade or assimilate polymers (Kale et al., 2015; Gajendiran et al., 2016; Yoshida et al., 2016; Skariyachan et al., 2018) (Table 7.10). According to Sen and Raut (2015), the biodegradation of polymers can be divided into four steps (Figure 7.9):

1. microbial adherence
2. fragmentation
3. assimilation
4. mineralization.

7.4.10.1 Microbial Adherence

This is the first step in microbial degradation by forming highly synchronized, specific mechanisms of the interface between the substrate and the microbe, resulting in colonization and biofilm formation on plastic surface (Mohan, 2011). The dearth of any type of carbon source in the surroundings of microbial cells other than the plastics allows them to use plastic as food material, thus developing a plastic bioremediation technology with engineered microorganisms toward the hydrophobic

TABLE 7.10
Microbial Enzymes Involved in Polymer Degradation

Hydrolases (attacks carboxylic groups)	Reference	Oxidoreductases (forms easily fragmentable alcohol or peroxyl groups by adding O_2 atoms)	References
Esterases	(Lucas et al., 2008; Kawai, 2010)	Monooxygenases	(Mishra et al., 2001)
Lipases		Dioxygenases	
Cutinase		Laccases (Ca atoms)	

Source: (Muller et al., 2005; 2006).

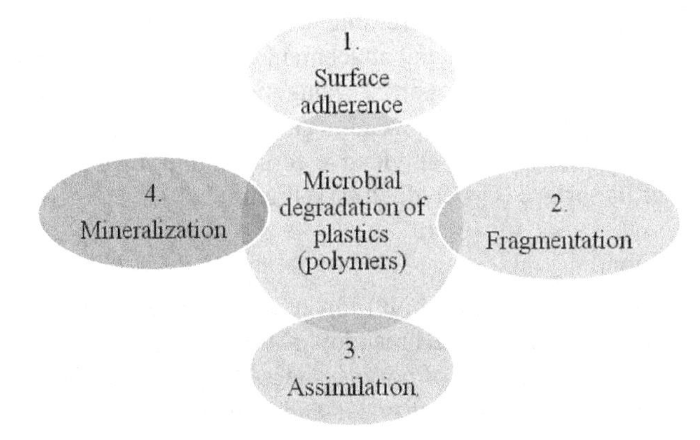

FIGURE 7.9 Process of microbial degradation of plastics.

Source: B. Vajramma.

plastic (carbon) assimilation. Many abiotic and biotic factors influence the degradation process. The surrounding abiotic conditions, including topography, surface roughness, salinity, temperature, hydrophobicity, pH and presence of oxygen, etc., change the polymer morphologically and help in microbial colonization (Shah et al., 2008; Nauendorf et al., 2016; Pathak and Navneet, 2017). The biotic conditions may be exogenous or endogenous in nature. Endogenous factors depend on nutrient sources available to the microbe, which helps in adherence (initial physical interaction of bacteria toward substrate) and biofilm formation (Alshehrei, 2017). Biofilm formation involves three steps (Garrett et al., 2008).

Surface colonization by microbes, alteration in physicochemical properties of surface, biofilm development, and maturation of functional colonies depend on specified events at the individual cell/population level (Dang and Lovell, 2016). The firmness of microbial adherence, and its colonization depend on the following factors and any variation may affect the rate of degradation:

1. Microbial abilities, polymer type and properties, pre-treatments, time, and physical properties of connection between both the surface of the polymer and the related microorganisms (Christensen et al., 1995).
2. Thermal properties, as well as surface properties, such as roughness, charge, and energy, have a essential influence on microbial cell attachment to the surface, which provides a clue to explain cellular activities in the bio-degradation operation (Tokiwa et al., 2009).

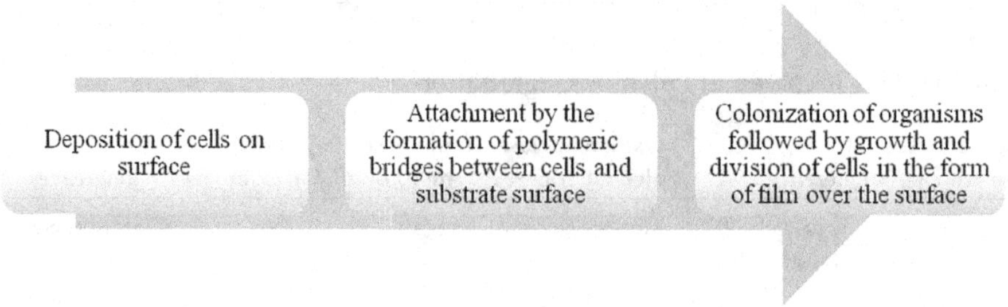

FIGURE 7.10 Steps involved in biofilm formation.

Source: B. Vajramma.

Conditioning or optimization is crucial for increasing the degradation rate. The plasticizers (thermal or chemical-based) affect the colonization and attachment capability of the microbial cells (Marini et al., 2007). The complete conversion or partial change in the polymer chemical structure helps in biodegradation by reducing its molecular weight (Gu, 2003). Bacteria like *Rhodococcus ruber* adhere and colonize on the surface of polyethylene to form biofilm, then gradually deteriorate and reduce hydrophobicity of its surface (Orr et al., 2004) (Figure 7.10).

7.4.10.2 Fragmentation

Several surrounding abiotic (chemical, dynamic) forces and pretreatments increase the hydrophilicity of polymers, which favors the microbial breakdown of the large polymeric structure into smaller fragments (da Luz et al., 2014). Initiation of polymer degradation starts with terminal branches (at the edges) by extracellular depolymerizing enzymes and formed oligomers ingested intracellularly (Mueller, 2006).

The oxidation of polymer depends on the exposure time and the kind of additives used in the film (Singh and Sharma, 2008; Tokiwa et al., 2009). Metabolizing of oligomers is influenced by the microbial communities. Ex. Actinomycetes have enormous capabilities in polymerization-free polyester, but are unable to metabolize its by-products (Mor and Sivan, 2008; Sivan, 2011).

7.4.10.3 Assimilation

Low molecular weight particles formed on the surface of the plastic by separation with external microbial work pass via the cell membranes and gain entry to the metabolic pathway during the assimilation activity (Sen and Raut, 2015; Alshehrei, 2017). Polymer biodegradation (PD) is basically a transfer of electrons (Kyrikou and Briassoulis, 2007). The released electrons from the oxidation process absorb energy and enter by the respiratory path through different intermediated compounds to the final electron acceptors, i.e., O_2 (aerobic), No_3^- and So_4^- (anaerobic). Of both aerobic and anaerobic biodegradation, aerobic process is more efficient (Gottschalk, 2012).

Some enzymes from microbes are identified to hydrolyze different xenobiotics, i.e. laccase, lipases, cutinases, and esterases with substrates of ester, lipids, and phenolic compounds (Kawai, 2010). The majority of scientific findings have revealed that exogenous depolymerization is a sequence of oxidative reactions preceded by β oxidation as required in the degradation of polyolefins and poly-ethers (Kawai, 2010; Wilkes and Aristilde, 2017). Polymer chains degradation depends on pro-oxidants which catalyze and form free radicals; these can react with O_2 which results in upcoming/next oxidation (Chiellini et al., 2006; Eubeler et al., 2010). When the microbe has polymer as the only source of carbon, its assimilation for the carbon movement could be approximated by the metabolites (Co_2) and progression of microbial populations (Skariyachan et al., 2018). Polymer bio-degradation is a surface erosion process as extracellular enzymes entered into the polymer structure work just on its exterior part. Tian et al. (2017) studied polystyrene degradation by using FTIR analysis where carbon was tracked using ^{14}C-polystyrene and concluded that polystyrene-generated carbonyl groups, when incubated with *Penicillium variabile*, had decreased ozonation and also reduced molecular weight.

7.4.10.4 Mineralization

In mineralization, all the solid biomass is converted into gaseous (Co_2 in aerobic and CH_4 in anaerobic respiration), water, minerals, and residual biomass (Vert et al., 2012; Sen and Raut, 2015). The microbes consume energy and carbon arising from the substrate and are converted into biomass for progress and respiration whereas the carbon is emitted as Co_2 and H_2O into the environment. This simple information helps in the study of the contribution of microbes toward changing the atmosphere. Yang et al., (2015) studied mealworms (*Tenebrio molitor*) and reported that they were 95% reliable in mineralizing styrofoam to Co_2, in a biomass when they were fed with ^{13}C labeled styrofoam. Polystyrene was labeled with ^{14}C to check its mineralization after 16 weeks with *P. variabile* and Tian et al. (2017) studied those fungi that mineralized low molecular weight labeled polymers easily, with a faster rate of mineralization.

7.5 CONCLUSION

The current global society is extremely dependent on the various industries and their products have transformed millions of lives. But the main concern is the generation of waste materials and the dumping of untreated waste from these industries is ultimately deteriorating the environment in many underdeveloped countries, which are ultimately experiencing serious climatic change and loss of biodiversity. Microbial remediation strategies to treat waste generated by industries are the focus of research globally. Unlike traditional pollutants treatment strategies, microbial degradation is low cost, producing nontoxic by-products and acts as the best alternative in future. In nature, microbes have the potential to degrade waste coming from industrial effluents, sewage, and agrochemical run-offs, though at relatively slow pace, are less efficient and depend on the surrounding crucial environmental conditions (temperature, pH, oxygen and nutrients). Highly efficient microbial mixed consortia can remediate different contaminants of the polluted soil. It is necessary to understand the basic sciences, such as environmental microbiology and analytical geochemistry that drive the microbial degradation with biotechnology principles. Thus, it is necessary to incorporate improvements in the existing microorganisms involved in biodegradation. Advanced genome editing techniques, directed evolution and metabolic pathway engineering techniques could be applied to enhance the degradation capabilities of engineered microorganisms.

REFERENCES

Abatenh, E., Gizaw, B., Tsegaye, Z., and Wassie, M. 2017. The role of microorganisms in bioremediation: A review. *Open Journal of Environmental Biology*, 2 (1): 038–046.

Abdulsalam, S., Adefila, S.S., Bugaje, I.M., and Ibrahim, S. 2013. Bioremediation of soil contaminated with used motor oil in a closed system. *Bioremediation and Biodegradation*, 3: 100–172.

Abo-Amer, A. 2011. Biodegradation of diazinon by *Serratia marcescens* DI101 and its use in bioremediation of contaminated environment. *Journal of Microbiology and Biotechnology*, 21: 71–80.

Adebajo, S.O., Balogun, S.A., and Akintokun, A.K. 2017. Decolorization of vat dyes by bacterial isolates recovered from local textile mills in Southwest. *Microbiology Research Journal International*, 18: 1–8.

Agarry, S.E., and Solomon, B.O. 2008. Kinetics of batch microbial degradation of phenols by indigenous *Pseudomonas fluorescence*. *International Journal of Environmental Science and Technology*, 5 (2): 223–232.

Aktar, W., Sengupta, D., and Chowdhury, A. 2009. Impact of pesticides use in agriculture: Their benefits and hazards. *Interdisciplinary Toxicology*, 2: 1–12.

Al-Jawhari, I.F.H. 2014. Ability of some soil fungi in biodegradation of petroleum hydrocarbon. *Journal of Applied and Environmental Microbiology*, 2: 46–52.

Alshehrei, F. 2017. Biodegradation of synthetic and natural plastic by microorganisms. *Journal of Applied and Environmental Microbiology*, 5: 8–19.

Alvarenga, N., Birolli, W.G., Seleghim, M.H., and Porto, A.L. 2014. Biodegradation of methyl parathion by whole cells of marine-derived fungi *Aspergillus sydowii* and *Penicillium decaturense*. *Chemosphere*, 117: 47–52.

Amin, M.M., Khanahmad, H., Teimouri, F., Sadani, M., Karami, M.A., and Rahimmanesh, I. 2017. Improvement of biodegradability of explosives using anaerobic-intrinsic bioaugmentation approach. *Bulgarian Chemical Communications*, 49: 735–741.

Anahid, S., Yaghmaei, S., and Ghobadinejad, Z. 2011. Heavy metal tolerance of fungi. *Scientia Iranica*, 18 (3): 502–508.

Angayarkanni, J., Palaniswamy, M. and Swaminathan, K. 2003. Biotreatment of distillery effluent using *Aspergillus niveus*. *Bulletin of Environmental Contamination and Toxicology*, 70: 268–277.

Aranda, E., Ullrich, R., and Hofrichter, M. 2010. Conversion of polycyclic aromatic hydrocarbons, methyl naphthalenes and dibenzofuran by two fungal peroxygenases. *Biodegradation*, 21: 267–281.

Aziz, M.W., Sabit, H., and Tawakkol, W. 2014. Biodegradation of malathion by *Pseudomonas* spp. and *Bacillus* spp. isolated from polluted sites in Egypt. *American-Eurasian Journal of Agricultural and Environmental Sciences*, 14: 855–862.

Azubuike, C.C., Chikere, C.B., and Okpokwasili, G.C. 2016. Bioremediation techniques—classification based on site of application: Principles, advantages, limitations and prospects. *World Journal of Microbiology and Biotechnology*, 32: 180.

Barriault, D., Plante, M.M., and Sylvestre, M. 2002. Family shuffling of a targeted bphA region to engineer biphenyl dioxygenase. *Journal of Bacteriology*, 184 (14): 3794–3800.

Benito, G.G., Miranda, M.P., and Santos, D.R. 1997. Decolorization of wastewater from an alcoholic fermentation process with *Trametes versicolor*. *Bioresource Technology*, 61: 33–37.

Bennett, J.W., Hollirah, P., Waterhouse, A., and Horvath, K. 1995. Isolation of bacteria and fungi from TNT contaminated composts and preparation of ^{14C}-ring labeled TNT. *International Biodeterioration and Biodegradation*, 35 (4): 421–430.

Bernardes, M.F.F., Pazin, M., Pereira, L.C., and Dorta, D.J. 2015. Impact of pesticides on environmental and human health. In Andreazza, A.C. (Ed.), *Toxicology Studies: Cells, Drugs and Environment*. Croatia: InTech, pp. 195–233.

Bharagava, R.N., Chandra, R., and Rai, V. 2009. Isolation and characterization of aerobic bacteria capable of the degradation of synthetic and natural melanoidins from distillery effluent. *World Journal of Microbiology and Biotechnology*, 25: 737–744.

Boopathy, M.A., and Senthilkumar, S.N.S. 2014. Media optimization for the decolorization of distillery spent wash by biological treatment using *Pseudomonas fluorescence*. *International Journal of Innovations in Engineering and Technology*, 4: 1–8.

Boopathy, R., and Manning, J. 1998. A laboratory study of the bioremediation of 2,4,6-trinitrotoluenecontaminated soil using aerobic anaerobic soil slurry reactor. *Water Environment Research*, 70: 80–86.

Boopathy, R., Kulpa, C.F., and Wilson, M. 1993. Metabolism of 2,4,6-trinitrotoluene (TNT) by *Desulfovibrio* sp. (B strain). *Applied Microbiology and Biotechnology*, 39: 270–275.

Burghal, A.A., Najwa, M.J.A., and Al-Tamimi, W.H. 2016. Mycodegradation of crude oil by fungal species isolated from petroleum contaminated soil. *International Journal of Innovative Research in Science, Engineering and Technology*, 5: 1517–1524.

Burken, J.G., and Schnoor, J.L. 2003. Uptake and metabolism of atrazine by poplar trees. *Environment Science and Technology*, 31: 1399–1406.

Cabrera, J.A., Kurtz, A., Sikora, R.A., and Schouten, A. 2010. Isolation and characterization of fenamiphos degrading bacteria. *Biodegradation*, 21: 1017–1027.Caceres, T.P., Megharaj, M., Malik, S., Beer, M., and Naidu, R. 2009. Hydrolysis of fenamiphos and its toxic oxidation products by *Microbacterium* sp. in pure culture and groundwater. *Bioresource Technology*, 100: 2732–2736.

Caceres, T.P., Megharaj, M., and Naidu, R. 2008. Biodegradation of the pesticide fenamiphos by ten different species of green algae and cyanobacteria. *Current Microbiology*, 57: 643–646.

Cao, W., and Wang, Y. 2004. Compound-specific hydrogen and carbon isotopic fractionations during artificial enhanced bioremediation of petroleum hydrocarbons. In *Proceedings of 227th ACS National Meeting, Anaheim, CA, USA, March 28–April 1, GEOC-028*. Washington, DC: American Chemical Society.

Carrier, T.A., and Keasling, J.D. 1997a. Controlling messenger RNA stability in bacteria: Strategies for engineering gene expression. *Biotechnology Progress*, 13 (6): 699–708.

Carrier, T.A., and Keasling, J.D. 1997b. Engineering mRNA stability in *E. coli* by the addition of synthetic hairpins using a 50 cassette system. *Biotechnology and Bioengineering*, 55 (3): 577–580.

Chandra, R., Bharagava, R.N., Rai, V., and Singh, S.K. 2009. Characterization of sucrose-glutamic acid Maillard products (SGMPs) degrading bacteria and their metabolites. *Bioresource Technology*, 100: 6665–6668.

Chang, C.H., Liu, C.C., and Tseng, P.Y. 2013. Emissions inventory for rice straw open burning in Taiwan based on burned area classification and mapping using formosat-2 satellite imagery. *Aerosol and Air Quality Research*, 13 (2): 474–487.

Chaudhary, D.K., and Kim, J. 2016. *Sphingomonas naphthae sp. nov*., isolated from oil-contaminated soil. *International Journal of Systematic and Evolutionary Microbiology*, 66: 4621–4627.

Chen, C., and Wang, J.L. 2007. Characteristics of Zn2+ biosorption by *Saccharomyces cerevisiae*. *Biomedical and Environmental Sciences*, 20: 478–482.

Chen, W., Bruhlmann, F., Richins, R.D., and Mulchandani, A. 1999. Engineering of improved microbes and enzymes for bioremediation. *Current Opinion in Biotechnology*, 10 (2): 137–141.

Chiellini, E., Corti, A., Antone, S., and Baciu, R. 2006. Oxo-biodegradable carbon backbone polymers—oxidative degradation of polyethylene under accelerated test conditions. *Polymer Degradation and Stability*, 91: 2739–2747.

Chowdhary, P., Raj, A., and Bharagava, R.N. 2018. Environmental pollution and health hazards from distillery wastewater and treatment approaches to combat the environmental threats. *Chemosphere*, 194: 229–246.

Chowdhary, P., Sammi, S.R., Pandey, R., Kaithwas, G., Raj, A., Singh, J., and Bharagava, R.M. 2020. Bacterial degradation of distillery wastewater pollutants and their metabolites characterization and its toxicity evaluation by using *Caenorhabditis elegans* as terrestrial test models. *Chemosphere*, 261: 127689.

Christensen, G.D., Baldassarri, L., and Simpson, W.A. 1995. Methods for studying microbial colonization of plastics. *Methods in Enzymology*, 253: 477–500.

Couto, S.R., Sanroman, M.A., and Gubitz, G.M. 2005. Influence of redox mediators and metal ions on synthetic acid dye decolorization by crude laccase from *Trametes hirsuta*. *Chemosphere*, 58: 417–422.

Crameri, A., Raillard, S.A., Bermudez, E., and Stemmer, W.P. 1998. DNA shuffling of a family of genes from diverse species accelerates directed evolution. *Nature*, 391 (6664): 288.

Cresnar, B., and Petric, S. 2011. Cytochrome P450 enzymes in the fungal kingdom. *Biochimica et Biophysica Acta*, 1814: 29–35.

Crocker, F.H., Indest, K.J., and Fredrickson, H.L. 2006. Biodegradation of the cyclic nitramine explosives RDX, HMX, and CL-20. *Applied Microbiology and Biotechnology*, 73 (2): 274–290.

Cycon, M., Zmijowska, A., Wojcik, M., and Piotrowska-Seget, Z. 2013. Biodegradation and bioremediation potential of diazinon-degrading Serratia marcescens to remove other organophosphorus pesticides from soils. *Journal of Environmental Management*, 117: 7–16.

Dahiya, J., Singh, D., and Nigam, P. 2001. Decolourisation of molasses wastewater by cells of *Pseudomonas fluorescens* immobilized on porous cellulose carrier. *Bioresource Technology*, 78: 111–114

Dai, M., and Copley, S.D. 2004. Genome shuffling improves degradation of the anthropogenic pesticide pentachlorophenol by *Sphingobium chlorophenolicum* ATCC 39723. *Applied and Environmental Microbiology*, 70 (4): 2391–2397.

da Luz, J.M.R., Paes, S.A., Bazzolli, D.M.S., Totola, M.R., Demuner, A.J., and Kasuya, M.C.M. 2014. Abiotic and biotic degradation of oxo-biodegradable plastic bags *by Pleurotus ostreatus*. *PLOS One*, 9: e107438. Available from: https://doi.org/10.1371/journal. pone.0107438.

Damalas, C.A. 2009. Understanding benefits and risks of pesticide use. *Scientific Research and Essays*, 4: 945–949.

Damare, S., Singh, P., and Raghukumar, S. 2012. Biotechnology of marine fungi. *Progress in Molecular and Subcellular Biology*, 53: 277–297.

Dang, H., and Lovell, C.R. 2016. Microbial surface colonization and biofilm development in marine environments. *Microbiology and Molecular Biology Reviews*, 80: 91–138.

Das, A., Mishra, S., and Verma, V.K. 2015. Enhanced biodecolorization of textile dye remazol navy blue using an isolated bacterial strain *Bacillus pumilus* HKG212 under improved culture conditions. *Journal of Biochemical Technology*, 6: 962–969.

Das, A.P., and Mishra, S. 2010. Biodegradation of metallic carcinogen hexavalent chromium Cr (VI) by an indigenously isolated bacterial strain. *Journal of Carcinogenesis*, 9: 6.

Das, K., and Mukherjee, A.K. 2007. Crude petroleum-oil biodegradation efficiency of *Bacillus subtilis* and *Pseudomonas aeruginosa* strains isolated from a petroleum-oil contaminated soil from North-East India. *Bioresource Technology*, 98: 1339–1345.

Davison, J., Heusterspreute, M., Chevalier, N., Ha-Thi, V., and Brunei, F. 1987. Vectors with restriction site banks V. pJRD215, a wide-host-range cosmid vector with multiple cloning sites. *Gene*, 51 (2–3): 275–280.

Dehorter, B., and Blondeau, R. 1993. Isolation of an extracellular Mn-dependent enzyme mineralizing melanoidins from the white rot fungus *Trametes versicolour*. *FEMS Microbiology Letters*, 109: 117–122.

De la Cruz, E., Bravo-Duran, V., Ramirez, F., and Castillo, L.E. 2014. Environmental hazards associated with pesticide import into Costa Rica, 1977–2009. *Journal of Environmental Biology*, 35: 43–55.

De Lorenzo, V., Fernandez, S., Herrero, M., Jakubzik, U., and Timmis, K.N. 1993. Engineering of alkyl- and haloaromatic-responsive gene expression with mini-transposons containing regulated promoters of biodegradative pathways of *Pseudomonas*. *Gene*, 130 (1): 41–46.

Demnerova, K., Mackova, M., Spevakova, V., Beranova, K., and Kochankova, L. 2005. Two approaches to biological decontamination of groundwater and soil polluted by aromatics characterization of microbial populations. *International Microbiology*, 8: 205–211.

Deshmukh, R., Khardenavis, A.A., and Purohit, H.J. 2016. Diverse metabolic capacities of fungi for bioremediation. *Indian Journal of Microbiology*, 56: 247–264.

Dixit, R., Wasiullah, M., Malaviya, D., Pandiyan, K., Singh, U.B., Sahu, A., Shukla, R., ... Paul, D. 2015. Bioremediation of heavy metals from soil and aquatic environment: An overview of principles and criteria of fundamental processes. *Sustainability*, 7 (2): 2189–2212.

Dos Santos Bazanella, G.C., de Souza, D.F., Castoldi, R., Oliveira, R.F., Bracht, A., and Peralta, R.M. 2013. Production of laccase and manganese peroxidase by *Pleurotus pulmonarius* in solid-state cultures and application in dye decolorization. *Folia Microbiologica*, 58: 641–647.

Duffner, F.M., Kirchner, U., Bauer, M.P., and Muller, R. 2000. Phenol/cresol degradation by the thermophilic *Bacillus thermoglucosidasius* A7: Cloning and sequence analysis of five genes involved in the pathway. *Gene*, 256: 215–221.

Eissa, F.I., Mahmoud, H.A., Zidan, N.A., and Belal, E.B.A. 2006. Microbial, thermal and photo degradation of cadusafos and carbofuran pesticides. *Journal of Pest Control and Environmental Sciences*, 14: 107–130.

El-Borai, A.M., Eltayeb, K.M., Mostafa, A.R., and El-Assar, S.A. 2016. Biodegradation of industrial oil-polluted wastewater in Egypt by bacterial consortium immobilized in different types of carriers. *Polish Journal of Environmental Studies*, 25: 1901–1909.

Emmanuel, A., Jacob, E., and Liberty, T. 2013. Effluent's characteristics of some selected food processing industries in Enugu and Anambra states of Nigeria. *Journal of Environment and Earth Science*, 3 (9): 46–53.

Engling, G., Lee, J.J., Tsai, Y.W., Lung, S.C.C., Chou, C.C.K., and Chan, C.Y. 2009. Size resolved anhydrosugar composition in smoke aerosol from controlled field burning of rice straw. *Aerosol Science and Technology*, 43 (7): 662–672.

Erika, A.W., Vivian, B., Claudia, C and Jorge, F.G. 2013. Biodegradation of phenol in static cultures by *Penicillium chrysogenum* erk1: Catalytic abilities and residual phytotoxicity. *Revista Argentina de Mcrobiologia*, 44: 113–121.

Eubeler, J.P., Bernhard, M., and Knepper, T.P. 2010. Environmental biodegradation of synthetic polymers II: Biodegradation of different polymer groups. *Trends in Analytical Chemistry*, 29: 84–100.

Fournier, D., Halasz, A., Spain, J., Fiurasek, P., and Hawari, J. 2002. Determination of key metabolites during biodegradation of hexahydro-1,3,5-trinitro-1,3,5-triazine with *Rhodococcus* sp. strain DN22. *Applied and Environmental Microbiology*, 68 (1): 166–172.

Freitas, A.C., Ferreira, F., Costa, A.M., Pereira, R., and Antunes, S.C. 2009. Biological treatment of the effluent from a bleached kraft pulp mill using basidiomycete and zygomycete fungi. *The Science of the Total Environment*, 407: 3282–3289.

Friss, N., and Myers-Keith, P. 1986. Biosorption of uranium and lead by *Streptomyces long-woodensis*. *Biotechnology Bioengineering*, 28: 21–28.

Fu, Y. and Viraraghavan, T. 2001. Fungal decolorization of dye wastewater: Review. *Bioresource Technology*, 79: 251–262.

Fulekar, M.H., Sharma, J., and Tendulkar, A. 2012. Bioremediation of heavy metals using biostimulation in laboratory bioreactor. *Environmental Monitoring and Assessment*, 184 (12): 7299–7307.

Fulekar, M.H., Singh, A., and Bhaduri, A.M. 2009. Genetic engineering strategies for enhancing phytoremediation of heavy metals. *African Journal of Biotechnology*, 8 (4): 529–535.

Fuller, M.E., McClay, K., Hawari, J., Pauquet, L., Malone, T.E., Fox, B.G., and Steffan, R.J. 2009. Transformation of RDX and other energetic compounds by xenobiotic reductases XenA and XenB. *Applied Microbiology and Biotechnology*, 84 (3): 535–544.

Gajendiran, A., Krishnamoorthy, S., and Abraham, J. 2016. Microbial degradation of low-density polyethylene (LDPE) by *Aspergillus clavatus* strain JASK1 isolated from landfill soil. *Biotechnology*, 6: 52.

Gallardo, M.E., Ferrandez, A., De Lorenzo, V., Garcia, J.L., and Diaz, E.1997. Designing recombinant *Pseudomonas* strains to enhance biodesulfurization. *Journal of Bacteriology*, 179 (22): 7156–7160.

Gao, Y., Chen, S., Hu, M., Hu, Q., Luo, J., and Li, Y. 2012. Purification and characterization of a novel chlorpyrifos hydrolase from *Cladosporium cladosporioides* Hu-01. *PLOS One*, 7, e38137.

Garg, S., and Chaudhary, S. 2017. Treatment of wastewater of food industry by membrane bioreactors. *International Advanced Research Journal in Science, Engineering and Technology*, 4 (6): 153–156.

Garret, T.R., Bhako, M., and Zang, Z. 2008. Bacterial adhesion and biofilms on surfaces. *Progress in Natural Science*, 18 (9): 1049–1056.

Ghosh, M., Verma, S.C., Mengoni, A., and Tripathi, A.K. 2004. Enrichment and identification of bacteria capable of reducing chemical oxygen demand of anaerobically treated spent wash. *Journal of Applied Microbiology*, 6: 241–278.

Gomez-Acata, S., Esquivel-Rios, I., Perez-Sandoval, M.V., Navarro-Noya, Y., Rojas-Valdez, A., Thalasso, F., Luna-Guido, M., and Dendooven, L. 2016. Bacterial community structure within an activated sludge reactor added with phenolic compounds. *Applied Microbiology and Biotechnology*, 101 (8): 3405–3414.

Gottschalk, G. 2012. *Bacterial Metabolism*. New York: Springer.

Goyal, S., and Sindhu, S.S. 2011. Composting of rice straw using different inocula and analysis of compost quality. *Microbiology Journal*, 1 (4): 126–138.

Grabic-Galic, D., and Vogel, T.M. 1987. Transformation of toluene and benzene by mixed methanogenic culture. *Applied and Environmental Microbiology*, 53 (2): 254–260.

Gu, J.D. 2003. Microbiological deterioration and degradation of synthetic polymeric materials: Recent research advances. *International Biodeterioration and Biodegradation*, 52: 69–91.

Guermouche M'rassi, A., Bensalah, F., Gury, J., and Duran, R. 2015. Isolation and characterization of different bacterial strains for bioremediation of n-alkanes and polycyclic aromatic hydrocarbons. *Environmental Science and Pollution Research*, 22: 15332–15346.

Haleyur, N., Shahsavari, E., Taha, M., Khudur, L.S., Koshlaf, E., and Osborn, A.M. 2018. Assessing the degradation efficacy of native PAH-degrading bacteria from aged, weathered soils in an Australian former gasworks site. *Geoderma*, 321: 110–117.

Harayama, S. 1998. Artificial evolution by DNA shuffling. *Trends in Biotechnology*, 16 (2): 76–82.

Hassan, M.M., Alam, M.Z., and Anwar, M.N. 2013. Biodegradation of textile azo dyes by bacteria isolated from dyeing industry effluent. *International Research Journal of Biological Sciences*, 2: 27–31.

Hayashi, K., Ono, K., Kajiura, M.S., Sudo, S., Yonemura, S., Fushimi, A., and Tanabe, K. 2014. Trace gas and particle emissions from open burning of three cereal crop residues: Increase in residue moistness enhances emissions of carbon monoxide, methane and particulate. *Atmospheric Environment*, 95: 36–44.

Hesham, A., Khan, S., Tao, Y., Li, D., and Zhang, Y. 2012. Biodegradation of high molecular weight PAHs using isolated yeast mixtures: Application of metagenomic methods for community structure analyses. *Environmental Science and Pollution Research International*, 19: 3568–3578.

Hidayat, A., and Tachibana, S. 2012. Biodegradation of aliphatic hydrocarbon in three types of crude oil by *Fusarium* sp. F092 under stress with artificial sea water. *Journal of Environmental Science and Technology*, 5: 64–73. Available at: https://goo.gl/L73KgW.

Hisatsuka, K., Nakahara, T., Sano, N., and Yamada, K. 1971. Formation of rhamnolipid by *Pseudomonas aeruginosa* and its function in hydrocarbon fermentation. *Agricultural and Biological Chemistry*, 35: 686–692.

Holba, A.G., Dzou, I.L., Hickey, J.J., Franks, S.G., May, S.J., and Lenney, T. 1996. Reservoir geochemistry of South Pass 61 field. Gulf of Mexico: Compositional heterogeneities reflecting filling history and biodegradation. *Organic Geochemistry*, 24: 1179–1198.

Holloway, P., Knoke, K.L., Trevors, J.T., and Lee, H., 1998. Alteration of the substrate range of haloalkane dehalogenase by site-directed mutagenesis. *Biotechnology and Bioengineering*, 59 (4): 520–523.

Huang, C.C., Chien, M.F., and Lin, K.H. 2010. Bacterial mercury resistance of TnMERI1 and its application in bioremediation. *Interdisciplinary Studies on Environmental Chemistry*, 3 (11): 21–29.

Huang, Y., Xiao, L., Li, F., Xiao, M., Lin, D., and Long, X. 2018. Microbial degradation of pesticide residues and an emphasis on the degradation of cypermethrin and 3-phenoxy benzoic acid: A review. *Molecules*, 23: 2313.

Hussaini, S., Shaker, M., and Asef, M. 2013. Isolation of bacterial for degradation of selected pesticides. *Bulletin of Environment, Pharmacology and Life Sciences*, 2: 50–53.

Ibrahim, M.W., and Essa, M.A. 2010. Tolerance and utilization of organophosphorus insecticide by nitrogen fixing Cyanobacteria. *Egyptian Journal of Botany*, 27: 225–240.

Ibrahim, W., Mohamed, K., El-Shahat, R., and Asmaa, A. 2014. Biodegradation and utilization of organophosphorus pesticide malathion by cynobacteria. *BioMed Research International*, 2: 392682.

Infante, J.C., De Arco, R.D., and Angulo, M.E. 2014. Removal of lead, mercury and nickel using the yeast *Saccharomyces cerevisiae*. *Revista MVZ Cordoba*, 19: 4141–4149.

Iranzo, M., Sainz-Pardo, I., Boluda, R., Sanchez, J., Mormeneo, S., and Hemery, G.E. 2001. The use of microorganisms in environmental remediation. *Annals of Microbiology*, 51: 135–143.

Izumi, Y., Ohshiro, T., Ogino, H., Hine, Y., and Shimao, M. 1994. Selective desulfurization of dibenzothiophene by *Rhodococcus erythropolis* D-1. *Applied and Environmental Microbiology*, 60 (1): 223–226.

Jariyal, M., Jindal, V., Mandal, K., Gupta, V.K., and Singh, B. 2018. Bioremediation of organo phosphorus pesticide phorate in soil by microbial consortia. *Ecotoxicology and Environmental Safety*, 159: 310–316.

Jasinska, A., Bernat, P., and Paraszkiewicz, K. 2013. Malachite green removal from aqueous solution using the system rapeseed press cake and fungus *Myrothecium roridum*. *Desalination and Water Treatment*, 51: 7663–7671.

Jasinska, A., Paraszkiewicz, K., Sip, A., and Dlugon, J. 2015. Malachite green decolorization by the filamentous fungus *Myrothecium roridum*: Mechanistic study and process optimization. *Bioresource Technology*, 194: 43–48.

Jasinska A., Rozalska, S., Bernat, P., Paraszkiewicz, K., and Dlugonski, J. 2012. Malachite green decolorization by non-basidiomycete filamentous fungi of *Penicillium pinophilum* and *Myrothecium roridum*. *International Biodeterioration and Biodegradation*, 73: 33–40.

Jenkins, B.M., Mehlschau, J.J., Williams, R.B., Solomon, C., Balmes, C.J., Kleinman, M., and Smith, N. 2003. Rice straw smoke generation system for controlled human inhalation exposures. *Aerosol Science and Technology*, 37 (5): 437–454.

Joshi, N., and Deepali, 2012. Study of ground water quality in and around Sidcul industrial area, Haridwar, Uttarakhand, India. *International Journal Applied Technology Environmental Sanitation*, 2 (2): 129–134.

Kadiyala, V., and Spain, J.C. 1998. A two-component monooxygenase catalyzes both the hydroxylation of p nitrophenol and the oxidative release of nitrite from 4-nitrocatechol in *Bacillus sphaericus* JS905. *Applied and Environmental Microbiology*, 64 (7): 2479–2484.

Kalavathi, D.F., Uma, L., and Subramanian, G. 2001. Degradation and metabolization of the pigment-melanoidin in distillery effluent by the marine cyanobacterium *Oscillatoria boryana* BDU 92181. *Enzyme and Microbial Technology*, 29, 246–251.

Kalderis, D., Hawthorne, S.B., Clifford, A.A., and Gidarakos, E. 2008. Interaction of soil, water and TNT during degradation of TNT on contaminated soil using subcritical water. *Journal of Hazardous Materials*, 159 (2–3): 329–334.

Kale, S.K., Deshmukh, A.G., Dudhare, M.S. and Patil, V.B. 2015. Microbial degradation of plastic: A review. *Journal of Biochemical Technology*, 6: 952–961.

Kambe, T.N., Shimomura, M., Nomura, N., Chanpornpong, T., and Nakahara, T., 1999. Decolourization of molasses wastewater by *Bacillus* sp. under thermophilic and anaerobic conditions. *Journal of Bioscience and Bioengineering*, 87, 119e121.

Karakaya, P., Christodoulatos, C., Koutsospyros, A., Balas, W., Nicolich, S., and Sidhoum, M. 2009. Biodegradation of the high explosive hexanitrohexaazaiso-wurtzitane (CL-20). *International Journal of Environmental Research and Public Health*, 6 (4): 1371–1392.

Karigar, C.S., and Rao, S.S. 2011. Role of microbial enzymes in the bioremediation of pollutants: A review. *Enzyme Research*, 2011: 805187.

Karpouzas, D.G., Fotopoulou, A., Menkissoglu-Spiroudi, U., and Singh, B.K. 2005. Non-specific biodegradation of the organophosphorus pesticides, cadusafos and ethoprophos, by two bacterial isolates. *FEMS Microbiology Ecology*, 53: 369–378.

Kavitha, G.V., and Beebi, S.K. 2003. Biodegradation of phenol in a packed bed reactor using peat media. *Asian Journal of Microbiology, Biotechnology and Environmental Sciences*, 5 (2): 157–159.

Kawai, F., 2010. The biochemistry and molecular biology of xenobiotic polymer degradation by microorganisms. *Bioscience Biotechnology and Biochemistry*, 74: 1743–1759.

Kawa-Rygielska, J., Chmielewska, J., and Plaskowska, E. 2007. Effect of raw material quality on fermentation activity of distillery yeast. *Polish Journal of Food and Nutrition Sciences*, 57: 275–279.

Kehinde, F.O., and Isaac, S.A. 2016. Effectiveness of augmented consortia of *Bacillus coagulans*, *Citrobacter koseri* and *Serratia ficaria* in the degradation of diesel polluted soil supplemented with pig dung. *African Journal of Microbiology Research*, 10: 1637–1644.

Khongkhaem, P., Intasiri, A., and Luepromchai, E. 2011. Silica-immobilized *Methylobacterium* sp. NP3 and Acinetobacter sp. PK1 degrade high concentrations of phenol. *Letters in Applied Microbiology*, 52: 448–452.

Kort, M.J. 1979. Colour in the sugar industry. In de Birch, G.G., and Parker, K.J. (Eds), *Science and Technology*. London: Applied Science, pp. 97–130.

Kuchner, O., and Arnold, F.H. 1997. Directed evolution of enzyme catalysts. *Trends in Biotechnology*, 15 (12): 523–530.

Kujan, P., Prell, A., Safar, H., Sobotka, M., Rezanka, T., and Holler P. 2006. Use of the industrial yeast *Candida utilis* for cadmium sorption. *Folia Microbiologica*, 51 (4): 257–260.

Kumamaru, T., Suenaga, H., Mitsuoka, M., Watanabe, T., and Furukawa, K. 1998. Enhanced degradation of polychlorinated biphenyls by directed evolution of biphenyl dioxygenase. *Nature Biotechnology*, 16 (7): 663.

Kumar, A., Kumar, S., and Kumar, S. 2005. Biodegradation kinetics of phenol and catechol using *Pseudomonas putida* MTCC 1194. *Biochemical Engineering Journal*, 22 (2), 151–159.

Kumar, S., Chaurasia, P., and Kumar, A. 2016. Isolation and characterization of microbial strains from textile industry effluents of Bhilwara, India: Analysis with bioremediation. *Journal of Chemical and Pharmaceutical Research*, 8: 143–150.

Kumar, S., Kaushik, G., Dar, M.A., Nimesh, S., Lopez-Chuken, U.J., and Villarreal-Chiu, J.F. 2018. Microbial degradation of organophosphate pesticides: A review. *Pedosphere*, 28: 190–208.

Kyrikou, I., and Briassoulis, D. 2007. Biodegradation of agricultural plastic films: A critical review. *Journal of Polymers and Environment*, 15: 125–150.

Lee, T.H., Aoki, H., Sugano, Y., and Shoda, M. 2000. Effect of molasses on the production and activity of dye-decolourizing peroxidase from *Geotrichum candidum*. *Journal of Bioscience and Bioengineering*, 89: 545–549.

Lin, C., Gan, L., and Chen, Z.L. 2010. Biodegradation of naphthalene by strain Bacillus fusiformis (BFN). *Journal of Hazardous Materials*, 182: 771–777.

Liu, K., Han, W., Pan, W.P., and Riley, J.T. 2001b. Polycyclic aromatic hydrocarbon (PAH) emissions from a coal-fired pilot FBC system. *Journal of Hazardous Materials*, 84: 175–188.

Liu, Y.H., Chung, Y.C., and Xiong, Y. 2001a. Purification and characterization of a dimethoate degrading enzyme of *Aspergillus niger* ZHY256, isolated from sewage. *Applied and Environmental Microbiology*, 67: 3746–3749.

Logeshwaran, P., Megharaj, M., Chadalavada, S., Bowman, M., and Naidu, R. 2018. Petroleum hydrocarbons (PH) in groundwater aquifers: An overview of environmental fate, toxicity, microbial degradation and risk-based remediation approaches. *Environmental Technology and Innovation*, 10: 175–193.

Loukidou, M.X., Matis, K.A., Zouboulis A.I., and Liakopoulou-Kyriakidou, M. 2003. Removal of As (V) from wastewaters by chemically modified fungal biomass. *Water Research*, 37 (18): 4544–4552.

Lucas, N., Bienaime, C., Belloy, C., Queneudec, M., Silvestre, F., and Nava-Saucedo, J.E. 2008. Polymer biodegradation: Mechanisms and estimation techniques: A review. *Chemosphere*, 73: 429–442.

Machado, M.D., Soares, E.V., Soares, H.M. 2010. Removal of heavy metals using a brewer's yeast strain of Saccharomyces cerevisiae: Chemical speciation as a tool in the prediction and improving of treatment efficiency of real electroplating effluents. *Journal of Hazardous Materials*, 180 (1–3): 347–353.

Madaeni, S.S., and Eslamifard, M.R. 2010. Recycle unit wastewater treatment in petrochemical complex using reverse osmosis process. *Journal of Hazardous Materials*, 124: 404–409.

Madhuri, R., and Rangaswamy, V. 2009. Biodegradation of selected insecticides by *Bacillus* and *Pseudomonas* sps in groundnut fields. *Toxicology International*, 16: 127.

Madukasi, E.I., Dai, X., He, C., and Zhou, J., 2010. Potentials of phototrophic bacteria in treating pharmaceutical wastewater. *International Journal of Environmental Science and Technology*, 7 (1), 165–174.

Mahjoubi, M., Cappello, S., Souissi, Y., Jaouani, A., and Cherif, A. 2018. Microbial bioremediation of petroleum hydrocarbon-contaminated marine environments. In Zoveidavianpoor, M. (Ed.), *Recent Insights in Petroleum Science and Engineering*. London: IntechOpen. https://doi.org/10.5772/intechopen.72207.

Maliji, D., Olama, Z., and Holail, H. 2013. Environmental studies on the microbial degradation of oil hydrocarbons and its application in Lebanese oil polluted coastal and marine ecosystem. *International Journal of Current Microbiology and Applied Sciences*, 2: 1–18.

Margesin, R., Fonteyne, P.A., and Redl, B. 2005. Low-temperature biodegradation of high amounts of phenol by *Rhodococcus* spp. and basidiomycetous yeasts. *Research in Microbiology*, 156: 68–75.

Marimuthu, T., Rajendran, S., and Manivannan, M. 2013. A review on fungal degradation of textile dye effluent. *Acta Chimica and Pharmaceutica Indica*, 3 (2): 192–200.

Marini, M., De Niederhausern, S., Iseppi, R., Bondi, M., Sabia, C., and Toselli, M. 2007. Antibacterial activity of plastics coated with silver-doped organic-inorganic hybrid coatings prepared by sol-gel processes. *Biomacromolecules*, 8: 1246–1254.

Mattuschka, B., Junghaus, K., and Straube, G. 1993. Biosorption of metals by waste biomass. *Biohydrometallurgical Technologies*, 2: 125–132.

Maulin, P., Shah., Patel, K.A., Nair, S.S., and Darji, A.M. 2013. Microbial degradation of textile dye (Remazol Black B) by *Bacillus* spp. ETL-2012. *Journal of Bioremediation and Biodegradation*, 4: 1–5.

McIllroy, S., Saunders, A.M., and Albertsen, M. 2015. MiDAS: A field guide to the microbes of activated sludge. *Database (Oxford)*, 2015: bav062.

Megharaj, M., Singh, N., Kookana, R.S., Naidu, R., and Sethunathan, N. 2003. Hydrolysis of fenamiphos and its oxidation products by a soil bacterium in pure culture, soil and water. *Applied Microbiology and Biotechnology*, 61: 252–256.

Megharaj, M., Venkateswarlu, K., and Rao, A.S. 1987. Metabolism of monocrotophos and quinalphos by algae isolated from soil. *Bulletin of Environmental Contamination and Toxicology*, 39: 251–256.

Mercimek, H.A., Dincer, S., Guzeldag, G., Ozsavli, A., Matyar, F., Arkut, A., Kayis, F., and Ozdenefe, M.S. 2015. Degradation of 2,4,6-trinitrotoluene by *P. aeruginosa* and characterization of some metabolites. *Brazilian Journal of Microbiology*, 46 (1): 103–111.

Mishra, V., Lal, R., and Srinivasan. 2001. Enzymes and operons mediating xenobiotic degradation in bacteria. *Critical Reviews in Microbiology*, 27: 133–166.

Miyata, N., Iwahori, K., and Fujita, M. 1998. Manganese-independent and dependent decolourization of melanoidin by extracellular hydrogen peroxide and peroxidases from *Coriolus hirsutus* pellets. *Journal of Fermentation and Bioengineering*, 85: 550–553.

Mohamed, A.T., El Hussein, A.A., El Siddig, M.A., and Osman, A.G. 2011. Degradation of oxyfluorfen herbicide by soil microorganisms: Biodegradation of herbicides. *Biotechnology*, 10: 274–279.

Mohammad, P., Azarmidokht, H., Fatollah, M., and Mahboubeh, B. 2006. Application of response surface methodology for optimization of important parameters in decolorizing treated distillery wastewater using *Aspergillus fumigates* UB260. *International Biodeterioration and Biodegradation*, 57 (4): 195–199.

Mohan, K. 2011. Microbial deterioration and degradation of polymeric materials. *Journal of Biochemical Technology*, 2: 210–215.

Mohan, N., Balasubramanian, N., and Ahmed Basha, C. 2007. Electrochemical oxidation of textile wastewater and its reuse. *Journal of Hazardous Materials*, 147: 644–651.

Mohanty, G., and Mukherji, S. 2008. Biodegradation rate of diesel range n-alkanes by bacterial cultures *Exiguobacterium aurantiacum* and *Burkholderia cepacia*. *International Biodeterioration and Biodegradation*, 61 (3): 240–250.

Moller, S., Pedersen, A.R., Poulsen, L.K., Arvin, E., and Molin, S. 1996. Activity and three-dimensional distribution of toluene-degrading *Pseudomonas putida* in a multispecies biofilm assessed by microscopy. *Applied and Environmental Microbiology*, 12: 4632–4640.

Monica, P., Darwin, R.O., Manjunatha, B., Zuniga, J.J., and Diego, R. 2016. Evaluation of various pesticides-degrading pure bacterial cultures isolated from pesticide-contaminated soils in Ecuador. *African Journal of Biotechnololgy*, 15: 2224–2233.

Mor, R., and Sivan, A. 2008. Biofilm formation and partial biodegradation of polystyrene by the actinomycete *Rhodococcus ruber*: Biodegradation of polystyrene. *Biodegradation*, 19: 851–858.

Mueller, R.J. 2006. Biological degradation of synthetic polyesters: Enzymes as potential catalysts for polyester recycling. *Process Biochemistry*, 41: 2124–2128.

Muller, R.J., Schrader, H., Profe, J., Dresler, K., and Deckwer, W.D. 2005. Enzymatic degradation of poly (ethylene terephthalate): Rapid hydrolyse using a hydrolase from *T.* fusca. *Macromolecular Rapid Communications*, 26: 1400–1405.

Nagy, B., Manzatu, C., Maicaneanu, A., Indolean, C., Lucian, B.T., and Majdik, C. 2017. Linear and nonlinear regression analysis for heavy metals removal using *Agaricus bisporus* macrofungus. *Arabian Journal of Chemistry*, 10 (2): S3569–S3579.

Nair, C.N., Jayachandran, K., and Shashidhar, S. 2008. Biodegradation of phenol. *African Journal of Biotechnology*, 7 (25): 4951–4958.

Naranjo-Briceno, L., Pernıa, B., Perdomo, T., Gonzalez, M., Inojosa, Y., and De Sisto, A. 2019. Potential role of extremophilic hydrocarbonoclastic fungi for extra-heavy crude oil bioconversion and the sustainable development of the petroleum industry. In Tiquia Arashiro, S.M., and Grube, M. (Eds), *Fungi in Extreme Environments: Ecological Role and Biotechnological Significance*. Cham: Springer, pp. 559–586.

Nauendorf, A., Krause, S., Bigalke, N.K., Gorb, E.V., Gorb, S.N., and Haeckel, M. 2016. Microbial colonization and degradation of polyethylene and biodegradable plastic bags in temperate fine-grained organic-rich marine sediments. *Marine Pollution Bulletin*, 103: 168–178.

Niti, C., Sunita, S., and Kamlesh, K. 2013. Bioremediation: An emerging technology for remediation of pesticides. *Research Journal of Chemistry and Environment*, 17: 88–105.

Norton, J.M. 2012. Fungi for bioremediation of hydrocarbon pollutants. *Hohonu*, 10: 18–21.

Ntuli, F., Omoregbe, I., Kujipa, P., Muzenda, E., and Belaid, M. 2009. *Characterization of Effluent from Textile Wet Finishing Operations*, vol. 1. San Francisco, CA: WCECS.

Ohmoho, S., Kainuma, M., Kamimura, K., Sirianuntapiboon, S., and Aoshima, I. 1988. Adsorption of melanoidin to the mycelia of *Aspergillus oryazae* Y-2–32. *Agricultural and Biological Chemistry Tokyo*, 52: 381–386.

Orr, I.G., Hadar, Y., and Sivan, A. 2004. Colonization, biofilm formation and biodegradation of polyethylene by a strain of *Rhodococcus ruber*. *Applied Microbiology and Biotechnology*, 65: 97–104.

Ortiz-Hernandez, M.L., Sanchez-Salinas, E., Dantan-Gonzalez, E., and Castrejon-Godinez, M.L. 2013. Pesticide biodegradation: Mechanisms, genetics and strategies to enhance the process. In Chamy, R. (Ed.), *Biodegradation—Life of Science*. London: IntechOpen, pp. 251–287.

Ou, L.T., Thomas, J.E., and Dickson, D.W., 1994. Degradation of fenamiphos in soil with a history of continuous fenamiphos applications. *Soil Science Society of America Journal*, 58: 1139–1147.

Pandey, B., Baghel, P.S., and Shrivastava, S. 2014. To study the bioremediation of monocrotophos and to analyze the kinetics effect of Tween 80 on fungal growth. *Indo-American Journal of Pharmaceutical Research*, 4: 925–930.

Panikov, N.S., Sizova., M.V., Ross, D., Christodoulatos, C., Balas, W., and Nicolich, S. 2007. Biodegradation kinetics of the nitramine explosive CL-20 in soil and microbial cultures. *Biodegradation*, 18: 317–332.

Pant, D., and Adholeya, A. 2007. Identification, ligninolytic enzyme activity and decolorization potential of two fungi isolated from a distillery effluent contaminated site. *Water, Air, and Soil Pollution*, 183: 165–176.

Pant, D., and Adholeya, A. 2010. Development of a novel fungal consortium for the treatment of molasses distillery wastewater. *Environmentalist*, 30: 178–182.

Paranthaman, S.R., and Karthikeyan, B. 2015. Bioremediation of heavy metal in paper mill effluent using *Pseudomonas* spp. *International Journal of Microbiology*, 1: 1–5.

Parte, S.G., Mohekar, A.D., and Kharat, A.S. 2017. Microbial degradation of pesticide: A review. *African Journal of Microbiology Research*, 11: 992–1012.

Pathak, V.M., and Navneet. 2017. Review on the current status of polymer degradation: A microbial approach. *Bioresources and Bioprocessing*, 4: 15.

Patowary, K., Patowary, R., Kalita, M.C., and Deka, S. 2016. Development of an efficient bacterial consortium for the potential remediation of hydrocarbons from contaminated sites. *Frontiers in Microbiology*, 7: 1092.

Pawlak, Z., Rauckyte, T., and Oloyede, A. 2008. Oil, grease and used petroleum oil management and environmental economic issues. *Journal of Achievements in Materials and Manufacturing Engineering*, 26: 11–17.

Pedro, P., Francisco, J.E., Joao, F., and Ana, L. 2014. DNA damage induced by hydroquinone can be prevented by fungal detoxification. *Toxicology Reports*, 1: 1096–1105.

Pena-Montenegro, T.D., Lozano, L., and Dussan, J. 2015. Genome sequence and description of the mosquitocidal and heavy metal tolerant strain *Lysinibacillus sphaericus* CBAM5. *Standards in Genomic Sciences*, 10: 1–10.

Perez-Martin, J., and de Lorenzo, V. 1996. VTR expression cassettes for engineering conditional phenotypes in *Pseudomonas*: Activity of the Pu promoter of the TOL plasmid under limiting concentrations of the XylR activator protein. *Gene*, 172 (1): 81–86.

Phulpoto, H., Qazi, M.A., Mangi, S., Ahmed, S., and Kanhar, N.A. 2016. Biodegradation of oil-based paint by Bacillus species monocultures isolated from the paint warehouses. *International Journal of Environmental Science and Technology*, 13: 125–134.

Phutela, U.G., and Sahni, N. 2012. Effect of *Fusarium* sp. on paddy straw digestibility and biogas production. *Journal of Advanced Laboratory Research in Biology*, 3 (1): 12–15.

Prescott, L.M., Harley, J.P., and Klein, D.A. 2002. *Microbiology*. 5th edn. New York: McGraw Hill, pp. 10–14.

Priyalaxmi, R., Murugan, A., Raja, P., and Raj, K.D. 2014. Bioremediation of cadmium by *Bacillus safensis* (JX126862), a marine bacterium isolated from mangrove sediments. *International Journal of Current Microbiology and Applied Sciences*, 3: 326–335.

Pushkar, B.K., Sevak, P.I., and Singh, A. 2015. Isolation and characterization of potential microbe for bio remediating heavy metal from Mithi River. *Annals of Applied Bio Sciences*, 2 (2): 20–27.

Qiu, X., Zhong, Q., Li, M., Bai, W., and Li, B. 2007. Biodegradation of p-nitrophenol by methyl parathion-degrading *Ochrobactrum* sp. B2. *International Biodeterioration Biodegradation*, 59: 297–301.

Qu, Y., Ma, Q., Deng, J., Shen, W., Zhang, X., He, Z., Van Nostr, J.D., Zhou, J., and Zhou, J. 2015. Responses of microbial communities to single walled carbon nano tubes in phenol wastewater treatment system. *Environmental Science and Technology*, 49: 4627–4635.

Ramasamy, R.K., Congeevaram, S., and Thamaraiselvi, K. 2011. Evaluation of isolated fungal strain from e-waste recycling facility for effective sorption of toxic heavy metal Pb (II) ions and fungal protein molecular characterization-a Mycoremediation approach. *Asian Journal of Experimental Biological Sciences*, 2 (2): 342–347.

Ramos, J.L., Dı´az, E., Dowling, D., de Lorenzo, V., Molin, S., and O'Gara, F. 1994. The behavior of bacteria designed for biodegradation. *Bio Technology*, 12 (12): 1349.

Ravi, R.K., Pathak, B., and Fulekar, M.H. 2015. Bioremediation of persistent pesticides in rice field soil environment using surface soil treatment reactor. *International Journal of Current Microbiology and Applied Sciences*, 4: 359–369.

Safiyanu, I., Isah, A.A., Abubakar, U.S., and Rita Singh, M. 2015. Review on comparative study on bioremediation for oil spills using microbes. *Research Journal of Pharmaceutical, Biological and Chemical Sciences*, 6: 783–790.

Salman, J.M., and Abdul-Adel, E. 2015. Potential use of cyanophyta species *Oscillatoria limnetica* in bioremediation of organophosphorus herbicide glyphosate. *Mesopotamia Environmental Journal*, 1: 15–26.

Sanghvi, G., Thanki, A., Pandey, S., and Singh, N.K. 2020. Bioremediation of pollutants. In Pandey, V.C., and Singh, V. (Eds), *Engineered Bacteria for Bioremediation*, Oxford: Elsevier, pp. 359–374.

Sangwan, P., Celin, S.M., and Hooda, L. 2015. Response surface methodological approach for optimizing process variables for biodegradation of 2,4,6-Trinitrotoluene using *Acinetobacter noscomialis*. *European Journal of Advances in Engineering and Technology*, 4: 51–56.

Sani, I., Safiyanu, I., and Rita, S.M. 2015. Review on bioremediation of oil spills using microbial approach. *International Journal of Educational Science and Research*, 3: 41–45.

Sarang, B., Richa, K., and Ram, C. 2013. Comparative study of bioremediation of hydrocarbon fuel. *International Journal of Biotechnology and Bioengineering Research*, 4: 677–686.

Sasikumar, C.S., and Papinazath, T. 2003. Environmental management: Bioremediation of polluted environment. In Bunch, M.J., Suresh, V., Madha, T. and Kumaran, V. (Eds), *Proceedings of the Third International Conference on Environment and Health*, pp. 15–17.

Sato, S.I., Nam, J.W., Kasuga, K., Nojiri, H., Yamane, H., and Omori, T. 1997. Identification and characterization of genes encoding carbazole 1, 9a-dioxygenase in *Pseudomonas* sp. strain CA10. *Journal of Bacteriology*, 179 (15): 4850–4858.

Sayler, G.S., Hooper, S.W., Layton, A.C., and King, J.H. 1990. Catabolic plasmids of environmental and ecological significance. *Microbial Ecology*, 19 (1): 1–20.

Sen, S.K., and Raut, S. 2015. Microbial degradation of low density polyethylene (LDPE): A review. *Journal of Environmental Chemical Engineering*, 3: 462–473.

Shah, A.A., Hasan, F., Hameed, A., and Ahmed, S. 2008. Biological degradation of plastics: A comprehensive review. *Biotechnology Advances*, 26: 246–265.

Shao, Z., Zhao, H., Giver, L., and Arnold, F.H. 1998. Random-priming in vitro recombination: An effective tool for directed evolution. *Nucleic Acids Research*, 26 (2): 681–683.

Sharma, I. 2021. *Bioremediation Techniques for Polluted Environment: Concept, Advantages, Limitations and Prospects*. London: IntechOpen. DOI: 10.5772/intechopen.90453.

Shedbalkar, U., and Jadhav, J. 2011. Detoxification of malachite green and textile industrial effluent by *Penicillium ochrochloron. Biotechnology and Bioprocess Engineering*, 16: 196–204.

Shiomi, N., 2015. Bioremediation of polluted waters using microorganisms. In Shiomi, N. (Ed.), *Advances in Bioremediation of Waste Water and Polluted Soil*. London: IntechOpen.

Shokrollahzadeh, S., Azizmohseni, F., Golmohammad, F., Shokouhi, H., and Khademhaghighat, F. 2008. Biodegradation potential and bacterial diversity of a petrochemical wastewater treatment plant in Iran. *Bioresource Technology*, 99 (14): 6127–6133.

Sidhu, G.K., Singh, S., Kumar, V., Dhanjal, D.S., Datta, S., and Singh, J. 2019. Toxicity, monitoring and biodegradation of organophosphate pesticides: A review. *Critical Reviews in Environmental Science and Technology*, 49: 1135–1187.

Simarro, R., Gonzalez, N., Bautista, L.F., and Molina, M.C. 2013. Assessment of the efficiency of in situ bioremediation techniques in a creosote polluted soil: Change in bacterial community. *Journal of Hazardous Materials*, 262: 158–167.

Singh, A., Kumar, A., and Srivastava, J.N. 2013. Assessment of bioremediation of oil and phenol contents in refinery waste water via bacterial consortium. *Journal of Petroleum and Environmental Biotechnology*, 4: 1–4.

Singh, B., and Sharma, N. 2008. Mechanistic implications of plastic degradation. *Polymer Degradation and Stability*, 93: 561–584.

Singh, B.K., and Walker, A. 2006. Microbial degradation of organophosphorus compounds. *FEMS Microbiology Reviews*, 30: 428–471.

Singh, H. 2006. *Mycoremediation: Fungal Bioremediation*. Hoboken, NJ: John Wiley & Sons.

Singh, R., Singh, P., and Sharma, R. 2014. Microorganism as a tool of bioremediation technology for cleaning environment: A review. *International Academy of Ecology and Environmental Science*, 4 (1): 1–6.

Sinha, S.N., and Biswas, K. 2014. Bioremediation of lead from river water through lead-resistant purple-nonsulfur bacteria. *Global Journal of Microbiology and Biotechnology*, 2: 11–18.

Sinha, S.N., Biswas, M., Paul, D., and Rahaman, S. 2011. Biodegradation potential of bacterial isolates from tannery effluent with special reference to hexavalent chromium. *Biotechnology Bioinformatics and Bioengineering*, 1: 381–386.

Sinha, S.N., and Paul, D. 2014. Heavy metal tolerance and accumulation by bacterial strains isolated from waste water. *Journal of Chemical, Biological and Physical Sciences*, 4: 812–817.Sirianuntapiboon, S., Phothilangka, P., and Ohmomo, S. 2004. Decolourization of molasses wastewater by a strain no. BP103 of acetogenic bacteria. *Bioresource Technology*, 92: 31–39.

Sisco, E., Najarro, M., Samarov, D., and Lawrence, J. 2015. Quantifying the degradation of TNT and RDX in a saline environment with and without UV-exposure. *Forensic Science International*, 251: 124–131.

Sivakumar, G., Xu, J., Thompson, R.W., Yang, Y., and Randol-Smith, P. 2012. Integrated green algal technology for bioremediation and biofuel. *Bioresource Technology*, 107: 1–9.

Sivan, A. 2011. New perspectives in plastic biodegradation. *Current Opinion in Biotechnology*, 22: 422–426.

Skariyachan, S., Manjunath, M., Shankar, A., Bachappanavar, N., and Patil, A.A. 2018. Application of novel microbial consortia for environmental site remediation and hazardous waste management toward low- and high-density polyethylene and prioritizing the cost-effective, eco-friendly, and sustainable biotechnological intervention. In Hussain, C.M. (Ed.), *Handbook of Environment Materials Management*. Cham: Springer, pp. 431–478.

Soleimani, N., Fazli, M.M., Mehrasbi, M., Darabian, S., and Mohammadi, J. 2015. Highly cadmium tolerant fungi: Their tolerance and removal potential. *Journal of Environmental Health Science and Engineering*, 13: 1–9.

Spina, F., Anastasi, A., Prigione, V., Tigini, V., and Varese, G.C. 2012. Biological treatment of industrial wastewater: A fungal approach. *Chemical Engineering Transactions*, 27: 175–180.

Srivastava, S., and Thakur, I.S. 2007. Evaluation of biosorption potency of *Acinetobacter* sp. for removal of hexavalent chromium from tannery effluent. *Iranian Journal of Environmental Health Science and Engineering*, 5 (3): 195–200.

Stapleton, R.D., and Sayler, G.S. 1998. Assessment of the microbiological potential for the natural attenuation of petroleum hydrocarbons in a shallow aquifer system. *Microbial Ecology*, 36 (3–4): 349–361.

Stemmer, W.P. 1994. Rapid evolution of a protein in vitro by DNA shuffling. *Nature*, 370 (6488): 389.

Stevenson, J.A., Bearpark, J.K., and Wong, L.L. 1998. Engineering molecular recognition in alkane oxidation catalyzed by cytochrome P450 cam. *New Journal of Chemistry*, 22: 551–552.

Strong, P.J., and Burgess, J.E. 2008. Treatment methods for wine-related ad distillery wastewaters: A review. *Bioremediation Journal*, 12: 70–87.

Su, C., Jiang, L., and Zhang, W. 2014. A review on heavy metal contamination in the soil worldwide: Situation, impact and remediation techniques. *Environmental Skeptics and Critics*, 3: 24–38.

Sukumar, S., and Nirmala P. 2016. Screening of diesel oil degrading bacteria from petroleum hydrocarbon contaminated soil. *International Journal of Advanced Research in Biological Sciences*, 3: 18–22.

Suresh, A., and Abraham, J. 2018. Bioremediation of hormones from waste water. In Hussain, C. (Ed.), *Handbook of Environmental Materials Management*, Cham: Springer, pp. 1–31.

Takeo, M., Abe, Y., Negoro, S., and Heiss, G. 2003. Simultaneous degradation of 4-nitrophenol and picric acid by two different mechanisms of *Rhodococcus* sp. PN1. *Journal of Chemical Engineering of Japan*, 36 (10): 1178–1184.

Talos, K., Pager, C., Tonk, S., Majdik, C., and Kocsis, B. 2009. Cadmium biosorption on native *Saccharomyces cerevisiae* cells in aqueous suspension. *Acta Universitatis Sapientiae, Agriculture and Environment*, 1: 20–30.

Tang, C.Y., Criddle, C.S., and Leckie, J.O. 2007. Effect of flux (trans membrane pressure) and membranes properties on fouling and rejection of reverse osmosis and nano filtration membranes treating perfluoro octane sulfonate containing waste water. *Environmental Science and Technology*, 41: 2008–2014.

Tang, M., and You, M. 2012. Isolation, identification and characterization of a novel triazophos degrading *Bacillus* sp. (TAP-1). *Microbiological Research*, 167: 299–305.

Tastan, B.E., Ertugrul, S., and Donmez, G. 2010. Effective bioremoval of reactive dye and heavy metals by Aspergillus versicolor. *Bioresource Technology*, 101 (3): 870–876.

Thakkar, A.P., Dhamankar, V.S., and Kapadnis, B.P. 2006. Biocatalytic decolourisation of molasses by *Phanerochaete chrysosporium*. *Bioresource Technology*, 97: 1377–1381.

Thangaraj, K., Kapley, A., and Purohit, H.J. 2007. Characterization of diverse *Acinetobacter* isolates for utilization of multiple aromatic compounds. *Bioresource Technology*, 99: 2488–2498.

Thompson, K.T., Crocker, F.H., and Fredrickson, H.L. 2005. Mineralization of the cyclic nitramine explosive hexahydro-1,3,5-trinitro-1,3,5-triazine by *Gordonia* and *Williamsia* spp. *Applied and Environmental Microbiology*, 71 (12): 8265–8272.

Tian, L., Kolvenbach, B., Corvini, N., Wang, S., Tavanaie, N., and Wang, L. 2017. Mineralization of 14C-labelled polystyrene plastics by *Penicillium variabile* after ozonation pre-treatment. *New Biotechnology*, 38: 101–105.

Tigini, V., Prigione, V., Giansanti, P., Mangiavillano, A., and Pannocchia, A. 2010. Fungal biosorption, an innovative treatment for the decolourisation and detoxification of textile effluents. *Water*, 2: 550–565.

Tokiwa, Y., Calabia, B., Ugwu, C., and Aiba, S. 2009. Biodegradability of plastics. *International Journal of Molecular Sciences*, 10: 3722–3742.

U.S. Environmental Protection Agency, 1997. Profile of the textile industry, Office of Compliance Sector Notebook Project. Washington, DC: US Environmental Protection Agency.

Valderrama, L.T., Del Campo, C.M., Rodriguez, C.M., Bashan, L.E., and Bashan, Y. 2002. Treatment of recalcitrant wastewater from ethanol and citric acid using the microalga *Chlorella vulgaris* and the macrophyte *Lemna minuscula*. *Water Research*, 36: 4185–4192.

Verma, S., Prasad, B., and Mishra, I.M. 2010. Pretreatment of petrochemical wastewater by coagulation and flocculation and the sludge characteristics. *Journal of Hazardous Materials*, 178: 1055–1064.

Vert, M., Doi, Y., Hellwich, K.H., Hess, M., Hodge, P., and Kubisa, P. 2012. Terminology for biorelated polymers and applications (IUPAC Recommendations 2012). *Pure and Applied Chemistry*, 84: 377–410.

Viji, J., and Neelanarayanan, P. 2015. Efficacy of lignocellulotytic fungi on the biodegradation of paddy straw. *International Journal of Environmental Research*, 9 (1): 225–232.

Wanapaisan, P., Laothamteep, N., Vejarano, F., Chakraborty, J., Shintani, M., and Muangchinda, C. 2018. Synergistic degradation of pyrene by five culturable bacteria in a mangrove sediment-derived bacterial consortium. *Journal of Hazardous Materials*, 342: 561–570.

Wang, D., Boukhalfa, H., Marina, O., Ware, D.S., Goering, T.J., Sun, F., Daligault, H.E., ... Starkenburg, S. 2016. Biostimulation and microbial community profiling reveal insights on RDX transformation in groundwater. *Microbiology Open*, 6 (2): 1–4.

Ward, D.M., and Brock, T. 1978. Hydrocarbon biodegradation in hypersaline environments. *Applied and Environmental Microbiology*, 35 (2): 353–359.

Watanabe, K., Kodoma, Y., Stutsubo, K., and Harayama, S. 2001. Molecular characterization of bacterial populations in petroleum contaminated groundwater discharge from undergoing crude oil storage cavities. *Applied and Environmental Microbiology*, 66: 4803–4809.

Watanabe, Y., Sugi, R., Tanaka, Y., and Hayashida, S. 1982. Enzymatic decolourization of melanoidin by *Coriolus* sp. *Agricultural and Biological Chemistry*, 46 (20): 1623–1630.

Wilkes, R.A., and Aristilde, L. 2017. Degradation and metabolism of synthetic plastics and associated products by *Pseudomonas* sp: Capabilities and challenges. *Journal of Applied Microbiology*, 123: 582–593.

Wilkie, A.C., Riedesel, K.J., and Ownes, J.M. 2000. Spillage characterization and anaerobic treatment of ethanol stillage from conventional and cellulosic feedstocks. *Biomass and Bioenergy*, 19: 63–102.

Wu, Y.H., Zhou, P., Cheng, H., Wang, C.S., and Wu, M. 2015. Draft genome sequence of *Microbacterium profundi* Shh49T, an *Actinobacterium* isolated from deep-sea sediment of a polymetallic nodule environment. *Genome Announcements*, 3: 1–2.

Yadav, I.S., and Devi, N.L. 2017. Pesticides classification and its impact on human and environment. *Environmental Science and Engineering*, 6: 140–158.

Yadav, M., Singh, S., Sharma, J., and Deo Singh, K. 2011. Oxidation of polyaromatic hydrocarbons in systems containing water miscible organic solvents by the lignin peroxidase of *Gleophyllum striatum* MTCC-1117. *Environmental Technology*, 32: 1287–1294.

Yan, J., Niu, J., Chen, D., Chen, Y., and Irbis, C. 2014. Screening of *Trametes* strains for efficient decolorization of malachite green at high temperatures and ionic concentrations. *International Biodeterioration and Biodegradation*, 87: 109–115. Available at: https://goo.gl/mJedH2.

Yang, Q., Liang, Y., and Zhou, T. 2008. TNT determination based on its degradation by immobilized HRP with electrochemical sensor. *Electrochemistry Communications*, 10 (8): 1176–1179.

Yang, Q., Yang, M., Pritsch, K., Yediler, A., and Kettrup, A. 2003. Decolorisation of synthetic dyes and production of manganese dependent peroxidase by new fungal isolates. *Biotechnology Letters*, 25: 709–713.

Yang, Y., Yang, J., Wu, W.M., Zhao, J., Song, Y., and Ga, L. 2015. Biodegradation and mineralization of polystyrene by plastic-eating mealworms: Part 1. Chemical and physical characterization and isotopic tests. *Environmental Science and Technology*, 49: 12080–12086. https://doi.org/10.1021/acs.est.5b02661.

Yeruva, D.K., Jukuri, S., Velvizhi, G., Kumar, N.A., Swamy, Y.V., and Mohan, V.S. 2015. Integrating sequencing batch with bio electrochemical treatment for augmenting remediation efficiency of complex petrochemical wastewater. *Bioresource Technology*, 188: 33–42.

Yogesh, P., and Akshaya G. 2016. Evaluation of bioremediation potential of isolated bacterial culture YPAG-9 (*Pseudomonas aeruginosa*) for decolorization of sulfonated di-azodye reactive red HE8B under optimized culture conditions. *International Journal of Current Microbiology and Applied Sciences*, 5: 258–272.

Yoshida, S., Hiraga, K., Takehana, T., Taniguchi, I., Yamaji, H., and Maeda, Y. 2016. A bacterium that degrades and assimilates poly (ethylene terephthalate). *Science*, 351: 1196–1199.

Zang, W.J., Jiang, F.B., and Ou, J.F. 2011. Global pesticide consumption and pollution; with china as focus. *Proceedings of the International Academy of Ecology and Environmental Sciences*, 1 (2): 125–144.

Zhang, T., Shao, M.F., and Ye, L. 2012. 454 pyrosequencing reveals bacterial diversity of activated sludge from 14 sewage treatment plants. *The ISME Journal*, 6 (6): 1137–1147.

Zhu, X., Ni, X., Waigi, M.G., Liu, J., Sun, K., and Gao, Y., 2016. Biodegradation of mixed PAHs by PAH-degrading endophytic bacteria. *International Journal of Environmental Research and Public Health*, 13: E805.

8 Microbial Bioformulation Technology for Applications in Bioremediation

*Jia May Chin and Adeline Su Yien Ting**
School of Science, Monash University Malaysia, Bandar Sunway, Selangor, Malaysia
*Corresponding author Email: adeline.ting@monash.edu

CONTENTS

8.1 Introduction ... 155
8.2 Environmental Pollution and Remediation ... 157
8.2.1 Bioremediation ... 160
8.2.2 Bioformulation ... 161
8.3 Types of Bioformulations ... 162
8.3.1 Solid-Based Bioformulations ... 162
8.3.2 Liquid-Based Bioformulations ... 163
8.4 Bioformulation in Non-Bioremediation Applications ... 164
8.5 Bioformulation in Bioremediation Applications ... 166
8.6 Benefits and Limitations of Bioformulations ... 169
8.7 Commercial Bioformulations on the Market and Their Challenges ... 170
8.8 Conclusion ... 174
Acknowledgments ... 174
References ... 174

8.1 INTRODUCTION

Over the last century, industrialization and urbanization have rapidly progressed. This puts huge pressure on the environment as large quantities of toxic pollutants seeped into the environment. Moreover, these pollutants are often found to exceed the permissible limits in soils and wastewater (Chandra and Kumar, 2015; Agrawal et al., 2021; Singh et al., 2021; Machhirake et al., 2022). The common pollutants found include organic and inorganic wastes, radioactive wastes, toxic gaseous, heavy metals, oils, textile dyes and others from various industries (Chandra and Kumar 2017a, 2017b, 2017c; Bartholameuz et al., 2020; Arya et al., 2021; Kumar et al., 2021a, 2021b). Consequently, toxic pollutants that persist in the environment pose a threat, including the loss of biodiversity, crop losses, and health hazards (Frumkin and Haines, 2019). To manage these pollutants and reduce the impact of environmental hazards, several techniques have been adopted. The remediation strategy proposed and implemented to decontaminate soil and water includes physical, biological and chemical methods (Figure 8.1). This approach is varied: from the conventional groundwater "pump-and-treat" system method to wastewater purification in permeable reactive barriers (PRBs), to the recent new remediation technologies such as nano-remediation and bioremediation (Mulligan, Yong, and Gibbs, 2001; Rao et al., 2001; Wang and Mulligan, 2004; Khalid et al., 2017; Maitra, 2019; Ye et al., 2019).

DOI: 10.1201/9781003239956-10

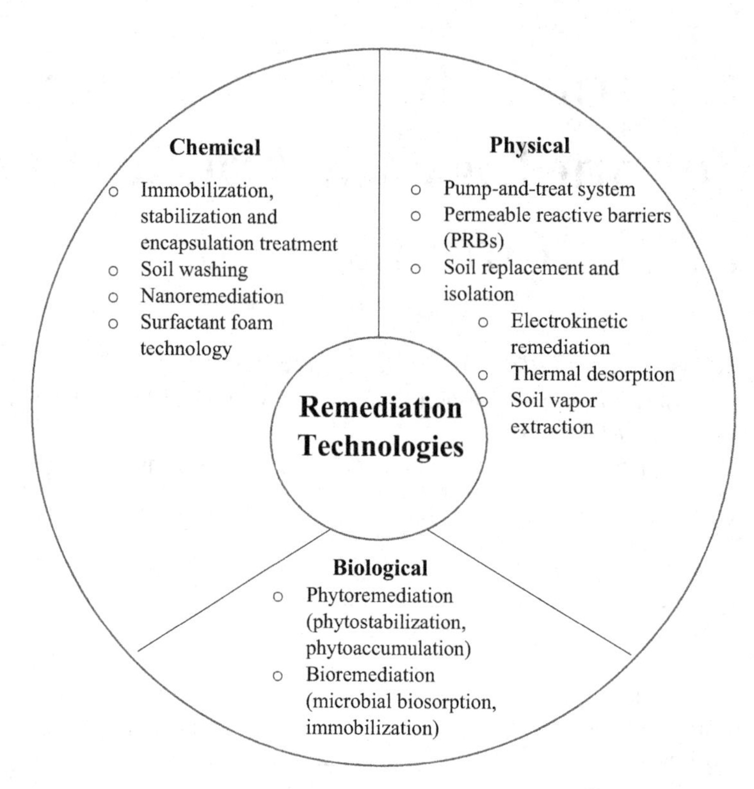

FIGURE 8.1 Summary of conventional and recent remediation technologies.

The conventional technologies, however, have several limitations. The 'pump-and-treat' system, for example, was the most common and popular remediation method used to treat polluted groundwater in the past. However, this method is costly and time-consuming as it may take 5–50 years to completely clean the entire contaminated site. In addition, its efficiency is less compared to recent technologies like bioremediation (Khan, Husain, and Hejazi, 2004). Soil washing is another common permanent treatment in soil remediation but has poor efficacy as well (Dermont et al., 2008). Other conventional technologies such as high-temperature incineration and chemical decomposition have also been introduced to transform toxic contaminants into innocuous substances. These technologies are generally more complex, costly, and are a health hazard to workers due to long-term exposure to the pollutants (Vidali, 2001).

In recent decades, bioremediation has emerged as an effective alternative technology that can treat both contaminated soils and groundwater. This option incurs a relatively lower cost, fewer technology demands, and provides a more environmentally friendly approach that has higher public acceptance (Khan, Husain, and Hejazi, 2004; Kumar et al., 2018). Bioremediation is a technology which involves pollutant degradation by using natural biological activity. This technology generally applies biological agents (microorganisms or microbial enzymes) to the contaminated site to reduce, degrade or immobilize the pollutants, which then may be destroyed or transformed into innocuous or less toxic substances (Kumar et al., 2011; Adams et al., 2015; Kumar et al., 2022). These processes are typically regulated by plants or naturally occurring bacteria and fungi. These organisms may be indigenous or adventive species to the contaminated site. In most cases, the introduction of microbial species into a contaminated site (via bioaugmentation) will lead to the increase in microbial diversity, subsequently accelerating the degradation of the pollutant (Kumar et al., 2011). The bioagents can be a single type of microorganisms or mixed culture, with studies showing successful bioremediation is more likely with the treatment using mixed bacterial culture

(Jabbar, Kadhim, and Mohammed, 2018; Chowdhary et al., 2020a). The biodegradation process of organisms is facilitated by favorable environmental conditions. Hence, most bioremediation technologies would treat the environment with different ingredients which are favorable for the growth of organisms and to enhance microbial activities. Biopiling is a well-known bioremediation technology that promotes biological processes by providing aeration, sufficient water and nutrients, and control of the pH and temperature of the contaminated soil. Biosparging is another bioremediation technology, which treats contaminated groundwater on-site (*in situ* bioremediation). This approach injects air and nutrients into the soil, indirectly increasing the biological activity of indigenous microorganisms (Khan, Husain, and Hejazi, 2004).

Nevertheless, bioremediation technologies, particularly the use of microorganisms, have several challenges limiting their application. As the agents used are biological, several factors affecting the viability of the bioagents have been identified to consequently impact the bioremediation efficacy. This includes the number of toxic contaminants and environmental growth factors in contaminated sites (e.g. nutrient availability, pH, temperature, soil type, aeration),which will influence the shelf-life of the bioagents (Vidali, 2001; Kumar et al., 2011; Chowdhary et al., 2020b). Nutrient availability is essential for microbial growth and to stimulate metabolic activity. Nutrients are critical as they are the basic building blocks of chemical enzymes and the carbon, oxygen, nitrogen, and hydrogen elements from nutrients constitute approximately 92% of the total element composition of a microbial cell (Vidali, 2001). Other known optimal environmental conditions that favour efficient degradation of contaminants are soil moisture (30–90%), oxygen content (10–40%), soil pH (6.5–8.0), temperature (20–30°C), and the concentration of contaminants (5–10% w:w hydrocarbon or <700ppm heavy metals) (ibid.).

To ensure a more comprehensive and effective bioremediation process, improvements can be made. One such strategy is to apply bioformulation to the bioagents (microorganisms). Bioformulation enhances their viability, increases their tolerance to adverse environmental conditions, and subsequently has greater efficacy in removing pollutants. Bioformulations are advantageous as they provide a microenvironment for bioagents, imitating the distinct living requirement of bioagents, thus permitting growth and viability, before application. Bioformulations can be in the form of dust, a semi-solid or liquid formulations. There are generally three main constituents to bioformulation: (1) the microbial agent (the active ingredient); (2) the carrier or matrix to hold or physically protect the biological agent; and (3) the additive to preserve the biological agent (Arora et al., 2016). By providing a suitable microenvironment to these biological agents, viability is enhanced when bioagents are introduced into the soil environment (Bashan et al., 2014). As such, bioformulation is a feasible approach and has been extensively studied for applications in bioremediation, as well as in the development of biofertilizers and biofungicides. This chapter will summarize the current progress in bioformulation in bioremediation. The application of bioformulation in both bioremediation and non-bioremediation technologies (i.e. biofertilizers, biofungicides) will also be discussed, as well as the status of bioformulations in the current market, and its benefits and challenges in these green approaches.

8.2 ENVIRONMENTAL POLLUTION AND REMEDIATION

Pollution, of the air, water, and soil, is the biggest global cause of environmental deterioration. The Lancet Commission on Pollution and Health revealed that these pollutions caused more than 9 million premature deaths in 2015 (Landrigan et al., 2018). It is believed that pollution-related diseases cause more deaths than smoking, infectious diseases or natural disasters (ibid.). Another major concern about the impacts of pollution is the loss of biodiversity and ecosystems due to degradation. According to the World Water Development report in 2017, 80% of global wastewater is most likely released back into our environment without proper treatment (WWAP, 2017). Moreover, water pollution has worsened since the 1990s, attributed to rapid industrialization (WWAP, 2018).

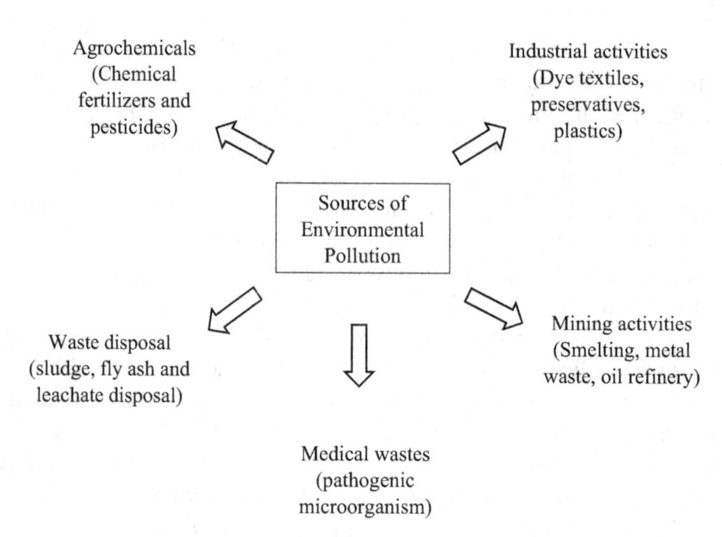

FIGURE 8.2 The main sources of environmental pollution.

With water pollution, soils are inevitably polluted as well, as water is one of the major compositions in soil and the polluted water permeates easily into the soil via leaching into the groundwater or through surface run-off.

The commonest type of environmental pollutant is chemical pollutants, which are typically from agriculture, industrial, waste disposal, and mining activities (Figure 8.2) (Chowdhary et al., 2018; Landrigan et al., 2018; Koli and Hussain, 2019). Chemical pollutants can be further classified into either organic pollutants (e.g. petroleum, organic acids, cellulose bark, chlorobenzene) or heavy metal pollutants (e.g. lead, mercury, metal nitrates, chromium) (Koli and Hussain, 2019).

The pollutants in wastewater differ according to the industries (Table 8.1). With the discharge of contaminated wastewater and the overuse of agrochemicals, soil pollution is inevitable as the hazardous chemicals enter the soil and are adsorbed by soil organic matter. The pollutants are then retained and accumulate in the soil environment (Ren et al., 2018). The presence of heavy metal and organic pollutants influences the composition of the soil organic matter, and this will impact the soil microbial diversity and their ability to decompose organic matter (Maron et al., 2018). The poor rate of decomposition of organic matter consequently affects the nutrient availability to crops, where the land becomes infertile and unsuitable for farming activities.

There is a high health risk when these pollutants come into contact with living organisms. Skin allergies, asthma, eye infections, as well as more severe cases such as cancer diseases, may develop upon exposure to pollutants (Koli and Hussain, 2019). The pollutants are also toxic to plants and animals, especially aquatic animals (Singh and Chandra, 2019). Furthermore, a recent study revealed that the wastewater from the textile industry from Kanpur city in India is phytotoxic, mutagenic, and genotoxic (Khan, Anas, and Malik, 2019). In the agriculture industry, wastewater is found to contain remarkably high levels of pesticides' residues. Pesticides such as organochlorine and carbamate are chemicals used to kill pests or control any pest outbreak that might destroy crops. It will remain in the environment after field application, subsequently leaching into the soil and groundwater, thus polluting the surrounding water sources. These toxic pesticides and their degraded products have an adverse effect on human health (respiratory, neurological, cardiovascular diseases), biodiversity loss and soil degradation (infertile land) (Boudh and Singh, 2019; Koli and Hussain, 2019). Besides organic pollutants, heavy metal contaminants are also toxic to humans, flora and fauna, due to their cytotoxic, carcinogenic, mutagenic and genotoxic properties (Mishra et al., 2019). As such, the removal of pollutants is critical to ensure a sustainable and safe living environment. By identifying

TABLE 8.1
General Types of Pollutants Found in Wastewater From Different Industries

Industry	Pollutants
Food	Organic sugars, residual pesticide, fats, oils and grease, nitrogen, phosphorus, chlorides, sulphates, microorganism
Petroleum and oil refinery	Solvents (benzene, toluene, xylene), nitrogen and sulphur compounds
Paper and pulp	Sulphides, bleaching liquors, organic acids, fibers, cellulose bark, wood sugars, calcium oxide, hydrochloric acid, chlorine dioxide, bisulfites, sodium hydroxide, sodium carbonates, terpenes, methanol, phenol
Chemical	Asbestos, sulfur, chlorinated biphenyl, heavy metals, volatile organic solvents, silica, acids
Textiles	Heavy metal salts, sodium hydrogen peroxide, inorganic chlorine compounds, synthetic polymers, organic compounds (phenols, alkanes, ethers, esters)
Metal	Cyanogens, phenols, ore, coke, limestone, alkalis, oils, heavy metals
Mining	Hydrogen sulphide, chlorides, heavy metals, acids, radioactive uranium
Tannery	Heavy metal, organic waste, lime, sulphides, microoganisms
Agriculture	Chlorinated hydrocarbons, organophosphates, heavy metals, manure and other residual organic compounds from fertilizers and pesticides
Nuclear power plant	Radioactive isotopes

Source: Khan, Anas, and Malik (2019); Koli and Hussain (2019); Singh and Chandra (2019).

the specific pollutants present in wastewater, a solution to decontaminate a polluted site can be rendered. The remediation process is more efficient and cost-effective with suitable treatment.

In the past, conventional waste disposal was by land filling, slowly evolving into more advanced methods such as physical, chemical and biological treatment. Remediation technologies with physical treatment include the removal of pollutants using naturally occurring forces or physical barriers (e.g. soil isolation, vitrification), whereas remediation technologies with chemical treatments apply the use of chemicals to elicit a reaction (e.g. coagulation or oxidation processes) to remove pollutants or convert non-biodegradable substances to biodegradable compounds (Koli and Hussain, 2019). The biological approach allows biodegradable substances to be broken down or detoxified via biological remediation techniques (e.g. aerobic or anaerobic biological processes) (Azubuike, Chikere, and Okpokwasili, 2016; Koli and Hussain, 2019). Remediation technologies have evolved in the last decades and Table 8.2 summarizes the common remediation techniques used since the 1980s.

Most polluted sites are treated via a combination of several techniques. For example, soil washing requires both physical soil isolation and chemical extraction of contaminants. These techniques can also be modified, adjusted and designed specifically based on the properties of contaminants. Therefore, the selection of remediation technologies is a crucial step in achieving successful remediation (Khan, Husain, and Hejazi, 2004; Ahmad et al., 2021). Conventional methods usually require a longer period of time and have a lower efficiency rate. For example, soil washing does not degrade pollutants completely and is less efficient in soil containing clay particles (Khan, Husain, and Hejazi, 2004). The pump-and-treat system is sturdy and effective but requires a large area to install a piece of costly equipment and demands maintenance costs and special handling techniques. The limitations of chemical techniques are that they produce chemical by-products. With increasing requirements in environmental sustainability, the biological approach is explored as this approach is known to have higher efficiency and have less demand for facilities compared to chemical and physical treatments (Maitra, 2019). Biological treatments such as biopiles and bioslurry are relatively simple, easy to install and manage, and only require a short remediation period (Khan, Husain, and Hejazi, 2004). Furthermore, bioremediation is strongly advocated nowadays due to its high sustainability and potential to rehabilitate the environment.

TABLE 8.2
The Various Remediation Technologies and Their Specific Use on Target Contaminants

Remediation techniques	Target contaminants	Description
Soil		
Soil washing (Dermont et al., 2008)	Petroleum residues, heavy metals, pesticides, semi-volatile organic compounds (SVOCs)	*Ex situ* technique. Concentrates pollutants from soil by physical separation or/and solubilizes pollutants with surfactants (chemical extraction)
Soil flushing (Atteia, Estrada, and Bertin, 2013)	Inorganic compounds, pesticides, fuels	Flush (injection or infiltration) the contaminated soils with extraction reagent, then fluid extracted is captured for further treatment before disposal
Thermal desorption (Navarro et al., 2009)	Hydrocarbon petroleum, polychloro biphenyls (PCBs), cyclic aromatic hydrocarbons (PAHs), pesticide, heavy metals	Heat soils at a temperature range of (100–600°C) according to the boiling point of contaminants, separate contaminants from soil after vaporization
Biopiles (Jabbar, Kadhim, and Mohammed, 2018)	Hydrocarbon petroleum, VOCs, pesticides	Contaminated soils are collected into a pile, giving heat, nutrients, water and aeration to stimulate microbial activity. Pile is covered with an impermeable liner to prevent leaching.
Bioslurry (Robles-González, Fava, and Poggi-Varaldo, 2008)	Non-halogenated SVOCs, VOCs, pesticides, PCBs	Contaminated soils are placed into bioreactors, mixed with selected microorganism as suspension under controlled conditions. Treated soil is properly disposed of after water removal.
Phytoremediation (Peng et al., 2009)	Heavy metals, radionuclides, petroleum hydrocarbons, chlorinated solvents, surfactants, etc.	Naturally occurring plants able to accumulate, absorb and degrade pollutants are used on site
Groundwater		
Pump-and-treat (Mackay and Cherry, 1989)	Dissolved metals, VOCs, SVOCs, fuels	Contaminants are pumped out with water into extraction wells and treated with other technologies
Permeable reactive barrier (Maitra, 2019)	Inorganic compounds, VOCs, SVOCs	A barrier is submerged into the area of polluted groundwater, contaminants are trapped when the water flows through
Ultraviolet-oxidation treatment (Hirvonen, Tuhkanen, and Kalliokoski, 1996)	VOCs, aromatics, alcohols, pesticides, dioxins, PCBs and other forms of organic carbons	Contaminated groundwater is treated in a reactor, containing oxygen-based oxidant and UV light. The chemical bonds in organic compounds will be loosened and oxidized.
Biosparging(Kao et al., 2008)	Petroleum hydrocarbon	Contaminants are degraded by naturally occurring organisms with enhancement in aeration and nutrients on site

Source: Khan, Husain, and Hejazi (2004); Khalid et al. (2017).

Note: VOCs: Volatile organic compounds; SVOCs: Semi-volatile organic compounds; PCBs: polychloro biphenyls; PAHs: polycyclic aromatic hydrocarbons.

8.2.1 Bioremediation

Bioremediation is a biological approach to restore the contaminated soils or waters to their natural state by using naturally occurring plants, microorganisms or their derivatives (Kumar et al., 2011; Khalid et al., 2017). Under controlled conditions, the pollutants are accumulated, immobilized,

TABLE 8.3
Advantages and Limitations of Bioremediation Approaches

Advantages	Disadvantages
• Natural process with harmless by-products generated from microbes • Complete destruction of targeted pollutants • Carry out *in situ* indicating less cost in transportation and labor • Able to customize based on specific site	• Limited to biodegradable pollutants and heavy metal pollutants. • Inconsistent speed of biodegradation • Robust engineering and high cost of maintenance • Acclimation of biological organisms • Lab-bench studies require longer time to extrapolate to effectively apply on a full-scale polluted site.

Source: Kumar et al. (2011); Bhatnagar and Kumari (2013).

reduced, transformed and/or degraded, resulting in pollutants achieving a permissible limit in the environment (Adams et al., 2015). This environmentally friendly technology provides a safer and non-invasive solution that aids in the re-establishment of the environment. The first reported microbial agent used in bioremediation was *Pseudomonas putida* (in 1974), due to its ability to degrade petroleum, with more species of bacteria and fungi discovered in the 1990s (Kumar et al., 2011; Bhatnagar and Kumari, 2013). Some of these microorganisms could accumulate, immobilize, or mobilize the pollutants, or induce phytoremediation by plants. Other mechanisms include the production of beneficial compounds such as siderophores, organic acids or biosurfactants, that can solubilize or degrade the pollutants (Ma et al., 2011). The microorganisms can be incorporated into the site for on-site decontamination (*in situ* bioremediation) or applied in bioreactors (*ex situ* bioremediation). Examples of *in situ* bioremediation techniques are bioventing, bioslurping, biosparging and phytoremediation. *Ex situ* bioremediation techniques include excavation of polluted samples, such as biopile, bioreactor, windrows and land farming (Azubuike, Chikere, and Okpokwasili, 2016). The advantages and limitations of the bioremediation approaches are summarized in Table 8.3.

The success of bioremediation, thus, depends on several key factors. This includes the presence of other microbial populations, the metabolic potential of the microorganisms, the availability of nutrients and the growth conditions that may affect the survival of microbes (pH, temperature, moisture level, carbon sources, oxidation-reduction condition) (Vidali, 2001; Bhatnagar and Kumari, 2013). To overcome some of these drawbacks, bioformulation is explored as an alternative to enhance the viability of microbial agents and their subsequent bioremediation efficacy. It has been proven that with a suitable formulation, bioremediation agents could perform better, even with a longer storage period. For example, Viegas et al. (2019) reported good bioremediation activity of lyophilized bacteria cells in a vermiculite carrier after one-month storage at 4°C. This has propelled research on the formulation of bioremediation agents as good storage stability and efficient degradation activity have been observed (Bjerketorp et al., 2018; Viegas et al., 2019).

8.2.2 Bioformulation

Bioformulation is an emerging green technology to enhance the bioremediation process, which can also be called bioinoculation, bioencapsulation or seeding (John et al., 2011; Bhatnagar and Kumari, 2013). Bioformulation comprises biological organisms, substrates, that help to preserve the organisms and a suitable carrier or matrix (Arora et al., 2016). The final products are presented in different forms, either solid bioformulation or liquid bioformulation (Figure 8.3).

Generally, bioformulation can be developed according to both scientific and commercial requirements, whereby the aim is to obtain efficient bioformulations with uniform microbial

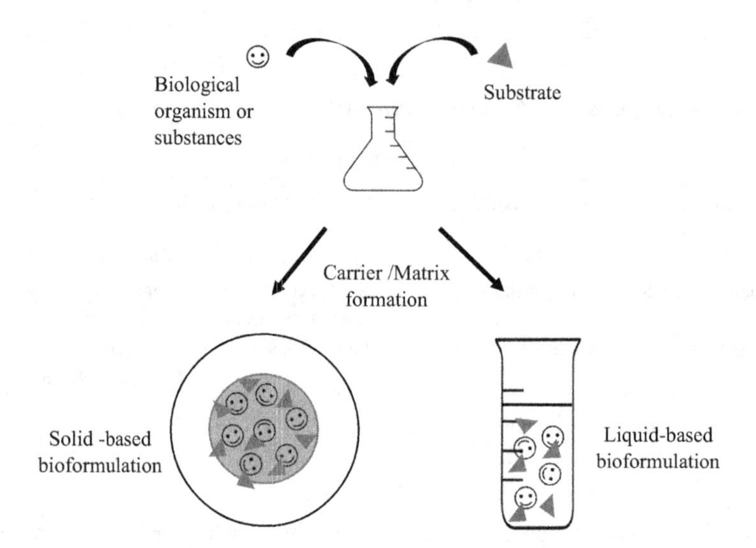

FIGURE 8.3 The schematic diagram of the concept of bioformulation.

coverage and high storage stability (Burges, 1998; Arora et al., 2016). The concept of bioformulation was discussed as early as 1998, when Burges (1998) explained the definition and reviewed the formulation of microbial biopesticides from the 1980s to the 1990s. The biological organisms can be microorganisms (bacteria, fungi) or biomolecules (metabolites or antibodies) or a consortium of different microorganism, depending on the application. Some common microorganisms bioformulated are plant growth-promoting rhizobacteria (PGPRs), biocontrol agents, plant endophytes and rhizobia (Khan et al., 2008; Ashwini and Srividya, 2014; Falcão et al., 2014).

Traditionally, bioformulation was used in agricultural industries to improve the performances of biofertilizers and biopesticides. Bioformulation is also used to inoculate microbes onto seeds and in seed biopriming to enhance crop performances and increase crop resistance to environmental stresses (O'Callaghan, 2016). For example, gum arabic, carboxymethyl cellulose and raffinose were reported as effective additives to enhance the adhesion and shelf-life of *Serratia plymuthica* on rape seeds (Abuamsha, Salman, and Ehlers, 2014). In animal feed development, bioformulation benefits by rendering higher stability of active ingredients when additives are added (Salmon et al., 2011), such as in the case of the enzyme phytase (extracted from *Aspergillus niger*) used in animal feed production (Spier et al., 2011). Bioformulation is an interesting approach as it can be applied in different methods by modifying the formulation. The substrates in bioformulation can be replaced by other bioadditives or biosurfactants according to the natural attributes of the bioagents. This ready-to-use bioformulation provides a handy green solution to enhance the adsorption of bioagents, improve target delivery and the acclimation of the bioagents. Moreover, target-specific bioformulation is cost-effective, non-hazardous and has longer shelf-life compared to the normal chemical formulation.

8.3 TYPES OF BIOFORMULATIONS

There are two major types of bioformulations: the solid and liquid bioformulations. They can be further classified according to the type of carrier materials used and the type of substrates added (e.g. additives, surfactants).

8.3.1 Solid-Based Bioformulations

Solid-based bioformulations use a solid carrier or matrix as carriers of the inoculants. The products appear in dry conditions and can be in the form of powder, dust, granules, pellets, capsules or

TABLE 8.4
Properties of Different Solid-Based Bioformulations

Products description	Granules	Powder	Dust	Pellet/Briquette
Size range	100–1000μm	50–100μm	5–50μm	Pellet: 6–8mm Briquette: 75–90mm
Carrier matrix	Starch, gluten, sugars, semolina, kaolin, talc, polymers	Sand, inert charcoal, fly ash, biopolymers	Clays, lactose, silica, granite, rock dust, talc	Peat, charcoal
Method of application	Direct application	Mix well in aqueous solution with additives before spraying	Direct application or mix in aqueous solution	Direct application
Example	Biopesticides on durum wheat (Mejri, Gamalero, and Souissi, 2013)	Biopesticides on pre- and post-harvest peach (Guijarro et al., 2007)	Biopesticide on cabbage (Faraone, MacPherson, and Hillier, 2018)	Fertilizer-pelleted seed treatment (Grahić et al., 2016)

Source: Burges (1998); Arora et al. (2016); Singh and Arora (2016).

briquette forms (Burges, 1998; Arora et al., 2016). Table 8.4 summarizes the types of solid bioformulations, based on their sizes, coating materials and methods of application. Powder-like and dust formulations are readily miscible and are therefore easily mixed with aqueous solution for spray application (Singh and Arora, 2016). Granules and pelleted formulations are also easily applied and can disintegrate easily upon contact with water. Overall, solid-based bioformulations benefit by protecting the bioagents from the diverse soil and storage environment, as bioagents are stabilized in the immobilized state and therefore have a longer survival period. Bioformulation also enables the slow release of bioagents into the environment, depending on the carrier material, thereby promoting higher sustainability of the bioformulation. However, the cost in applying solid bioformulation is higher due to the need for specialized equipment and tools required in loading solid-based bioformulations. Additional steps may be needed to dissolve solid-based bioformulations and large amounts may be required to achieve the best possible results. Solid-based bioformulations also occupy more space in storage and during transportation (ibid.). Moreover, there is high risk of inhaling fine dust while handling the powdered bioformulations, which is hazardous to human health.

8.3.2 Liquid-Based Bioformulations

Liquid-based bioformulations include bioinoculant concentrates, additives or adjuvants, and are usually in the form of suspension, emulsion or oil dispersion (Manikandan et al., 2010; Nagachandrabose, 2018). Liquid-based bioformulations can be oil-based or aqueous-based (water or polymer suspensions), amended with different inert components such as surfactants, additives and adhesives to improve the adherence and homogeneity of the bioagent (Singh and Arora, 2016). Table 8.5 summarizes the types of liquid-based bioformulations, coating materials and methods of application. Suspension concentrates are relatively cheaper to produce and are the most popular among all types of liquid-based bioformulation, as they only need water or a culture medium as their carrier matrix. Suspension concentrates are fine ground particles in an aqueous solution, not readily dissolved but well dispersed after agitation; therefore, easy to apply and no specialized tools are required. The liquid droplets of emulsion (0.1–10μm) are slightly smaller than suspension

TABLE 8.5
Properties of Different Liquid-Based Bioformulations

Product description	Suspension concentrates	Emulsions	Oil dispersions
Carrier matrix	Water; amended culture medium, polymer suspensions	Tween 20, plant oils, glycerine, water-soluble wax, sterile distilled water	Mineral or plant oils (addition of emulsifier)
Method of application	Dispersion of fine particulate suspensions with addition of dispersing agent	Dispersion of suspension concentrates with another immiscible liquid (oil)	Dispersion of particulate in oil, dilution might be needed
Example	Biopesticide against *Fusarium* wilt in tomato (Wong et al., 2019)	Biopesticide against almond bark beetle (Batta, 2007)	Oil-based liquid bioformulation of biopesticide against cacao black pod disease (Mbarga et al., 2014)

Source: Singh and Arora (2016).

concentrates (1–10μm) and are less evaporated due to the presence of the oil. Oil dispersions are comprised of thick concentrates (prepared from solid form bioagents) in oil and are diluted before application. They can be used to deliver hydrophobic bioagents, enhance retention and adherence (Singh and Arora, 2016). However, a low-temperature condition is needed for long-term storage due to the low stability of the liquid-based bioformulations.

8.4 BIOFORMULATION IN NON-BIOREMEDIATION APPLICATIONS

Bioformulation technology finds major uses in agriculture, followed by food and the pharmaceutical industry. Agricultural bioformulation is comprised mostly of beneficial microbial inoculants that are used to improve soil fertility in the field, to enhance plant growth or to control (biocontrol) phytopathogens (Arora et al., 2016). Instead of using agrochemicals which are known to impact soil microbial composition and fertility, the bioagents in bioformulations allow more nitrogen fixation and can enhance the nutrients' solubility and uptake by plants (Kumar, 2014). Bioformulation of bioagents is therefore an alternative to agrochemicals and is highly favored due to their non-toxic and biodegradable nature (ibid.). Bioformulations have also been found to successfully improve the shelf-life and viability of the bioagents, and also allow for better delivery of the bioagents. For example, a talc-based bioformulation and a wood flour bioformulation were compared on their efficacy in prolonging the shelf-life of bacterial bioagent, *Bacillus subtilis*, *Bacillus cereus* and *Pseudomonas fluorescens*. Results showed that wood flour bioformulations significantly reduced the disease incidence rate of cantaloupe root rot caused by *Fusarium solani*, and both talc-based and wood flour bioformulations successfully prolonged the shelf-life of the bioagents (Sallam et al., 2013). The viability of all bacterial formulation was prolonged to at least five months (approximately 10^5–10^6 CFU/mL) before a significant reduction was found (zero viability in the sixth month).

The microorganisms most typically selected for bioformulation include plant-growth-promoting rhizobacteria (PGPRs), endophytic bacteria (EBs) and fungi, that are capable of promoting crop performance and protecting plants from diseases via induced systemic resistance (Hardoim, van Overbeek, and van Elsas, 2008; Harish et al., 2008). Bioformulation for agricultural use is often developed according to the requirement from farmers, as well as cost and environmental considerations (Arora, Khare, and Maheshwari, 2010). In one study, the biopriming of banana plantlets were performed using endophytic bacteria (bioagent) in a talc-based formulation, which consists of talcum powder, calcium carbonate and carboxymethyl cellulose, and this successfully suppressed the banana bunchy top virus incidence (Harish et al., 2008). Mbarga et al. (2014) conducted a study to compare oil-based and aqueous-based bioformulations used in the formulation of the biological control agent *Trochoderma* spp. It was discovered that soybean oil-based bioformulation could decelerate the desiccation process and retained a longer shelf-life of *Trichoderma* conidia than the aqueous suspension (Mbarga et al., 2014). Bioformulation has also been extended to seed treatment, where the use of beneficial microorganisms on seeds has a positive effect on the germination rate and the stress tolerance (O'Callaghan, 2016). This approach uses seed biopriming techniques, whereby seeds are coated with bioagents under moist conditions in different formulations to promote uniform seed germination and defense responses against diseases (Ferrigo et al., 2014; Singh et al., 2016). For example, a commercial bioformulation of *Trichoderma harzanium* (strain T22) in the form of granules was used in biopriming maize seeds and results reported a significant reduction in *Fusarium verticillioides* disease incidence (58%) and mycotoxin contamination (53%). Other studies reported that *Pseudomonas fluorescens* and *Trichoderma harzanium* bioformulated with talcum powder had better seed germination rate, better yield and a significant reduction in *Fusarium* wilt disease incidence rate and severity when used in tomato seed treatment (Srivastava et al., 2010).

Other than for agricultural applications, bioformulation is also used in the food, pharmaceutical and cosmetic industries. In the food industry, bioformulation is mainly used for food preservation. Formulated bioagents are applied to pre- and post-harvest commodities to prevent rapid deterioration, in addition to maintaining the nutritional and economic value of post-harvest fruits/vegetables. One such example is *Pseudomonas syringe* that was bioformulated as a bioprotectant to reduce the risk of post-harvest diseases on citrus fruits (Singh and Arora, 2016). In another study, a wettable powder formulation of *Penicillium frequentas* enhanced conidial adherence on peach fruits to suppress brown rot disease of *Monilinia* spp. (Guijarro et al., 2007). In the cosmetics industry, bioformulations of active compounds are used in various products such as lotions, serums, body scrubs, masks, moisturizers and facial cream. Algae extract is one of the most common bioagents used in cosmetics production. Algae are a rich source of phytochemicals that can be refined and used in formulation to replace chemical additives (Michalak, Chojnacka, and Saeid, 2017). For example, extracts from the marine microalgae *Isochrysis* were used to replace chemical waxes in lipstick and lip balm, and they had comparable results with other waxes used in cosmetics (McIntosh et al., 2018). In the animal feed industry, bioactive substances are formulated into the animal diet to improve animal growth. A study on shrimp feed reported that a supplement bioformulation of algae extracts and minerals in dry pellet form promoted the shrimp growth rate (Ju, Forster, and Dominy, 2009). Similarly, the biotechnology industry has developed pharmaceutical formulas, such as ointments, solutions or creams, whereby bioformulation of natural products are used. For example, Buzia et al. (2018) successfully formulated bioactive compounds (capsaicin and piperine) from black pepper (*Piper nigrum*) and hot pepper (*Capsicum annuum*) with water, azotic substances, extractive substances and nitrogen, fats, cellulose and ashes. These bioformulated ointments and creams have antibacterial, anti-inflammatory and anti-rheumatic properties; and it is reported that the bioformulations prolonged the anti-rheumatic effect (ibid.). In recent years, pharmaceutical bioformulation has been expanding from internal administration to topical medication. One study focused on developing new bioformulations of the essential oil (EO) as alternatives to synthetic insect repellents. EO bioformulations (extracted from *Piper aduncum*) were made in the form of ointment, cream and gel and results showed that ointment (78.41%) could retain EO more effectively to repel *Aedes aegypti* mosquito, followed by cream (71.92%) and gel formulation (67.71%) (Mamood et al., 2017).

Bioformulation is therefore sought as a promising green approach to enhance environmental bioremediation. Bioremediation has higher public acceptance due to its low toxicity and other beneficial biological properties. Thus, bioformulation is gaining popularity for applications in bioremediation, as it reduces cost and enhances the efficacy of bioremediation technologies.

8.5 BIOFORMULATION IN BIOREMEDIATION APPLICATIONS

Most of the existing bioformulations for bioremediation can be applied in both contaminated soils and water. Table 8.6 describes a few recent studies on bioformulation in bioremediation based on different types of pollutants. Bioremediation of soils contaminated with agrochemicals has been widely studied using various microorganisms (bacteria, fungi, algae and yeast), and may include their metabolites. For example, biosurfactants produced by the bacteria strain, *Pseudomonas cepacia* were formulated with potassium sorbate (a stabilizer salt), and were successful in removing hydrophobic hydrocarbon on coral reefs (Da Silva et al., 2017). In their study, potassium sorbate acts as a preservative in this cell-free liquid bioformulation to stabilize the biological properties of biosurfactant. The biosurfactant then helps in solubilizing the hydrocarbon, thereby increasing the surface area for more biodegradation by the surrounding microorganisms (Rocha e Silva et al.,

TABLE 8.6
Bioformulation of Various Bioagents for the Bioremediation of Various Pollutants

Target pollutants	Bioagent	Carrier materials	Application method	Reference
Agrochemicals				
Organochlorine pesticide	Actinobacteria Consortium (*Streptomyces* sp. A2-A5-A11-M7-AC5-AC7)	Sodium alginate, calcium chloride	Apply alginate beads containing immobilized actinobacteria	(Fuentes et al., 2013)
Organophosphorus pesticide	Consortium (*Streptomyces* sp. AC5, AC9, GA11 and ISP13)	Liquid minimal medium supplemented with L-asparagine, magnesium sulphate, iron sulphate and dipotassium hydrogen phosphate	Drench mixed culture into contaminated soil (equal volume of each strains)	(Briceño et al., 2018)
Terbuthylazine (TBA)	*Arthrobacter aurescens*	Vermiculite	Resuspend formulated vermiculite in sterile saline solution before applying	(Viegas et al., 2019)
Fipronil	*Streptomyces rochei*	Fly ash, soil, 5% molasses	Apply suspension of pelleted microorganism	(Abraham and Gajendiran, 2019)
Hydrocarbon				
Pyrene (hydrocarbon)	*Pseudoxanthomonas*sp. RN402	Phosphate buffer supplemented with 1% glycerol	Drench suspension concentrates	(Nopcharoenkul, Pinphanichakarn, and Pinyakong, 2011)
Pyrene (hydrocarbon)	*Chlorella* ssp. MM3 (microalgae)	Sodium alginate, calcium chloride	Apply alginate beads containing immobilized microalga	(Subashchandrabose et al., 2017)

TABLE 8.6 (Continued)
Bioformulation of Various Bioagents for the Bioremediation of Various Pollutants

Target pollutants	Bioagent	Carrier materials	Application method	Reference
Petroleum hydrocarbon	*Pseudomonas cepacia*	Liquid culture medium supplemented with canola oil, corn steep liquor, sodium nitrate	Biosurfactant produced by bacteria is applied as emulsion form with potassium sorbate	(Da Silva et al., 2017)
Petroleum hydrocarbon	*Sphingomonaschangbaiensis, Pseudomonas stutzeri*	Biosurfactant alkyl polyglycosides	Apply in a reactor with contaminated soils	(Li et al., 2020)
Metals				
Chromium	*Agrobacterium* sp.	Liquid culture medium	Drench suspension concentrate	(Johanto, Cahyani, and Rosariastuti, 2018)
Copper, sulfate	*Desulfovibrio* sp.	Polyvinyl alcohol, sodium alginate, silica sand, calcium chloride	Apply immobilized beads (containing microalgae and bacteria)	(Li, Yang, and Geng, 2018)
Copper, zinc, nickel, lead	*Streptomyces fradiae*	Glutaraldehyde	Apply immobilized beads	(Simeonova, Godjevargova, and Ivanova, 2008)
Heavy metals and silver nanoparticles	*Phanerochaetechrysosporium*	Liquid culture medium with sodium hydrosulfide	Apply fungal pellet	(Huang et al., 2019)
Synthetic dyes				
Triphenylmethane dyes	*Trichoderma asperellum*	Sodium alginate, calcium chloride	Apply immobilized beads	(Chew and Ting, 2016)
Synthetic polymer				
Polymer (low density polyethylene and polyvinyl chloride)	Consortia (*Microbacterium* sp., *Pseudomonas putida, P. aeruginosa, P. otitidis, Bacillus aerius, B. cereus, Acanthopleurobacter pedis*)	Talc	Talc-based formulation is dissolved in sterile distilled water before application	(Sah et al., 2011)
Polyethylene microplastics	*Zalerionmaritimum*	Liquid minimal medium with glucose, malt extract, peptone, sea salts	Apply fungal biomass into bioreactors with liquid minimal medium	(Paço et al., 2017)
Low density polyethylene (LDPE)Consortia	(*Enterobacter* spp. and *Pantoea* spp.)	Liquid minimal medium	Drench liquid suspension	(Skariyachan et al., 2016)

2014; Da Silva et al., 2017). Some bioformulation is comprised of more than one type of microorganism and these are applied as mixed cultures. For example, Briceño et al. (2018) mixed four different strains of *Streptomyces* sp. in a modified minimal medium consisting of L-asparagine, to promote growth and subsequently enhance the biodegradation of organophosphate pesticides in soil. The liquid (free microorganism cells) and solid (immobilized cells) forms may also yield different levels of effectiveness in bioremediation. A study reported that alginate immobilized a consortium of *Streptomyces* sp. and had a significantly higher pesticide removal rate than free cells in liquid form, but both were able to remove organochlorine pesticides (pentachlorophenol and chlorpyrifos) in soil effectively (Fuentes et al., 2013). The better removal rate by alginate-immobilized *Streptomyces* sp. was attributed to the alginate beads providing a shield to protect the microbial cell from a harsh environment.

In bioremediation of the herbicide terbuthylazine in soil, a comparison between solid-based and liquid-based bioformulation was done using *Arthrobacter aurescens* strain TC1. This strain was formulated into powder (lyophilized cells supplemented with sucrose, glycerol and calcium chloride) form, vermiculite (powdered vermiculite carrier mixed with cell suspension, saline and calcium chloride) and cell paste (without any substrates) liquid form (Viegas et al., 2019). Viegas et al. claimed that the lyophilized powdered cells had 100% efficacy in bioremediation but the lowest (≥65%) in recovering viable cells, followed by vermiculite-adsorbed cells (81% bioremediation efficacy; ≥78% cell recovery) and cell paste (68% bioremediation efficacy; ≥78% cell recovery). This suggested that the lyophilization process may be detrimental to bacterial cell membrane and intercellular proteins, resulting in lower viability rate. However, the lyophilized cells performed better in bioremediation, possibly due to the composition in bioformulation that provided good stability and rich medium for microbial growth (ibid.). The preference in choosing the powder form bioformulation is also evident in a study of agrochemical degradation by *Streptomyces rochei.* Abraham and Gajendiran (2019) reported the effectiveness of *Streptomyces rochei* in degrading fipronil, a pesticide in soil and it showed an increase in cell number when formulated into powder form.

In certain bioformulations, additives were added to aid in rendering protection to the microbial cells and prolong their viability. Nopcharoenkul, Pinphanichakarn, and Pinyakong (2011) used a phosphate buffer containing glycerol in their bioformulation of *Pseudoxanthomonas* sp. RN402, which performed better in degrading polycyclic aromatic hydrocarbon (PAH), as bacterial cells were protected from abiotic stress and osmotic pressure was maintained. Other than bacteria, the microalga isolate, *Chlorella* ssp. MM3 also benefited from the addition of 0.005% Tween 80 into the alginate solid bioformulation, with an increase in 20% degradation of pyrene (Subashchandrabose et al., 2017). The addition of substrates such as hydrogen peroxide, phosphate and nitrates into liquid formulations of microbial consortia (*Pseudomonas aeruginosa* and *Bacillus subtilis*) were also able to enhance the degradation of hydrocarbon pollutants (benzene, toluene, and mixture of xylenes (BTX)) (Mukherjee and Bordoloi, 2012). Their study discovered that the substrates served as food sources for the bacteria, thereby enhancing microbial growth and the biodegradation process.

In bioremediating metal-polluted wastewater, it was discovered that metal immobilization, reduction and biosorption are the common principles considered in the development of bioformulation of bacterial isolates. Bioformulations of *Desulfovibrio* sp. (Li, Yang, and Geng, 2018), *Streptomyces* sp. (Simeonova, Godjevargova, and Ivanova, 2008), *Agrobacterium* sp. (Johanto, Cahyani, and Rosariastuti, 2018), *Aspergillus* sp. (Price, Classen, and Payne, 2001); and fungi such as *Pleurotus ostreatus*(Wang et al., 2019) and *Phanerochaete chrysosporium* (Chen et al., 2019; Huang et al., 2019) have been attempted. In one study, *Desulfovibrio* sp. (a sulfate-reducing bacteria, SRB) was immobilized in beads supplemented with microalgae (as the nutrient source), and this produced higher copper and sulfate removal rate than liquid suspended SRB (Li, Yang, and Geng, 2018). Synthetic dyes are also one of the common pollutants that are bioremediated by microorganisms. Immobilization of *Trichoderma asperellum* (Chew and Ting, 2016) and *Coriolopsis* sp (Chen and Ting, 2015) has been attempted. *T. asperellum* immobilized in alginate beads was reported to remove triphenylmethane dyes more effectively than free cells, due to a higher binding site present on alginate surface for dye molecules (Chew and Ting, 2016).

8.6 BENEFITS AND LIMITATIONS OF BIOFORMULATIONS

Bioformulation is aimed at benefitting the bioagents. The use of carrier materials in bioformulation aids in the dispersion of the bioagents and improves the overall stability of the bioagents. The carrier materials also serve as a coating matrix to shield the bioagents from adverse abiotic factors in soils or water, such as extreme pH, and from temperature and moisture stresses (Arora, Khare, and Maheshwari, 2010). Bioagents can be formulated into different forms, each having their own advantages as well as limitations, as summarized in Table 8.7. The liquid-based bioformulations are cost-effective as they are easier to handle and store, more uniform in dispersal and with higher efficiency. On the other hand, solid-based bioformulations are cheaper, and able to stabilize the bioagent and provide a longer shelf-life (Maheshwari et al., 2015). This was evident in a study whereby different types of carrier bioformulations of two common plant growth-promoting rhizobacteria (*Bacillus* and *Pseudomonas* sp.) were compared based on their shelf-life. Vermiculite, sawdust, sand and liquid effluent were four carrier matrixes chosen, vermiculite (solid-based bioformulation) was the best carrier with the lowest reduction in bacterial count. Vermiculite was claimed to have better water-holding capacity, optimum pH and moisture content that are more compatible with the bioagent, whereas the liquid effluent carrier matrix (liquid-based bioformulation) could only sustain the viability of the bioagent for a shorter period of time due to the faster rate of nutrient depletion (Maheshwari et al., 2015). Overall, both solid- and liquid-based bioformulations protect the bioagent from a harsh enironment, promote bioagent growth and prolong the survival rate during storage.

In fact, the presence of additives in bioformulation has also benefited the bioagent, with evidence showing positive changes in the growth, physiological and biochemical aspects of the bioagent (Singh and Arora, 2016). This kind of bioformulation stabilizes the bioagent with more nutrients and thereby improves its bioremediation efficacy. Several additives such as glucose, sucrose, xanthan, gum arabic, glycerol polyvinyl and polyethylene glycol, are used as carbon and nutrient sources for the growth of bioagents (Nopcharoenkul, Pinphanichakarn, and Pinyakong, 2011). The growth of the bioagent was evident in a study of a liquid bioformulation consisting of pyrene-degrading bacteria, *Pseudomonas aeruginosa* LP5 (with addition of 1.5% corn steep liquor). Corn steep liquor is a rich source of organic nitrogen, mineral, carbon and other essential nutrients, and produced higher biomass than bioformulation without additives. This liquid-based

TABLE 8.7
The Advantages and Limitations of Bioformulations

Advantages	Limitations
• Extend shelf-life of biological inoculant • Low cost in each bioformulation ingredient • Uniform spreading and additives increase bioremediation efficacy • Overcome acclimation of microbial inoculant in harsh environment • Stabilize inoculants during storage and transportations	• Inconsistent performances due to extrinsic factors • High cost in acquiring instrument for mass production • Special storage requirement (e.g. low humidity and temperature)

Source: Arora, Khare, and Maheshwari (2010); John et al. (2011).

bioformulation also significantly removed pyrene better than bioformulation without additives by 37.05% (Obayori et al., 2010).

Bioformulation is, unfortunately, not without its limitations. Several challenges to this approach have surfaced, such as the inconsistent performances due to the constantly changing environment and other extrinsic factors such as the indigenous bacterial population and the potentially toxic by-products of degraded pollutants. Also, a specific storage condition is needed to further prolong the shelf=life of the bioformulation. This is evident in a study focused on the storage temperature and use of sugar protectants in maintaining the viability of yeast bioagents (*Crytococcus laurentii* and *Pichia membranaefaciens*) in liquid formulation. Results showed the liquid bioformulation retained a higher survival rate of bioagents when stored at 4°C (with the addition of sugar) than at room temperature (Liu et al., 2009). This emphasizes the limitations of bioformulation whereby every parameter is important in optimizing a bioformulation, to ensure the optimum bioremediation efficacy.

8.7 COMMERCIAL BIOFORMULATIONS ON THE MARKET AND THEIR CHALLENGES

To date, there have been several commercially available bioformulations on the market. More research has been done on agricultural bioformulation than bioremediation (as cited in Table 8.8). Most of the products only revealed the active ingredients (bioagents) but not the carrier materials and additives. In the agricultural industry, biopesticide and biofertilizers are commonly used in agro-economically valuable crops. They are published and commercialized as organic use. For example, Actinovate® AG is a commercialized biofungicide in the United States, containing *Streptomyces*

TABLE 8.8
Commercially Available Bioformulations on the Market

Products	Bioagents	Formulation	Application	Function	Reference
Agricultural					
Actinovate® AG	*Streptomyces lydicus* WYEC 108	Solid-based	Solubilize in water for soil drenching, foliar spray or seed treatment	Control or suppress soil-borne plant diseases	(Himmelstein, Maul, and Everts, 2014)
Bio-Save	*Pseudomonas syringae*	Solid-based	Soil drench/spray	Prevent post-harvest disease of potatoes and fruits	(Al-Mughrabi et al., 2013)
Mycostop®	*Streptomyces griseoviridis* Strain K61	Solid-based	Soil spray/drench; seed treatment	Suppress *Botrytis* gray mold and root rot disease	(Peng et al., 2011)
RootShield Plus⁺ Granules	*Trichoderma harzanium* Rifai strain T-22, *Trichoderma virens* strain G-41	Solid-based	In-furrow soil treatment	Suppress soil-borne diseases	(Peng et al., 2011)
Sonata	*Bacillus pumilus* QST 2808	Liquid-based	Soil drench/spray	Control powdery mildew, downy mildew, rust, scab	(Borriss, 2011)
Serenade ASO	*Bacillus subtilis*	Liquid-based	Direct foliar spray	Suppress fruit damage caused by gray mold	(Peng et al., 2011)

TABLE 8.8 (Continued)
Commercially Available Bioformulations on the Market

Products	Bioagents	Formulation	Application	Function	Reference
Bioremediation					
BIOLEN IG30	Mixture of microorganism (not disclosed)	Solid-based	Solubilize in water, spray application	Biodegradation of hydrocarbon compound	(Bergueiro et al., 1998)
PRP® Powder	Mixture of microorganism and nutrients (not disclosed)	Solid-based	Direct application or solubilize in water	Absorption and biodegradation of hydrocarbon compound	unireminc.com/portfolio/prp-powder/
Oil Buster ™	Corncob and nutrients	Solid-based	Direct application	Absorption and aid in biodegradation of hydrocarbon compound	unireminc.com/portfolio/oil-buster
Enretech-1	Adsorbent with microorganism	Solid-based	Direct application	Biodegradation of petroleum contaminated soil	enretech.com.au/need-help-with-a-spill/remediation/
Inipol EAP22	Oleic acid and urea	Liquid-based	Spray application	Accelerate biodegradation of hydrocarbon compound	(Nikolopoulou and Kalogerakis, 2018)
RemActiv™	Mixture of nutrients and microorganism	Liquid-based	Spray application	Biodegradation of hydrocarbon compound	(Asquith et al., 2012)
Recycler Fuel Oil	Bacteria, actinomycetes and fungi	Liquid-based	Spray application	Biodegradation of aliphatic and aromatic hydrocarbons	(Asquith et al., 2012)
AgriRem®(A100, A90, A80, A70, A60, A50)	Mixture of microorganism	Liquid-based	Direct application	Biodegradation of hydrocarbon compounds, oils and greases, VOCs, PAHs, alcohols, heavy metals	www.agriremearthcare.com.au/AgriRem_Organic_Soil_Products.html
S-200 OilGone®	Oleic acid	Liquid-based	Spray application	Accelerate biodegradation of hydrocarbon compound	iepusa.com/s200.html
EOS 100	Vegetable oil	Liquid-based	Direct application	Biodegradation of long-lasting hydrocarbon compounds	www.eosremediation.com/eos-100-product-information

Note: VOCs: Volatile organic compounds; PAHs: polycyclic aromatic hydrocarbons.

lydicus WYEC 108 which is developed in wettable powder forms (Himmelstein, Maul, and Everts, 2014). It can be dissolved easily for use by soil drenching or via seed treatment. Both liquid-based and solid-based bioformulations are equally popular on the market. A biofungicide, *Bacillus subtilis* strain QST 713, called Serenade ASO (Bayer, the United Kingdom) is formulated in suspension concentrate, which can be directly used as a foliar spray for strawberry crops. Serenade ASO was also reported as effective as fungicides fluazinam and cyazofamid in reducing canola club-root disease (Peng et al., 2011). Most of the existing bioformulations in agriculture are solid-based formulations which can be solubilized in water for soil drenching or foliar spray application. Most bioformulations are compatible with other additives such as spreader-sticker, which can be used in conjunction with an application for better results.

In the bioremediation industry, the commercialized bioformulations are mostly found in powder forms (solid form) or liquid concentrates (liquid form). Both types are effective in bioremediating the corresponding pollutants. Products, such as BIOLEN IG30 and PRP® Powder, exist in powder form, contain a consortium of bacterium and nutrients that can be directly applied or solubilized in water, then sprayed or drenched to degrade hydrocarbon compounds. On the contrary, other solid-form bioformulation such as dry absorbent (Enretech-1) from Australia can only be directly applied in hydrocarbon-contaminated soil. Then, the hydrocarbon compounds are absorbed and encapsulated in the absorbent containing hydrocarbon-degrading microorganisms. On the contrary, a liquid-based bioformulation is easy to handle and quick to apply, such as RemActiv™. It is formulated as a liquid concentrate with selected naturally occurring microorganisms and a specific ratio of nutrients to degrade petroleum hydrocarbon in soil (Asquith et al., 2012). Similarly, Allcrobe Pty. Ltd. developed a series of bioremediation products in liquid bioformulations to degrade hydrocarbon compounds (Product name: Recycler Fuel Oil), organic compounds (Product name: Recycler Chemicals) and hydrogen sulphide (Product name: Recycler Grease Trap); and these commercially available liquid-based bioformulations are easily applied by diluting for spray or drench application. There is also non-microbial liquid-based product available on the market, with a study reporting the use of fertilizer bioformulation to improve the bioremediation efficacy. Inipol EAP 22 is a biodegradation accelerator, which acts as an oil-soluble additive that provides a rich carbon source (oleic acid) to the indigenous bacteria for bioaugmentation purposes (Bergueiro et al., 1998). It can be directly applied onto spill surfaces to enhance microbial growth and stimulate the hydrocarbon breakdown by indigenous hydrocarbon-degrading microorganisms. And contrary to other products, AgriRem® A100 from AgriRem Earth Care has multi-bioremediation properties in degrading hydrocarbons, VOCs, alcohols and heavy metals. The AgriRem bioformulations are also developed based on different pollutant sites, including hydrocarbon-based soils (A100), salinated land (A90), wastewater treatment (A80), sustainable turf management (A70), agricultural land (A60) and domestic garden and lawns (A50).

Despite the promising discovery of bioagents in bioremediation, very few are published as commercial products. The bioremediation industry has been facing challenges in maintaining consistency of bioremediation performances and developing a broad spectrum bioformulation. The bioremediation performances are dependent on the environmental conditions, where extreme environments are usually the harshest environment to support microbial growth. The change in intrinsic physiological and biological characteristic of the bioagent after application is often less known, due to the complexity in the interaction between indigenous microbes, bioagents and pollutants. Besides that, a broad spectrum bioformulation is currently unattainable due to the single function each bioremediating agent. To combine all beneficial bioagents in one, a lot more research efforts are needed to overcome the possibility of bioagent incompatibility. By applying multiple types of bioagents, the bioformulation must always be re-evaluated in a recycle loop manner according to the different situations, and typically follows the pathway in Figure 8.4 (Koli and Hussain, 2019). The selection of a carrier matrix and the addition of substrates are therefore crucial in the bioformulation design as they prolong the shelf-life of the bioagent.

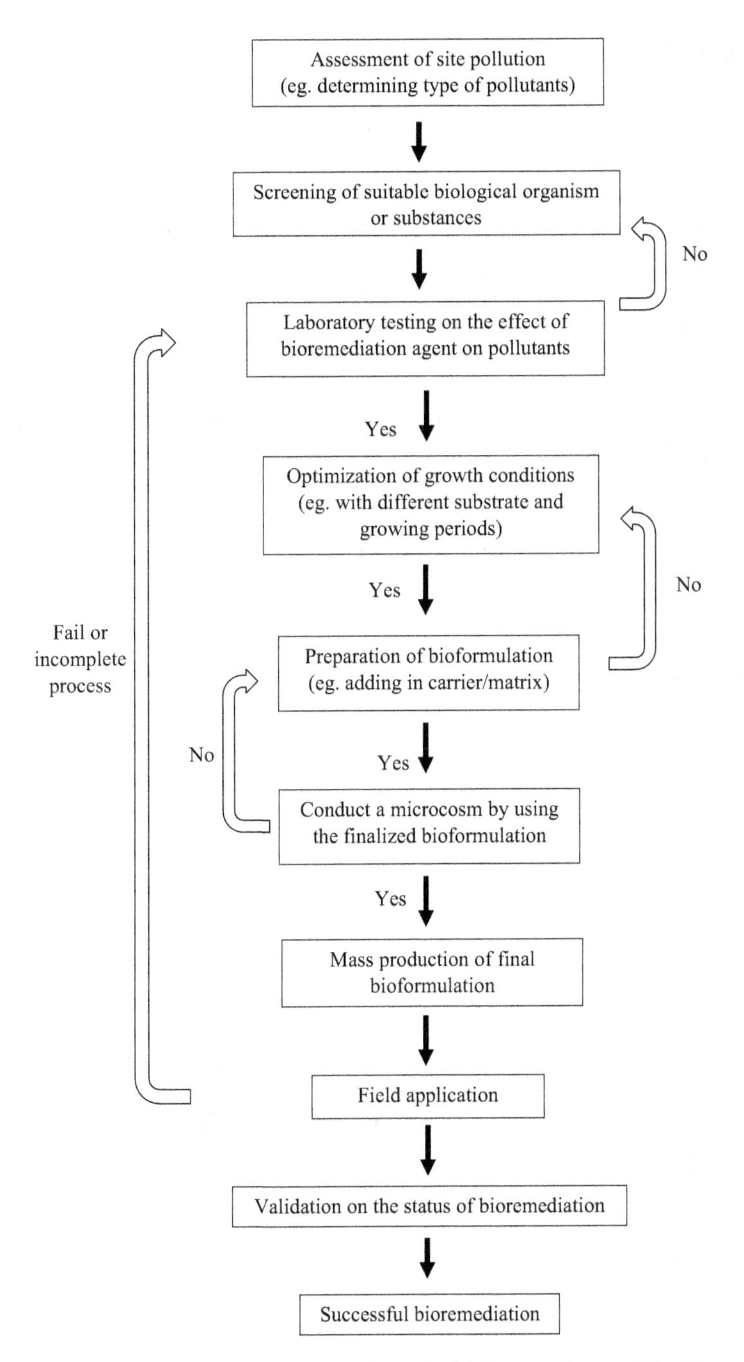

FIGURE 8.4 The proposed flow chart of customization of a bioformulation modified from the agricultural waste management method.

Source: Koli and Hussain (2019).

8.8 CONCLUSION

Bioformulation is undoubtedly a unique strategy in maximizing the bioremediation efficacy, making this green technology a better option among all the remediation technologies. The development of a suitable bioformulation for bioagents has brought many advantages in enhancing the efficacy of the bioremediation, with a higher prospect of environmental restoration and sustainability compared to the conventional bioremediation approach. A properly optimized bioformulation possesses great potential in the complete removal of pollutants. Moreover, the merit of the bioformulation technology is the formulation attributes, and the application methods that simplify the whole process for the end users. However, there is much to be explored as the consistency in the performance of bioformulations is not definite and is dependent on various extrinsic factors. The use of additives is expected to further stabilize the bioformulation. The development of suitable bioformulations may benefit from comprehensive interdisciplinary studies (involvement of soil chemistry, microbiology, plant biology, engineering, genomics, ecology) to evaluate the bioformulation under multiple environmental conditions. This can subsequently, improve our understanding of the microbial interaction with other organisms and with the environment.

ACKNOWLEDGMENTS

The authors acknowledge support from the Malaysia Toray Science Foundation for providing research funding and Monash University Malaysia for the funds and research facilities.

REFERENCES

Abraham, J., and Gajendiran, A. 2019. Biodegradation of fipronil and its metabolite fipronil sulfone by *Streptomyces rochei* strain AJAG7 and its use in bioremediation of contaminated soil. *Pesticide Biochemistry and Physiology* 155: 90–100.

Abuamsha, R., Salman, M., and Ehlers, R.-U. 2014. Role of different additives on survival of *Serratia plymuthica* HRO-C48 on oilseed rape seeds and control of *Phoma lingam*. *British Microbiology Research Journal* 4 (7): 737.

Adams, G.O., Fufeyin, P. T., Okoro, S. E., and Ehinomen, I. 2015. Bioremediation, biostimulation and bioaugmention: A review. *International Journal of Environmental Bioremediation & Biodegradation* 3 (1): 28–39.

Agrawal, N., Kumar, V., and Shahi, S. K., 2021. Biodegradation and detoxification of phenanthrene in in-vitro and in-vivo conditions by a newly isolated ligninolytic fungus *Coriolopsis byrsina* strain APC5 and characterization of their metabolites for environmental safety. *Environmental Science and Pollution Research.* https://doi.org/10.1007/s11356-021-15271-w

Ahmad, A., Singh, A. P., Khan, N., Chowdhary, P., Giri, B. S., Varjani, S., and Chaturvedi, P. (2021). Bio-composite of Fe-sludge biochar immobilized with Bacillus Sp. in packed column for bio-adsorption of Methylene blue in a hybrid treatment system: Isotherm and kinetic evaluation. *Environmental Technology & Innovation* 23: 101734.

Al-Mughrabi, K., Vikram, A., Peters, R., et al. 2013. Efficacy of *Pseudomonas syringae* in the management of potato tuber diseases in storage. *Biological Control* 64 (3): 315–322.

Arora, N., Khare, E., and Maheshwari, D. 2010. *Plant Growth Promoting Rhizobacteria: Constraints in Bioformulation, Commercialization, and Future Strategies*. Microbiology monograph. Cham: Springer.

Arora, N. K., Mehnaz, S., and Balestrini, R. (Eds.) 2016. *Bioformulations: For Sustainable Agriculture*. New Delhi: Springer (India) Pvt. Ltd.

Arya, S., Patel, A., and Kumar, S. 2021. Urban mining of obsolete computers by manual dismantling and waste printed circuit boards by chemical leaching and toxicity assessment of its waste residues. *Environmental Pollution* 283: 117033. https://doi.org/10.1016/j.envpol.2021.117033

Ashwini, N., and Srividya, S. 2014. Potentiality of *Bacillus subtilis* as biocontrol agent for management of anthracnose disease of chilli caused by *Colletotrichum gloeosporioides* OGC1. *3 Biotech* 4 (2): 127–136.

Asquith, E. A., Geary, P. M., Nolan, A. L., and Evans, C. A. 2012. Comparative bioremediation of petroleum hydrocarbon-contaminated soil by biostimulation, bioaugmentation and surfactant addition. *Journal of Environmental Science and Engineering A* 1 (5A): 637–640.

Atteia, O., Estrada, E. D. C., and Bertin, H. 2013. Soil flushing: A review of the origin of efficiency variability. *Reviews in Environmental Science and Bio/Technology* 12 (4):379–389.

Azubuike, C. C., Chikere, C. B., and Okpokwasili, G. C. 2016. Bioremediation techniques-classification based on site of application: Principles, advantages, limitations and prospects. *World Journal of Microbiology & Biotechnology* 32 (11): 180.

Bartholameuz, E. M., Hettiaratchi, J. P., Steele, M., and Kumar, S. 2020. Reaction kinetic analysis of manganese peroxidase augmented aerobic waste degradation. *Journal of Hazardous, Toxic, and Radioactive Waste* 24 (4): 04020043. https://doi.org/10.1061/(ASCE)HZ.2153-5515.0000536

Bashan, Y., de-Bashan, L., Prabhu, S., and Hernandez, J.-P. 2014. Advances in plant growth-promoting bacterial inoculant technology: Formulations and practical perspectives (1998–2013). *An International Journal on Plant-Soil Relationships* 378 (1): 1–33.

Batta, Y. 2007. Biocontrol of almond bark beetle (Scolytus amygdali Geurin-Meneville, Coleoptera: Scolytidae) using *Beauveria bassiana* (Bals.) Vuill. (Deuteromycotina: Hyphomycetes). *Journal of Applied Microbiology* 103 (5): 1406–1414.

Bergueiro, J., Dominguez, F., Guzman, E., Morales, N., and Perez-Navarro, A. 1998. FINASOL OSR 51 biodegradation by the biological activators BIOLEN IG 30 and BIOLEN IC 10. In *Arctic and Marine Oilspill Program Technical Seminar, 1998.* Ministry of Supply and Services, Canada, pp. 729–744.

Bhatnagar, S., and Kumari, R. 2013. Bioremediation: A sustainable tool for environmental management: A review. *Annual Research & Review in Biology 3:* 974–993.

Bjerketorp, J., Röling, W. F., Feng, X.-M., Garcia, A. H., Heipieper, H. J., and Håkansson, S. 2018. Formulation and stabilization of an *Arthrobacter* strain with good storage stability and 4-chlorophenol-degradation activity for bioremediation. *Applied Microbiology and Biotechnology* 102 (4): 2031–2040.

Borriss, R. 2011. Use of plant-associated *Bacillus* strains as biofertilizers and biocontrol agents in agriculture. In Maheshwari, D. K. (Ed.) *Bacteria in Agrobiology: Plant Growth Responses.* Cham: Springer.

Boudh, S., and Singh, J. S. 2019. Pesticide contamination: Environmental problems and remediation strategies. In Bharagava, R. N. and Chowdhary, P. (Eds.) *Emerging and Eco-Friendly Approaches for Waste Management*. Cham: Springer.

Briceño, G., Vergara, K., Schalchli, H., et al. 2018. Organophosphorus pesticide mixture removal from environmental matrices by a soil *Streptomyces* mixed culture. *Environmental Science and Pollution Research* 25 (22): 21296–21307.

Burges, H. D. 1998. *Formulation of Microbial Biopesticides Beneficial Microorganisms, Nematodes and Seed Treatments*. Dordrecht: Springer Netherlands.

Buzia, O. D., Mardare, N., Florea, A., Diaconu, C., Dinica, R. M., and Tatu, A. L. 2018. Formulation and preparation of pharmaceuticals with anti-rheumatic effect using the active principles of *Capsicum Annuum* and *Piper Nigrum. Rev. Chim.(Bucharest)* 69: 2854–2857.

Chandra, R., and Kumar, V., 2015. Biotransformation and biodegradation of organophosphates and organohalides. In Chandra, R. (Ed.) *Environmental Waste Management*. Boca Raton, FL: CRC Press. DOI:10.1201/b19243-17.

Chandra, R., and Kumar, V., 2017a. Phytoextraction of heavy metals by potential native plants and their microscopic observation of root growing on stabilised distillery sludge as a prospective tool for in-situ phytoremediation of industrial waste. *Environmental Science and Pollution Research* 24: 2605–2619. https://doi.org/10.1007/s11356-016-8022-1

Chandra, R., and Kumar, V., 2017b. Detection of *Bacillus* and *Stenotrophomonas* species growing in an organic acid and endocrine-disrupting chemicals rich environment of distillery spent wash and its phytotoxicity. *Environmental Monitoring and Assessment* 189: 26. https://doi.org/10.1007/s10661-016-5746-9

Chandra, R., and Kumar, V., 2017c. Detection of androgenic-mutagenic compounds and potential autochthonous bacterial communities during in-situ bioremediation of post methanated distillery sludge. *Frontiers in Microbiology* 8: 87. https://doi.org/10.3389/fmicb.2017.00887

Chen, S. H., and Yien Ting, A. S. 2015. Biodecolorization and biodegradation potential of recalcitrant triphenylmethane dyes by *Coriolopsis* sp. isolated from compost. *Journal of Environmental Management* 150 (Suppl. C): 274–280.

Chen, Y., Chen, Y., Li, Y., et al. 2019. Changes of heavy metal fractions during co-composting of agricultural waste and river sediment with inoculation of *Phanerochaete chrysosporium*. *Journal of Hazardous Materials* 378: 120757.

Chew, S. Y., and Ting, A. S. Y. 2016. Common filamentous *Trichoderma asperellum* for effective removal of triphenylmethane dyes. *Desalination and Water Treatment* 57 (29): 13534–13539.

Chowdhary, P., Raj, A., and Bharagava, R. N. 2018. Environmental pollution and health hazards from distillery wastewater and treatment approaches to combat the environmental. *Chemosphere* 194: 229–246.

Chowdhary, P., Raj, A., Verma, D., and Akhter, Y. (Eds.) 2020a. *Contaminants and Clean Technologies*, Boca Raton, FL: CRC Press.

Chowdhary, P., Raj, A., Verma, D., and Akhter, Y. (Eds.) 2020b. *Microorganisms for Sustainable Environment and Health*. Oxford: Elsevier.

Da Silva, R. S., Almeida, D., Brasileiro, P., Rufino, R., Luna, J., and Sarubbo, L. 2017. Biosurfactant formulation of *Pseudomonas cepacia* and application in the removal of oil from coral reef. *Chemical Engineering Transactions* 57: 649–654.

Dermont, G., Bergeron, M., Mercier, G., and Richer-Laflèche, M. 2008. Soil washing for metal removal: A review of physical/chemical technologies and field applications. *Journal of Hazardous Materials* 152 (1): 1–31.

Falcão, L. L., Silva-Werneck, J. O., Vilarinho, B. R., da Silva, J. P., Pomella, A. W. V., and Marcellino, L. H. 2014. Antimicrobial and plant growth-promoting properties of the cacao endophyte *Bacillus subtilis* ALB629. *Journal of Applied Microbiology* 116 (6): 1584–1592.

Faraone, N., MacPherson, S., and Hillier, N. K. 2018. Evaluation of repellent and insecticidal properties of a novel granite dust product in crop protection. *Journal of Pest Science* 91 (4): 1345–1352.

Ferrigo, D., Raiola, A., Rasera, R., and Causin, R. 2014. *Trichoderma harzianum* seed treatment controls *Fusarium verticillioides* colonization and fumonisin contamination in maize under field conditions. *Crop Protection* 65 (Suppl. C): 51–56.

Frumkin, H., and Haines, A. 2019. Global environmental change and noncommunicable disease risks. *Annual Review of Public Health* 40: 261–282.

Fuentes, M. S., Briceño, G. E., Saez, J. M., Benimeli, C. S., Diez, M. C., and Amoroso, M. J. 2013. Enhanced removal of a pesticides mixture by single cultures and consortia of free and immobilized *Streptomyces* strains. *BioMed Research International*. vol. 2013. http://dx.doi.org/10.1155/2013/392573

Grahić, J., Đikić, M., Gadžo, D., et al. 2016. Assessment of organic pelleted fertilizers influence on the chemical composition of seeds and yield components of buckwheat. *Radovi Poljoprivrednog Fakulteta Univerziteta u Sarajevu* [Works of the Faculty of Agriculture University of Sarajevo] 61 (66(2)): 31–47.

Guijarro, B., Melgarejo, P., Torres, R., Lamarca, N., Usall, J., and De Cal, A. 2007. Effects of different biological formulations of *Penicillium frequentans* on brown rot of peaches. *Biological Control* 42 (1): 86–96.

Hardoim, P. R., van Overbeek, L. S., and van Elsas, J. D. 2008. Properties of bacterial endophytes and their proposed role in plant growth. *Trends in Microbiology* 16 (10): 463–471.

Harish, S., Kavino, M., Kumar, N., Saravanakumar, D., Soorianathasundaram, K., and Samiyappan, R. 2008. Biohardening with plant growth promoting rhizosphere and endophytic bacteria induces systemic resistance against *Banana bunchy top* virus. *Applied Soil Ecology* 39 (2): 187–200.

Himmelstein, J., Maul, J., and Everts, K. 2014. Impact of five cover crop green manures and Actinovate on *Fusarium* wilt of watermelon. *Plant Disease* 98 (7): 965–972.

Hirvonen, A., Tuhkanen, T., and Kalliokoski, P. 1996. Treatment of TCE-and PCE contaminated groundwater using UV/H_2O_2 and O3/H_2O_2 oxidation processes. *Water Science and Technology* 33 (6): 67–73.

Huang, Z., He, K., Song, Z., et al. 2019. Alleviation of heavy metal and silver nanoparticle toxicity and enhancement of their removal by hydrogen sulfide in *Phanerochaete chrysosporium*. *Chemosphere* 224: 554–561.

Jabbar, N. M., Kadhim, E. H., and Mohammed, A. K. 2018. Bioremediation of soil contaminated with diesel using biopile system. *Al-Khwarizmi Engineering Journal* 14 (3): 48–56.

Johanto, A., Cahyani, V. R., and Rosariastuti, R. 2018. Bioremediation of paddy field contaminated by liquid waste from textile industry in order to reduce chromium in soil and rice plant. Paper presented at AIP Conference Proceedings.

John, R. P., Tyagi, R. D., Brar, S. K., Surampalli, R. Y., and Prévost, D. 2011. Bio-encapsulation of microbial cells for targeted agricultural delivery. *Critical Reviews in Biotechnology* 31: 211–226.

Ju, Z., Forster, I., and Dominy, W. 2009. Effects of supplementing two species of marine algae or their fractions to a formulated diet on growth, survival and composition of shrimp (*Litopenaeus vannamei*). *Aquaculture* 292 (3–4): 237–243.

Kao, C., Chen, C., Chen, S., Chien, H., and Chen, Y. 2008. Application of in situ biosparging to remediate a petroleum-hydrocarbon spill site: Field and microbial evaluation. *Chemosphere* 70 (8): 1492–1499.

Khalid, S., Shahid, M., Niazi, N., Murtaza, B., Bibi, I., and Dumat, C. 2017. A comparison of technologies for remediation of heavy metal contaminated soils. *Journal of Geochemical Exploration.* 182 (PB): 247–268.

Khan, F. I., Husain, T., and Hejazi, R. 2004. An overview and analysis of site remediation technologies. *Journal of Environmental Management* 71 (2): 95–122.

Khan, S., Anas, M., and Malik, A. 2019. Mutagenicity and genotoxicity evaluation of textile industry wastewater using bacterial and plant bioassays. *Toxicology Reports* 6: 193–201.

Khan, S. A., Hamayun, M., Yoon, H., et al. 2008. Plant growth promotion and *Penicillium citrinum. BMC Microbiology* 8 (1): 231.

Koli, S. K., and Hussain, A. 2019. Status of electronic waste management in India: A review. In Hussain, A. and Ahmed, S. (Eds.) *Advanced Treatment Techniques for Industrial Wastewater.* Hershey, PA: IGI Global.

Kumar, A., Bisht, B., Joshi, V., and Dhewa, T. 2011. Review on bioremediation of polluted environment: A management tool. *International Journal of Environmental Sciences* 1 (6): 1079.

Kumar, V. 2014. Characterization, bio-formulation development and shelf-life studies of locally isolated biofertilizer strains. *Octa Journal of Environmental Research* 2 (1).

Kumar, V., Agrawal, S., Shahi, S. K., Singh, S., and Ramamurthy, P. C., 2022. Bioremediation potential of newly isolated *Bacillus albus* strain VKDS9 for decolourization and detoxification of biomethanated distillery effluent and its metabolites characterization for environmental sustainability. *Environmental Technology & Innovation* 26: 102260. https://doi.org/10.1016/j.eti.2021.102260

Kumar, V., Ferreira, L. F. R., Sonkar, M., and Singh, J. 2021a. Phytoextraction of heavy metals and ultrastructural changes of *Ricinus communis* L. grown on complex organometallic sludge discharged from alcohol distillery. *Environmental Technology & Innovation* 22: 101382. https://doi.org/10.1016/j.eti.2021.101382

Kumar, V., Kaushal, A., Singh, K., and Shah, M. P., 2021b. Phytoaugmentation technology for phytoremediation of environmental pollutants: Opportunities, challenges and future prospects. In Kumar, V., Saxena, G., and Shah, M. P., (Eds) *Bioremediation for Environmental Sustainability: Approaches to Tackle Pollution for Cleaner and Greener Society.* Oxford: Elsevier. https://doi.org/10.1016/B978-0-12-820318-7.00016-2

Kumar, V., Shahi, S. K., and Singh, S., 2018. Bioremediation: An eco-sustainable approach for restoration of contaminated sites. In Singh J., Sharma D., Kumar G., and Sharma N. (Eds.) *Microbial Bioprospecting for Sustainable Development.* Singapore: Springer. https://doi.org/10.1007/978-981-13-0053-0_6

Landrigan, P. J., Fuller, R., Acosta, N. J. R., et al. 2018. The Lancet Commission on pollution and health. *Lancet* 391 (10119): 462–512.

Li, Q., Huang, Y., Wen, D., Fu, R., and Feng, L. 2020. Application of alkyl polyglycosides for enhanced bioremediation of petroleum hydrocarbon-contaminated soil using *Sphingomonas changbaiensis* and *Pseudomonas stutzeri. Science of the Total Environment* 719: 137456.

Li, Y., Yang, X., and Geng, B. 2018. Preparation of immobilized sulfate-reducing bacteria-microalgae beads for effective bioremediation of copper-containing wastewater. *Water, Air, & Soil Pollution* 229 (3): 54.

Liu, J., Tian, S.-P., Li, B.-Q., and Qin, G.-Z. 2009. Enhancing viability of two biocontrol yeasts in liquid formulation by applying sugar protectant combined with antioxidant. *BioControl* 54 (6): 817.

Ma, Y., Prasad, M. N. V., Rajkumar, M., and Freitas, H. 2011. Plant growth promoting rhizobacteria and endophytes accelerate phytoremediation of metalliferous soils. *Biotechnology Advances* 29 (2): 248–258.

Machhirake, N. P., Yadav, S., Krishna, V. and Kumar, S. 2022. Toxicity-removal efficiency of *Brassica juncea, Chrysopogon zizanioides* and *Pistia stratiotes* to decontaminate biomedical ash under non-chelating and chelating conditions: A pilot-scale phytoextraction study. *Chemosphere*, 287: 132416. https://doi.org/10.1016/j.chemosphere.2021.132416

Mackay, D. M., and Cherry, J. A. 1989. Groundwater contamination: Pump-and-treat remediation. *Environmental Science & Technology* 23 (6): 630–636.

Maheshwari, D., Dubey, R., Agarwal, M., Dheeman, S., Aeron, A., and Bajpai, V. 2015. Carrier based formulations of biocoenotic consortia of disease suppressive *Pseudomonas aeruginosa* KRP1 and *Bacillus licheniformis* KRB1. *Ecological Engineering* 81: 272–277.

Maitra, S. 2019. Permeable reactive barrier: A technology for groundwater remediation: A mini review. *Biodegradation* 80: 9.

Mamood, S. N. H., Hidayatulfathi, O., Budin, S. B., Ahmad Rohi, G., and Zulfakar, M. H. 2017. The formulation of the essential oil of *Piper aduncum Linnaeus* (Piperales: *Piperaceae*) increases its efficacy as an insect repellent. *Bulletin of Entomological Research* 107 (1): 49–57.

Manikandan, R., Saravanakumar, D., Rajendran, L., Raguchander, T., and Samiyappan, R. 2010. Standardization of liquid formulation of *Pseudomonas fluorescens* Pf1 for its efficacy against *Fusarium* wilt of tomato. *Biological Control* 54 (2): 83–89.

Maron, P.-A., Sarr, A., Kaisermann, A., et al. 2018. High microbial diversity promotes soil ecosystem functioning. *Applied and Environmental Microbiology* 84 (9): e02738-17.

Mbarga, J. B., Begoude, B., Ambang, Z., et al. 2014. A new oil-based formulation of *Trichoderma asperellum* for the biological control of cacao black pod disease caused by *Phytophthora megakarya*. *Biological Control* 77 : 15–22.

McIntosh, K., Smith, A., Young, K. L., et al. 2018. Alkenones as a promising green alternative for waxes in cosmetics and personal care products. *Cosmetics* 5 (2).

Mejri, D., Gamalero, E., and Souissi, T. 2013. Formulation development of the deleterious rhizobacterium *Pseudomonastrivialis* X33d for biocontrol of brome (Bromus diandrus) in durum wheat. *Journal of Applied Microbiology* 114 (1): 219–228.

Michalak, I., Chojnacka, K., and Saeid, A. 2017. Plant growth biostimulants, dietary feed supplements and cosmetics formulated with supercritical CO_2 algal extracts. *Molecules* 22 (1): 66.

Mishra, S., Bharagava, R. N., More, N., et al. 2019. Heavy metal contamination: An alarming threat to environment and human health. In Mishra, C. K. (Ed.) *Environmental Biotechnology: For Sustainable Future*. Cham: Springer.

Mukherjee, A. K., and Bordoloi, N. K. 2012. Biodegradation of benzene, toluene, and xylene (BTX) in liquid culture and in soil by *Bacillus subtilis* and *Pseudomonas aeruginosa* strains and a formulated bacterial consortium. *Environmental Science and Pollution Research* 19: 3380–3388. https://doi.org/10.1007/s11356-012-0862-8

Mulligan, C., Yong, R., and Gibbs, B. 2001. Remediation technologies for metal-contaminated soils and groundwater: an evaluation. *Engineering Geology* 60 (1–4): 193–207.

Nagachandrabose, S. 2018. Liquid bioformulations for the management of root-knot nematode, *Meloidogyne hapla* that infects carrot. *Crop Protection* 114: 155–161.

Navarro, A., Cañadas, I., Martinez, D., Rodriguez, J., and Mendoza, J. 2009. Application of solar thermal desorption to remediation of mercury-contaminated soils. *Solar Energy* 83 (8): 1405–1414.

Nikolopoulou, M., and Kalogerakis, N. 2018. Biostimulation strategies for enhanced bioremediation of marine oil spills including chronic pollution. In Steffan, R. (Ed.) *Consequences of Microbial Interactions with Hydrocarbons, Oils, and Lipids: Biodegradation and Bioremediation. Handbook of Hydrocarbon and Lipid Microbiology*. Cham: Springer, pp. 1–10.

Nopcharoenkul, W., Pinphanichakarn, P., and Pinyakong, O. 2011. The development of a liquid formulation of *Pseudoxanthomonas* sp. RN402 and its application in the treatment of pyrene-contaminated soil. *Journal of Applied Microbiology* 111 (1): 36–47.

Obayori, O. S., Adebusoye, S. A., Ilori, M. O., Oyetibo, G. O., Omotayo, A. E., and Amund, O. O. 2010. Effects of corn steep liquor on growth rate and pyrene degradation by *Pseudomonas* strains. *Current Microbiology* 60 (6): 407–411.

O'Callaghan, M. 2016. Microbial inoculation of seed for improved crop performance: issues and opportunities. *Applied Microbiology and Biotechnology* 100 (13): 5729–5746.

Paço, A., Duarte, K., da Costa, J. P., et al. 2017. Biodegradation of polyethylene microplastics by the marine fungus *Zalerion maritimum*. *Science of the Total Environment* 586: 10–15.

Peng, G., McGregor, L., Lahlali, R., et al. 2011. Potential biological control of clubroot on canola and crucifer vegetable crops. *Plant Pathology* 60 (3): 566–574.

Peng, S., Zhou, Q., Cai, Z., and Zhang, Z. 2009. Phytoremediation of petroleum contaminated soils by *Mirabilis Jalapa L.* in a greenhouse plot experiment. *Journal of Hazardous Materials* 168 (2–3): 1490–1496.

Price, M. S., Classen, J. J., and Payne, G. A. 2001. *Aspergillus niger* absorbs copper and zinc from swine wastewater. *Bioresource Technology* 77 (1): 41–49.

Rao, P. S. C., Jawitz, J. W., Enfield, C. G., Falta Jr, R., Annable, M. D., and Wood, A. L. 2001. Technology integration for contaminated site remediation: Clean-up goals and performance criteria. *Groundwater Quality: Natural and Enhanced Restoration of Groundwater Pollution* 275: 571–578.

Ren, X., Zeng, G., Tang, L., et al. 2018. Sorption, transport and biodegradation—an insight into bioavailability of persistent organic pollutants in soil. *Science of the Total Environment* 610: 1154–1163.

Robles-González, I. V., Fava, F., and Poggi-Varaldo, H. M. 2008. A review on slurry bioreactors for bioremediation of soils and sediments. *Microbial Cell Factories* 7 (1): 5.

Rocha e Silva, N. M. P., Rufino, R. D., Luna, J. M., Santos, V. A., and Sarubbo, L. A. 2014. Screening of *Pseudomonas* species for biosurfactant production using low-cost substrates. *Biocatalysis and Agricultural Biotechnology* 3 (2): 132–139.

Sah, A., Negi, H., Kapri, A., Anwar, S., and Goel, R. 2011. Comparative shelf life and efficacy of LDPE and PVC degrading bacterial consortia under bioformulation. *Ekologija* 57 (2).

Sallam, N. A., Riad, S. N., Mohamed, M. S., and El-eslam, A. S. 2013. Formulations of *Bacillus* spp. and *Pseudomonas fluorescens* for biocontrol of cantaloupe root rot caused by *Fusarium solani*. *Journal of Plant Protection Research* 53 (3): 295–300.

Salmon, D. N., Piva, L. C., Binati, R. L., et al. 2011. Formulated products containing a new phytase from *Schyzophyllum* sp. phytase for application in feed and food processing. *Brazilian Archives of Biology and Technology* 54 (6): 1069–1074.

Simeonova, A., Godjevargova, T., and Ivanova, D. 2008. Biosorption of heavy metals by dead *Streptomyces fradiae*. *Environmental Engineering Science* 25 (5): 627–634.

Singh, A. K., and Chandra, R. 2019. Pollutants released from the pulp paper industry: Aquatic toxicity and their health hazards. *Aquatic Toxicology* 211: 202–216.

Singh, R., and Arora, N. K. 2016. Bacterial formulations and delivery systems against pests in sustainable agro-food production. *Food Science* 1: 1–11.

Singh, S., Anil, A. G., Khasnabis, S., Kumar, V., Nath, B., Sunil Kumar Naik, T. S., Subramanian, S., ... Ramamurthy, P. C. 2021. Sustainable removal of Cr(VI) using graphene oxide-zinc oxide nanohybrid: Adsorption kinetics, isotherms, and thermodynamics. *Environmental Research* 203: 111891. https://doi.org/10.1016/j.envres.2021.111891

Singh, V., Upadhyay, R. S., Sarma, B. K., and Singh, H. B. 2016. Seed bio-priming with *Trichoderma asperellum* effectively modulate plant growth promotion in pea. *International Journal of Agriculture, Environment and Biotechnology* 9 (3): 361–365.

Skariyachan, S., Manjunatha, V., Sultana, S., Jois, C., Bai, V., and Vasist, K. S. 2016. Novel bacterial consortia isolated from plastic garbage processing areas demonstrated enhanced degradation for low density polyethylene. *Environmental Science and Pollution Research* 23 (18): 18307–18319.

Spier, M., Fendrich, R., Almeida, P., et al. 2011. Phytase produced on citric byproducts: Purification and characterization. *World Journal of Microbiology and Biotechnology* 27 (2): 267–274.

Srivastava, R., Khalid, A., Singh, U. S., and Sharma, A. K. 2010. Evaluation of arbuscular mycorrhizal fungus, fluorescent *Pseudomonas* and *Trichoderma harzianum* formulation against *Fusarium oxysporum* f. sp. *lycopersici* for the management of tomato wilt. *Biological Control* 53 (1): 24–31.

Subashchandrabose, S. R., Logeshwaran, P., Venkateswarlu, K., Naidu, R., and Megharaj, M. 2017. Pyrene degradation by *Chlorella* sp. MM3 in liquid medium and soil slurry: Possible role of dihydrolipoamide acetyltransferase in pyrene biodegradation. *Algal Research* 23: 223–232.

Vidali, M. 2001. Bioremediation: An overview. *Pure and Applied Chemistry* 73 (7): 1163–1172.

Viegas, C. A., Silva, V. P., Varela, V. M., Correia, V., Ribeiro, R., and Moreira-Santos, M. 2019. Evaluating formulation and storage of *Arthrobacter aurescens* strain TC1 as a bioremediation tool for terbuthylazine contaminated soils: Efficacy on abatement of aquatic ecotoxicity. *Science of the Total Environment* 668: 714–722.

Wang, S., and Mulligan, C. N. 2004. An evaluation of surfactant foam technology in remediation of contaminated soil. *Chemosphere* 57 (9): 1079–1089.

Wang, Y., Yi, B., Sun, X., et al. 2019. Removal and tolerance mechanism of Pb by a filamentous fungus: A case study. *Chemosphere* 225: 200–208.

Wong, C. K. F., Saidi, N. B., Vadamalai, G., Teh, C. Y., and Zulperi, D. 2019. Effect of bioformulations on the biocontrol efficacy, microbial viability and storage stability of a consortium of biocontrol agents against *Fusarium* wilt of banana. *Journal of Applied Microbiology* 127 (2): 544–555.
WWAP (United Nations World Water Assessment Programme). 2017. *The United Nations World Water Development Report 2017: Wastewater, the Untapped Resource.* Paris: UNESCO.
WWAP (United Nations World Water Assessment Programme). 2018. *The United Nations World Water Development Report 2018: Nature-Based Solutions for Water.* Paris: UNESCO.
Ye, J., Chen, X., Chen, C., and Bate, B. 2019. Emerging sustainable technologies for remediation of soils and groundwater in a municipal solid waste landfill site: A review. *Chemosphere* 227: 681–702.

9 The Role of Microbes in the Degradation of Plastics and Directions Toward Greener Bioplastic

A.K. Priya,[*,1] *D. Balaji,*[2] *J. Vijayaraghavan,*[3] *J. Thivya,*[4] *and R. Anand*[5]

[1]Department of Chemical Engineering, KPR Institute of Engineering and Technology, Coimbatore, Tamil Nadu, India
[2]Department of Mechanical Engineering, KPR Institute of Engineering and Technology, Coimbatore, Tamil Nadu, India
[3]Department of Civil Engineering, University College of Engineering, Ramanathapuram, Tamil Nadu, India
[4]Department of Civil Engineering, University College of Engineering, Dindigul, Tamil Nadu, India
[5]Department of Electrical and Electronics Engineering. Nehru Institute of Engineering and Technology, Coimbatore, Tamil Nadu, India
*Corresponding author Email: akpriy@gmail.com

CONTENTS

9.1 Introduction 182
9.2 Plastics Origins and Pollution 185
  9.2.1 Bioplastics 186
  9.2.2 Recycling Routes 186
  9.2.3 Biopolymers 187
9.3 Classification of Plastics Based on Biodegradability 187
  9.3.1 Non-Biodegradable Plastics 188
  9.3.2 Biodegradable Plastics 188
    9.3.2.1 Types of Biodegradable Plastics 188
9.4 Microbes and Their Mechanisms for Plastic Biodegradation 190
  9.4.1 Degradation Routes 190
    9.4.1.1 Biodegradation 190
    9.4.1.2 Pyrolyse 192
    9.4.1.3 Solvolysis 192
    9.4.1.4 Biodegradability and Compostability 192
    9.4.1.5 Bioplastic as a Packaging Material 192
  9.4.2 Factors Affecting Plastic Biodegradation 192
    9.4.2.1 Revelation States Moisture 193
    9.4.2.2 pH and Temperature 193
    9.4.2.3 Enzyme Characteristics 193
    9.4.2.4 Polymer Characteristics 193

DOI: 10.1201/9781003239956-11

9.5 Current Research on Plastic Biodegradation 194
9.5.1 Sustainability and End-of-Life Options for Bioplastics 194
9.6 Future Prospects 195
9.7 Conclusion 195
References 196

9.1 INTRODUCTION

"Bio" is one of the dictionary's most widely used codes. The bio beginning of different terms is complemented by the fact that disciplines, incidents or items are defined as highly valuable, helpful and even nutritious news. Here it is used accurately, even incorrectly, when bio is used for deceptive purposes, similar to the instance of supposed "bio-dynamic cultivating," where most analysts are dubious of the idea that this process has been set up through logical techniques, fearing instead it is due to other ways. In reality, biodynamic approaches cannot be checked and affirmed and there is no evidence of the improvement in plant or soil quality evaluated by the horticultural administration framework (Ruse, 2013). It is not enough, in any case, to describe anything as "biological" to make it safe and beneficial.

This wording additionally incorporates bioplastics (Prata et al., 2019; Yang et al., 2019): innovative, eager materials. These have supplanted another material, plastics, which, since World War II, have progressively infiltrated the environment and which are an intrinsic part of people's everyday lives, found in their homes, in the workplace, in the means of transport, in virtually all the instruments and things we use daily and, in any event, even in our clothes.

The twenty-first century thrives on unprecedented economic development, but it is also facing irreparable environmental damage (Moharir and Kumar, 2019; Singh,, Kumar, Mishra et al., 2020; Singh, Kumar, Khapre et al., 2020; Kumar et al., 2021). Plastic contamination has recently been highlighted at all levels from development to disposal and incineration as a global crisis (Scalenghe, 2018; Chowdhary et al., 2020). Ordinary and artificially produced bioplastics from limitless or oil-based resources are intended to have an inconsequential carbon impression, high reusable worth and hard and fast biodegradability or compostability.

It is hard to assess the overall amount of bioplastics in the world due to the constantly expanding scope of bio-based and biodegradable polymers and expanded speculation interests in the bioplastics field. The Nova Institute's most recent study estimates that worldwide bioplastic creation potential has risen fundamentally from some 2.11 million tons in 2018 to an expected 2.62 million tons in 2023 (European Bioplastics, 2018). Europe is leading bioplastic innovative work and, other than Asia, is the main center for the production and use of bioplastics (Bio-based Building Blocks and Polymers, 2018). With numerous noteworthy bioplastics joining the expanded applications market sectors, industry is interested in raising the production potential. In fact, the importance of sustainability and the circulatory economy in the bioplastics industry has influenced significant growth and technological maturity with multiple production routes.

Degradable plastics using microbes are attracting the main attention as an alternative to artificial plastics (Chowdhary, Hare, et al., 2020). The role of microbes in the degradation of plastics and directions toward greener bioplastic are shown in Figure 9.1. Artificial plastics are increasing as hazardous waste in the environment which severely disturbs the environmental balance. Hence, most of the plastic manufacturers regularly enhance the degradable harvests using microbes suitable for domestic, industrial, and agricultural purposes. While large-scale manufacturing of plastics only goes back to the 1950s, they have become essential materials for a wide range of everyday uses. The plastics business has quickly expanded because of the assorted variety of accessible plastics and the moderately modest oil use. Plastics ordinarily have fantastic mechanical qualities with low mass thickness and idleness, making them suitable for a wide assortment of uses. In a 2014 review, a bewildering 311 MTs of plastic were produced in a single year, expending 6% of the world's oil

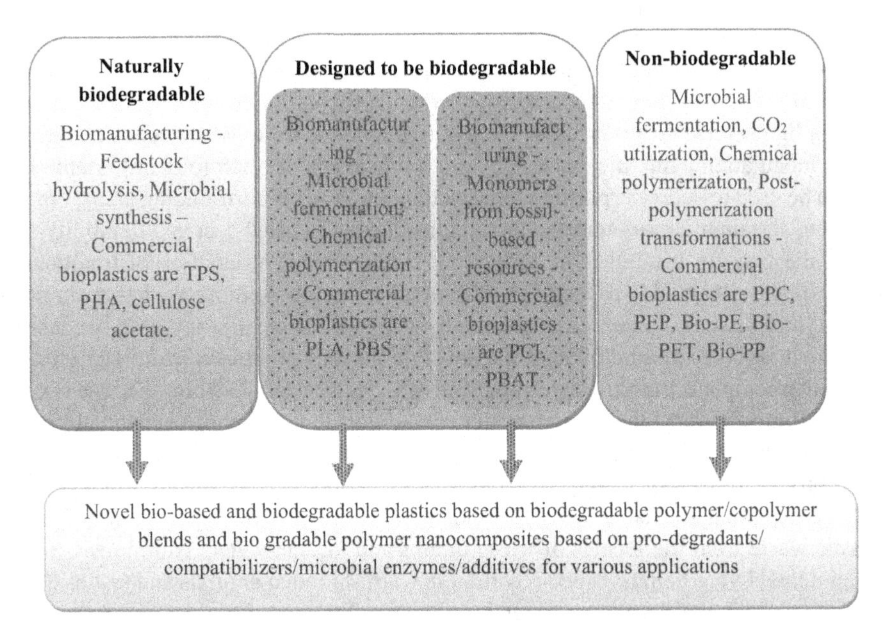

FIGURE 9.1 Role of microbes in degradation of plastics and directions toward greener bioplastic.

production as a polymer amalgamation feedstock. Some 26% by volume were used in packaging applications, of which only 5% were reused for additional purposes, costing £62–92 billion. Annual plastic production is supposed to double by 2034, reaching 124 MTs in 2050 and using 20% of the world's oil production.

Polyethylene (PE) (36%), polypropylene (PP) (21%), polyvinyl chloride (PVC) (12%), polyurethane (PUR), poly-ethylenetrephthalate (PET) and polystyrene (PS) (under 10% each) account for worldwide non-fiber plastics production: Fiber is overwhelmingly made of polyester, polyamide and polyacrylic strands (Helanto et al., 2019), with PET involving 70%. These six non-fiber processes and the filaments comprise 92% of all plastics ever produced. The accompanying greatest market share is advances in building, which consumed 69% of all PVC and 19% of the full-scale non-fiber plastics, and 42% of all non-fiber plastics (by and large, PP, PE and PET). Between 2010 and 2025, the plastic waste estimated to reach the oceans is projected at 100 MTs (Elvers et al., 2016; Nakajima et al., 2017). After plastic reaches the sea, it debases gradually and breaks into smaller and smaller pieces before they become microplastics. The fact that microplastics are eaten by zooplankton and phytoplankton (green growth) produces widespread contamination in marine environments and is having a negative impact there. Microplastics have shown that they can inhibit algae growth, and since about 70% of global oxygen is generated by photosynthesizing marine plants, this can have significant implications for climate change and global warming. In the end, microplastics bio-accumulate and once they release monomers, additives and pollutants in the body, they cause physiological damage ranging from oxidation to carcinogenic behavior. Annual human microplastic intake is calculated to be between 74,000 and 121,000 particles. The bioplastic industry emerges as a potential option for the substitution of fossil fuel-based polymers to resolve such concerns.

Biopolymer precursors are provided by biomass, rendering them sustainable. Via photosynthesis, biomass growth absorbs CO_2 (European Bioplastics, 2018). As a result, total bioplastics manufacturing has a lower carbon footprint than petro plastics. Besides, some bioplastics have high biodegradability which additionally makes them more ecologically acceptable. Worldwide biopolymer production is expected to reach 2.62 MTs by 2023 (Bio-based Building Blocks and Polymers, 2018). Despite this quick market development, bioplastics still account for less than 1% of plastics production. The more costly processing and generally less mechanical properties of bioplastics are the key reasons why they are used in smaller quantities than petro plastics.

Reduction of waste volumes is accomplished when sufficient microorganisms are added during the disposal of bioplastics, thus increasing the availability of space in waste deposits. For bioplastic degradation, inoculation with the necessary microorganisms is therefore important. Bioplastic degraders can be used to decrease plastic waste accumulation in the surrounding areas. For enhanced biodegradation, certain microorganisms may be introduced. The low cost and flexibility of plastics have opened the way for a number of applications. The annual production of traditional (petrochemical) plastics is expected to exceed 350 million tons. Their boundless use has prompted serious ecological pollution issues, nonetheless, essentially due to the huge amount of expendable plastics produced, which involves around 50% of the absolute plastics produced. From 1950 to 2015, less than 10% of the complete plastic waste produced was really reused, while the rest was disposed of or delivered into the air. On account of the concoction mixture and mechanical properties of plastics and especially their sturdiness, these polymers represent an ecological danger. The release of polymers produced from oil specifically emits CO_2 emanations into the environment and hence adds an ozone-depleting substance to the development of greenhouse gases. In addition, the long-lasting plastic waste from the locales of waste dumps and marine waters enormously affects these biological systems' living beings, especially marine warm-blooded animals and turtles that mistake nourishment for plastic build-up, sometimes leading to suffocation.

Microplastics, i.e., plastic particles with a breadth of under 5 mm, obtained from the decay of full-scale plastics were found not just in staples, such as drinking water, soil, and air, yet additionally in unforeseen zones, for example, the depths of Antarctica and the Arctic ice residue. Also, microplastics, which represent approximately 13% of plastic waste mass, and macroplastics can be mistakenly taken for food by living animals, which is fatal, as they cause harm or mischief to human health by ingestion. In addition, the dependency on fossil fuels must be reduced and the carbon footprint produced by petroleum-based plastics must decrease. The words "polymers" and "plastics" are used interchangeably in this chapter. However, it is explained that plastics are usually polymers to which plasticizers have been added. The drawn-out presence and problem of plastics in the air gave impetus to the creation of new materials, i.e., biodegradable bioplastics. These materials can be broken down by the same abiotic processes that disintegrate traditional plastics such as photodegradation, oxidation, hydrolysis and microbial degradation. Latest estimates show that the global production of bioplastics slightly exceeds 2.1 million tons, equal to 0.6% of the global generation of plastics.

Bio-based biodegradable plastics are becoming progressively significant, particularly in packaging, health care and agriculture, yet their contribution to plastics is still extremely low. Furthermore, their ecological insurance and materials are centered around compelling waste administration and network training plans (Rujnić-Sokele et al., 2017). A few investigations have indicated the astounding ability of fossil-based biodegradable polymers to quickly debase naturally from the effect of certain microspheric species, including microbes and parasites under distressing conditions, through the improvement of the exoenzymes (Mohanty et al., 2000; Sharma et al., 2003; Ghosh et al., 2013). Proteases, lipases and cutinases are fundamental microbial chemicals suitable for polymer biodegradation (Tokiwa et al., 2009; Muhamad et al, 2015). Also, Rhizopus delemar produces compounds, for example, esterases and lipases. The research on complex polymers, for example, poly (ethylene-adipate) and PCL (Jin et al., 2000; Lam et al., 2009) was demonstrated using arrhizous plants, such as *cylindraceous candida*. In business applications, the share of biodegradable plastic polymers is very restricted primarily on the grounds that they have an unpredictable structure and there is a lack of data on advanced conditions for fast corruption (Rujnić-Sokele et al., 2017). In any event, biodegradable plastics may eventually supplant non-biodegradable polymers if systems, for example, legitimate waste administration, trash control, network training and the development of modern biodegradation offices are established.

Miniature creatures use an assortment of strategies, including the dynamic use as a wholesome source or the roundabout activity of various microbial catalysts, to break up complex polymers. *Pseudomonas fluorescens*, *P. aeruginosa*, and *Penicillium simplicissimum* (Norman et al., 2002; Raziya Fathima et al., 2016; Singh et al., 2017) are the most frequently used bacterial and fungal strains for polymer biodegradation. The focus of this review is on classifying plastics, especially biodegradable plastics, and on established microbial polymer degradation mechanisms. The rationality of use in environmental protection of biodegradable polymers is also debated. Finally, the chapter also presents existing insights into biodegradable plastics research accompanied by prospects of plastic biodegradation. The key aims of this chapter are to examine and investigate the biodegradation of bioplastics under different conditions. Likewise, since bioplastics are generally seen as practical by definition, this investigation features their application potential with the end goal that their inefficient use is avoided.

9.2 PLASTICS ORIGINS AND POLLUTION

The origins of the modern plastics period can be dated to the mid-nineteenth century, to Christian Schönbein's discovery of nitrocellulose. He combined cotton, nitric acid and sulfuric acid. Guncotton was obtained, a material which at first proved very hazardous, as it was readily inflammable. Hilaire Bernigaud de Chardonnet produced a semi-manufactured fiber called "counterfeit silk," more generally known after 1924 as rayon, by the disintegration of cellulose in watery sodium hydroxide by the occurrence of carbon disulfide. In 1908, the packaging film, called cello protein, was variously treated with a similar colloidal arrangement by Jacques Branden Berger. At that point, in 1907, Leo Baekeland produced the first completely manufactured polymer, Bakelite, by mixing, at controlled temperatures and weight, phenol and formaldehyde. Bakelite was regularly used to produce numerous machines and the key weapons manufactured throughout World War II. First, PVC (polyvinyl chloride) was developed, which supplanted the materials used for water pipes, electric links and floors, and afterwards polystyrene or nylon, created by Dow Chemical and DuPont, which were used for new packaging and material strands. In any case, in the second half of the twentieth century a genuine "plastic revolution" started after Rex Whinfield and James Tennant Dickson developed PET (polyethylene terephthalate) in 1941, and, following two years of ethylene polymerization, Karl Ziegler improved it. At first, PET succeeded colossally in making the tertial tissue known as "heap," then it produced plastic jugs. Notwithstanding, PET entered the universe of food packaging in 1973, when Nathaniel Wyeth licensed the primary jug to be used as a soda pop container. In the last part of the twentieth century, melamine formaldehyde was found capable of replacing box wood in school workshops, and isotatic polypropylene was discovered, for which the Italian Giulio Natta, along with the German Karl Ziegler, won the 1963 Nobel Prize. The Moplen brand fabricated polypropylene industrially, at that point it became part of the story of the Italian "financial boost" in the 1960s.

The ensuing decades incorporated the reformist development and production of plastic materials in progressively intricate and capricious applications because of the alleged techno-polymers, which likewise were demonstrated to be better than uncommon metals, because of their warm and mechanical opposition qualities. Techno-polymers have a glass-like or nebulous structure, generally are thermoplastics, and can be made clear whenever spread out (Plastics Europe, 2019).

It has been estimated that since the beginning of the introduction of plastics that in excess of 6 billion tons of plastic waste have been produced overall (Ellen McArthur Foundation, n.d.). Less than 10% of the 300 million tons of plastic waste produced overall are reused yearly, while the rest ends up in waste dumps and the seas, thus, after some time, producing smaller and unsafe petro-polymers ingested by oceanic creatures and zooplankton (Geyer et al., 2017; Plastic Oceans, n.d.; UNEP-UN Environment, n.d.). It has been estimated that more than 5 trillion plastic particles of around 300,000 tons, including bisphenol, are found on the seacoasts and deliver different poisonous substances. More than 100,000 marine animals are killed annually and millions of birds and fish die from ingesting these microplastics. Around a million of the gooney birds in the Midway Islands are confirmed to have plastics in their stomach (Eriksen et al., 2014). The largest collection areas of ocean plastics now are found in the subtropical waters between California and the Hawaii Islands, called the Great Pacific Garbage Patch (GPGP). The GPGP consists of over 80,000 tons of floating plastic waste, a trash island covering over 1.6 million square km and many times bigger than France (Guern, 2019). Besides, just 26% of the huge amounts of plastic used every year in the EU is reused, 38% is kept in dumps and 36% is destroyed by ignition (Plastics Europe Annual Review, 2019).

9.2.1 Bioplastics

Bioplastics first developed back in the 1950s and reappeared in the 1980s, yet only recently have they become an important consideration; specifically, their assembly started during the 2000s. The enthusiasm for bioplastics in logical articles (Scopus diaries) has expanded essentially with the quantity of distribution in the last five years. In 1952, the first logical distribution regarding this matter was composed and applied in a clinical approach. Bioplastics is the typical term depicting bio-based plastics such as biodegradable plastics, including those made of biogenic materials; but not all bio-based plastics are biodegradable. Then again, a petrochemical starting point might be available for some biodegradable plastics. A bioplastic is thus the plastic which is naturally acquired, biodegradable or both. The annual production of bioplastics is moderately small, representing only one out of a hundred of around 360 million tons of plastics produced every year. In 2019, total production was 2.11 million tons. Of these, 44.5% were non-biodegradable bioplastics and 55.5% were biodegradable bioplastics. Bioplastics will be the ideal alternative to customary plastics, since they have practically the same applications. They are likewise a reasonable way to deal with the natural environment, due to both their biodegradability and their use in the treatment of biogenic crude materials. These two variables accord with the circular economy's targets. For example, while bioplastics can be produced from manufactured mixes using diverse waste products, standard plastics are created by irreversible cycles reliant on oil-based goods.

The prospect of manufacturing bioplastics from waste is very important, as it would mean replacing traditional plastics with bio-based polymers, requiring approximately 0.02% of the total arable land on Earth used for agricultural products. This percentage is expected to raise the projected demand for plastics by more than 50% by 2050, prompting an urgent inquiry whether land should be used to develop food crops or as a plastic tank, which was the situation for the original biofuels that require significant soil crops. Without a doubt, most bioplastics are still produced using crop-based feedstocks (starches and plant materials), which both increase water use and lower the area for the creation of food. In such a manner, progress would be made from a petroleum product-driven society and unreasonable waste administration to a waste administration dependent on a sustainable power source, a reduced non-renewable energy source and reuse-orientated waste administration by making new advances in bioplastics. Finally, this accords with the circular economy.

9.2.2 Recycling Routes

After plastic is collected, sorted and washed, four reuse paths are conceivable (Figure 9.2). (1) Primary reuse is a closed circle which can be accomplished only by the demonstrated foundation of a large plastic piece. This means either reuse or the mechanical reuse of scrap plastic by the closed circle so as to make things with their unique structure. (2) Secondary reuse alludes to the mechanical handling/minimizing, by mechanical methods of changing waste plastics into a less demanding item. The accompanying advantages have been accomplished in mechanical reuse over production reuse: lower production costs, lower global climate change potential, less use of non-sustainable assets, less fermentation and eutrophication. (3) Tertiary reuse means production techniques that

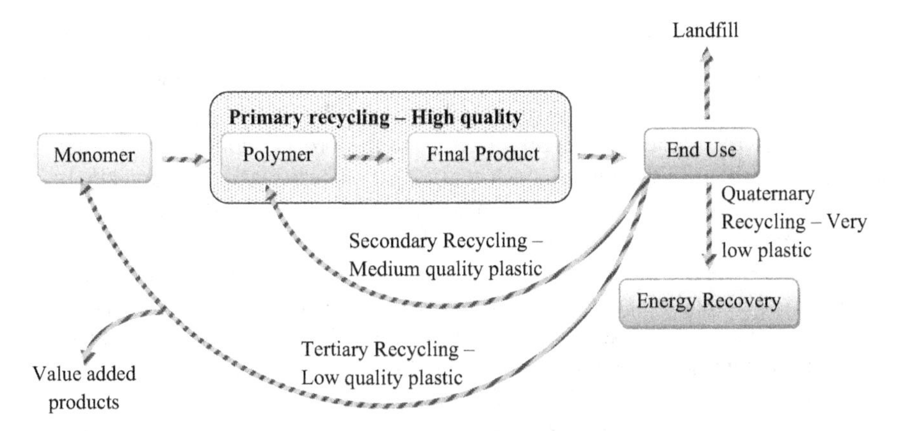

FIGURE 9.2 Options for plastic waste treatment.

depolymerize and corrupt plastic waste into monomeric units or legitimately into other valuable materials. Chemical reuse strategies are different and every method will have its focal points and its downsides, depending upon the polymer type. Compound reuse profits by mechanical reuse: the possibility of assembling quality materials, a round polymer produced as virgin monomer can be repolymerized for interminable numbers of reuses. (4) Quaternary reuse alludes to the vitality recuperation from second-rate plastic waste cremation. At the point when the heat is generated, the subsequent vitality is used to deliver steam and power. This sort of reuse must be done as a final process since it not only destroys the coordinated vitality of the polymers, but also discharges into nature harmful synthetic substances and dioxins.

9.2.3 Biopolymers

Bioplastic is bio-based, biodegradable or both. Bio-based polymer is sourced from carbon biomass, for example, vegetable fats, cellulose or maize starch. The biopolymer might be bio-situated in full or to a limited extent. Bio-based polymers are materials which are more reasonable than oil-based polymers. The plants that produce biomass feedstocks for the amalgamation of biopolymers assimilate CO_2 as they extend and subsequently lessen the net carbon impression of bioplastic production. Likewise, when discarded in an incinerator, the biopolymers discharge lower levels of CO_2 compared to conventional petro-plastics. There is a significant difference between biomass of the first and second eras. The former concerns food biomass (sugar stick, for example) while the latter includes unappetizing biomass (lignocellulosic). Using a second-era biomass, for example, backwoods build-ups, to integrate bioplastic materials that can supplant petro-plastics, will reduce the effect of environmental change significantly because of a lower carbon footprint. An assortment of issues must be settled with the end goal of bioplastics supplanting customary petro-plastics. Cost decrease in their advance ought to be significant, their thermo-mechanical and hindrance properties should be improved, biodegradability should be improved and their accessibility expanded. With the continuation of its rapid growth, the overall output of the bioplastic industry will rise and output prices will decrease. The main question is whether the properties of bioplastics can be strengthened to compete with petro-plastics.

9.3 CLASSIFICATION OF PLASTICS BASED ON BIODEGRADABILITY

There are two categories of plastics, i.e., non-biodegradable plastics and biodegradable plastics. Certain polymer plastics (biodegradable and non-biodegradable) are chemical structures.

9.3.1 Non-Biodegradable Plastics

Fossil-based and bio-based polymers incorporate non-biodegradable plastics. Most of non-biodegradable plastics used really are fossil-based engineered polymers sourced from hydrocarbon and oil subsidiaries (petrochemicals). Inferable from the broad duplication of small monomer gatherings, their atomic weight is high (Ghosh et al., 2013). These plastics are amazingly steady and don't readily join the biosphere debasement measures (Vijaya and Reddy, 2008). The majority of the ware polymers presently used are either non-biodegradable or their pace of debasement is excessively delayed to completely crumble. A large number of the regularly used plastics, for example, PVC, PP, PS, PET, PUR and PE are non-biodegradable plastics. Due to the lack of waste administration, they gather in enormous amounts in the environment and are a risk to the planet (Krueger et al., 2015). Polyolefin-determined plastics, for example, PE, are as of now used for the production of plastic films for different plastic merchandise such as sheets for packaging, transport and shopping bags and packs and mugs. Because of their life span and ecological dependability, polyolefin presents issues because of an absence of waste administration. In this manner, the executive plans of such non-biodegradable polymers must be empowered (Shah et al., 2008). What's more, idleness and opposition to microbial assault in a portion of these polymers are deliberately decreased through the presentation of starch and support of oxidants to prevent fracture (Vijaya and Reddy, 2008). In light of an absence of persuasive evidence regarding corruption, oxo-biodegradable plastics are esteemed to be non-biodegradable (Reddy, 2008).

9.3.2 Biodegradable Plastics

Biodegradable plastics comprise both biodegradable and fossil-based polymers, depending on the degree of biodegradability and the microbial acclimatization. Enzyme hydrolysis requires the biodegradation of plastics (Wackett et al., 2001). A living being type, a pretreatment plan and polymer attributes are some of the components that impact the adequacy of biodegradation measures. What's more, a portion of the critical qualities of plastics debasement incorporates versatility, crystallinity, type of useful gatherings, physicality, substance components, sub-atomic weight and added substances present in polymers (Artham et al., 2008). Microorganisms discharge exoenzymes which make polymer structures decay into smaller atoms, for example, dimers and monomers. During debasement, little particles are likewise a lot smaller than the semi-porous layers of a bacterial cell, to be used for both vitality and carbon source (Gu, 2003; Jayasekara et al., 2005). Both oxygen-consuming and anaerobic pathways are engaged with biodegradation responses (Reddy, 2008).

9.3.2.1 Types of Biodegradable Plastics

9.3.2.1.1 Bio-Based Biodegradable Plastics

Natural biodegradable plastics come from renewable resources. In certain industrial applications, bio-based biodegradable polymers have an environmental value because they are entirely biodegradable (Kale et al., 2007). Bio-based, biodegradable polymers, such as cellulose and starch, are ingested by microorganisms persistently because they reduce their molecular weight by enzyme action in an extracellular manner. Starch is the supreme mutual bio-based polymer used in biodegradable plastics production. The wide application of starch for the synthesis of bio-based plastics is due to its abundance, it is ready to use, its low cost and biodegradability under certain environmental conditions (Kyrikou et al., 2007; Nanda et al., 2010; Chattopadhyay et al., 2011). Starch is a suitable alternative, consisting of amylopectin and amylase polymers. In addition to several more bio-based biodegradable content in packaging, starch-based polymers are divided into two types: (1) polymer filled with starch, and (2) polymer based on starch (Jayasekara et al., 2003). These polymers can be totally decayed by microorganisms (bacteria, fungi and algae) and various environmental factors (Kasirajan et al., 2012). Different microorganisms of separated soil are claimed to corrupt bio-based polyps in both anaerobic and vigorous conditions (Shah et al., 2008), such as *Variovorax paradoxus*, *Comamonas sp*., *Aspergillus fumigatus*, *Acidovoraxfaecilis*, and *P. lemoignei*. Bio-based biodegradable plastics are widely known as PHA and PLA (Elbanna et al., 2004).

9.3.2.1.2 Polyhydroxyalkanoates

This is bio-based polyester that is biodegradable and framed by bacterial aging as sugars and lipids (Shimao, 2001). Because of their biodegradability, PHA polymers can be used in the packaging, clinical, and drug industries (Philip et al., 2007). Food materials, disposable clinical instruments, packaging materials and certain paints (Flieger et al., 2003) are additionally well-known PHA items. The microbial biodegradation of PHA relies upon dirt and nature. Microorganisms can corrupt PHAs under restricted vitality and use them as a carbon and vitality source (Chen and Patel, 2011). Some delegated bacterial genera are *Bacillus*, *Burkholderia, Nocardiopsis* and *Cupriavidus* for PHA biodegradation. Also, contagious agents like Mycobacterium and Micromycetes, by using vigorous and anaerobic instruments, are known to acclimatize PHA (Boyandin et al., 2013).

9.3.2.1.3 Polylactic Acid

Polylactic corrosive is a biodegradable plastic created by Nature Works in the USA as a business venture (manufacturing 150,000 tons each year). It originates from sustainable assets, for example, maize starch, custard roots or sugar stick. Inferable from the limit of the polymer to be coordinated in human and animal bodies, it has been broadly used in medication. Due to its accessibility, biodegradability and great mechanical credits, PLA is generally fundamental among bio-based biodegradable plastics (Liu et al., 2000). Hydrolytic corruption of PLA items has recently been said to be completely possible by microorganisms, *Amycolatopsis sp*. Soil-confined licheniformis has been claimed to corrupt PLA (Fukushima et al., 2009). And *Cryptococcus sp*, a lipase segregated from a growth strain S-2, indicated a distant homology of the cutinase family proteins and exhibited fruitful PLA natural debasement (Masaki et al., 2005). Poly is another bio-based biodegradable polymer used in medication and food supplies that has miniature creatures (Anderson et al., 2012).

9.3.2.1.4 Fossil-Based Biodegradable Plastics

For some reason, fossil-based biodegradable plastics have been used particularly in the packaging industry. But most fossil plastics are non-biodegradable and present significant waste management issues (Hoshino et al., 2003). The scraping of non-biodegradable fossil-based plastics in humus is a key problem in the management of pollution (Goldstein, 2005). These plastics are ordinarily used in drug packaging, different food items, make-up and various synthetics. The corruption of non-biodegradable polymers dependent on fossils is a very moderate cycle. A corruption component incorporates various natural operators, including organisms and their compounds (Chen, 2009; Mir et al., 2017). Studies of the corruption of plastics at present are zeroing in on the portrayal of microorganisms equipped to debase fossil environmental plastic, growing new debasement systems based upon chemicals, and combining duplicates of the qualities encoding biodegradation proteins (Vijaya and Reddy, 2008). Be that as it may, for productive and expedient biodegradation, microorganism-based biodegradation measures must be customized for various ecological conditions. A few instances of biodegradable polymers dependent on fossils are depicted here.

9.3.2.1.5 Polyethylene Succinate

Polyethylene succinate (PES) is one of the polyethylene esters shaped either by copolymerization of ethylene oxide and succinic anhydride or by ethylene glycol and succinic corrosive poly-build-up (Hoang et al., 2007). The plastic industry uses PES as a paper covering specialist to form films for cultivating and for shopping bags. This polymer has been viably debased by a mesophilic bacterial strain called *Pseudomonas sp.* AKS2 (Tribedi et al., 2014). In comparison to the diversity of PCL degrading microbes, there is a small distribution of microbes that degrade PES. Another thermophilic

PES-degrading strain is *Bacillus sp.* in soil isolation of TT96 (Tokiwa et al., 2009). Moreover, *Bacillus* and *Paenibacillus genera* phylogenetically include multiple mesophilic microbes isolated with the intrinsic capacity to degrade PES (Tezuka et al., 2004).

9.3.2.1.6 Polycaprolactone

Polycaprolactone is a biodegradable polymer dependent on fossils, which can be debased rapidly by vigorous and anaerobic microorganisms. This almost glasslike polyester is blended with different copolymers to prepare packaging material in biomedicinal items, catheters and blood packs. While exorbitant, its adaptability and biodegradability have stood out (Wu, 2005). Microbial lipases and esterases can corrupt PCL (Karakus, 2016). The bacteria responsible for the degradation of PCL are commonly present in the atmosphere (Shimao, 2001), such as *Aspergillus* sp., a fungal strain. The ST-01 degradation of PCL into a broad range of products, including butyric, succinic, caproic and valeric acids, was stated to be effective (Sanchez et al., 2000).

9.4 MICROBES AND THEIR MECHANISMS FOR PLASTIC BIODEGRADATION

Organisms (principally microorganisms and parasites) regularly harvest extracellular catalysts which corrupt bio- and fossil-based plastics of various sorts (Shah, 2014). Microscopic organisms and growths work using digestion and protein to disintegrate these polymers into CO_2 and H_2O. The presence and reactant action of these compounds vary as indicated by and even inside the microbial species. Because of this accuracy, various catalysts are known to debase different types of polymers. *Bacillus sp.*, for example, *Brevi bacillus* and *Brevi bacillus* sp. assembling proteases engaged with various polymers' corruption (Sivan, 2011). Lignin-debased organisms likewise contain laccases that are used by the oxidation cycle to catalyze sweet-smelling and non-fragrant mixes (Mayer and Staples, 2002). These microbial compounds frequently change in a viable and ecologically amicable way the biodegradation pace of polymers. The linkage to different microorganisms and compounds is claimed to be biodegradable like non-biodegradable polymers, for example, PHA, PLA, PET, PHB, PVC and PBS. Microorganisms with polymers, joined by surface colonization, are the fundamental pathways associated with plastic biodegradation. Plastic chemical-based hydrolysis requires two stages: first, the catalysts attach themselves to the polymer substrate and, then, the division into hydrolytes (Figure 9.3). Polymers, for example, oligomers, dimers and monomers, suffer little debasement in their atomic weight and basically have been changed by mineralization to CO_2 and H_2O. Under vigorous conditions, microbes and smaller natural mixes use oxygen as an electron receptor, and along these lines CO_2 and water are the finished results. Under anaerobic conditions (Figure 9.3), microorganisms pulverize polymers without oxygen. As electron accepters of anaerobic microscopic organisms, sulfates, nitrate, iron, carbon dioxide and manganese are used (Priyanka and Archana, 2011). To streamline the conditions under which polymers can be successfully corrupted, new microbial chemicals and pathways should be investigated. Non-biodegradable polymers (fossil and bio-based) are called non-biodegradables, but once new microbial strains and pathways to spoil these polymers have been recognized, they may be totally biodegradable. New polymers are earnestly required for their ecologically agreeable and sustainable use in the packaging and healthcare industry in the class of biodegradable polymers.

9.4.1 DEGRADATION ROUTES

9.4.1.1 Biodegradation

Polymer biodegradation is a blend of the metabolic items (H_2O, CO_2, biomass, and so forth) of the microorganism that is answerable to catalysts, of the abiotic responses (photodegradation, oxidation, hydrolysis) and the compound's cleavage of its polymer chains. The microbial qualities that influence the biodegradation rate include: the kinds of microorganisms, the conveyance of

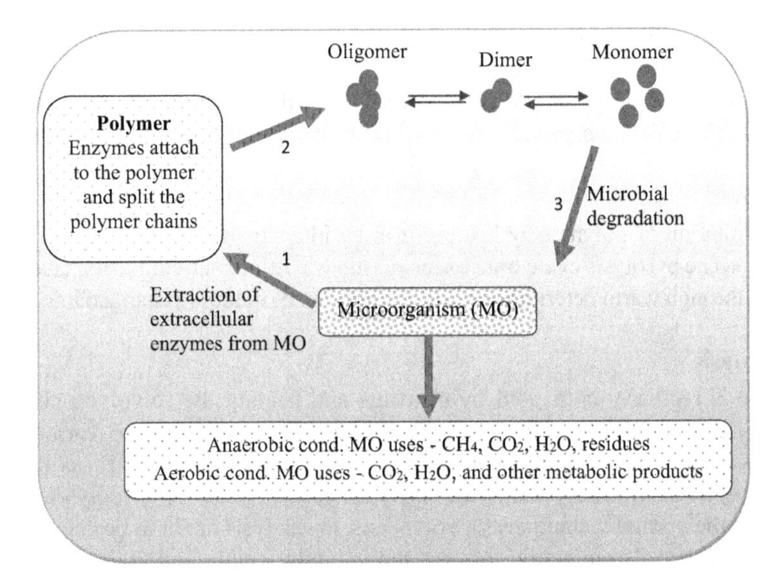

FIGURE 9.3 Plastic biodegradation: mechanism under aerobic and anaerobic conditions.

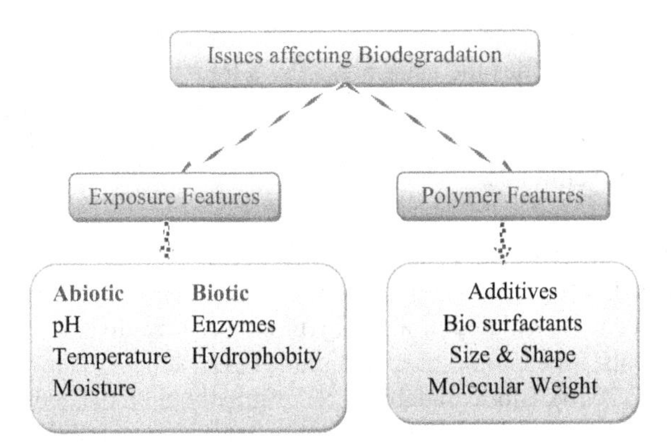

FIGURE 9.4 Factors affecting rate of biodegradation.

the microorganisms, the microbial development conditions (pH, temperature, dampness content, oxygen, supplements) and the organism types of chemicals (intracellular, extracellular) (Figure 9.4). The structure of the first order (chemical construction, molecular weight) and the structure of the higher order (Tg, tm, crystallinity) are the plastic characteristics affecting the pace. Biodegradation takes place in two distinct phases. In the beginning phase, polymer chains are both debased and abbreviated by abiotic and extracellular protein assaults; chemicals corrupt the nebulous regions of a plastic ideally. The crystallinity of a plastic will accordingly quickly increase and the total will reach almost 100%. The subsequent stage happens when polymer ties are small enough to convey microorganisms into cells where they can bio-absorb and mineralize. There are numerous international requirements that, under particular circumstances, each describes the biodegradability of bioplastics. For example, norm EN 13432:2000 states that in order to be compostable, bioplastics have to decompose by 90% and mineralize in industrial composters between 50 and 60°C in a period within 3 months. Biodegradable plastics should in principle not be collected as they can be left to biodegrade in the air under a few conditions, including expanding soil richness, lower plastic collection in waste destinations, and reducing waste administration costs. Indeed, in any case, the debasement rate may not exactly be what is desired, as it relies upon the type of bioplastic and the conditions. Another downside to biodegradation is that all meaning in the molecular structure of the polymer is lost. Instead, we should work toward a circular economy solution where bioplastics are industrially recycled and chemically recycled to recapture their commodity value.

9.4.1.2 Pyrolyse

This is the exploitation of polymer by heating it in an idle situation through or without impetuses (warm breaking). The pyrolysis cycle transforms plastic waste into natural fumes, charcoal and gases without oxygen through warm deterioration. The natural fumes made are changed into oil by build-up.

9.4.1.3 Solvolysis

Substance reuse is regularly cultivated by cleavage and heating the solvolytic chain. Solvolysis includes a few depolymerization methods and fractional depolymerization. Various solvents and heated elements are used to depolymerize a polymer into its monomers or to break the polymer into its oligomers or other manufactured substances. Hydrolysis, alcoholysis, glycolysis, aminolysis and ammonolysis are the normal techniques for solvolysis. Hydrolysis ought to be considered as having two components: water dissemination through the polymer weight and the concurrent hydrolysis response. On the off chance that water dissemination is moderate, the corruption of polymer happens fundamentally on a superficial level (heterogeneous disintegration) when contrasted with the hydrolysis rate, the debasement happens through the entire polymer mass (homogeneous disintegration). Alcoholysis is a response where liquor is the nucleophile. Regarding producing reused polyesters, the liquor bunch divides the ester bonds by a transesterification response. The polymer depolymerizes its monomers or added value goods. For the chemistry of polyesters, glycolysis refers to the incorporation of a glycole into the polymeric chains, the breaking of the ester bonds and their substitution with hydroxyl terminals.

9.4.1.4 Biodegradability and Compostability

The European standard EN 16575 portrays bio-based as obtained from biomass. Biodegradable mixes are materials that can be isolated in water by miniature creatures, for example, microbes and organisms; gaseous petrols, for example, carbon dioxide (CO_2) and methane (CH_4) and biomass (Van der Zee, 1997; Wu et al., 2016). Biodegradability relies greatly upon the ecological conditions, such as the temperature, the presence of miniature creatures and the accessibility of oxygen and water. Compostable materials are composite materials. Mechanically fertilizing the soil conditions needs high temperatures (55–60°C), joined with high mugginess and oxygen presence, and are really the best when contrasted with other soil, ocean and surface corruption conditions. Consistency with EN 13432 is viewed as a decent mechanical packaging material compostability measure. Plastic packaging must be identified as compostable as per the EN 13432 standard (Bos, 2016).

9.4.1.5 Bioplastic as a Packaging Material

Polylactic acid (PLA) is a 100% bio-based material presently used for packaging. PLA is exceptionally suited to the creation of compostable packaging materials. Pertinent advantages of PLA are its clear nature, clarity, unbending nature, printability, measure capacity and phenomenal ability to prevent smells which enhance its packaging applications. PLA is endorsed for direct contact with food and is used in various packaging materials. PLA is additionally used in relation to other bio-based and biodegradable polymers to upgrade inflexibility and quality and limit costs (ibid.).

9.4.2 Factors Affecting Plastic Biodegradation

During the biodegradation cycle, numerous variables impact the polymer properties, such as the states of introduction and the attributes of the compound (Figure 9.4). Some of these factors can be found below.

9.4.2.1 Revelation States Moisture

Dampness can influence polymer biodegradation in different ways on account of the important water interest for microbial development and duplication. Therefore, the speed of degradation is increased due to rapid microbial activity in the presence of ample moisture (Ho et al., 1999). Moisture-rich conditions further help the hydrolysis process by producing more reactions to chain splitting.

9.4.2.2 pH and Temperature

The pH can change the movement of hydrolysis by changing the acidic or key conditions. The hydrolysis centralizations of PLA cases are, for example, ideal at pH 5 (Auras et al., 2004; Henton et al., 2005). The adulterating consequences of different polymers change the pH conditions, followed by the movement of corruption and the microbial turn of events. Moreover, impetus degradability is essentially influenced by the temperature of the polymer unwinding. Polyester is less biodegradable with a higher dissolving point. Possible protein degradability reduces as the temperature increases, for example, scrubbed R lipase, or ground-breaking polyesters such as PCL with low relaxing centers (Tokiwa et al., 2004).

9.4.2.3 Enzyme Characteristics

Various catalysts have explicit dynamic locales and are equipped for biodegrading various types of polymers. For instance, straight chain polyesters obtained from 6–12 C-particle dicorrosive monomers were quickly debased by proteins created by contagious species A. An and flavus. Niger contrasted with polyesters obtained from other monomer straight chains. Extracellular proteins engaged with the depolymerization of PHB (depolymerases) have been found to debase PHB by various components, relying upon the specific depolymerase framed by the microbial substances. As a result of their hydrophobicity and 3D structure, plastics obtained from petrochemical sources cannot be swiftly corrupted in nature (Yamada-Onodera et al., 2001). Also, the hydrophobic feature of PE intervenes in the making of a microorganism biofilm to reduce biodegradation (Hadad et al., 2005).

9.4.2.4 Polymer Characteristics

9.4.2.4.1 Molecular Weight

Meanwhile, from the perspective of biodegradability, atomic weight has a significant influence in deciding numerous polymer properties. Degradability is brought down as sub-atomic weight increases. Higher sub-atomic weight PCL (> 4000) was consistently debased by R strain lipase delemar conversely with a low sub-atomic polymer weight. Microbial chemicals will undoubtedly focus on a low sub-atomic weight foundation.

9.4.2.4.2 Shape and Size

The properties, for example, polymer shape and size, assume a noteworthy role during the time spent in debasement. Polymers with a huge surface can be quickly corrupted in contrast to polymers with a restricted surface area (Stevens, 2003; Kijchavengkul et al., 2008; Tan, 2015). There is a normalized structure and size basis for the biodegradation of various sorts of plastics (ASTM D., 2004).

9.4.2.4.3 Additives

Non-polymeric contamination, for example, the colors (impetus waste or scrap used for transformation results of polymerization and added substances) or fillers influence the limit of debasement. As the dialect cellulosic filler increases in the example, the warm steadiness is diminished and the debris content expands. The dispersal and interfacial bond between the dialect cellulosic filler and the thermoplastic polymer are the key factors that influence the composite framework's warm dependability (Yang et al., 2005). Metals likewise work as solid pro-oxidants in the polyole fine production of thermo-oxidative corruption of polymers.

9.4.2.4.4 *Biosurfactants*

Biosurfactants are amphiphilic intensifiers that remain basically framed on living surfaces. Polymer biodegradation (fossil-based and bio-based) is improved because of its low poisonousness and high biodegradability (Orr et al., 2004). Inferable from the presence of exceptionally useful collections, biosurfactants make it conceivable to biodegrade under extraordinary temperature, pH and saltiness conditions (Kawai et al., 2002; Kawai et al., 2004).

9.5 CURRENT RESEARCH ON PLASTIC BIODEGRADATION

The use of naturally neighborly biodegradable bio- and fossil-based plastics together is increasing. Biodegradable plastics are an achievement aimed at appropriate waste administration and littering control methodologies in order to make a sustainable society (Iwata 2015). The Cat, a non-biodegradable fossil-based polymer, is used globally in plastic items and has increased overall enthusiasm for nature. Another bacterial strain, Ideonellasakaiensis 201-F6, has as of late remained detached, using PET as a wellspring of vitality and carbon. The strain produces two proteins for hydrolyzing PET and terephthalic corrosive mono (2-hydroxyethyl) into two ecologically damaging monomers, to be specific, ethylene glycol and terephthalic corrosive (Yoshida et al., 2016). In this way, more streamlining in cells of the PET corruption measure is primarily required to locate its numerous other eco-accommodating applications. Another exploration shows that the Pantoea spp. microbial consortium, *Enterobacter* spp., has LDPE debasement potential (Skariyachan et al., 2016). Researchers have recognized a few microorganisms can change natural styrene (mechanical plastic waste material) into PHA. A viable styrene-debasing bacterium is the disengaged strain, P. putida NBUS12. Another bacterial strain, *Achromobacterxylosoxidans*, has recently been found to affect the structure of high-thickness polyethylene (Kowalczyk et al., 2016). Also, *Anoxy bacillus rupiensis* Ir3 (JQ912241), a serious thermophilic bacterium, was secluded from hydrocarbon-dirtied Iraqi land and exhibited its potential for use as carbon-contaminated fragrant mixes and their debasement (Mahdi et al., 2016). There are accordingly extensive examination studies around the globe to set up techniques for debasement of fossil- and bio-based polymers and discover their new earth neighborly applications and waste administration plans.

9.5.1 Sustainability and End-of-Life Options for Bioplastics

The worldwide pattern of fossil-based plastic production and the burning of plastic waste adds up to around 400 million tons of CO_2 consistently (Rahimi et al., 2017). Replacing fossil-based plastics with biodegradable plastics would reduce the level of their carbon footprint. Notwithstanding, it is important for bioplastics to be used as a naturally reasonable choice to assess their supportability regarding their end-of-life results. Not all bio-based polymers are viewed as biodegradable, and a portion of the biodegradable polymers can be made from fossil crude materials. Certainly, groups of notable bioplastics, for example, PHB, PCL, and starch and their blends have ended up being biodegraded in both regulated and interesting unmanaged conditions (Narancic et al., 2018), however, the absence of control of their removal will create unregulated natural corruption adding to current plastic contamination (Saibuatrong et al., 2017). It is therefore of the utmost importance that the properties and processing conditions of each individual bioplastic, rather than standardized waste management plans, are regarded in real end-of-life management. Life cycle analysis (LCA) is a vital method in calculating and quantifying, subject to border conditions and evaluation considerations, the benefits or consequences of any bioplastic.

The most favored method for handling bioplastic waste is recycling as close to traditional plastic waste (Hong et al., 2019). However, recycling, depending on the biodegradable existence of the bioplastics and/or whether the polymeric content is considered biodegradable only under controlled conditions, can be mechanical, chemical/catalytic or organic. The separate recycling options shown in Figure 9.2 reflect the most advanced closed-loop management of bioplastic waste after consumption. Notable heterogeneity, low market volumes, different sources and high potential for the pollution of plastic waste are the primary difficulties of post-buyer bioplastic waste reuse. These issues plainly exhibit the requirement for more viable substance and biochemical cycles for the recycling of bioplastic waste into reusable valued items (Tang et al., 2019; Chowdhary et al., 2020). Both reuse and recuperation standards, including the extraction of high-value compound/monomer items through synthetic reuse, dissolvable extraction and coage of biofuels and unstable unsaturated fats through anaerobic assimilation, will definitely have a positive effect on the circular economy. Some management methods do not recycle bioplastics directly into their initial monomer (Albertsson and Hakkarainen, 2017). However, it is worth investing in the recovery of bioplastic waste from post-use and offer incentives for recycling or energy recycling to contribute to the circular economy and sustainable bioplastic waste management (Garcia and Robertson, 2017).

9.6 FUTURE PROSPECTS

The most innovative and environmentally reasonable approach to address garbage removal from an assortment of sources is to use biodegradable plastics in numerous applications, for example, packaging, farming and health care. Whenever used, biodegradable and fossil-based polymers are viably corrupted in nature and in cells.. Right now, petrochemical non-biodegradable items used for the production of plastics represent an extraordinary ecological risk, especially without littering control and waste administration frameworks. In some applications, the market for environmentally friendly polymers is constantly growing. In future, the use of such materials should be concentrated in the manufacture of packaging materials, food packaging and disposable medical devices. Biodegradable plastics are also useful in the field, such as agricultural films, fishing materials (fishing nets), biomass-absorbable materials in therapeutics, operational frames and sterile products. In addition, biodegradable plastics should be used where there is an inevitable diffusion into the atmosphere or where it is difficult to remove the waste. Then again, it is imperative to appropriately deal with the waste and control litter to exploit these polymers in the earth. Biodegradable plastics of the next decade will create a more sustainable community for particular purposes. These plastics can likewise be biodegraded and recycled to make their reuse conceivable in a manageable way. Thus, a careful comprehension of the biomass structure commonly produced is expected to create new biodegradable plastic polymers through slight auxiliary changes. Biomass researchers, synthetic chemists, process engineers and microbiologists should make the individual and collective use of their skills to make society more sustainable through the production of environmentally-friendly materials.

9.7 CONCLUSION

Using plastic materials is unavoidable if we want to satisfy our regular daily existence. Plastics use is an ever-expanding field. This must be firmly connected with waste administration and littering control and the use of certain applications for a reasonable ecological security of bio-based and fossil-based biodegradable materials. Distinctive microbial networks may bio-vert such plastic polymers through vigorous and anaerobic systems into less complex items. There is an urgent need to combine and additionally order new naturally based biodegradable polymers to debase fossil-based polymers. Moreover, the amount of biodegradable materials must be increased in some modern applications for natural security and sustainability to be guaranteed, as well as the creation of streamlined mechanical debasement plants and littering control systems.

REFERENCES

Albertsson, A.C., and Hakkarainen, M., 2017. Designed to degrade. *Science*, 358(6365), 872–873.

Anderson, J.M., and Shive, M.S., 2012. Biodegradation and biocompatibility of PLA and PLGA microspheres. *Advanced Drug Delivery Reviews*, 64, 72–82.

Artham, T., and Doble, M., 2008. Biodegradation of aliphatic and aromatic polycarbonates. *Macromolecule Bioscience* 8(1), 14–24.

ASTM D, 2004. 6400–04 Standard Specification for Compostable Plastics. West Conshohocken, PA: ASTM International.

Auras, R., Harte, B., and Selke, S., 2004. An overview of polylactides as packaging materials. *Macromolecular Bioscience*, 4(9), 835–864.

Bio-based Building Blocks and Polymers, 2018. Global capacities, production and trends, 2018–2023. Available at: www.liebertpub.com/doi/10.1089/ind.2019.29179.rch

Bos, 2016. Bioplastics and food—enemies or allies. Paper presented at Sustainable Plastics 2016 Conference, Köln, March 1–2.

Boyandin, A.N., Prudnikova, S.V., Karpov, V.A., Ivonin, V.N., Đỗ, N.L., Nguyễn, T.H., Lê, T.M.H., … Volova, T.G., 2013. Microbial degradation of polyhydroxyalkanoates in tropical soils. *International Biodeterioration & Biodegradation*, 83, 77–84.

Chattopadhyay, S.K., Singh, S., Pramanik, N., Niyogi, U.K., Khandal, R.K., Uppaluri, R., and Ghoshal, A.K., 2011. Biodegradability studies on natural fibers reinforced polypropylene composites. *Journal of Applied Polymer Science*, 121(4), 2226–2232.

Chen, G.G.Q. (Ed.), 2009. *Plastics from Bacteria: Natural Functions and Applications* Vol. 14. Cham: Springer Science & Business Media.

Chen, G.Q., and Patel, M.K., 2012. Plastics derived from biological sources: Present and future: A technical and environmental review. *Chemical Reviews*, 112(4), 2082–2099.

Chowdhary, P., Hare, V., Singh, A.K., Pandit S., and Chaturvedi, P., 2020. Emerging environmental contaminants: Sources, consequences, and future challenges. In Chowdhary, P., and Raj, A. (Eds.), *Contaminants and Clean Technologies*, Boca Raton, FL: CRC Press.

Elbanna, K., Lütke-Eversloh, T., Jendrossek, D., Luftmann, H., and Steinbüchel, A., 2004. Studies on the biodegradability of polythioester copolymers and homopolymers by polyhydroxyalkanoate (PHA)-degrading bacteria and PHA depolymerases. *Archives of Microbiology*, 182(2–3), 212–225.

Ellen McArthur Foundation, n.d. The new plastics economy: Rethinking the future of plastics. Available at: www.ellenmacarthurfoundation.org/publications/the-new-plastics-economy-rethinking-the-future-of-plastics

Elvers, D., Song, C.H., Steinbüchel, A., and Leker, J., 2016. Technology trends in biodegradable polymers: Evidence from patent analysis. *Polymer Reviews*, 56(4), 584–606.

Eriksen, M., Lebreton, L.C., Carson, H.S., Thiel, M., Moore, C.J., Borerro, J.C., Galgani, F., Ryan, P.G., and Reisser, J., 2014. Plastic pollution in the world's oceans: More than 5 trillion plastic pieces weighing over 250,000 tons afloat at sea. *PloS One*, 9(12), e111913.

European Bioplastics, 2018. Bioplastics market data. Available at: https://european-bioplastics.org/wp-content/uploads/2016/02/Report_Bioplastics-Market-Data_2018.pdf

Flieger, M., Kantorova, M., Prell, A., Řezanka, T., and Votruba, J., 2003. Biodegradable plastics from renewable sources. *Folia Microbiologica*, 48(1), 27.

Fukushima, K., Abbate, C., Tabuani, D., Gennari, M., and Camino G., 2009. Biodegradation of poly (lactic acid) and its nanocomposites. *Polymer Degradation Stability*, 94(10), 1646–1655.

Garcia, J.M., and Robertson, M.L., 2017. The future of plastics recycling. *Science*, 358(6365), 870–872.

Geyer, R., Jambeck, J.R., and Law, K.L., 2017. Production, use, and fate of all plastics ever made. *Science Advances*, 3(7), e1700782.

Ghosh, S.K., Pal, S., and Ray, S., 2013. Study of microbes having potentiality for biodegradation of plastics. *Environmental Science and Pollution Research*, 20(7), 4339–4355.

Goldstein, N., 2005. Source separated MSW composting in the US. *BioCycle*, 46(12), 20.

Gu, J.D., 2003. Microbiological deterioration and degradation of synthetic polymeric materials: Recent research advances. *International Biodeterioration & Biodegradation*, 52(2), 69–91.

Guern, C.L., 2019. When the mermaids cry: The great plastic tide. Available at: www.plastic-pollution.org

Hadad, D., Geresh, S., and Sivan, A., 2005. Biodegradation of polyethylene by the thermophilic bacterium *Brevibacillusborstelensis*. *Journal of Applied Microbiology*, 98(5), 1093–1100.

Helanto, K.E., Matikainen, L., Talja, R., and Rojas, O.J., 2019. Bio-based polymers for sustainable packaging and biobarriers: A critical review. *BioResources*, 14(2), 4902–4951.

Henton, D.E., Gruber, P., Lunt, J., and Randall, J., 2005. Polylactic acid technology. *Natural Fibers, Biopolymers, and Biocomposites*, 16, 527–577.

Ho, K.L.G., Pometto, A.L., and Hinz, P.N., 1999. Effects of temperature and relative humidity on polylactic acid plastic degradation. *Journal of Environmental Polymer Degradation*, 7(2), 83–92.

Hoang, K.C., Tseng, M., and Shu, W.J., 2007. Degradation of polyethylene succinate (PES) by a new thermophilic Microbispora strain. *Biodegradation*, 18(3), 333–342.

Hong, M., and Chen, E.Y.X., 2019. Future directions for sustainable polymers. *Trends in Chemistry*, 1, 148–151.

Hoshino, A., Tsuji, M., Ito, M., Momochi, M., Mizutani, A., Takakuwa, K., Higo, S., Sawada, H., and Uematsu, S., 2003. Study of the aerobic biodegradability of plastic materials under controlled compost. In Chiellini, E. and Solaro, R. (Eds.), *Biodegradable Polymers and Plastics*, Boston, MA: Springer, pp. 47–54.

Iwata, T., 2015. Biodegradable and bio-based polymers: Future prospects of eco-friendly plastics. *Angewandte Chemie International Edition*, 54(11), 3210–3215.

Jayasekara, R., Harding, I., Bowater, I., Christie, G.B., and Lonergan, G.T., 2003. Biodegradation by composting of surface modified starch and PVA blended films. *Journal of Polymers and the Environment*, 11(2), 49–56.

Jayasekara, R., Harding, I., Bowater, I., and Lonergan, G. T., 2005. Biodegradability of a selected range of polymers and polymer blends and standard methods for assessment of biodegradation. *Journal of Polymers and the Environment*, 13(3), 231–251.

Jin, H.J., Lee, B.Y., Kim, M.N., and Yoon, J.S., 2000. Thermal and mechanical properties of mandelic acid-copolymerized poly (butylene succinate) and poly (ethylene adipate). *Journal of Polymer Science Part B Polymer Physics*, 38(11): 1504–1511. https://doi.org/10.1002/(SICI)1099-0488(20000601)38:113.0.CO;2-4.

Kale, G., Kijchavengkul, T., Auras, R., Rubino, M., Selke, S.E., and Singh, S.P., 2007. Compostability of bioplastic packaging materials: An overview. *Macromolecular Bioscience*, 7(3): 255–277. https://doi.org/10.1002/mabi. 200600168

Karakus, K., 2016. Polycaprolactone (PCL) based polymer composites filled wheat straw flour. *Kastamonu Üniversitesi Orman Fakültes iDergisi*, 16(1), 264–268.

Kasirajan, S., and Ngouajio, M., 2012. Polyethylene and biodegradable mulches for agricultural applications: A review. *Agronomy for Sustainable Development*, 32(2), 501–529.

Kawai, F., Watanabe, M., Shibata, M., Yokoyama, S., and Sudate, Y., 2002. Experimental analysis and numerical simulation for biodegradability of polyethylene. *Polymer Degradation and Stability*, 76(1), 129–135.

Kawai, F., Watanabe, M., Shibata, M., Yokoyama, S., Sudate, Y., and Hayashi, S., 2004. Comparative study on biodegradability of polyethylene wax by bacteria and fungi. *Polymer Degradation and Stability*, 86(1), 105–114.

Kijchavengkul, T., and Auras, R., 2008. Compostability of polymers. *Polymer International*, 57(6), 793–804.

Kowalczyk, A., Chyc, M., Ryszka, P., and Latowski, D., 2016. Achromobacterxylosoxidans as a new microorganism strain colonizing high-density polyethylene as a key step to its biodegradation. *Environmental Science and Pollution Research*, 23(11), 11349–11356.

Krueger, M.C., Harms, H., and Schlosser, D., 2015. Prospects for microbiological solutions to environmental pollution with plastics. *Applied Microbiology and Biotechnology*, 99(21), 8857–8874.

Kumar, V., Srivastava, S., and Thakur, I.S., 2021. Enhanced recovery of polyhydroxyalkanoates from secondary wastewater sludge of sewage treatment plant: Analysis and process parameters optimization. *Bioresource Technology Reports*, 15, 100783. https://doi.org/10.1016/j.biteb.2021.100783

Kyrikou, I., and Briassoulis, D., 2007. Biodegradation of agricultural plastic films: A critical review. *Journal of Polymers and the Environment*, 15(2), 125–150.

Lam, C.X., Hutmacher, D.W., Schantz, J.T., Woodruff, M.A., and Teoh, S.H., 2009. Evaluation of polycaprolactone scaffold degradation for 6 months in vitro and in vivo. *Journal of Biomedical Materials Research Part A: An Official Journal of the Society for Biomaterials, the Japanese Society for Biomaterials, and the Australian Society for Biomaterials and the Korean Society for Biomaterials*, 90(3), 906–919.

Liu, L., Li, S., Garreau, H., and Vert, M., 2000. Selective enzymatic degradations of poly (L-lactide) and poly (ε-caprolactone) blend films. *Biomacromolecules*, *1*(3), 350–359.

Mahdi, M.S., Ameen, R.S., and Ibrahim, H.K., 2016. Study on degradation of nylon 6 by thermophilic bacteria Anoxybacillusrupiensis Ir3 (JQ912241). *International Journal of Advanced Research in Biological Sciences*, 3(9), 200–209.

Masaki, K., Kamini, N.R., Ikeda, H., and Iefuji, H., 2005. Cutinase-like enzyme from the yeast *Cryptococcus* sp. strain S-2 hydrolyzes polylactic acid and other biodegradable plastics. *Applied and Environmental Microbiology*, 71(11), 7548–7550.

Mayer, A.M., and Staples, R.C., 2002. Laccase: New functions for an old enzyme. *Phytochemistry*, 60(6): 551–565. https://doi.org/10.1016/ S0031-9422(02)00171-1

Mir, S., Asghar, B., Khan, A.K., Rashid, R., Shaikh, A.J., Khan, R.A., and Murtaza, G., 2017. The effects of nanoclay on thermal, mechanical and rheological properties of LLDPE/chitosan blend. *Journal of Polymer Engineering*, 37(2), 143–149.

Mohanty, A.K., Misra, M.A., and Hinrichsen, G.I., 2000. Biofibres, biodegradable polymers and biocomposites: An overview. *Macromolecular Materials and Engineering*, 276(1), 1–24.

Moharir, R. V., and Kumar, S., 2019. Challenges associated with plastic waste disposal and allied microbial routes for its effective degradation: A comprehensive review. *Journal of Cleaner Production*, 208, 65–76. https://doi.org/10.1016/j.jclepro.2018.10.059

Muhamad, W.N.A.W., Othman, R., Shaharuddin, R.I., and Irani, M.S., 2015. Microorganism as plastic biodegradation agent towards sustainable environment. *Advance Environmental Biology*, 9, 8–14.

Nakajima, H., Dijkstra, P., and Loos, K., 2017. The recent developments in biobased polymers toward general and engineering applications: Polymers that are upgraded from biodegradable polymers, analogous to petroleum-derived polymers, and newly developed. *Polymers*, 9(10), 523.

Nanda, S., Sahu, S., and Abraham, J., 2010. Studies on the biodegradation of natural and synthetic polyethylene by Pseudomonas spp. *Journal of Applied Sciences and Environmental Management*, 14(2), 57–60.

Narancic, T., Verstichel, S., Reddy Chaganti, S., Morales-Gamez, L., Kenny, S.T., De Wilde, B., BabuPadamati, R., and O'Connor, K.E., 2018. Biodegradable plastic blends create new possibilities for end-of-life management of plastics but they are not a panacea for plastic pollution. *Environmental Science & Technology*, 52(18), 10441–10452.

Norman, R.S., Frontera-Suau, R., and Morris, P.J., 2002. Variability in Pseudomonas aeruginosa lipopolysaccharide expression during crude oil degradation. *Applied and Environmental Microbiology*, 68(10), 5096–5103.

Orr, I.G., Hadar, Y.. and Sivan, A., 2004. Colonization, biofilm formation and biodegradation of polyethylene by a strain of Rhodococcusruber. *Applied Microbiology and Biotechnology*, 65(1), 97–104.

Philip, S., Keshavarz, T., and Roy, I., 2007. Polyhydroxyalkanoates: Biodegradable polymers with a range of applications. *Journal of Chemical Technology & Biotechnology: International Research in Process, Environmental & Clean Technology*, 82(3), 233–247.

Plastic Oceans, n.d. Report. Available at: www.plasticoceans.org

Plastics Europe, 2019. Annual review. Available at: www.plasticseurope.org

Prata, J.C., Silva, A.L.P., Da Costa, J.P., Mouneyrac, C., Walker, T.R., Duarte, A.C., and Rocha-Santos, T., 2019. Solutions and integrated strategies for the control and mitigation of plastic and microplastic pollution. *International Journal of Environmental Research and Public Health*, 16(13), 2411.

Priyanka, N., and Archana, T., 2011. Biodegradability of polythene and plastic by the help of microorganism: A way for brighter future. *Journal of Environmental and Analytical Toxicology*, 1(4), 1000111.

Rahimi, A., and García, J.M., 2017. Chemical recycling of waste plastics for new materials production. *Nature Reviews Chemistry*, 1(6), 1–11.

Raziyafathima, M., Praseetha, P.K., and Rimal, I.R.S., 2016. Microbial degradation of plastic waste: A review. *Chemical and Biological Sciences*, 4, 231–242.

Reddy, M., 2008. Oxo-biodegradation of polyethylene. *Journal of Applied Polymer Science*, 111(3), 1426–1432.

Rujnić-Sokele, M., and Pilipović, A., 2017. Challenges and opportunities of biodegradable plastics: A mini review. *Waste Management & Research*, 35(2), 132–140.

Ruse, M., 2013. Popular science to professional science, in Pigliucci, M. and Boudry, M. (Eds.), *Philosophy of Pseudoscience: Reconsidering the Demarcation Problem*, Chicago: University of Chicago Press.

Saibuatrong, W., Cheroennet, N., and Suwanmanee, U., 2017. Life cycle assessment focusing on the waste management of conventional and bio-based garbage bags. *Journal of Cleaner Production*, 158, 319–334.

Sanchez, J.G., Tsuchii, A., and Tokiwa, Y., 2000. Degradation of polycaprolactone at 50° C by a thermotolerant Aspergillus sp. *Biotechnology Letters*, 22(10), 849–853.

Scalenghe, R., 2018. Resource or waste? A perspective of plastics degradation in soil with a focus on end-of-life options. *Heliyon*, 4(12), e00941.

Shah, A.A., Hasan, F., Hameed, A., and Ahmed, S., 2008. Biological degradation of plastics: A comprehensive review. *Biotechnology Advances*, 26(3), 246–265.

Shah, A.A., Kato, S., Shintani, N., Kamini, N.R., and Nakajima-Kambe, T., 2014. Microbial degradation of aliphatic and aliphatic-aromatic co-polyesters. *Applied Microbiology and Biotechnology*, 98(8), 3437–3447.

Sharma, S., Singh, I., and Virdi, J., 2003. A potential Aspergillus species for biodegradation of polymeric materials. *Current Science*, 84(11): 1399–1402.

Shimao, M., 2001. Biodegradation of plastics. *Current Opinion in Biotechnology*, 12(3), 242–247.Singh, E., Kumar, A., Khapre, A., Saikia, P., Kumar S.K., and Kumar, S., 2020. Efficient removal of arsenic using plastic waste char: Prevailing mechanism and sorption performance. *Journal of Water Process Engineering*, 33, 101095. https://doi.org/10.1016/j.jwpe.2019.101095

Singh, E., Kumar, A., Mishra, R., You, S., Singh, L., Kumar, S., and Kumar, R. 2020. Pyrolysis of waste biomass and plastics for production of biochar and its use for removal of heavy metals from aqueous solution. *Bioresource Technology*, 20-06197R2. https://doi.org/10.1016/j.biortech.2020.124278

Singh, N., Hui, D., Singh, R., Ahuja, I.P.S., Feo, L., and Fraternali, F., 2017. Recycling of plastic solid waste: A state of art review and future applications. *Composites Part B: Engineering*, 115, 409–422.

Sivan, A. 2011. New perspectives in plastic biodegradation. *Current Opinion in Biotechnology*, 22(3): 422–426. https://doi.org/10.1016/j.copbio.2011. 01.013

Skariyachan, S., Manjunatha, V., Sultana, S., Jois, C., Bai, V., and Vasist, K.S., 2016. Novel bacterial consortia isolated from plastic garbage processing areas demonstrated enhanced degradation for low density polyethylene. *Environmental Science and Pollution Research*, 23(18), 18307–18319.

Stevens, E.S., 2003. What makes green plastics green? *Biocycle*, 44(3), 24–27.

Tan, G.Y.A., Chen, C.L., Ge, L., Li, L., Tan, S.N., and Wang, J.Y., 2015. Bioconversion of styrene to poly (hydroxyalkanoate)(PHA) by the new bacterial strain *Pseudomonas putida* NBUS12. *Microbes and Environments*, 30(1), 76–85.

Tang, X., and Chen, E.Y.X., 2019. Toward infinitely recyclable plastics derived from renewable cyclic esters. *Chem*, 5(2), 284–312.

Tezuka, Y., Ishii, N., Kasuya, K.I., and Mitomo, H., 2004. Degradation of poly (ethylene succinate) by mesophilic bacteria. *Polymer Degradation and Stability*, 84(1), 115–121.

Tokiwa, Y., and Calabia, B.P., 2004. Review degradation of microbial polyesters. *Biotechnology Letters*, 26(15), 1181–1189.

Tokiwa, Y., Calabia, B.P., Ugwu, C.U., and Aiba, S., 2009. Biodegradability of plastics. *International Journal of Molecular Sciences*, 10(9), 3722–3742.

Tribedi, P., and Sil, A.K., 2014. Cell surface hydrophobicity: A key component in the degradation of polyethylene succinate by Pseudomonas sp. AKS 2. *Journal of Applied Microbiology*, 116(2), 295–303.UNEP-UN Environment, n.d. Report. Available at: web.unep.org

Van der Zee, M., 1997. Structure-biodegradability relationships of polymeric materials. PhD thesis. Universiteit Twente.

Vijaya, C. and Reddy, R.M., 2008. Impact of soil composting using municipal solid waste on biodegradation of plastics. *Indian Journal of Biotechnology*, 7, 235–239.

Wackett, L.P., and Hershberger, C.D., 2001. *Biocatalysis and Biodegradation: Microbial Transformation of Organic Compounds* (No. QP517. B5 W33). Washington, DC: ASM Press.

Wu, C.S., 2005. A comparison of the structure, thermal properties, and biodegradability of polycaprolactone/chitosan and acrylic acid grafted polycaprolactone/chitosan. *Polymer*, 46(1), 147–155.

Wu, J., Zhu, Q., Chu, J., Liu, H., and Liang, L., 2016. Measuring energy and environmental efficiency of transportation systems in China based on a parallel DEA approach. *Transportation Research Part D: Transport and Environment*, 48, 460–472.

Yamada-Onodera, K., Mukumoto, H., Katsuyaya, Y., Saiganji, A., and Tani, Y., 2001. Degradation of polyethylene by a fungus, Penicillium simplicissimum YK. *Polymer Degradation and Stability*, 72(2), 323–327.

Yang, H.S., Wolcott, M.P., Kim, H.S., and Kim, H.J., 2005. Thermal properties of lignocellulosic filler-thermoplastic polymer bio-composites. *Journal of Thermal Analysis and Calorimetry*, 82(1), 157–160.
Yang, J., Ching, Y.C., and Chuah, C.H., 2019. Applications of lignocellulosic fibers and lignin in bioplastics: A review. *Polymers*, 11(5), 751.
Yoshida, S., Hiraga, K., Takehana, T., Taniguchi, I., Yamaji, H., Maeda, Y., Toyohara, K., … Oda, K., 2016. A bacterium that degrades and assimilates poly (ethylene terephthalate). *Science*, 351(6278), 1196–1199.

10 Microalgae
The Role of Phycoremediation in Treated Chrome Sludge from the Electroplating Industry and in Biomass Production

M. Muthukumaran
PG and Research Department of Botany, Ramakrishna Mission Vivekananda College, Chennai, Tamil Nadu, India
Corresponding author Email: mmuthukumaran@rkmvc.ac.in

CONTENTS

10.1 Introduction 201
10.2 Effect of Heavy Metals Contamination on the Environment 202
10.3 Methodology 202
10.4 Results and Discussion 202
 10.4.1 Phycoremediation: Techniques and Mechanisms 203
 10.4.2 Microalgae for Wastewaters Remediation 203
 10.4.3 Phytoremediation for Heavy Metals Using Microalgae 204
10.5 Phycoremediation of Chrome Sludge from the Electroplating Industry 205
 10.5.1 The Experiment Raceway Pond Study 205
 10.5.2 Potential of Microalga *Desmococcus Olivaceus* Grown in Chrome Sludge of Electroplating Industry in the Open Raceway Pond Study 206
 10.5.3 Growth and Chrome-Sludge Reduction by *Desmococcus Olivaceus* 206
 10.5.4 Phycoremediation of Chromium Heavy Metal from the Electroplating Industry 206
 10.5.5 Phycoremediation: Algal Biomass Production and Its Applications 207
10.6 Conclusion 210
Acknowledgements 211
References 211

10.1 INTRODUCTION

Inorganic and organic substances are delivered into the environment because of agricultural, domestic and industrial uses that frequently lead to serious contamination (Randrianarison and Ashraf, 2017; Agrawal et al., 2021; Dhote et al., 2021; Kumar et al., 2021). The current global ecological issues raise unavoidable problems for our use of natural resources. Providing the human population with clean water is turning into a global issue (Wollmann et al., 2019). The far-reaching results of different types of contamination, including noxious heavy metals and other emerging risky contaminants, is of grave concern (Bilal et al., 2018; Dutta et al., 2021; Kumar et al., 2021). Looking after the biological status of water sources is an emerging problem for some developing countries,

DOI: 10.1201/9781003239956-12

specifically, how to reduce nitrogen and phosphorus in industrial wastewater (Mohsenpour et al., 2021). Emerging types of contamination are attracting attention because of their possible danger to human well-being and the environment. The previous decade saw the re-established significance of microalgae to bioremediate various pollutants from wastewater (Maryjoseph and Ketheesan, 2020).

10.2 EFFECT OF HEAVY METALS CONTAMINATION ON THE ENVIRONMENT

Water contamination occurs because of the arrival of an unreasonable amount of dangerous heavy metals in metropolitan wastewater, and this problem is expanding the hazard to aquatic ecosystems (Moharir and Kumar, 2020). Treatment of wastewater using algae is called phycoremediation, which is a special, modest, technique to treat the contaminated water using a natural strategy (Kumar et al., 2018; Prathiba et al., 2020). The heavy metals are delivered into streams because of different anthropogenic uses. Treating them is vital, since heavy metals are non-biodegradable and endure in the environment, representing a danger to the biota because of their harmful elements (Kaplan, 2013; Kumar, 2018). Figure 10.1 shows the toxicological properties of heavy metals.

10.3 METHODOLOGY

This chapter discusses how microalgae can be used in phycoremediation to treat chrome sludge from the electroplating industry and to produce biomass for feedstock applications. Sources used for the data were Scielo, PubMed, Web of Science, Science Direct and scholastic papers.

10.4 RESULTS AND DISCUSSION

Contamination of surface water has now become one of the major problems of the natural world. There are numerous studies by different authors underscoring the connection of algae to clean wastewater

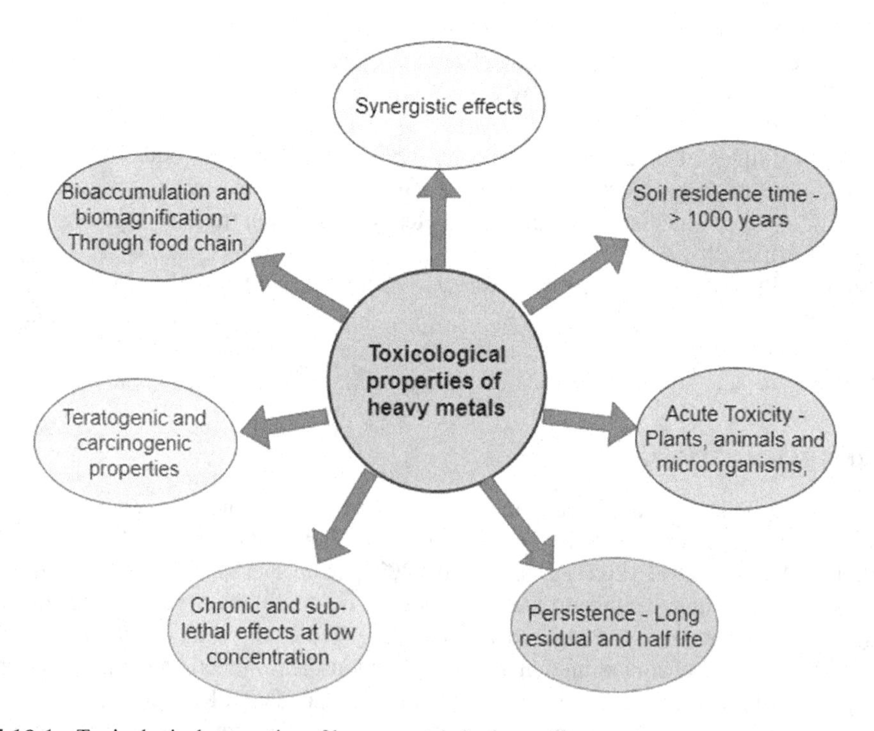

FIGURE 10.1 Toxicological properties of heavy metals in the environment.

(Randrianarison and Ashraf, 2017). The possibility of microalgae-based effluent treatment, including the techniques for strain choice, the impact of wastewater types, the photobioreactor plan, the monetary possibility appraisal, and other main points of contention that impact the treatment execution, was presented by (Wang et al., 2016). The incorporation of wastewater and microalgae and microalgal properties with respect to biofuel creation was studied by Salama et al. (2017). The coupling of wastewater treatment by microalgae is a successful method of remediation and creates a financially clever microalgal biofuel (Bhatt et al., 2014). Phycoremediation can be characterized as the use of microalgae for expulsion or biotransformation of contaminants and CO_2 from wastewater treatment with the associated biomass production (Ahmad et al., 2021).

10.4.1 Phycoremediation: Techniques and Mechanisms

Algae are an assorted collection of photosynthetic creatures with probable significant bioremediation and industrial applications that can reduce the expense of using energy. Algae have been involved in different applications, such as biofuel, biofertilizer and medicine to wastewater treatment. Numerous types of algae have been used for their ability to treat wastewater (Bansal et al., 2018; Chowdhary et al., 2018). The mechanism and elements associated with using algae to eliminate supplements and contaminants from wastewater were discussed prior to discovering the capability of algal-based methods to remediate lake eutrophication. Procedures used in the phycoremediation of wastewater incorporate algal turf scrubbers, algal biofilm, a high rate algal pond (HRAP) and algal immobilized. The mechanisms of harvesting, the factors affecting the phycoremediation and strain choice in the phycoremediation of eutrophicated lakes were additionally considered by Razak and Sharip (2020). Their review demonstrates microalgae have fostered a wide range of adsorption (intracellular) and retention (extracellular) components to adapt to heavy metal toxicity (Kaplan, 2013).

Heavy metal expulsion systems incorporate flocculation, sedimentation, absorption and cations and anion trade, precipitation, complexation, reduction/oxidation/decrease, microbiological movement and take-up. Microalgae eliminate heavy metals straightforwardly from contaminated water through two significant mechanisms: (1) a digestion subordinate take-up into their cells at low fixations; and (2) biosorption which is a non-dynamic adsorption measure (Chekroun and Baghour, 2013; Chowdhary et al., 2020b). The mechanisms of contaminants evacuation by microalgae and the phycoremediation capability of various sorts of wastewaters, the likely uses of wastewater that has developed a microalgal biomass, the existing difficulties, and the future course of microalgal applications in treatment of wastewater were discussed by Al-Jabri et al. (2021). Travieso et al. (1999) studied the resistance and take-up of Zn, Cd and Cr of *Scenedesmus acutus* by immobilization. Biosorption strategies can be an option in contrast to the normal industrial techniques for evacuation of heavy metals as they are harmless to the ecosystem and sustainable (Ali Redha, 2020).

10.4.2 Microalgae for Wastewaters Remediation

The sustainable use of a wastewater treatment framework should be technologically attainable, harmless to the ecosystem and financially practical (Kumar, 2021). The proof is whether incorporating miniature microalgae as an option for natural wastewater treatment is innovatively feasible for the environment (Chowdhary and Raj, 2020a; Mohsenpour et al., 2021). Microalgae cultures offer a rich answer as a tertiary treatment because of the capacity of microalgae to use inorganic mixtures for their growth development and, furthermore, for their ability to eliminate heavy metals, and other toxic natural mixtures (Randrianarison and Ashraf, 2017). The microalgae are effective at expelling various kinds of wastewaters and can be investigated for the phycoremediation of these polluted sites (Renuka et al., 2015). Microalgae can use supplements like phosphorus or nitrogen substances in wastewater (Li et al., 2019). The possible use of *Chlorella vulgaris* for phycoremediation of textile wastewater (TW) was examined using four clusters of cultures in a high-rate algal pond (HRAP) containing textile dye (Lim et al., 2010). The microalgae *Scenedesmus* sp., *Chlorella sorokiniana* and *Chlorella variabilis* developed on tannery wastewater (TWW) were investigated as potential biological agents to decrease the defilement burden and safely release the effluents (Nagi et al., 2020). The microalgae *Chlorella sorokiniana* may have certain attributes permitting it to be a better organism for treating wastewater in mass cultivation (Bohutskyi et al., 2015). The green microalgae, *Scenedesmus* sp. and *Chlorella* sp. were isolated and developed on crude industrial wastewater for synchronous supplement expulsion and biomass production with lipid creation (Ansari et al., 2017). The bioremediation of wastewater containing chlorides was performed with local microalgae (Ramírez et al., 2018).

The microalgal species, such as *Scenedesmus obliquus* and *Chlorella vulgaris*, at different strengths adequately reduce the COD, BOD and TDS levels in leather-processing industrial effluents (Elumalai et al., 2014). The absorption of Zn by *Spirulina platensis* in the biomass has been used to eliminate numerous contaminants from effluents by retention (Palaniswamy and Veluchamy, 2017). The adaptability of microalgae to the wastewaters/effluents was deliberate under laboratory conditions by choosing the microalgae, such as green algae (Chlorophyceae) *Chlamydomonas pertusa, Chlorococcum humicola, Chlorococcum vitiosum, Chlorella vulgaris, Dactylococcopsis raphioides, Scenedesmus dimorphus, Desmococcus olivaceus, Scenedesmus incrassatulus* and *Scenedesmus accuminatus* and the diatom (Bacillariophyceae) *Amphiprora paludosa, Amphora leaives, Navicula pennata, Amphora turgida, Thallasosira weissflogii* and *Synedra tabulate* and the blue-green algae (Cyanophyceae), *Chroococcus turgidus*, etc. The correct choice of microalgae could treat effluents from different industries, like textile dyeing, soft drinks production, chemical production, oil drilling, detergents, alginate and petrochemical and leather processing industrial effluents. After choosing the microalgal strain, it was taken into the field together with a pilot test and scaled up to treat a particular industrial effluent. The microalgae, *Desmococcus olivaceus, Chlorococcum humicola, Chroococcus turgidus, Amphora turgida* and *Chlorella vulgaris*, were included in a huge revision of pH, COD, BOD, TDS, in the removal of colour, phosphate, and nitrate (about 60–80%). The immobilized cells of *Scenedesmus accuminatus* could eliminate TDS (28–54%) and phosphate, nitrate (44–76%) which are important substances of industrial effluents. The use of microalgal biomass reaped from the treated effluents for business applications is described in (Muthukumaran, 2005, 2009; Muthukumaran and Sivasubramanian, 2017).

10.4.3 Phytoremediation for Heavy Metals Using Microalgae

Heavy metals can be remediated from a microalgal point of view and this chapter now gives an outline of the different basic ways of using this innovative technology (Suresh Kumar et al., 2015). To decide the capacity of *Chlorella vulgaris* in retaining lead (Pb) and the impact of the variety of Pb heavy metal, focus on the development of *Chlorella vulgaris* (Dewi and Nuravivah, 2018). Phycoremediation potentially can work to expel heavy metals from sewage wastewater (Kaur et al., 2018). *Chlorella* sp. was the ruling algal species in the adjustment pond water during the whole period and was read for its Pb^{2+} and Zn^{2+} metals expulsion productivity (Kumar and Goyal, 2010). *Scenedesmus* sp. and *Chlorella* sp. radically reduce heavy metals like lead, sulfate, nickel, zinc and copper in the effluent of the pharmaceutical industry (Prathiba et al., 2020). A similar take-up investigation of the poisonous components boron, arsenic, manganese, zinc and copper in monometallic and multimetallic arrangements by four green microalgae, *Chlorella vulgaris*, *Scenedesmus almeriensis* and *Chlamydomonas reinhardtii*, and a native Chlorophyceae, assessed the impact of pH and conductivity (Saavedra et al., 2018). Three microalgal isolates *Scenedesmus quadricauda* and *Pseudochlorococcum typicum* (Chlorophyta) and *Phormidium ambiguum* (Cyanophyta), were tested for resistance and evacuation of cadmium (Cd^{2+}), lead (Pb^{2+}) and mercury (Hg^{2+}) and in aqueous solutions as a solitary metal species group at a concentration of 5–100mg/L under controlled laboratory research facility conditions (Shanab et al., 2012). Bioremediation of Cu by xeric green

TABLE 10.1
Phycoremediation of Heavy Metals by Microalgae in Wastewaters/Effluents Treatment

Microalgae	Sources of heavy metals (HMs)	Heavy metals	Reduction of HMs (%)	References
Desmococcus olivaceus	Electroplating industry chrome sludge	Cr Zn Cu	33.33 16.49 08.38	Muthukumaran, 2009; Muthukumaran et al., 2012; Muthukumaran and Sivasubramanian, 2017
Chlorella vulgaris	Electroplating and galvanizing industry effluent Monometallic solution (1000mg/L) Wastewater Domestic wastewater	Cr[IV] Mn Zn Cu Pb Ni Pb Zn	81.3 mg/L 99.40 91.90 88.00 99.40 99.00 72.00 73.00	Sibi, 2016; Saavedra et al., 2018; Manzoor et al., 2019; Kumar and Goyal, 2010
Chlorella miniate	Aqueous solution (30mg/L)	Ni	60.00–73.00	Chong et al., 2000
Spirulina platensis	Aqueous solution (06.06% $HgCl_2$) Electroplating Industrial Effluent	Hg Zn	03.41 100mg/L	Sharma et al., 2019; Palaniswamy R. and Veluchamy C., 2017
Scenedesmus almeriensis	Monometallic solution (1000mg/L) Wastewater	As B Cd	40.70 38.60 73.00	Saavedra et al., 2018; Travieso et al., 1999
Pseudochlorococcum typicum	Aqueous solution (5–100mg/L)	Hg Pb Cd	97.00 86.00 70.00	Shanab et al., 2012
Synechocystis sp.	Aqueous solution (30mg/L	Zn	40.00	Chong et al., 2000
Synechocystis salina	Wastewater	Fe Cr Ni Hg	66.00 60.00 70.00 77.00	Worku and Sahu, 2014

algae bioremediator, *Chlorococcum hemicolum* was explored (Harish et al., 2011). The *Chlorella vulgaris* was extremely successful at removing different heavy metals and is suitable to eliminate over 90% of heavy metals from wastewater, by and large (Manzoor et al., 2019). The results are shown in Table 10.1.

10.5 PHYCOREMEDIATION OF CHROME SLUDGE FROM THE ELECTROPLATING INDUSTRY

10.5.1 The Experiment Raceway Pond Study

A field trial experiment was directed in an open raceway cultivation pond built on the premises of the Ramakrishna Mission Vivekananda College, Chennai. The pond (a reinforced cement tank) was built with the tank wall thickness of 25cm. The dimensions of the inner pond were built to such an extent so that the length was 365cm and the width was 180cm. The pond depth was kept at 44cm, remembering the adequate light needed for the development of the growth of the microalgae.

A segment divider was built in the pond with a length of 246cm and width of 12cm. The floor was built with a slight slant on one or the other side of the partition in the other direction to empower the appropriate microalgal culture mixing. On the wall of the pond, white tiles were fixed to prevent leakage of the medium/effluent into the divider and for better visual perception of algal development. It additionally helps in reflecting the daylight, in this way enabling more light to reach the microalgal development. The tank was furnished with a tap water association for arrangement of the medium and two outlets to enable the tank to be cleaned. The tank substance was manually mixed thrice per day to guarantee the appropriate scattering of supplements and better air circulation (Muthukumaran et al., 2005; Muthukumaran, 2009; Muthukumaran et al., 2012; Muthukumaran and Sivasubramanian, 2017).

10.5.2 Potential of Microalga *Desmococcus Olivaceus* Grown in Chrome Sludge of Electroplating Industry in the Open Raceway Pond Study

In the starter lab preliminaries, *Desmococcus olivaceus* showed better endurance in 0.2% chrome sludge concentration than other microalgae. So it was chosen for additional field tests in the open raceway pond study and the phycoremediation of sludge was performed. At first, 1 KL of altered CFTRI medium was poured into the tank (alongside EDTA to chelate the metal particles according to Monika et al. (2000), in the proportions of 3mg/g of sludge) into which *Desmococcus olivaceus* was introduced to the pond. The sludge and total dissolved solid (TDS) showed a slow decrease and the algal cell check expanded. When both the raw or untreated and the algae-treated chrome sludge of the electroplating industry were investigated, the potential phytoremediator of the sludge effluent showed a significant decrease in all the parameters examined. A surprising decrease was found in the total dissolved solid (TDS) about (74.24%), potassium (60%), sodium (50%), phosphate (86.16%). COD and BOD showed decrease levels of 23.80% and 22.70% respectively. The heavy metals like zinc, copper and chromium showed decrease levels of 16.49%, 8.38% and 33.33% individually (Muthukumaran, 2009; Muthukumaran et al., 2012; Muthukumaran and Sivasubramanian, 2017). The results are shown in Figure 10.2 and Table 10.2.

10.5.3 Growth and Chrome-Sludge Reduction by *Desmococcus Olivaceus*

This experimental trial was performed essentially to break down the development of the microalga, *Desmococcus olivaceus* in the chrome sludge and measure the synchronous decrease of sludge in the open pilot tank. In the examination carried out over 12 days, 0.2% chrome sludge was blended in with an altered CFTRI medium alongside the metal chelating specialist EDTA (60g/KL) in an open raceway pond. On the first day of the examination, 2 litres of *Desmococcus olivaceus* culture were added. The parameters, such as TDS, pH, growth rate and chrome sludge decrease, were observed. There was an impressive sludge decrease (20.95%), when the cell count of *Desmococcus olivaceus* was additionally discovered to be greatest (308×10^4 cells/ml) on the 10th day (Muthukumaran, 2009; Muthukumaran et al., 2012; Muthukumaran and Sivasubramanian, 2017). The results are shown in Table 10.3.

10.5.4 Phycoremediation of Chromium Heavy Metal from the Electroplating Industry

The tanning process produces high strength wastewater comprising heavy metals, and inorganic and organic pollutants, which may negatively influence general well-being and the environment. The wastewater contains significant amounts of heavy metals including Cr, which is cancer-causing, teratogenic and mutagenic and determined in the environment. A few physico-synthetic treatment methods have been used to treat the wastewater of the tannery industry and were demonstrated to significantly reduce the degree of harmful Cr and reduce different poisons to low levels. Treatment

FIGURE 10.2 Heavy metals reduction and algal biomass production by *Desmococcus Olivaceus* grown in chrome sludge of electroplating industry in an open raceway pond.

Source: Muthukumaran (2009), Muthukumaran et al. (2012), Muthukumaran and Sivasubramanian (2017).

of wastewater by microalgal frameworks can play a significant part in the phycoremediation of tannery wastewater and is considered a minimal expense, and an effective treatment choice that can conceivably eliminate inorganic and organic pollutants and heavy metals. Microalgae have been able to reduce the harmful Cr (VI) to a considerably less poisonous Cr (III), particularly when combined with initiated carbon (Mohammed and Dabai, 2020). A huge number of various types of algae can expel chromium from wastewater (Jyoti and Awasthi, 2014). Biosorption of chromium (VI) offers a possible option in contrast to ordinary metal expulsion techniques. The microalga *Chlorella vulgaris* biomass could be viewed as a promising minimal expense biosorbent for the evacuation of Cr (VI) from electroplating and galvanizing industrial effluents (Sibi, 2016). The microalga *Desmococcus olivaceus* achieved a considerable reduction of chromium (33.33%) in the treatment of chrome sludge from the electroplating industry (Muthukumaran, 2009; Muthukumaran et al., 2012; Muthukumaran and Sivasubramanian, 2017).

10.5.5 Phycoremediation: Algal Biomass Production and Its Applications

Microalgae can assume a significant role in a circular bio-economy by producing excellent items, like lipids, proteins, and colorants, in the biomass delivered by the wastewater treatment cleaning

TABLE 10.2
Comparison of Physico-Chemical Parameters of Raw and Algae-Treated Electroplating Industrial Chrome Sludge in an Open Raceway Pond

Parameters	Raw sludge effluent	Algae-treated sludge effluent	% of reduction
Physical examination			
Turbidity NTU	69.0	18.9	41.6
Total solids mg/L	3022	1955	35.30
Total dissolved solids (TDS) mg/L	66	17	**74.24**
Total suspended solids (TSS) mg/L	2956	1938	34.43
Electrical conductivity (micro mho/cm)	4702	2745	41.42
Chemical examination			
pH	7.95	8.91	
Alkalinity pH (as Ca CO_3) mg/L	8	16	
Alkalinity total (as Ca CO_3) mg/L	1393	1027	26.27
Total hardness (as Ca CO_3) mg/L	680	520	23.52
Calcium (as Ca) mg/L	176	198	
Magnesium (as Mg) mg/L	58	48	17.24
Sodium (as Na) mg/L	680	340	**50**
Potassium (as K) mg/L	50	20	**60**
Iron (as Fe) mg/L	6.48	4.70	27.46
Manganese (as Mn) mg/L	Nil	Nil	Nil
Free ammonia (as NH_3) mg/L	38.08	23.52	**38.23**
Nitrite (as NO_2) mg/L	0.84	0.36	**41.14**
Nitrate (as NO_3) mg/L	45	23	**48.88**
Chloride (as Cl) mg/L	351	176	**49.85**
Fluoride (as F) mg/L	1.12	0.99	**11.60**
Sulphate (as SO_4) mg/L	186	129	**30.64**
Phosphate (as PO_4) mg/L	43.51	6.02	**86.16**
Tidy's test (as O) mg/L	67.2	58.4	13.09
Silica (as SiO_2) mg/L	40.12	47.68	15.85
BOD mg/L	643	497	**22.70**
COD mg/L	210	160	**23.80**
Total Kjeldhal Nitrogen mg/L	51.52	59.36	
Copper (as Cu) mg/L	0.01371	0.01256	08.38
Zinc (as Zn) mg/L	0.285	0.238	16.49
Chromium (as Cr) mg/L	0.024	0.016	**33.33**

Source: Muthukumaran (2009), Muthukumaran et al. (2012), Muthukumaran and Sivasubramanian (2017).

Note: Some of the chemical parameters were significantly reduced, shown by numbers highlighted in bold.

measure (Wollmann et al., 2019). Microalgae cultures offer an intriguing advance for the treatment of wastewater, since they provide a tertiary biotreatment combined with the creation of possibly significant biomass, which can be used for several purposes (Abdel-Raouf et al., 2012). The microalgae grow well in supplement-rich wastewater by engrossing the natural supplements and changing them into helpful biomass. The reaped biomass can be used as animal feed, biofertilizer, and an elective fuel source for the production of biodiesel (Amenorfenyo et al., 2019). Microalgae have emerged as a possible source of reasonable energy as of late in light of their higher biomass efficiency and capacity to destroy air and water contamination by means of bio-extraction (Hakeem et al., 2014). Electroplating industrial chrome sludge supported algal growth development when the open

TABLE 10.3
Growth and Sludge Reduction by *Desmococcus Olivaceus* Grown in a Pilot Tank Amended with Chrome Sludge from Electroplating Industry

Day	Cell count (X10^4 cells/ml)	% sludge reduction
1st	3	0
2nd	8	–
3rd	12	04.98
4th	40	–
5th	82	09.41
6th	128	–
7th	143	13.87
8th	161	–
9th	260	–
10th	308	20.95
11th	298	–
12th	280	11.21

Source: Muthukumaran (2009), Muthukumaran et al. (2012), Muthukumaran and Sivasubramanian (2017).

TABLE 10.4
Algal Biomass Production by *Desmococcus Olivaceus* Grown in Open Raceway Pond Amended with Chrome Sludge from Electroplating Industry

Day	Algal biomass in dry wgt (X10^4 gm/KL)
1st	001.0766
2nd	002.8712
3rd	004.3068
4th	014.3560
5th	029.4298
6th	045.9392
7th	051.3227
8th	057.7829
9th	093.3140
10th	110.5412
11th	106.9522
12th	100.4920

raceway pond chrome sludge was treated with the microalga, *Desmococcus olivaceus*. There was a great decrease in slop and the production of biomass in the open raceway pond changed with the chrome sludge (Muthukumaran et al., 2012). The results are shown in Table 10.4.

Microalgae can play a part in energy-efficient carbon-unbiased wastewater treatment. This is achievable by coupling carbon catch with wastewater treatment and the resulting biomass creation (Mohammed and Mota, 2019). To reduce costs and work on natural sustainability, an ideal microalgal source framework should be based on agricultural, domestic, municipal and industrial wastewaters

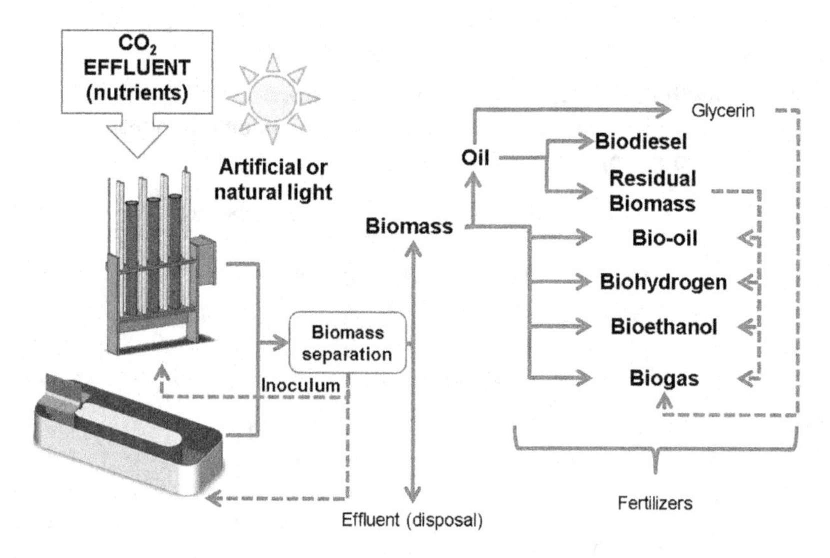

FIGURE 10.3 Flow chart showing effluent/wastewater treatment using microalgae and the biomass harvested and utilized in various applications of biofuel.

Source: Muthukumaran and Sivasubramanian (2017).

as a principal wellspring of water and supplements. However, the microalgae should likewise be kind to fluctuating wastewater quality, while yet creating sufficient biomass and lipids (Stemmler et al., 2016). The development of algae that create high biomass in wastewaters is an alluring interaction on account of the double use of microalgae for phycoremediation and the production of cheap biomass as a value-added by-product (Ummalyma and Anamika Singh, 2021). Creation of the microalgae source can be coordinated with wastewater and industrial wellsprings of carbon dioxide. It was shown that a blended culture of wild algal species could effectively be developed on wastewater supplements and conceivably scaled to be commercially useful (Dalrymple et al., 2013). The production and outflow of CO_2 from various sources have caused critical changes in the environment, which is of significant concern and identified as global warming. Compared to other CO_2 evacuation methods, microalgae can efficiently eliminate CO_2 through the fast creation of algal biomass (Molazadeh et al., 2019). The result is shown in Figure 10.3.

10.6 CONCLUSION

As the number of polluting industries is continually expanding, contamination is turning into a significant global worry. It is agreed that heavy metals contamination is one of the significant dangers to water bodies, with unique reference to chromium, which is released into the waterway systems by industries. Considering the limitations of conventional chemical treatment strategies, the utilization of organic entities straightforwardly or by implication to eliminate pollutants from aquatic or ground environments is a significant, inventive remediation innovation, conceivably relevant to treat an assortment of impurities. The global recognition of the necessity of biological equilibrium, reduction in contamination, and a safe environment requires an innovation. Water pollution is one of the biggest environmental concerns and clearly bioremediation can offer an unrivalled response to deal with this issue. Bioremediation to deal with heavy metals is viewed as a financially encouraging option in contrast to customary techniques for remediation. Phycoremediation is a practical green innovation with eco-friendly technology. This chapter discussed microalgae's role in the phycoremediation of chrome sludge from the electroplating industry. There is a huge number of various types of algae that can expel chromium from sludge and the effluents of various industries. The microalga, *Chlorella vulgaris* biomass could be viewed as a promising, very low-cost, biosorbent for the evacuation of Cr from electroplating industrial effluents. An open raceway pond of chrome sludge was treated with microalga, *Desmococcus olivaceu* and there was a considerable decrease in the sludge, and biomass was produced in the open raceway pond with amended chrome sludge. A wonderful decrease was found in TDS, phosphate, sodium and potassium. In doing this, the microalgae essentially reduced some heavy metals like copper, chromium and zinc. The microalgae cultivation for delivering high biomass in wastewater treatment is an appealing move due to the double use of algae for phycoremediation and to produce cheap biomass for value-added by-products.

ACKNOWLEDGEMENTS

The author is very grateful to the Secretary, Principal and HOD of Botany, Ramakrishna Mission Vivekananda College (Autonomous), Chennai, Tamil Nadu, India, for providing the facilities to finish this research work.

REFERENCES

Abdel-Raouf, N., Al-Homaidan, A. A., and Ibraheem, I. B. M. (2012). Microalgae and wastewater treatment. *Saudi Journal of Biological Sciences*, 19(3), 257–275. https://doi.org/10.1016/j.sjbs.2012.04.005

Agrawal, N., Kumar, V., and Shahi, S. K. (2021). Biodegradation and detoxification of phenanthrene in in-vitro and in-vivo conditions by a newly isolated ligninolytic fungus *Coriolopsis byrsina* strain APC5 and characterization of their metabolites for environmental safety. *Environmental Science and Pollution Research*. https://doi.org/10.1007/s11356-021-15271-w

Ahmad, I., Abdullah, N., Koji, I., Yuzir, A., and Mohamad, S. E. (2021). Potential of microalgae in bioremediation of wastewater. *Bulletin of Chemical Reaction Engineering and Catalysis*, 16(2), 413–429. https://doi.org/10.9767/bcrec.16.2.10616.413-429

Ali Redha, A. (2020). Removal of heavy metals from aqueous media by biosorption. *Arab Journal of Basic and Applied Sciences*, 27(1), 183–193.

Al-Jabri, H., Das, P., Khan, S., Thaher, M., and Abdulquadir, M. (2021). Treatment of wastewaters by microalgae and the potential applications of the produced biomass—a review. *Water (Switzerland)*, 13(27), 1–26. https://doi.org/10.3390/w13010027

Amenorfenyo, D. K., Huang, X., Zhang, Y., Zeng, Q., Zhang, N., Ren, J., and Huang, Q. (2019). Microalgae brewery wastewater treatment: Potentials, benefits and the challenges. *International Journal of Environmental Research and Public Health*, 16(11), 1910. https://doi.org/10.3390/ijerph16111910

Ansari, A. A., Khoja, A. H., Nawar, A., Qayyum, M., and Ali, E. (2017). Wastewater treatment by local microalgae strains for CO_2 sequestration and biofuel production. *Applied Water Science*, 7(7), 4151–4158. https://doi.org/10.1007/s13201-017-0574-9

Bansal, A., Shinde, O., and Sarkar, S. (2018). Industrial wastewater treatment using phycoremediation technologies and co-production of value-added products. *Journal of Bioremediation and Biodegradation*, 9(1), 1–10. https://doi.org/10.4172/2155-6199.1000428

Bhatt, N. C., Panwar, A., Bisht, T. S., and Tamta, S. (2014). Coupling of algal biofuel production with wastewater. *Scientific World Journal*, 2014, 210504. https://doi.org/10.1155/2014/210504

Bilal, M., Rasheed, T., Sosa-Hernández, J. E., Raza, A., Nabeel, F., and Iqbal, H. M. N. (2018). Biosorption: An interplay between marine algae and potentially toxic elements—A review. *Marine Drugs*, 16(2), 1–16. https://doi.org/10.3390/md16020065

Bohutskyi, P., Liu, K., Nasr, L. K., Byers, N., Rosenberg, J. N., Oyler, G. A., Betenbaugh, M. J., and Bouwer, E. J. (2015). Bioprospecting of microalgae for integrated biomass production and phytoremediation of unsterilized wastewater and anaerobic digestion centrate. *Applied Microbiology and Biotechnology*, 99(14), 6139–6154. https://doi.org/10.1007/s00253-015-6603-4

Chekroun, K. B., and Baghour, M. (2013). The role of algae in phytoremediation of heavy metals: A review. *Journal of Materials and Environmental Science*, 4(6), 873–880.

Chong, A. M. Y., Wong, Y. S., and Tam, N. F. Y. (2000). Performance of different microalgal species in removing nickel and zinc from industrial wastewater. *Chemosphere*, 41(1–2), 251–257. https://doi.org/10.1016/S0045-6535(99)00418-X

Chowdhary, P., Raj, A., and Bharagava, R. N. (2018) Environmental pollution and health hazards from distillery wastewater and treatment approaches to combat the environmental. *Chemosphere*, 194, 229–246.

Chowdhary, P., Raj, A., Verma, D., and Akhter, Y. (eds) (2020a). *Microorganisms for Sustainable Environment and Health*. Oxford: Elsevier.

Chowdhary, P., Raj, A., Verma, D., and Akhter, Y. (eds) (2020b). *Contaminants and Clean Technologies*, Boca Raton, FL: CRC Press.

Dalrymple, O. K., Halfhide, T., Udom, I., Gilles, B., Wolan, J., Zhang, Q., and Ergas, S. (2013). Wastewater use in algae production for generation of renewable resources: A review and preliminary results. *Aquatic Biosystems*, 9(1), 1–11. https://doi.org/10.1186/2046-9063-9-2

Dewi, E. R. S., and Nuravivah, R. (2018). Potential of microalgae chlorella vulgaris as bioremediation agents of heavy metal Pb (lead) on culture media. *E3S Web of Conferences*, 31, 3–6. https://doi.org/10.1051/e3sconf/20183105010

Dhote, L., Pandey, R. A., Middey, A., Mandal, N., and Kumar, S. (2021). Co-combustion of distillery sludge and coal for application in boiler and subsequent utilization of the generated bottom ash. *Environmental Science and Pollution Research*, 28(27): 36742–36752. https://doi.org/10.1007/s11356-021-13277-y

Dutta, D., Arya, S., and Kumar, S. (2021). Industrial wastewater treatment: Current trends, bottlenecks, and best practices. *Chemosphere*, 285, 131245. https://doi.org/10.1016/j.chemosphere.2021.131245

Elumalai, S., Ramganesh, S., and Sangeetha, T. (2014). Phycoremediation for leather industrial effluent: Treatment and recycling using green microalgae and its consortia. *International Journal of Current Biotechnology*, *2*(10), 1–9.

Hakeem, K. R., Jawaid, M., and Rashid, U. (2014). Biomass and bioenergy: Processing and properties. *Biomass and Bioenergy: Processing and Properties*, 9783319076, 1–367. https://doi.org/10.1007/978-3-319-07641-6

Harish, M., Sundaramoorthy, S., Kumar, D., and Vaijapurkar, S. G. (2011). Cu2+ uptake by Chlorococcum hemicolum: a xeric chlorophycean alga. *Nature Precedings*, 1–7. https://doi.org/10.1038/npre.2011.5657.1

Jyoti, J., and Awasthi, M. (2014). Bioremediation of wastewater chromium through microalgae: A review. *International Journal of Engineering Research and Technology*, 3(6), 1210–1215.

Kaplan, D. (2013). Absorption and adsorption of heavy metals by microalgae. In Richmond, A. (Ed.) *Handbook of Microalgal Culture: Applied Phycology and Biotechnology*. 2nd edn, Oxford: Blackwell Science, pp. 602–611. https://doi.org/10.1002/9781118567166.ch32

Kaur, H., Rajor, A., and Kaleka, A. S. (2018). Role of phycoremediation to remove heavy metals from sewage water: Review article. *Journal of Environmental Science and Technology*, 12(1), 1–9. https://doi.org/10.3923/jest.2019.1.9

Kumar, R., and Goyal, D. (2010). Waste water treatment and metal (Pb2+, Zn2+) removal by microalgal based stabilization pond system. *Indian Journal of Microbiology*, 50(S1), 34–40. https://doi.org/10.1007/s12088-010-0063-4

Kumar, V. (2018). Mechanism of microbial heavy metal accumulation from polluted environment and bioremediation. In Sharma, D., and Saharan, B. S. (Eds) *Microbial Fuel Factories*. Boca Raton, FL: CRC Press.

Kumar, V. (2021). Phytoremediation of distillery effluent: Current progress, challenges, and future opportunities. In Saxena, G., Kumar, V., and Shah, M. P. (Eds), *Bioremediation for Environmental Sustainability: Toxicity, Mechanisms of Contaminants Degradation, Detoxification and Challenges*. Oxford: Elsevier. https://doi.org/10.1016/B978-0-12-820524-2.00014-6

Kumar, V., Shahi, S. K., Ferreira, L. F. R., Bilal, M., Biswas, J. K., and Bulgariu, L. (2021). Detection and characterization of refractory organic and inorganic pollutants discharged in biomethanated distillery effluent and their phytotoxicity, cytotoxicity, and genotoxicity assessment using *Phaseolus aureus* L. and *Allium cepa* L. *Environmental Research*, 201, 111551. https://doi.org/10.1016/j.envres.2021.111551

Kumar, V., Shahi, S. K., and Singh, S. (2018). Bioremediation: An eco-sustainable approach for restoration of contaminated sites. In Singh, J., Sharma, D., Kumar, G., and Sharma, N. (Eds) *Microbial Bioprospecting for Sustainable Development*. Singapore: Springer. https://doi.org/10.1007/978-981-13-0053-0_6Li, G., Bai, X., Li, H., Lu, Z., Zhou, Y., Wang, Y., Cao, J., and Huang, Z. (2019). Nutrients removal and biomass production from anaerobic digested effluent by microalgae: A review. *International Journal of Agricultural and Biological Engineering*, 12(5), 8–13. https://doi.org/10.25165/j.ijabe.20191205.3630

Lim, S. L., Chu, W. L., and Phang, S. M. (2010). Use of *Chlorella vulgaris* for bioremediation of textile wastewater. *Bioresource Technology*, 101(19), 7314–7322. https://doi.org/10.1016/j.biortech.2010.04.092

Manzoor, F., Karbassi, A., and Golzary, A. (2019). Removal of heavy metal contaminants from wastewater by using *Chlorella vulgaris* Beijerinck: A review. *Current Environmental Management*, 6(3), 174–187. https://doi.org/10.2174/2212717806666190716160536

Maryjoseph, S., and Ketheesan, B. (2020). Microalgae based wastewater treatment for the removal of emerging contaminants: A review of challenges and opportunities. *Case Studies in Chemical and Environmental Engineering*, 2(August), 100046. https://doi.org/10.1016/j.cscee.2020.100046

Mohammed, K., and Dabai, A. I. (2020). Chromium removal from tannery wastewater: A review. *PLATFORM: A Journal of Science and Technology*, 3(1), 63–71. https://researchgate.net/publication/343609834

Mohammed, K., and Mota, C. (2019). Microalgae and sustainable wastewater treatment: A review. *Bayero Journal of Pure and Applied Sciences*, 11(1), 408. https://doi.org/10.4314/bajopas.v11i1.65s

Moharir, R. V., and Kumar, S. (2020). Structural characterization of LDPE films to analyse the impact of heavy metals and effect of UV pre-treatment on polymer degradation. *Journal of Cleaner Production*. 298. https://doi.org/10.1016/j.jclepro.2021.126670

Mohsenpour, S. F., Hennige, S., Willoughby, N., Adeloye, A., and Gutierrez, T. (2021). Integrating micro-algae into wastewater treatment: A review. *Science of the Total Environment*, 752(September), 142168. https://doi.org/10.1016/j.scitotenv.2020.142168

Molazadeh, M., Ahmadzadeh, H., Pourianfar, H. R., Lyon, S., and Rampelotto, P. H. (2019). The use of microalgae for coupling wastewater treatment with CO_2 biofixation. *Frontiers in Bioengineering and Biotechnology*, 7(March). https://doi.org/10.3389/fbioe.2019.00042

Monika, A. M., Kedziorek, A. and Bourg, C. M. (2000). Solubilization of lead and chromium during the percolation of EDTA through a soil polluted by smelting activities. *Journal of Contaminant Hydrology*, 40(4), 381–392.

Muthukumaran, M. (2009). Studies on the phycoremediation of industrial effluents and utilization of algal biomass. PhD thesis. University of Madras.

Muthukumaran, M., Raghavan, B. G., Subramanian, V. V., and Sivasubramanian, V. (2005). Bioremediation of industrial effluent using microalgae. *Indian Hydrobiology*, 7(suppl.), 105–122.

Muthukumaran. M., and Sivasubramanian, V. (2017). Microalgae cultivation for biofuels: Cost, energy balance, environmental impacts and future perspectives. In Singh, R. S., Pandey, A., and Gnansounou, E. (Eds) *Biofuels: Production and Future Perspectives*. Boca Raton, FL:CRC Press, pp. 363–411.

Muthukumaran, M., Thirupathi, P., Chinnu, K., and Sivasubramanian, V. (2012). Phycoremediation efficiency and biomass production by micro alga *Desmococcusolivaceus* (Persoon et Acharius) J. R. Laundon treated on chrome-sludge from an electroplating industry—An open raceway pond study. *International Journal of Current Science* (Special Issue), 52–62.

Nagi, M., He, M., Li, D., Gebreluel, T., Cheng, B., and Wang, C. (2020). Utilization of tannery wastewater for biofuel production: New insights on microalgae growth and biomass production. *Scientific Reports*, 10(1), 1–14. https://doi.org/10.1038/s41598-019-57120-4

Palaniswamy, R., and Veluchamy, C. (2017). Biosorption of heavy metals by *Spirulina platensis* from electroplating industrial effluent. *Environmental Science: An Indian Journal*, 13(4), 139.

Prathiba, S., Muninathan, N., Ponnulakshmi, R., and Nalini, D. (2020). Phycoremediation of heavy metal removal from pharmaceutical industrial effluents, Kandigai, Kanchipuram District, Tamil Nadu. *Medico-Legal Update*, 20(4), 2328–2331. https://doi.org/10.37506/mlu.v20i4.2193

Ramírez, M. E., Vélez, Y. H., Rendón, L., and Alzate, E. (2018). Potential of microalgae in the bioremediation of water with chloride content. *Brazilian Journal of Biology*, 78(3), 472–476. https://doi.org/10.1590/1519-6984.169372

Randrianarison, G., and Ashraf, M. A. (2017). Microalgae: A potential plant for energy production. *Geology, Ecology, and Landscapes*, 1(2), 104–120.

Razak, S. B. A., and Sharip, Z. (2020). The potential of phycoremediation in controlling eutrophication in tropical lakes and reservoirs: A review. *Desalination and Water Treatment*, 180, 164–173. https://doi.org/10.5004/dwt.2020.25078

Renuka, N., Sood, A., Prasanna, R., and Ahluwalia, A. S. (2015). Phycoremediation of wastewaters: A synergistic approach using microalgae for bioremediation and biomass generation. *International Journal of Environmental Science and Technology*, 12(4), 1443–1460. https://doi.org/10.1007/s13762-014-0700-2

Saavedra, R., Muñoz, R., Taboada, M. E., Vega, M., and Bolado, S. (2018). Comparative uptake study of arsenic, boron, copper, manganese and zinc from water by different green microalgae. *Bioresource Technology*, 263(July), 49–57. https://doi.org/10.1016/j.biortech.2018.04.101

Salama, E. S., Kurade, M. B., Abou-Shanab, R. A. I., El-Dalatony, M. M., Yang, I. S., Min, B., and Jeon, B. H. (2017). Recent progress in microalgal biomass production coupled with wastewater treatment for biofuel generation. *Renewable and Sustainable Energy Reviews*, 79(November), 1189–1211. https://doi.org/10.1016/j.rser.2017.05.091

Shanab, S., Essa, A., and Shalaby, E. (2012). Bioremoval capacity of three heavy metals by some microalgae species (Egyptian isolates). *Plant Signaling and Behavior*, 7(3), 392–399.

Sharma, A., Kaur, K., Chatli, A. S., and Town, M. (2019). Bioremediation of heavy metal ($HgCl_2$) from water by *Spirulina platensis*. *International Journal of Scientific Development and Research*, 4(3), 461–466.

Sibi, G. (2016). Biosorption of chromium from electroplating and galvanizing industrial effluents under extreme conditions using *Chlorella vulgaris*. *Green Energy and Environment*, 1(2), 172–177. https://doi.org/10.1016/j.gee.2016.08.002

Stemmler, K., Massimi, R., and Kirkwood, A. E. (2016). Growth and fatty acid characterization of microalgae isolated from municipal waste-treatment systems and the potential role of algal-associated bacteria in feedstock production. *PeerJ*, 2016(3), e1780. https://doi.org/10.7717/peerj.1780

Suresh Kumar, K., Dahms, H. U., Won, E. J., Lee, J. S., and Shin, K. H. (2015). Microalgae: A promising tool for heavy metal remediation. *Ecotoxicology and Environmental Safety*, 113, 329–352. https://doi.org/10.1016/j.ecoenv.2014.12.019

Travieso, L., Cañizares, R. O., Borja, R., Benítez, F., Domínguez, A. R., Dupeyrón, R., and Valiente, Y. V. (1999). Heavy metal removal by microalgae. *Bulletin of Environmental Contamination and Toxicology*, 62(2), 144–151. https://doi.org/10.1007/s001289900853

Ummalyma, S. B., and Singh, A. (2021). Importance of algae and bacteria in the bioremediation of heavy metals from wastewater treatment plants. In Shah, M., Rodriguez-Couto, S., and Kumar, V. (Eds) *New Trends in Removal of Heavy Metals from Industrial Wastewater*, Oxford: Elsevier, pp. 343–357. https://doi.org/10.1016/b978-0-12-822965-1.00014-3

Wang, Y., Ho, S. H., Cheng, C. L., Guo, W. Q., Nagarajan, D., Ren, N. Q., Lee, D. J., and Chang, J. S. (2016). Perspectives on the feasibility of using microalgae for industrial wastewater treatment. *Bioresource Technology*, 222, 485–497. https://doi.org/10.1016/j.biortech.2016.09.106

Wollmann, F., Dietze, S., Ackermann, J. U., Bley, T., Walther, T., Steingroewer, J., and Krujatz, F. (2019). Microalgae wastewater treatment: Biological and technological approaches. *Engineering in Life Sciences*, 19(12), 860–871. https://doi.org/10.1002/elsc.201900071

Worku, A., and Sahu, O. (2014). Reduction of heavy metal and hardness from ground water by algae. *Journal of Applied and Environmental Microbiology*, 2(3), 86–89.

11 Environmental Sensing and Detection of Toxic Heavy Metals by Metal Organic Frameworks-Based Electrochemical Sensors

Komal Rizwan[1,*] *and Muhammad Bilal*[2*]
[1]Department of Chemistry, University of Sahiwal, Sahiwal, Pakistan
[2]Institute of Chemical Technology and Engineering, Faculty of Chemical Technology, Poznan University of Technology, Berdychowo, Poznan, Poland.
[*]Corresponding authors Email: komal.rizwan45@yahoo.com; muhammad.bilal@put.poznan.pl

CONTENTS

11.1 Introduction 215
11.2 Types of MOF Composites 216
    11.2.1 Structural Design of MOF Materials for Electrochemical Sensing 217
    11.2.2 MOF-Based Sensors for Toxic Heavy Metal Ions Sensing 217
11.3 Conclusion 220
References 223

11.1 INTRODUCTION

Heavy metals are naturally occurring elements, but they are responsible for environmental pollution globally. Industries, agriculture, the military, medical, and the domestic sectors are mainly responsible for the increase in heavy metal pollution in the ecosystem due to the improper release of heavy metals in water. Some heavy metals are important for life, but excessive consumption of these can cause various chronic and acute diseases, including infertility, cardiovascular issues, hypertension, hyperglycemia, and various neurological disorders. These toxic heavy metal ions include Hg (II), As (III and V), Pb (II), Cr (III and VI), Cd (II) and Cu (II). Materials for sensing heavy metals need to be developed, even if the heavy metals are at low concentrations, so that after sensing the heavy metals, proper efforts may be made to remove them from the environment (Boix et al., 2020; Ru, Wang, Wang et al., 2021).

Metal organic frameworks (MOFs) are porous hybrid nanomaterials generated by coordinating a connection among organic ligands, metal ions, and metal ion clusters (Rasheed, Hassan, Bilal, Hussain, and Rizwan, 2020; Rasheed, Rizwan, Bilal, and Iqbal, 2020). MOF-based derivatives have an extensive range of applications for storing energy, medicines, gas, and catalysis, etc. (Liu, Li, and Pang, 2020; Rasheed, Hassan, et al., 2020). Recently, sensing applications of MOF materials have been explored in various fields. MOFs possess a potential sensing application because of their huge surface area, periodic structure, active coordination sites, high porosity, and tunable structures,

DOI: 10.1201/9781003239956-13

and these characteristics bestow the MOFs with great catalytic potential and make them promising materials for electrode coatings which are used in sensing applications (Xu, Li, Xue, and Pang, 2018). Their huge surface area and porosity are advantageous in efficiently concentrating and in the mass transfer of analytes, which may amplify the response of the signal effectively and increase the sensitivity of sensing. The specific size and shape of MOF cavities enable them to contain great selectivity regarding particular analytes. Furthermore, many MOFs have weak electrical conductivity by nature, and the small pore size of MOFs is responsible for limiting their full potential.

Incorporation of other materials such as biomolecules, polymers, aptamers, etc. in MOFs has generated hybrid MOFs (Rasheed and Rizwan, 2022). These materials enhance the sensing potential of pristine MOFs (Liu et al., 2020). Incorporation of carbon-based materials in MOFs results in augmented conductivity of the pristine MOFs, which leads to an enhancement in electro-activity of the developed sensing devices (Xue, Zheng, Xue, and Pang, 2019). These composites can integrate the benefits and overcome the disadvantages of a single component. Thus, due to the synergistic effect, novel attributes are obtained which a single component can never achieve. These MOF-derived nanostructures have been determined widely for their sensing potential applications and attempts have been made by researchers to regulate the key features of sensors, such as accuracy, stability, sensitivity, selectivity, etc. (J. Zhang, Wang, and Li, 2019). This chapter focuses on applications of MOF-based sensors in heavy metal sensing.

11.2 TYPES OF MOF COMPOSITES

Large numbers of guest materials are hosted by MOF structures due to their large surface area (Railey, Song, Liu, and Li, 2017). MOF composites play a vital role in overcoming the deficiencies of pristine MOFs. MOF composites are widely used in electrochemical applications. MOF composites are categorized into four types, based on the composite material dimension as: (1) MOF/0-D; (2) MOF/1D; (3) MOF/2D; and (4) MOF/3D. MOFs' intrinsic functions and properties can be enhanced by 0D materials, which are tiny and a have large surface energy. 0D MOF composites can be made in a variety of ways, the most common of which involve encapsulating prefabricated nanoparticles in MOFs or using MOFs as templates to fabricate nanoparticles on the outer edges (Deng, Bo, and Guo, 2018). The one-step solvothermal reduction method improves the synthetic efficacy and makes fabrication easy (Deng et al., 2018). A double solvent method (Pang, Zhang, Cheng, Lai, and Huang, 2015) can prevent MNP accumulation on the outermost part of MOFs, and the colloidal deposition strategy avoids the use of more reactive, expensive organic solvents and precursors (Zheng et al., 2018).

1D/MOF includes MOFs enveloped around the 1D materials, 1D material on the surface of MOFs or MOFs with 1D nano-belts generated on the carbon cloth. Through *in-situ* polymerization, the MOFs can be combined with a 1D nano-belt by chemical oxidation. To fabricate 1D composites, the self-template approach is used, in which 1D materials are the template to absorb metal ions through solvents, thereby commencing the generation of MOFs without any kind of surface modification. In addition, the spontaneous phase separation approach is suitable to produce MOFs like ZIF-8 powder and it is also suitable for a wide range of supported semiconductor 1D materials application (Wee et al., 2014). An easy hydrothermal strategy has been embraced to synthesize CNTs/Mn-MOF, which overcomes the restrictions of previously studied Mn-based materials (Y. Zhang et al., 2015).

MOF/2D materials can also be produced by growing MOFs on 2D nanosheets and encapsulating the 2D materials in the MOFs. The solvothermal approach is commonly used to combine the 2D nanosheet materials, such as graphene oxide with MOFs. An ultrasound approach can also be used to fabricate 2D composites which provides uniform products and has an easy reaction in comparison to the solvothermal strategy (W. Zhang et al., 2016). MOF/3D composites, which include core-shell-like and cubic kinds of frameworks, can be fabricated via the melt-diffusion approach

(Z. Wang et al., 2015) and the stepwise method. To construct 3D composites, the layer-by-layer (LBL) assembly approach may be used, which can carefully manage the shell thickness of MOFs such as 3D core shell composites, in which the core provides a reaction surface and the porous shell acts as a boundary layer to inhibit the surface from activating and aggregating with the neighboring components during the reaction (Ke, Qiu, Yuan, Jiang, and Zhu, 2012).

11.2.1 Structural Design of MOF Materials for Electrochemical Sensing

Different metal-based MOFs, including copper, nickel, chromium, cobalt, etc. have been synthesized and employed as modifiers for electrodes used for the sensing of hydrogen peroxide, glucose, and various organic molecules (Lopa et al., 2018). These electroactive MOFs can effectively enhance the response signals because of their advantageous electrocatalytic activities, wide surface areas, and ultra-high pore size. Huang reported a ground-breaking study by merging Cu-MOF and Co-MOF into flexible sensors for nutrient detection applications (Ling et al., 2018). The massive MOF particles were crushed into nanoparticles and uniformly diffused in a Nafion solution, which allowed the MOFs to be fixed on the flexible PET substrate using a drop casting approach.

A single or multichannel adjustable sensor was fabricated by integrating the MOF-coated flexible PET substrate with a microfluidic device to evaluate essential nutrients like ascorbic acid, glycine, and L-tryptophan with detection resolutions of 14.97, 0.71, and 4.14μM, respectively. This study highlighted MOFs' huge potential for electrical sensing applications. Conventional MOFs, on the other hand, are generally micrometer-sized crystals or agglomerated powders with three-dimensional structures (Jiao, Seow, Skinner, Wang, and Jiang, 2019), which is inefficient in producing effective mass/electron transport and efficiently immobilizing MOFs materials on the electrode surfaces. As a result, a weak response was noted: the typical MOF-based electrochemical sensors showed a sensitivity of 100 $\mu A\ mM^{-1}cm^{-2}$, the minimum limit of detection (LOD) was at a micromolar concentration, and the linear detection range was in the milli-molar range (Liu et al., 2020). To overcome this constraint, the rational design and controllable synthesis of MOFs with small sizes, as well as other novel techniques, have been developed to improve their sensing performances (Khan, Hasan, and Jhung, 2018).

11.2.2 MOF-Based Sensors for Toxic Heavy Metal Ions Sensing

MOF-based sensors are extensively employed in different fields for sensing environmental contaminants. The discovery of electrochemical sensors is now the main method for sensing pollutants. The release of toxic heavy metals directly into the environment is extremely dangerous to human health. The heavy metal ions are non-degradable and possess a long half-life. They accumulate in the human body and long-term exposure to them can cause cancer. Therefore, it is necessary to develop highly sensitive, cost-effective, fast, and miniaturized electrochemical sensing devices to locate heavy metals.

A unique MOF structure, NH_2-MIL-53(Cr), was fabricated via a facile reflux strategy by reacting $Cr(NO_3)_3.9H_2O$ with 2-aminoterephthalic-acid in solvent DMF/H_2O. The fabricated MOF electrode was used for lead (Pb^{+2}) ions sensing in an aqueous solution. An NH_2-MIL-53(Cr) modified electrode exhibited great stability for sensing lead ions and this may be attributed to the interaction of the amino moiety of the MOF with the lead ions. An interference study was carried out on the existence of other co-existing ions in real water samples. The as-synthesized MOF electrode exhibited high selectivity toward Pb^{+2} ions found in cadmium, copper, and zinc ions. The detection linear range was found to be $4x10^{-7}$ to $8x10^{-5}$ M, while the LOD was $3.05x10^{-8}$ M (Guo et al., 2016).

In the latest study, a nanocomposite, comprised of Zn-based MOF and graphene nanosheets (Gr/MOF), was fabricated through the solvo-thermal strategy. The synthesized GO/MOF was cast onto GCE to sense As (III) ions, and displayed prodigious efficiency with a low LOD value of 0.06 μg/

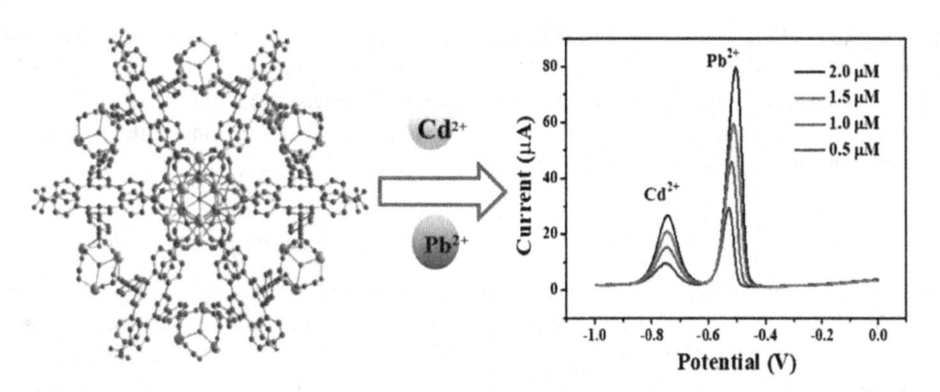

FIGURE 11.1 Ln-MOF-based sensor for recognition of Pb^{+2} and Cd^{+2} ions.

Source: Ye et al. (2020).

L. This fabricated MOF system was applied to determine the As(III) ions in the aqueous samples (Baghayeri, Ghanei-Motlagh, Tayebee, Fayazi, and Narenji, 2020). Ye and his coworkers designed an MOF-coated GCE to detect cadmium (Cd^{+2}) and lead (Pb^{+2}) ions in trace amounts. Lanthanide MOF (ZJU-27) was fabricated through the solvo-thermal approach by using $Yb(NO_3)_3.5H_2O$ with 1,3,5-benzene-trisbenzoic acid found in 2-FBA and was further applied to manipulate GCE. This synthesized sensor showed great stability and sensitivity in real water samples to sense $Pb^{+2}$and Cd^{+2}ions. The LOD for Pb^{+2} was found to be 0.228 ppb (Figure 11.1) (Ye et al., 2020).

Mercury is a toxic metal and exists in different forms as inorganic salts, elemental and organic molecules. The harmful effects caused by mercury metal are dangerous to living beings, and the main concern of scientists is to sense the trace level of mercury through electrochemical approaches (Devaraj, Yesudass, Rajendran, and Ponce, 2021). Recently, another sensor was constructed for Hg^{+2} ions sensing in tuna fish and tap water samples. The sensor was synthesized by using the Cu-MOF nanoparticles, which were synthesized via an easy single-step hydrothermal reaction of $CuNO_3.3H_2O$ with NH_2-BDC in a mixture of dimethylformamide/ethanol and PVP (polyvinylpyrrolidone). The limit of detection of $Hg^{+2}$was found to be 0.0633 nM. The linear detection range was found to be 0.1–50 nM. The fabricated sensor efficiently analyzed the mercury ions (Hg^{+2}) in the tuna fish samples and tap water. The fabricated sensor showed great sensitivity, reproducibility, repeatability and great applicability for Hg^{+2} ions detection (Figure 11.2) (Singh et al., 2020).

It is important to simultaneously detect the various metals with great ease, sensitivity, and reliability. Wang and colleagues developed a new sensing modality for concurrent detection of different metal ions (Cd^{+2}, Cu^{+2} and Pb^{+2}). The sensor has been synthesized by compositing ferrocene-carboxylic acid-based MOF (Fc-NH_2-UiO-66) with trGNO. The porous framework and huge surface area made it promising for adsorption and pre-concentration of heavy metals. Incorporation of Fc and reduced graphene enhanced the electrochemical potential and conductivity as well. The sensor trGNO/Fc-NH_2-UiO-66 efficiently detected the metals. The linear range of detection was found to be 0.8 nM, 8.5 nM and 0.6 nM for Cu^{2+}, Cd^{2+} and Pb^{2+} respectively. The sensor was applied to find the metal ions in tap water (X. Wang et al., 2020).

Lu and coworkers developed multi-layer modified electrodes for sensing applications. The surface of GCE was modified with many multi-layer films which were synthesized through repeated alternating deposition of MWCNTs-COOH and UiO-66-NH_2 (Zr-based metal organic frameworks). A novel fabricated three-layered MWCNTs–COOH/UiO-66-NH_2/MWCNTs-COOH/GCE was used for simultaneous sensing of Pb^{2+} and Cd^{2+} through square-wave anodic stripping voltammetry. The high electro-catalytic potential of the fabricated sensor was due to a combination of various adsorption sites at Zr-MOF with MWCNTs-COOH and chelation mechanisms between NH_2 moiety and

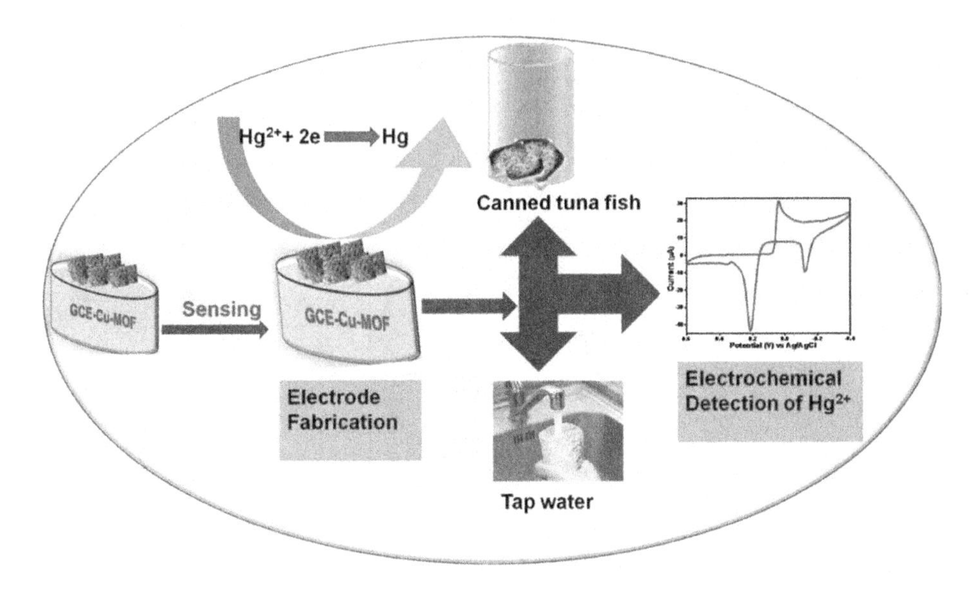

FIGURE 11.2 Cu-MOF-based electrochemical senor for detection of Hg^{+2} ions in tap water and tuna fish.

Source: Singh et al. (2020).

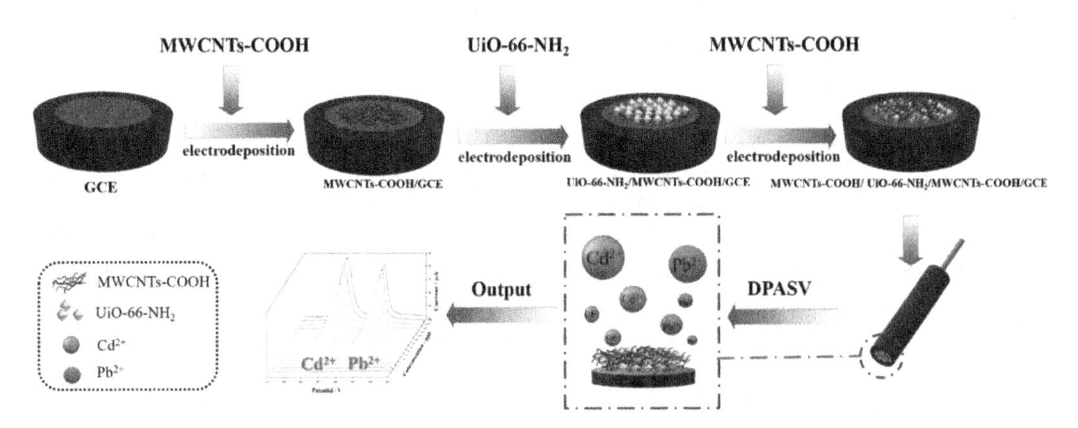

FIGURE 11.3 Fabrication of the MWCNTs–COOH/UiO-66-NH_2/MWCNTs–COOH/GCE for electrochemical sensing of Cd^{2+} and Pb^{2+}.

Source: Lu et al. (2021).

metals ions are also important. The synthesized sensor showed a detection performance at high sensitivity to Cd^{+2} and Pb^{+2}. The lower detection limit was obtained for Cd^{+2} (0.090 ppb) and Pb^{+2} (0.071 ppb) respectively (Figure 11.3). This synthetic strategy may prove promising for applications in energy, the environment, and health-related fields (Z. Lu et al., 2021).

The NH_2-MIL-53(Al) and polypyrrole (PPy)-based nanocomposite was fabricated to sense trace amounts of Pb^{+2} and Cu^{+2}. The fabricated nanocomposite (NH_2-MIL-53(Al)/PPy) was used to modify the gold electrode. Polypyrrole nanowires were synthesized via a chemical polymerization strategy. NH_2-MIL-53(Al) was deposited onto nanowires through an electrochemical approach. X-ray, TEM, attenuated total reflection infrared and field emission SEM were used to analyze the features of the nanocomposite. The cooperative outcome of PPy and NH_2-MIL-53(Al) endowed

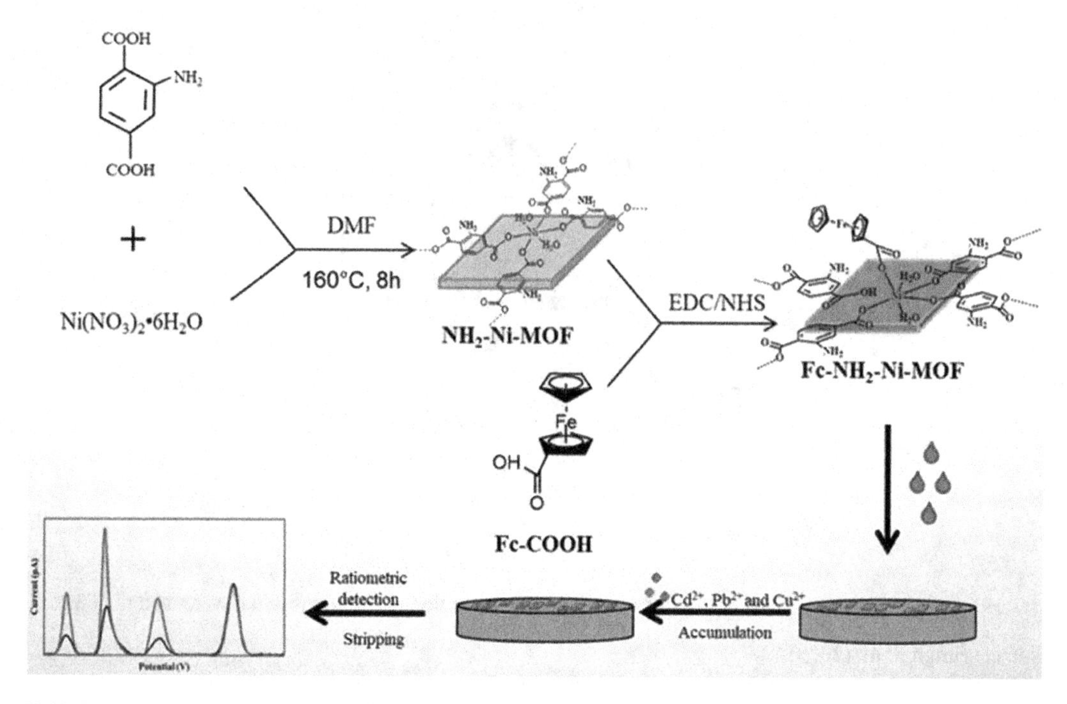

FIGURE 11.4 Nickel-based MOF functionalized with ferrocene for sensing of copper, lead and cadmium metal ions.

Source: Wan et al. (2021).

the modified electrode with great capability to sense Pb^{+2} and Cu^{+2} in the range of 1–400 μg/L. The detection limit was found to be 0.315 μg/L and 0.244 μg/L for Pb^{+2} and Cu^{+2} respectively. The synthesized, nanocomposite modified electrode showed enhanced performance compared to individual modified electrodes (N. Wang et al., 2020).

To electrochemically monitor mercury ions in dairy products, copper-based MOFs (Cu-MOFs) have been used (X. Zhang, Zhu, et al., 2020). Recently, a nickel-based MOF was fabricated (NH_2-Ni-MOF) and through post-synthetic modification it was functionalized with ferrocene (Fc). Ferrocene is known to be an electrochemically active compound. This ferrocene-based unique electrochemical sensor (Fc-NH_2-Ni-MOF) was used to detect different heavy metal ions. Synthesized NH_2-Ni-MOF showed a nanoplate-like structure, which provides an improved electrode area and larger active regions and leads to enhanced adsorption of the heavy metal ions. Incorporation of ferrocene on MOFs increases the MOF material's conductivity and also gives a reference signal for ratiometric analysis. This fabricated sensor showed a wide linear range for lead, copper, and cadmium metals. The sensor showed LOD of 0.2 nM, 6.3 nM and 7.1 nM for lead, copper and cadmium metal ions respectively (Figure 11.4). The sensor showed improved reproducibility with high sensitivity, reliability, and selectivity. So this sensor can be effectively used for the detection of toxic metals in real samples on a wide scale (Wan et al., 2021). Various MOF-based sensors with their electrochemical sensing capacity for different heavy metals along with detection limits, detection range and sensor sensitivity are presented in detail in Table 11.1.

11.3 CONCLUSION

The ecosystem is facing a serious threat, so there is a need for continuous monitoring and effective determination of pollutants that are important for life. Environmental protection has encouraged

TABLE 11.1
Various MOF-Based Sensors for Electrochemical Sensing of Heavy Metal Ions

MOF-based materials	Target	LOD	Linear range	Sensitivity	Reference
Gr/Zn-MOF	As (III)	0.06 μg/L	0.2–25 ppb (μg/L)	0.6237 μA ppb-1	(Baghayeri, Ghanei-Motlagh, et al., 2020)
MOF-867 nanocrystals/pyrrole monomers	Cd^{2+}	0.29 μg/L	–	–	(Li, Cai, Shao, Chen, and Wang, 2021)
Lanthanide MOF (ZJU-27)/GCE	Pb^{+2}	0.228 ppb	–	–	(Ye et al., 2020)
Cu-MOF/GCE	Hg^{+2}	0.0633 nM	0.1–50 nM	0.062 μA nM^{-1}	(Singh et al., 2020)
trGNO/Fc-NH_2-UiO-66	Cd^{+2}	8.5 nM	0.01–2 μM	–	(X. Wang et al., 2020)
	Pb^{+2}	0.6 nM	0.001–2 μM	–	
	Cu^{+2}	0.8 nM	0.001–2 μM	–	
UiO-66- NH_2/ GaOOH	Cd^{+2}	0.016 μM	0.35–1.60 μM	–	(Ru, Wang, Cui, et al., 2021)
	Pb^{+2}	0.028 μM	0.55–2.50 μM	–	
	Cu^{+2}	0.019 μM	0.30–1.40 μM	–	
	Hg^{+2}	0.006 μM	0.10–0.45 μM	–	
MWCNTs–COOH/UiO-66-NH_2/MWCNTs–COOH/GCE	Pb^{+2}	0.071 ppb	–	0.3012 μA ppb^{-1}	(Lu et al., 2021)
	Cd^{+2}	0.090 ppb	–	0.2528 μA ppb^{-1}	
Fc-NH_2-Ni-MOF	Cd^{+2}	7.1 nM	0.01 μM- 2.0 μM		(Wan et al., 2021)
	Pb^{+2}	0.2 nM	0.001 μM- 2.0 μM	–	
	Cu^{+2}	6.3 nM	0.01 μM-2.0 μM	–	
Co/Mo-MOF	As (V)	0.04 ppb	–	–	(Vaid, Dhiman, Kumar, Kim, and Kumar, 2021)
Cu-MOF/GCE	Ti (I)	0.11 ppb	0.5–700 ppb	–	(Baghayeri, Amiri, Safapour Moghaddam, and Nodehi, 2020)
	Hg^{+2}	0.17 ppb	1–400 ppb	–	
Mn-PCN-222 / ITO	Cu^{+2}	0.013μM	0.06–100μM	–	(Fischer et al., 2021)
	Cd^{+2}	0.100μM	0.30–100μM	–	
porph@MOF	Pb^{+2}	5 ppm	–	–	(X. Zhang, Huang, et al., 2020)
Au/SWNTs@MOF-199	Pb^{+2}	25 pM/L	0.1 mM–1 pM	5.46 μA/M	(Bodkhe et al., 2021)
Graphene aerogel (GA)/UiO-66-NH_2	Cd^{+2}	0.02 μM	0.06–3 μM	–	(Lu et al., 2019)
	Pb^{+2}	1.5 nM	0.01–4 μM	–	
	Cu^{+2}	7 nM	0.1–3.5 μM	–	
	Hg^{+2}	2 nM	0.005–3	–	
MOFs HKUST-1/GCE	Pb+2	2 nM	10 nM - 10 μM	–	(Hu, Zhang, Chi, Yang, and Yang, 2020)
	Cd^{+2}	5 nM	10 nM - 10 μM	–	
SPAN/UiO-66-NH_2	Cd^{+2}	0.17 μg/L	0.5 μg/L - 100 μg/L	–	(Zhao, Bai, Yan, Wang, and Zhang, 2019)
Ca-MOF	Pb^{+2}	0.64 μg/L	–	–	(Pournara et al., 2019)
	Cu^{2+}	1.4 μg/L	–	–	
	Zn^{2+}	1.1 μg/L	–	–	
	Cd^{2+}	1.3 μg/L	–	–	
Yb-MOF/GCE	Cd^{2+}	3 ppb	0–50 ppb	–	(Nguyen et al., 2021)
	Pb^{2+}	1.6 ppb	0–50 ppb	–	
Graphite paste/Ca-MOF mixture	Hg^{+2}	0.6 μg/L	–	–	(Kokkinos et al., 2020)

(continued)

TABLE 11.1 (Continued)
Various MOF-Based Sensors for Electrochemical Sensing of Heavy Metal Ions

MOF-based materials	Target	LOD	Linear range	Sensitivity	Reference
Zn/Ni-ZIF-8/XC-72/Nafion/GCE	Pb^{+2}	0.0150 ppm	0.794–39.6 ppm	–	(D. Li et al., 2021)
	Cu^{+2}	0.0096 ppm	0.397–19.9 ppm	–	
NH_2-MIL-53(Al)/polypyrrole nanocomposite/gold electrode	Pb^{+2}	0.315 μg/L	1–400 μg/L	–	(N. Wang et al., 2020)
	Cu^{+2}	0.244 μg/L	–	–	
Bi-BTC/Rgo/GCE	Pb^{+2}	0.021 μg/L	0.062–20.72 μg/L		(Zou et al., 2020)
Cu(I)-MOF/acetylene black/GCE	Hg^{+2}	1.01 nM	2.0 μM to 64 μM	16.85 $\mu A\ \mu M^{-1}\ cm^{-2}$	(X.-Q. Wang, Tang, Ma, Wu, and Yang, 2021)
ZIF-8@dimethylglyoxime/ β-cyclodextrin/RGO	Ni (II)	0.005 μM	0.01–1.0 μM	–	(Cui et al., 2020)
Phytic acid/Polypyrrole)/ZIF-8@ZIF-67/GCE	Cu^{+2}	14.8 nM	0.2–200 μM, 200–600 μM	–	(W. Zhang et al., 2019)
	Pb^{+2}	2.9 nM	0.02–200 μM 200–600 μM		
Cu-MOF	Hg^{+2}	4.8 fM	–	–	(X. Zhang, Zhu, et al., 2020)
Ag-MOF/ITO	Cu^{+2}	0.68 μM	50.0–1.0 μM	–	(Kwon and Kim, 2021)
	Pb^{+2}	0.64 μM	50.0–1.0 μM	–	
MIL-47(as)/carbon paste electrode (CPE)	Cu^{+2}	87.80 nM	1–10 Mm	–	(Niu et al., 2021)
	Pb^{+2}	29.01 nM	1–10 μM	–	
	Hg^{+2}	67.67 nM	1–10 μM	–	
UiO-66/Bi/ screen-printed carbon electrodes (SPCE)	Cd+2	1.7310–50 μg/L	20–60 μg/L	–	(Ding et al., 2021)
	Pb^{+2}	1.0010–50 μg/L	20–60 μg/L	–	
CUiO-66/Bi/SPCE	Cd+2	4.67 μg/L	20–120 μg/L	–	
	Pb^{+2}	1.24 μg/L	20–120 μg/L	–	
UiO-66/Bi/GCE	Cd+2	0.94 μg/L	10–50 μg/L	–	
	Pb^{+2}	2.01 μg/L	10–50 μg/L	–	
CUiO-66/Bi/GCE	Cd+2	1.16 μg/L	10–50 μg/L	–	
	Pb^{+2}	1.14 μg/L	10–50 μg/L	–	
UiO-66-NH_2/MWCNTs	Pb^{+2}	0.5 nmol/L	0.001–0.8 μmol/L	–	(Sun et al., 2020)
Mn-MOF/Nafion/GCE	Cd^{+2}	0.12 ppb	–	–	(Li et al., 2019)
ZIF-67/expanded graphite/GCE	Cd^{2+}	1.13 nM	–	–	(Ma et al., 2020)
	Hg^{2+}	1.28 nM	–	–	
	Pb^{2+}	1.11 nM	–	–	
	Cu^{2+}	2.23 nM	–	–	
Zr-DMBD MOFs/3D-KSC (kenaf stem-derived carbon)	Hg^{+2}	0.05 μM	0.25 μM–3.5 μM	324.58 $\mu A\ \mu M^{-1}\ cm^{-2}$	(Yang, Peng, Han, Song, and Wang, 2020)
GO@Au/Cu(II)-MOFs	Hg^{2+}	0.001 aM	0.10 aM to 100 nM		(X. Zhang, Jiang, et al., 2020)
Bi/MIL-101(Cr)/carbon cloth electrode	Cd^{+2}	3 μM	–	–	(Shi et al., 2019)
UiO-66-NH_2@PANI	Cd^{+2}	3 μM	0.5–600 μg/L	–	(Y. Wang et al., 2017)
Zr(IV)-based MOFs with attached thioether side moities	Hg+2	7.3 fM	0.01 nM to 3 μM	–	(Fu et al., 2019)
Fe(III)-based MOF	Pb^{2+}	2.27 PM	0.01 to 10.0 nM	–	(Z. Zhang et al., 2017)
	As^{3+}	6.73 pM	0.01 to 10.0 nM	–	

wide investigations to develop practical and simple methods, valuable materials, and perfect sensors to determine toxic pollutants. Recently, MOFs have been employed effectively to enhance various techniques such as amperometry and voltammetry to sense different contaminants. MOFs possess novel structural features, such as great porosity, a vast surface area, good pore volume, and adjustable chemical functionalities and these features make them attractive for sensing applications. In recent years, great developments in electrochemical sensing applications of nanomaterials and MOFs have been achieved, but still various challenges in the detection of environmental pollutants through MOFs exist.

MOFs possess insulator-type features, therefore fabrication of redox active MOFs is a major challenge for MOF-based sensors. Compositing of MOFs with various materials as metal nanoparticles, polymers, or carbonaceous materials has received enormous attention. Developing MOF-derived sensors for the determination of toxic heavy metals, such as cadmium, mercury, lead, and arsenic has been presented in this chapter in detail. Different MOF-derived materials have robust connections and the binding of targets for metal ion sensing were described in this chapter. Synthesis of different MOFs and electrode generation were presented. We also have reported on the sensitivity, LOD, selectivity and linear range of MOF-based sensors for sensing heavy metal ions. Synthesizing MOF-derived nanocomposites and their deployment for electrochemical sensing of metals is still in its initial stages. MOF derivatives have great sensing applications, but this practice is still carried out on the lab scale only. So, there is a need to make MOF sensors on a commercial scale to be available in markets for easy access to sensing practice.

The controlled fabrication of MOF-based novel cues for mutual applications, including biosensing or imaging of drug release, is urgently required for the visualization and the quantification of therapeutics and diagnostics. MOF-based sensors will provide a great future in the fields of environmental and clinical analysis. For future outcomes, efforts should be made to increase the sensitivity of sensors. It is important to increase the redox capability of MOF-based materials by selecting multivalent ligands and metals. To understand the selectivity in sensing of various species, it is important to design and fabricate MOFs along with the desired pores and sites through control of conditions, modifications, and synthetic methodologies. A controlled immobilization approach on the surface of the electrode as an electro-chemical deposition, the layer-by-layer method and the Langmuir method should be considered due to the requirement of an organized structure for stable signals. In electrochemical biosensing, water solubility is very important, therefore functional moieties should be applied to MOFs through post-modification, *in-situ* synthesis, and physical methods. With the development of nanotechnology, MOF-based materials will show significant applications in electrochemical sensing.

REFERENCES

Baghayeri, M., Amiri, A., Safapour Moghaddam, B., and Nodehi, M. (2020). Cu-based MOF for simultaneous determination of trace Tl (I) and Hg (II) by stripping voltammetry. *Journal of the Electrochemical Society*, 167(16), 167522. doi:10.1149/1945-7111/abd1ee.

Baghayeri, M., Ghanei-Motlagh, M., Tayebee, R., Fayazi, M., and Narenji, F. (2020). Application of graphene/zinc-based metal-organic framework nanocomposite for electrochemical sensing of As(III) in water resources. *Analytica Chimica Acta*, 1099, 60–67. https://doi.org/10.1016/j.aca.2019.11.045

Bodkhe, G. A., Hedau, B. S., Deshmukh, M. A., Patil, H. K., Shirsat, S. M., Phase, D. M., … Shirsat, M. D. (2021). Selective and sensitive detection of lead Pb(II) ions: Au/SWNT nanocomposite-embedded MOF-199. *Journal of Materials Science*, 56(1), 474–487. doi:10.1007/s10853-020-05285-z

Boix, G., Troyano, J., Garzon-Tovar, L., Camur, C., Bermejo, N., Yazdi, A., … Imaz, I. (2020). MOF-beads containing inorganic nanoparticles for the simultaneous removal of multiple heavy metals from water. *ACS Applied Materials and Interfaces*, 12(9), 10554–10562.

Cui, X., Yang, B., Zhao, S., Li, X., Qiao, M., Mao, R., … Zhao, X. (2020). Electrochemical sensor based on ZIF-8@dimethylglyoxime and β-cyclodextrin modified reduced graphene oxide for nickel (II) detection. *Sensors and Actuators B: Chemical*, 315, 128091. https://doi.org/10.1016/j.snb.2020.128091

Deng, M., Bo, X., and Guo, L. (2018). Encapsulation of platinum nanoparticles into a series of zirconium-based metal-organic frameworks: Effect of the carrier structures on electrocatalytic performances of composites. *Journal of Electroanalytical Chemistry*, 815, 198–209.

Devaraj, M., Yesudass, S., Rajendran, S., and Ponce, L. C. (2021). Metal organic framework-based nanomaterials for electrochemical sensing of toxic heavy metal ions: Progress and their prospects. *Journal of the Electrochemical Society*, 168, 037513.

Ding, Y., Wei, F., Dong, C., Li, J., Zhang, C., and Han, X. (2021). UiO-66 based electrochemical sensor for simultaneous detection of Cd(II) and Pb(II). *Inorganic Chemistry Communications*, 131, 108785. https://doi.org/10.1016/j.inoche.2021.108785

Fischer, R. A., Zhou, Z., Mukherjee, S., Hou, S., Li, W., and Elsner, M. (2021). Porphyrinic MOF film for multifaceted electrochemical sensing. *Angewandte Chemie*. https://doi.org/10.1002/anie.202107860.

Fu, L., Xie, K., Wang, A., Lyu, F., Ge, J., Zhang, L., … Ruan, S. (2019). High selective detection of mercury (II) ions by thioether side groups on metal-organic frameworks. *Analytica Chimica Acta*, 1081, 51–58. https://doi.org/10.1016/j.aca.2019.06.055

Guo, H., Wang, D., Chen, J., Weng, W., Huang, M., and Zheng, Z. (2016). Simple fabrication of flake-like NH_2-MIL-53 (Cr) and its application as an electrochemical sensor for the detection of Pb^{2+}. *Chemical Engineering Journal*, 289, 479–485.

Hu, R., Zhang, X., Chi, K.-N., Yang, T., and Yang, Y.-H. (2020). Bifunctional MOFs-based ratiometric electrochemical sensor for multiplex heavy metal ions. *ACS Applied Materials and Interfaces*, 12(27), 30770–30778. doi:10.1021/acsami.0c06291

Jiao, L., Seow, J. Y. R., Skinner, W. S., Wang, Z. U., and Jiang, H.-L. (2019). Metal–organic frameworks: Structures and functional applications. *Materials Today*, 27, 43–68.

Ke, F., Qiu, L.-G., Yuan, Y.-P., Jiang, X., and Zhu, J.-F. (2012). Fe_3O_4@MOF core–shell magnetic microspheres with a designable metal–organic framework shell. *Journal of Materials Chemistry*, 22(19), 9497–9500.

Khan, N. A., Hasan, Z., and Jhung, S. H. (2018). Beyond pristine metal-organic frameworks: Preparation and application of nanostructured, nanosized, and analogous MOFs. *Coordination Chemistry Reviews*, 376, 20–45.

Kokkinos, C., Economou, A., Pournara, A., Manos, M., Spanopoulos, I., Kanatzidis, M., … Papaefstathiou, G. S. (2020). 3D-printed lab-in-a-syringe voltammetric cell based on a working electrode modified with a highly efficient Ca-MOF sorbent for the determination of Hg(II). *Sensors and Actuators B: Chemical*, 321, 128508. https://doi.org/10.1016/j.snb.2020.128508

Kwon, D., and Kim, J. (2021). Ag metal organic frameworks nanocomposite modified electrode for simultaneous electrochemical detection of copper (II) and lead (II). *Journal of Applied Electrochemistry*, 51(8), 1207–1216. doi:10.1007/s10800-021-01569-7.

Li, D., Qiu, X., Guo, H., Duan, D., Zhang, W., Wang, J., … Zhang, Z. (2021). A simple strategy for the detection of Pb(II) and Cu(II) by an electrochemical sensor based on Zn/Ni-ZIF-8/XC-72/Nafion hybrid materials. *Environmental Research*, 202, 111605. https://doi.org/10.1016/j.envres.2021.111605

Li, Y., Cai, Y., Shao, K., Chen, Y., and Wang, D. (2021). A free-standing poly-MOF film fabricated by post-modification and interfacial polymerization: A novel platform for Cd2+ electrochemical sensors. *Microporous and Mesoporous Materials*, 323, 111200. https://doi.org/10.1016/j.microm eso.2021.111200

Li, Y., Xia, T., Zhang, J., Cui, Y., Li, B., Yang, Y., and Qian, G. (2019). A manganese-based metal-organic framework electrochemical sensor for highly sensitive cadmium ions detection. *Journal of Solid State Chemistry*, 275, 38–42. https://doi.org/10.1016/j.jssc.2019.03.051

Ling, W., Liew, G., Li, Y., Hao, Y., Pan, H., Wang, H., … Huang, X. (2018). Materials and techniques for implantable nutrient sensing using flexible sensors integrated with metal–organic frameworks. *Advanced Materials*, 30(23), 1800917.

Liu, C.-S., Li, J., and Pang, H. (2020). Metal-organic framework-based materials as an emerging platform for advanced electrochemical sensing. *Coordination Chemistry Reviews*, 410, 213222.

Lopa, N. S., Rahman, M. M., Ahmed, F., Sutradhar, S. C., Ryu, T., and Kim, W. (2018). A base-stable metal-organic framework for sensitive and non-enzymatic electrochemical detection of hydrogen peroxide. *Electrochimica Acta*, 274, 49–56.

Lu, M., Deng, Y., Luo, Y., Lv, J., Li, T., Xu, J., … Wang, J. (2019). Graphene aerogel–metal–organic framework-based electrochemical method for simultaneous detection of multiple heavy-metal ions. *Analytical Chemistry*, 91(1), 888–895. doi:10.1021/acs.analchem.8b03764.

Lu, Z., Zhao, W., Wu, L., He, J., Dai, W., Zhou, C., … Ye, J. (2021). Tunable electrochemical of electrosynthesized layer-by-layer multilayer films based on multi-walled carbon nanotubes and metal-organic framework as high-performance electrochemical sensor for simultaneous determination cadmium and lead. *Sensors and Actuators B: Chemical*, 326, 128957. https://doi.org/10.1016/j.snb.2020.128957

Ma, L., Zhang, X., Ikram, M., Ullah, M., Wu, H., and Shi, K. (2020). Controllable synthesis of an intercalated ZIF-67/EG structure for the detection of ultratrace Cd^{2+}, Cu^{2+}, Hg^{2+} and Pb^{2+} ions. *Chemical Engineering Journal*, 395, 125216. https://doi.org/10.1016/j.cej.2020.125216

Nguyen, M. B., Nga, D. T. N., Thu, V. T., Piro, B., Truong, T. N. P., Yen, P. T. H., … Ha, V. T. T. (2021). Novel nanoscale Yb-MOF used as highly efficient electrode for simultaneous detection of heavy metal ions. *Journal of Materials Science*, 56(13), 8172–8185. doi:10.1007/s10853-021-05815-3.

Niu, B., Yao, B., Zhu, M., Guo, H., Ying, S., and Chen, Z. (2021). Carbon paste electrode modified with fern leaves-like MIL-47 (as) for electrochemical simultaneous detection of Pb (II), Cu (II) and Hg (II). *Journal of Electroanalytical Chemistry*, 886, 115121.

Pang, H., Zhang, Y., Cheng, T., Lai, W.-Y., and Huang, W. (2015). Uniform manganese hexacyanoferrate hydrate nanocubes featuring superior performance for low-cost supercapacitors and nonenzymatic electrochemical sensors. *Nanoscale*, 7(38), 16012–16019.

Pournara, A. D., Margariti, A., Tarlas, G. D., Kourtelaris, A., Petkov, V., Kokkinos, C., … Manos, M. J. (2019). A Ca^{2+} MOF combining highly efficient sorption and capability for voltammetric determination of heavy metal ions in aqueous media. *Journal of Materials Chemistry A*, 7(25), 15432–15443.

Railey, P., Song, Y., Liu, T., and Li, Y. (2017). Metal organic frameworks with immobilized nanoparticles: Synthesis and applications in photocatalytic hydrogen generation and energy storage. *Materials Research Bulletin*, 96, 385–394.

Rasheed, T., Hassan, A. A., Bilal, M., Hussain, T., and Rizwan, K. (2020). Metal-organic frameworks based adsorbents: A review from removal perspective of various environmental contaminants from wastewater. *Chemosphere*, 259, 127369.

Rasheed, T., and Rizwan, K. (2022). Metal-organic frameworks based hybrid nanocomposites as state-of-the-art analytical tools for electrochemical sensing applications. *Biosensors and Bioelectronics*, 199, 113867. doi:https://doi.org/10.1016/j.bios.2021.113867.

Rasheed, T., Rizwan, K., Bilal, M., and Iqbal, H. M. N. (2020). Metal-organic framework-based engineered materials: Fundamentals and applications. *Molecules*, 25(7), 1598.

Ru, J., Wang, X., Cui, X., Wang, F., Ji, H., Du, X., and Lu, X. (2021). GaOOH-modified metal-organic frameworks UiO-66-NH2: Selective and sensitive sensing four heavy-metal ions in real wastewater by electrochemical method. *Talanta*, 234, 122679. https://doi.org/10.1016/j.talanta.2021.122679

Ru, J., Wang, X., Wang, F., Cui, X., Du, X., and Lu, X. (2021). UiO series of metal-organic frameworks composites as advanced sorbents for the removal of heavy metal ions: Synthesis, applications and adsorption mechanism. *Ecotoxicology and Environmental Safety*, 208, 111577. https://doi.org/10.1016/j.ecoenv.2020.111577

Shi, E., Yu, G., Lin, H., Liang, C., Zhang, T., Zhang, F., and Qu, F. (2019). The incorporation of bismuth(III) into metal-organic frameworks for electrochemical detection of trace cadmium(II) and lead(II). *Microchimica Acta*, 186(7), 451. doi:10.1007/s00604-019-3522-6.

Singh, S., Numan, A., Zhan, Y., Singh, V., Van Hung, T., and Nam, N. D. (2020). A novel highly efficient and ultrasensitive electrochemical detection of toxic mercury (II) ions in canned tuna fish and tap water based on a copper metal-organic framework. *Journal of Hazardous Materials*, 399, 123042.

Sun, X., Chen, Y., Xie, Y., Wang, L., Wang, Y., and Hu, X. (2020). Preparation of a chemically stable metal–organic framework and multi-walled carbon nanotube composite as a high-performance electrocatalyst for the detection of lead. *Analyst*, 145(5), 1833–1840. doi:10.1039/C9AN02299F.

Vaid, K., Dhiman, J., Kumar, S., Kim, K.-H., and Kumar, V. (2021). Mixed metal (cobalt/molybdenum) based metal-organic frameworks for highly sensitive and specific sensing of arsenic (V): Spectroscopic versus paper-based approaches. *Chemical Engineering Journal*, 426, 131243. https://doi.org/10.1016/j.cej.2021.131243

Wan, J., Shen, Y., Xu, L., Xu, R., Zhang, J., Sun, H., … Wang, X. (2021). Ferrocene-functionalized Ni(II)-based metal-organic framework as electrochemical sensing interface for ratiometric analysis of Cu^{2+}, Pb^{2+} and Cd^{2+}. *Journal of Electroanalytical Chemistry*, 895, 115374. https://doi.org/10.1016/j.jelechem.2021.115374

Wang, N., Zhao, W., Shen, Z., Sun, S., Dai, H., Ma, H., and Lin, M. (2020). Sensitive and selective detection of Pb (II) and Cu (II) using a metal-organic framework/polypyrrole nanocomposite functionalized electrode. *Sensors and Actuators B: Chemical*, 304, 127286. https://doi.org/10.1016/j.snb.2019.127286

Wang, X., Qi, Y., Shen, Y., Yuan, Y., Zhang, L., Zhang, C., and Sun, Y. (2020). A ratiometric electrochemical sensor for simultaneous detection of multiple heavy metal ions based on ferrocene-functionalized metal-organic framework. *Sensors and Actuators B: Chemical*, 310, 127756. https://doi.org/10.1016/j.snb.2020.127756

Wang, X.-Q., Tang, J., Ma, X., Wu, D., and Yang, J. (2021). A novel copper(I) metal–organic framework as a highly efficient and ultrasensitive electrochemical platform for detection of Hg(II) ions in aqueous solution. *CrystEngComm*, 23(16), 3043–3051. doi:10.1039/D1CE00197C.

Wang, Y., Wang, L., Huang, W., Zhang, T., Hu, X., Perman, J. A., and Ma, S. (2017). A metal-organic framework and conducting polymer-based electrochemical sensor for high performance cadmium ion detection. *Journal of Materials Chemistry A*, 5(18), 8385–8393.

Wang, Z., Wang, B., Yang, Y., Cui, Y., Wang, Z., Chen, B., and Qian, G. (2015). Mixed-metal-organic framework with effective Lewis acidic sites for sulfur confinement in high-performance lithium–sulfur batteries. *ACS Applied Materials and Interfaces*, 7(37), 20999–21004.

Wee, L. H., Janssens, N., Sree, S. P., Wiktor, C., Gobechiya, E., Fischer, R. A., ... Martens, J. A. (2014). Local transformation of ZIF-8 powders and coatings into ZnO nanorods for photocatalytic application. *Nanoscale*, 6(4), 2056–2060.

Xu, Y., Li, Q., Xue, H., and Pang, H. (2018). Metal-organic frameworks for direct electrochemical applications. *Coordination Chemistry Reviews*, 376, 292–318.

Xue, Y., Zheng, S., Xue, H., and Pang, H. (2019). Metal-organic framework composites and their electrochemical applications. *Journal of Materials Chemistry A*, 7(13), 7301–7327.

Yang, H., Peng, C., Han, J., Song, Y., and Wang, L. (2020). Three-dimensional macroporous Carbon/Zr-2,5-dimercaptoterephthalic acid metal-organic frameworks nanocomposites for removal and detection of Hg(II). *Sensors and Actuators B: Chemical*, 320, 128447. https://doi.org/10.1016/j.snb.2020.128447

Ye, W., Li, Y., Wang, J., Li, B., Cui, Y., Yang, Y., and Qian, G. (2020). Electrochemical detection of trace heavy metal ions using a Ln-MOF modified glass carbon electrode. *Journal of Solid State Chemistry*, 281, 121032. https://doi.org/10.1016/j.jssc.2019.121032

Zhang, J., Wang, D., and Li, Y. (2019). Ratiometric electrochemical sensors associated with self-cleaning electrodes for simultaneous detection of adrenaline, serotonin, and tryptophan. *ACS Applied Materials and Interfaces*, 11(14), 13557–13563.

Zhang, W., Fan, S., Li, X., Liu, S., Duan, D., Leng, L., ... Qu, L. (2019). Electrochemical determination of lead(II) and copper(II) by using phytic acid and polypyrrole functionalized metal-organic frameworks. *Microchimica Acta*, 187(1), 69. doi:10.1007/s00604-019-4044-y.

Zhang, W., Tan, Y., Gao, Y., Wu, J., Hu, J., Stein, A., and Tang, B. (2016). Nanocomposites of zeolitic imidazolate frameworks on graphene oxide for pseudocapacitor applications. *Journal of Applied Electrochemistry*, 46(4), 441–450.

Zhang, X., Huang, X., Xu, Y., Wang, X., Guo, Z., Huang, X., ... Zou, X. (2020). Single-step electrochemical sensing of ppt-level lead in leaf vegetables based on peroxidase-mimicking metal-organic framework. *Biosensors and Bioelectronics*, 168, 112544. https://doi.org/10.1016/j.bios.2020.112544

Zhang, X., Jiang, Y., Zhu, M., Xu, Y., Guo, Z., Shi, J., ... Wang, D. (2020). Electrochemical DNA sensor for inorganic mercury(II) ion at attomolar level in dairy product using Cu(II)-anchored metal-organic framework as mimetic catalyst. *Chemical Engineering Journal*, 383, 123182. doi:https://doi.org/10.1016/j.cej.2019.123182

Zhang, X., Zhu, M., Jiang, Y., Wang, X., Guo, Z., Shi, J., ... Han, E. (2020). Simple electrochemical sensing for mercury ions in dairy product using optimal Cu^{2+}-based metal-organic frameworks as signal reporting. *Journal of Hazardous Materials*, 400, 123222. doi:https://doi.org/10.1016/j.jhazmat.2020.123222

Zhang, Y., Lin, B., Sun, Y., Zhang, X., Yang, H., and Wang, J. (2015). Carbon nanotubes@ metal–organic frameworks as Mn-based symmetrical supercapacitor electrodes for enhanced charge storage. *RSC Advances*, 5(72), 58100–58106.

Zhang, Z., Ji, H., Song, Y., Zhang, S., Wang, M., Jia, C., ... Liu, C.-S. (2017). Fe(III)-based metal-organic framework-derived core-shell nanostructure: Sensitive electrochemical platform for high trace determination of heavy metal ions. *Biosensors and Bioelectronics*, 94, 358–364. https://doi.org/10.1016/j.bios.2017.03.014

Zhao, X., Bai, W., Yan, Y., Wang, Y., and Zhang, J. (2019). Core-shell self-doped polyaniline coated metal-organic-framework (SPAN@UIO-66-NH2) screen printed electrochemical sensor for Cd2+ Ions. *Journal of the Electrochemical Society*, 166(12), B873–B880. doi:10.1149/2.0251912jes.

Zheng, S., Li, B., Tang, Y., Li, Q., Xue, H., and Pang, H. (2018). Ultrathin nanosheet-assembled $[Ni_3(OH)_2(PTA)_2(H_2O)_4]$ $2H_2O$ hierarchical flowers for high-performance electrocatalysis of glucose oxidation reactions. *Nanoscale*, 10(27), 13270–13276.

Zou, J., Zhong, W., Gao, F., Tu, X., Chen, S., Huang, X., … Yu, Y. (2020). Sensitive electrochemical platform for trace determination of Pb^{2+} based on multilayer Bi-MOFs/reduced graphene oxide films modified electrode. *Microchimica Acta*, 187(11), 603. doi:10.1007/s00604-020-04571-6.

12 Beneficial Functions of Vermiwash and Vermicompost for Sustainable Agriculture

Ankeet Bhagat,[1] *Sumit Singh,*[1] *Kasahun Gudeta,*[2,3]
Siddhant Bhardwaj,[4] *and Sartaj Ahmad Bhat*[5,*]
[1]Department of Zoology, Guru Nanak Dev University, Amritsar, Punjab, India
[2]Shoolini University Biotechnology and Management Sciences, School of Biological and Environmental Sciences, Solan, Himachal Pradesh, India
[3]Department of Biology, Adama Science and Technology University, Adama, Ethiopia
[4]Government Degree College (Boys), Kathua, Jammu and Kashmir, India
[5]River Basin Research Center, Gifu University, Yanagido, Gifu, Japan
*Corresponding author Email: sartajbhat88@gmail.com

CONTENTS

12.1 Introduction 229
12.2 Preparation of Vermiwash 231
12.2.1 Vermiwash Collection Method 231
12.2.2 Important Parameters for Vermicomposting 232
12.2.2.1 Selection of Earthworms 232
12.2.2.2 pH 232
12.2.2.3 Temperature 232
12.2.2.4 Moisture 232
12.2.2.5 Aeration 232
12.2.2.6 Site 232
12.3 Composition of Vermiwash and Its Specific Role 232
12.4 Efficacy as a Biopesticide/Pest Control Agent 233
12.5 Vermicompost and Vermiwash as Plant Growth Promoters 234
12.6 Advantages of Vermiwash over Synthetic Inorganic Chemicals 236
12.7 Factors Leading to Soil Health Improvement and Enhancement of Crop Productivity by Vermicompost and Vermiwash 236
12.8 Conclusion 237
References 237

12.1 INTRODUCTION

Multiple factors such as the population explosion, improvement in living standards and other anthropogenic activities such as urbanization, intensive agricultural practices, extensive exploitation of natural resources, etc. have led to the accumulation of a variety of waste products on land that adversely affect the environment, and waste management is a global problem these days (Singh et al., 2011; Khapre et al., 2021; Rena et al., 2022). Moreover, in order to meet the demands of the

DOI: 10.1201/9781003239956-14

rapidly increasing global population, agricultural practices throughout the world have increased intensively and chemical fertilizers are being used on the soil in a continuous and uncontrolled manner which ultimately leads to a reduction in the fertility of the soil in the long run (Verma et al., 2018). Enormous amounts of chemical synthetic fertilizers and pesticides were also used during the 'Green Revolution' in order to boost the total crop yield from agricultural fields, which resulted in better yield and productivity (Datta et al., 2016; Sharma and Singhavi, 2017). Synthetic pesticides, fertilizers, and weedicides include carbamates, organophosphates, organochlorines, pyrethroids, hexaconazole, benomyl, propiconazole, etc. (Chandra and Kumar, 2015; Patibanda and Ranganathswamy, 2018; Tudi et al., 2021). The indiscriminate and excessive use of all such chemicals has led to unprecedented problems for the normal functioning of our ecosystem, which include faster evolution of resistant pest varieties, harm to their natural predators, harm to non-target organisms, soil health contamination, and impairment of the global food web associated with the soil ecosystem, and ultimately leading to human health complications as well (Khater, 2012; Sharma and Singhvi, 2017; Chauhan et al., 2018; Khan, 2019; Tudi et al., 2021). As such, the global scientific and agricultural community is required to shift their focus from chemical farming to organic farming (Chauhan et al., 2018).

With the growing concerns and awareness regarding the adverse effects of agricultural chemicals on human and environmental health, organic farming has currently gained attention in both the developed as well as the developing nations (Makkar et al., 2017; Rahman and Zhang, 2018; Chowdhary and Raj, 2020). "Sustainable agriculture" can only be ensured and realized with the help of a variety of organic farming systems, including processes of biological origin, such as compost and vermin compost. It is an established fact that earthworms have the capability to consume different types of organic wastes, such as cattle dung, livestock excreta, oil palm waste, sewage sludge, agricultural residue, and many other agro-industrial residues (Bhat et al., 2018). Vermicomposting technology is the process of the decomposition of a variety of organic wastes into nutrient-enriched vermicast, relying on the combined activity of both earthworms as well as microorganisms, and during which the earthworms also increase in number, size and weight (Manyuchi et al., 2013). Vermicomposting, nowadays, is progressively becoming accepted as an organic farming and solid waste management (SWM) technique and two vital bio-fertilizers, namely, vermin compost and vermin wash, are produced by using this technology (Kaur et al., 2015).

The former, vermicompost, is a homogeneous, humus-like stable end product of vermicomposting formed after the mutual activity of both earthworms and microorganisms (Lim et al., 2015). The latter, vermiwash, is one of the important by-products of vermiculture and the vermicomposting industry, carrying a combination of secretions and wash of the earthworms. It is a type of organic fertilizer collected in the form of drainage from vermiculture/vermicompost units (Das et al., 2014). It is an excellent source of a variety of nutrients in organic farming (Zambare et al., 2008). The quality of vermiwash produced by earthworms relies on the vermicompost being used (Sreenivas et al., 2000). Its composition includes different soluble plant nutrients like N, K, Ca, P, and micronutrients, vitamins, hormones, such as cytokinins, auxin etc., amino acids, enzymes, such as phosphatases, proteases, ureases and amylases, useful microbes, such as heterotrophic bacteria, fungi, actinomycetes, including N_2 fixers, such as *Azotobacter* spp., *Rhizobium* spp., *Agrobacterium* spp. etc., PO_4^- solubilizers and many other worm secretions (Zambare et al., 2008; Das et al., 2014; Varghese and Prabha, 2014). These microbes help plants by making available inorganic N_2, amino acids, and inorganic phosphates via processes such as aminofication and nitrification.

Studies suggest vermiwash is an effective promoter of plant growth for the sustainable production of crops on a low-input, eco-friendly basis (Suthar, 2010). It carries excellent growth-promoting properties in addition to biopesticidal properties (Sundararasu and Jeyasankar, 2014). It helps in root initiation, plant development and endorses growth rate. Vermiwash and vermicompost, when added to the soil, enhance crop yield and growth, soil humification, microbial activity and enzyme production (Kibatu and Mamo, 2014). Being organically rich with primary fertilizer nutrients, such as nitrogen, potassium, phosphorus, carbon, etc., both the vermicompost and the vermiwash can be explored and used as bio-fertilizers for agricultural practices (Manyuchi et al., 2013; Das et al., 2014; Lim et al., 2015).

12.2 PREPARATION OF VERMIWASH

Vermiwash is the by-product of vermiculture and vermicomposting, which consists of wash and the secretion of worms. It is procured from a vermicomposting unit in the form of effluent. No tools or devices are required for the collection of vermiwash except for a valve, which is fixed at the base of the vermiwash unit (Das et al., 2014; Sundararasu and Jeyasankar, 2014). The vermiwash unit is constructed with the help of a plastic bucket or barrel. The base of the bucket/barrel is packed with broken bricks and gravel up to 20–25 cm with a covering of coarse sand (20–30 cm) on top. A layer of moist loamy soil (25 cm) is laid on the top of the coarse sand. Finally, the vermiwash unit is loaded up with pre-digested cow dung and various organic wastes. Around 700–800 individual earthworms are introduced into the vermiwash unit. With the assistance of a punctured plastic container, water is sprinkled on the unit and left overnight. Vermiwash is collected through the valve of the unit on a daily basis usually in the morning (Das et al., 2014; Chattopadhyay, 2015). The standard of vermiwash delivered by the worms relies upon the vermicompost which is used (Sreenivas et al., 2000). Vermiwash soaks up the earthworms' secretions, various enzymes, phytohormones and other nutrients from the different organic waste to use as organic fertilizer (Sundaravadivelan et al., 2011; Nath and Singh, 2012). The colour of the vermiwash changes from yellow to dark (Esakkiammal et al., 2015).

12.2.1 VERMIWASH COLLECTION METHOD

There are many methods for the collection of vermiwash but the most commonly and commercially used methods include the Ecoscience Research Foundation (ERF) method and Ismail's method (Meena et al., 2021). The ERF has developed a technique of vermiwash collection in which either barrels or buckets and sometimes earthen pots are used (Figure 12.1) (Srivastava and Beohar, 2005).

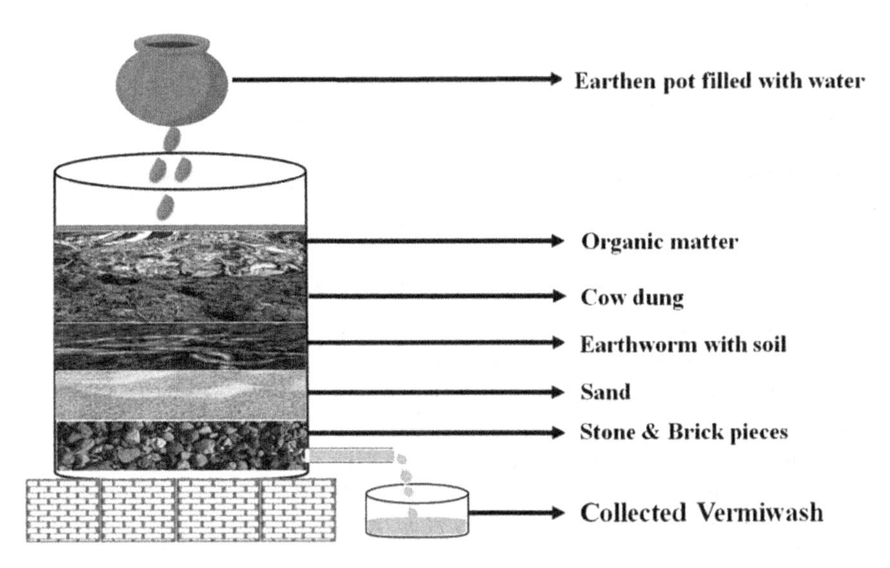

FIGURE 12.1 Set-up of ERF method for the collection of vermiwash.Source; Adopted from Gudeta et al. (2021).

12.2.2 Important Parameters for Vermicomposting

12.2.2.1 Selection of Earthworms

Epigeic earthworms that live on the soil surface and feed on litter are mainly used for the purpose of vermicomposting (Ansari and Ismail, 2012). Examples include *Eisenia fetida, Eudrilus eugeniae* and *Perionyx excavatus* (Chanu et al., 2018).

12.2.2.2 pH

The pH of both the pre-digested dung and the organic waste is a crucial factor affecting the process of vermicomposting. A pH around 7 has been reported to be optimum for the maximum growth and activity of the earthworms (Abd-Manaf et al., 2009). Some reports suggest that the earthworms can even survive at pH ranging from 4.3–6.9 (Hau et al., 2005; Khare et al., 2005).

12.2.2.3 Temperature

Temperature has a great effect on the activity, metabolism, growth, respiration and reproduction of earthworms (Kaur, 2020). Earthworms are highly susceptible to the harsh climate. They are negatively affected by very hot or very cold temperatures and direct sunlight also has a negative impact on the health of earthworm (Abd-Manaf et al., 2009). However, the temperature ranging from 20–35°C has been reported to be ideal for vermicomposting (Nagavallemma et al., 2004; Borah et al., 2007).

12.2.2.4 Moisture

The growth rate of the worms is strongly influenced by the moisture content of the organic waste. The moisture range of 60–65% at all temperature ranges of vermicomposting has been reported to be optimum (Yadav et al., 2010).

12.2.2.5 Aeration

Earthworms are aerobic organisms and their activity is maximized when their bedding material is highly porous as well as aerated (Abd-Manaf et al., 2009). Earthworms themselves help in this task as they aerate the bedding in which they reside by continuously moving through it. *Eisenia fetida* have been reported to move away from the anaerobic water-saturated substrate, or from high concentration of CO_2 or H_2S (Kaur, 2020).

12.2.2.6 Site

Vermicompost production can be done easily in shady areas having high humidity and low temperature. Sites such as an unused cattle shed or poultry shed, etc. can be used for the vermicompost set-up. Wastes heaped for vermicompost production are usually covered with moist gunny bags (Chanu et al., 2018).

12.3 COMPOSITION OF VERMIWASH AND ITS SPECIFIC ROLE

Vermiwash is important in the initiation of rooting, root growth, by raising the organic and nutrient content of the soil, all of which are available to the plants and results in a high crop production (Nandhini and Venmathi, 2017). Vermiwash consists of a large number of decomposing bacteria, mucus, vitamins, various accessible minerals, different phytohormones, a mixture of proteases, amylases, urease and phosphatase, different antimicrobial peptides, micro and macro nutrients, and a variety of helpful microbes, such as heterotrophic bacteria, fungi, actinomycetes, and nitrogen-fixing bacteria, which includes species of *Azotobacter*, *Agrobacterium*, and *Rhizobium*, phosphate solubilizers, etc. (Das et al., 2014; Varghese and Prabha, 2014; Chattopadhyay, 2015; Gudeta et al., 2021).

Vermiwash is enriched with many bioavailable nutrients and growth-promoting substances (phytohormones) and these components have great potential to germinate seed and seedling

TABLE 12.1
Functions of Different Vermiwash Components in Plants

S. no.	Component of vermiwash	Role in plants	Reference
1	Micronutrients (N, P, K, C)	Growth and development	Manyuchi et al., 2013
2	Pro-vitamin D	Growth promoter	Karthikairaj and Isaiarasu, 2013
3	Carbohydrates	Provide energy	Varghese and Prabha, 2014
4	Proteins	Provide immunity	Varghese and Prabha, 2014
5	Macronutrients (Fe, Cu, Zn, Mn)	Growth promoter	Varghese and Prabha, 2014
6	Auxin	Enhance root tip length	Rajasooriya and Karunarathna, 2020
7	Cytokinin and Gibbrellin	Rapid cell division	Rajasooriya and Karunarathna, 2020
8	Protease, Amylase, and Phosphatase	Development and growth	Verma et al., 2018
9	*Azotobacter* sp, *Agrobacterium* sp and *Rhizobium* sp	Improves seed germination and helps in the fixation of Nitrogen	Nandhini and Venmathi, 2017
10	Ca, Mg and Na	Increases the disease resistance, Ca promotes microbial activity	Verma et al., 2018; Manyuchi et al., 2013
11	Amino acids	Chlorophyll synthesis	Das et al., 2014

vigour (Chattopadhyay, 2015). The macronutrients found in vermiwash are required for optimal growth, cell enlargement, increased photosynthesis the formation of carbohydrates, and the translocation of solutes (Hiradeve et al., 2011). The development of leaves and stems is encouraged by nitrogen. Furthermore, N_2 enhances the dark colour of the plants and increases the standard of their foliage. Phosphorus (P) induces the growth of plants, their maturity and flower development, improves the disease resistance power of the plant, as well as its water usage efficiency (Manyuchi et al., 2013).

Secondary nutrients found in vermiwash, such as sulfur (S) and calcium (Ca) are required by the plants in smaller amounts than the basic nutrients. Ca enhances the microbial activity while improving the nitrogen metabolism and reducing plant respiration. S, on the other hand, is used by the plant to form chlorophyll and also to form a necessary component of a number of amino acids (ibid.).

Vermiwash also contains carbohydrates and proteins (Varghese and Prabha, 2014). The protein content in vermiwash was reported to be higher compared to the control. Proteolytic enzymes secreted from the guts of the earthworms may account for the increase in protein content in the vermiwash. It is also an innocuous and eco-friendly liquid, which supports the growth of bacteria by forming a defensive layer for their endurance and development (Das et al., 2014). The components and different parameters of vermiwash and their specific roles are presented in Table 12.1.

12.4 EFFICACY AS A BIOPESTICIDE/PEST CONTROL AGENT

Agriculture or farming plays a crucial part in sustaining global food security and also forms an important part of the global economy. Pest attacks along with other plant diseases can cause serious damage to the food crops, resulting in their lower yield. Therefore, proper control of these pests and diseases is urgently needed (Awad-Allah and Khalil, 2019). For better growth, insect-pest and disease management, vermiwash is physically used as foliar spray (10% solution) on the

TABLE 12.2
Use of Different Products of Vermicomposting for Pest Control

S. no.	Product of vermicomposting	Pest controlled	Host crop	Reference
1	Vermicompost extract	*Meloidogyne incognita*	Carrot	Rao et al., 2017
2	Vermicompost extract	*Meloidogyne incognita*	Tomato	Rao et al., 2015
3	Vermicompost	*Earias vitella*	Lady finger	Hussain et al., 2020
4	Vermiwash	*Helicoverpa armigera*	Chick pea	Haralu et al., 2018
5	Vermiwash	*Tetranychus urticae*	Bean	Aghamohammadi et al., 2016
6	Vermiwash	*Lipaphis erysimi*	Mustard	Nath and Singh, 2012
7	Vermiwash + Neem Extract+ Garlic extract+ Custard apple extract	*Leptocoryza varicornis*	Rice	Mishra et al., 2015

plants (Ravisankar and Gangaiah, 2016). Vermiwash is well known to show plant growth-promoting and pesticidal characteristics. Various vermiwash sprays expeditiously control multiple diseases and pest modalities, which demonstrate it as an eco-friendly bio-pesticide (Gudeta et al., 2021). Vermicompost and vermiwash may be employed together for a better outcome in controlling pests and disease as well as enhancing the quality of the soil and the desired crop growth (Sulaiman and Mohamad, 2020). A combination of these has been reported to reduce two crucial pests of chili, i.e. thrips and mites (Pathma and Sakthivel, 2012). Vermiwash on translocation into the plant system changed the physiological processes, including the enzymatic action which corresponded to a higher metabolite accumulation in the plants and thus resistance against insect-pests (Haralu et al., 2018).

Rao et al. (2015) reported that vermicompost extract inhibits the population of nematode, *Meloidogyne incognita.* Similarly, the application of vermicompost in carrot fields suppressed 10–24% of *Meloidogyne incognita* population and disease incidence (Rao et al., 2017). The vermicompost was also reported to be efficient in checking the growth rate of fruit borer in lady finger (Hussain et al., 2020). Haralu et al. (2018) reported that vermiwash was also efficient against *Helicoverpaarmigera*, which usually infects the pod of chickpeas. The combined application of vermiwash and acaricide azocyclotin has also been documented to show pesticidal activity against *Tetranychusurticae* in bean plants (Aghamohammadi et al., 2016). Table 12.2 lists various reports of pesticidal activity of vermicompost and its by-products.

The vermiwash obtained from the vermicompost made up of buffalo dung plus gram bran with neem oil and liquid extract of garlic (*Allium sativum*) was reported to be capable of controlling 97% of the total aphid infestation on the mustard, *Brassica campestris*, plants (Nath and Singh, 2012). The combination also promoted crop growth, induced early flowering and enhanced total yield of mustard up to 3.5-fold with respect to the control (ibid.). Vermiwash collected from different types of animal dung and municipal solid wastes in combination with neem oil (*Azadirachata indica*) and aqueous extract of garlic bulbs (*Allium sativum*) was reported to be a better alternative for the management of *Leptocoris avaricornis* infestation in rice (Mishra et al., 2015).

12.5 VERMICOMPOST AND VERMIWASH AS PLANT GROWTH PROMOTERS

Vermitechnology makes use of various types of organic wastes, improves the quality of raw waste, transforms these wastes, and the resultant vermicompost, when applied to different crops in the fields, facilitates plant growth by improving the nutritional status and physical properties of the soil (Eo and Park, 2019). Vermicompost carries high concentrations of humus, nutrients, phytohormones, various enzymes, microbes and a number of other components that have the potential to defend plants against a number of plant pests and diseases (Hanc and Pliva, 2013). Actually, vermicompost formed post digestion and excretion activities by worms is a form of nutrient-rich and organic soil amendment carrying considerable potential to improve crop productivity. Microorganisms and earthworms mutually transform it into a finely divided and peat-like matter with low C:N ratios and elevated porosity, water- and air-holding capacity (Edwards, 1997; Ali et al., 2015). Vermicompost includes a variety of nutrients including nitrates, phosphorus, potassium, calcium, and magnesium, in soluble forms that are readily available to be absorbed by the plants, indicating its great potential to be exploited as a medium for plant growth in the horticultural and agricultural sector industries (Atiyeh et al., 2000). Vermicomposting plays a key role in nitrogen recycling and augments various physico-chemical and biological properties of soil, thereby facilitating growth and quality as well as the yield of many field crops (Nagavallemma et al., 2004; Bharagava and Chowdhary, 2019). An application of vermicompost has been reported to significantly improve growth and flowering in the case of *Crossandra* compared to the untreated plants (Gajalakshmi and Abbasi, 2004).

Similarly, vermiwash has also been reported to significantly improve plant growth and development, as discussed above. Treatment of plants with vermiwash resulted in the extensive development of plant root systems carrying larger root volume with increased root length, whereas treatment with chemical fertilizers forms a weak, lean and poorly branched root system (Lazcano et al., 2009; Makkar et al., 2017). Humic acids have already been reported to enhance the nutrient uptake in plants by increasing the permeability of the root cell membranes, and the presence of this humic acid both in vermicompost as well as vermiwash might be the reason for this enhancement in root growth and developmental parameters (Tattini et al., 1990; Valdrighi et al., 1996; Makkar et al., 2017).

As vermiwash comprises significant amounts of mucus, excretory substances (urea, ammonia and phenols) and a number of beneficial microbes which directly or indirectly promote plant growth, it can act as an efficient plant growth promoter for sustainable crop productivity (Suthar, 2010). Vermiwash also carries enzymes and secretions of earthworms that stimulate the growth and yield of a variety of crops and also develop resistance in crops (Zambare et al., 2008). Vermiwash improves the production, growth and germination of some plants that are mentioned above.

Nath and Singh (2011) reported that the application of foliar spray of soya bean (*Glycine max* L.) with vermiwash, collected as a by-product of vermicomposts formed from a mixture of buffalo dung and agro/kitchen wastes, reflected its potential to enhance growth and productivity and, further, reduced the rate of crop infestation by pests. The growth-enhancing effect of vermiwash has also been reported on the crinkle red variety of the flamingo flower, *Anthurium andraeanum* (Karuna et al., 1999). Seedlings of plants such as *Vigna mungo*, *Vigna radiata* and *Sesamum indicum* when sprayed with vermiwash of *Perionyx excavates* also demonstrated up-regulation of various plant growth parameters, including root and shoot lengths, number of leaves and twigs, and the total plant biomass (Hatti et al., 2010). Similarly, in Okra or *Abelmoschus esculentus*, the exomorphological characters, including height of plant, length and diameter of the internodes, number of leaves, surface area of leaves, root length, dry and wet mass of plant shoot and root, were also positively affected with vermiwash treatment (Elumalai et al., 2013).

In a similar study by Gajjela and Chatterjee (2019), the application of vermiwash not only enhanced the yield in the case of organic bitter gourd but also improved the quality and shelf life of the fruits. As such, in the long run, the use of vermiwash has been suggested to make the production system more sustainable. The exposure of vermiwash and vermicompost (vermicompost @ 5 t/ha + vermiwash 5 sprays at 10 days interval after 30 DAS) to Okra var. *Kashi Pragati* showed a significant increase in fruit characters, such as length, girth, weight, fruiting span, fruits per plant and also yield per hectare (Sharma et al., 2014). Awadhpersad et al. (2021) treated tomato plants (*Lycopersicones culentum* Mill) with a vermicompost and vermiwash combination and at the time of harvesting their results indicated that the features such as plant height, wet and dry shoot mass, root mass, density and length, yield and fruit mass were enhanced compared to normal. Furthermore, the ratio of flowering and fruiting was also observed to be significantly enhanced when vermiwash was applied as a foliar spray. Table 12.3 presents the growth-promoting results of using vermiwash on different plants.

TABLE 12.3
Growth-Promoting Effects of Vermiwash on Crop Plants

S. No.	Crop	Results	References
1	Marigold	Increased production	Shivsubramanian and Ganeshkumar, 2004
2	Radish	Production increase by 7.3% after weekly application of vermiwash	Zambare et al., 2008
3	Rice	Growth improved and production increased after vermiwash application	Tharmaraj et al., 2011
4	Flax	Improvement in germination and yield after foliar application of vermiwash integrated with vermicompost	Makkar et al., 2017
5	Tabasco Pepper	Vermiwash increased growth rate due to increased macronutrients and micronutrients uptake	Varghese et al., 2014
6	Eggplant	High yield of brinjal due to increased availability of more exchangeable nutrients in the soil by the application of vermiwash	Sundararasu and Jeyasankar, 2014
7	Allium cepa	Combinations of vermiwash and humic acid (3:1.5% vermiwash: humic acid) increased growth, soft shoots, number of branches, total carbohydrate and total protein contents	Prasad et al., 2016

12.6 ADVANTAGES OF VERMIWASH OVER SYNTHETIC INORGANIC CHEMICALS

Currently, a major threat confronting the world is environmental degradation and a major part of this is attributed to the extensive use of chemical fertilizers that in the long run lead to deterioration of the environment, a decrease in soil fertility, reduced agricultural productivity, eutrophication and the degradation of the soil (Khan and Ishaq, 2011; Ray and Ray, 2011). Studies on the long-term and uncontrolled application of chemical fertilizers in an unbalanced manner indicate that these have detrimental effects on the physical, chemical and biological properties of the soil, inducing secondary and micronutrient deficiencies, or a nutrient imbalance in the soil as well as in plants. Further, the soil microbiota is also severely affected (Kumar et al., 2018). Vermicomposting presents an eco-friendly and cost-effective alternative to these as a biofertilizer that promises high value that increases plant growth and productivity by supplementing the essential nutrients (Suthar, 2007). The application of vermicompost facilitates the aggregation of the soil and stabilizes its structure, thereby improving the air and water retention capacity of the soil, and also encourages the root system of plants to extensively develop (ibid.; Parthasarathi et al., 2008). The nutrient mineralization of the soil is also enhanced, resulting in higher and better crop productivity. Furthermore, vermicomposts possess a higher capacity for base exchange and carry more exchangeable magnesium, potassium and calcium than normal soil (Suthar, 2007). A study by Karmegam and Daniel (2008) reported that the effect of vermicompost on plant growth was equal to that of chemical fertilizer applied either alone or partially supplemented with vermicompost (Karmegam and Daniel, 2008).

12.7 FACTORS LEADING TO SOIL HEALTH IMPROVEMENT AND ENHANCEMENT OF CROP PRODUCTIVITY BY VERMICOMPOST AND VERMIWASH

In simple words, vermicompost can be considered the excreta of earthworms which is enriched with plant nutrients, growth hormones and high quality humus (Adhikary, 2012). It can be used as a key source of organic matter added to the soil for sustainable crop yield. Its application usually results in an improvement in soil health and high crop productivity based on the following reasons:

- Vermicompost is enriched with beneficial micro flora, including cellulose decomposers, N_2-fixers, P-solubilizers, etc. that can improve the soil quality. It also carries the cocoons of the earthworms thereby increasing their soil population and activity (Mahanta et al., 2021; Chanu et al., 2018).
- It is devoid of pathogens, weed seeds, toxic elements, etc. (Ahirwar and Hussain, 2015)
- It reduces the chances of pest attacks and diseases. Vermiwash carries phytohormones and pesticidal features. The spray of vermiwash potentially controls many types of plant diseases and pests, and can be an ecofriendly bio-pesticide (Thakur and Sood, 2019). Two-months-old vermiwash was experimentally observed to be very effective in managing and curbing bacterial leaf spot and early leaf blight diseases in tomato, fusarium wilt in the case of brinjal, mustard damping of root disease and, hence, can be explored as an efficient biocontrol agent for these diseases (Pattnaik et al., 2015).
- A number of immobilized enzymes, including protease, amylase, cellulase, lipase, and chitinase, etc. existing in vermicompost biodegrade the agricultural residues present in the soil so that further decomposition by microbial attack is boosted (Rana, 2018).
- Production of many crops is reduced owing to an increase in the salinity of the agricultural soils (Rengasamy, 2010) but the humic matter present in vermicompost and vermiwash is associated with the organic constituents of earth and amplifies the tolerance capacity of the plant roots. The stress symptoms generated in the plants cultivated in saline conditions are also lessened by supplementing them with vermicompost and vermiwash (Ruiz-Lau et al., 2020).
- Superiority of vermicompost over other synthetic growth media is more pronounced in plant nurseries.
- Vermicompost and vermiwash can serve as rooting media for growing various saplings, particularly in nurseries.
- Vermiwash and vermicompost promote overall plant growth by encouraging the growth and formation of new shoots and leaves and also help in improving the overall quality as well as the shelf-life of the yield (Rana, 2018; Chanu et al., 2018).

12.8 CONCLUSION

It has been proved that vermiwash is an efficient fertilizer that contributes to the growth and yield of crops when applied directly or combined with fertilizer or manure in a definite ratio. It was also noted that the crops treated with vermiwash suffer less insect-pest infestation. Vermiwash functions as a liquid organic fertilizer a gentle biocide, a rhizospheric liquid fertilizer, a natural biopesticide, an excellent growth promoter, and as worm cast or worm manure and is popularly referred as black gold. It is used effectively in organic farming for soil health as well as disease management for a sustainable crop yield, and also to protect the environment and human health. Vermiwash shows potential applications in sustainable development in agriculture biotechnology with respect to its origin, cost effectiveness, easy availability, time saving, reproducibility, reliability and eco-friendliness.

REFERENCES

Abd Manaf, L., Jusoh, M. L. C., Yusoff, M. K., Ismail, T. H. T., Harun, R., Juahir, H., and Jusoff, K. (2009). Influences of bedding material in vermicomposting process. *International Journal of Biology*, 1(1), 81.

Adhikary, S. (2012). Vermicompost, the story of organic gold: a review, *Agricultural Sciences*, 3(7): 905–917.

Aghamohammadi, Z., Etesami, H., and Alikhani, H. A. (2016). Vermiwash allows reduced application rates of acaricide azocyclotin for the control of two spotted spider mite, *Tetranychusurticae* Koch, on bean plant (*Phaseolus vulgaris* L.). *Ecological Engineering*, 93, 234–241.

Ahirwar, C. S., and Hussain, A. (2015). Effect of vermicompost on growth, yield and quality of vegetable crops. *International Journal of Applied and Pure Science and Agriculture*, 1(8), 49–56.

Ali, U., Sajid, N., Khalid, A., Riaz, L., Rabbani, M. M., Syed, J. H., and Malik, R. N. (2015). A review on vermicomposting of organic wastes. *Environmental Progress and Sustainable Energy*, 34(4), 1050–1062.

Ansari, A. A., and Ismail, S. A. (2012). Role of earthworms in vermitechnology. *Journal of Agricultural Technology*, 8(2), 403–415.

Atiyeh, R. M., Subler, S., Edwards, C. A., Bachman, G., Metzger, J. D., and Shuster, W. (2000). Effects of vermicomposts and composts on plant growth in horticultural container media and soil. *Pedobiologia*, 44(5), 579–590.

Awad-Allah, S. F. A., and Khalil, M. S. (2019). Effects of vermicompost, vermicompost tea and a bacterial bioagent against *Meloidogyne incognita* on banana in Egypt. *Pakistan Journal of Nematology*, 37(1), 25–33.

Awadhpersad, V. R. R., Ori, L., and Ansari, A. (2021). Production and effect of vermiwash singly and in combination with vermicompost on the growth, development and productivity of tomato in the greenhouse in Suriname. *Asian Journal of Agriculture*, 5(1).

Bhat, S. A., Singh, J., and Vig, A. P. (2018). Earthworms as organic waste managers and biofertilizer producers. *Waste and Biomass Valorization*, 9(7), 1073–1086.

Borah, M. C., Mahanta, P., Kakoty, S. K., Saha, U. K., and Sahasrabudhe, A. D. (2007). Study of quality parameters in vermicomposting. *Indian Journal of Biotechnology*, 6, 410–413.

Chanu, L. J., Hazarika, S., Choudhury, B. U., Ramesh, T., Balusamy, A., Moirangthem, P., … and Sinha, P. K. (2018). A guide to vermicomposting-production process and socio economic aspects. *Extension Bulletin*, 81, 30.

Chattopadhyay, A. (2015). Effect of vermiwash of *Eisenia foetida* produced by different methods on seed germination of green mung, *Vigna radiata*. *International Journal of Recycling of Organic Waste in Agriculture*, 4(4), 233–237.

Chauhan, A., Ranjan, A., and Jindal, T. (2018). Biological control agents for sustainable agriculture, safe water and soil health. In Jindal, T. (ed.) *Paradigms in Pollution Prevention* (pp. 71–83). Cham: Springer.

Chandra, R., and Kumar, V. (2015). Biotransformation and biodegradation of organophosphates and organohalides. In Chandra, R. (ed.) *Environmental Waste Management*. Boca Raton, FL: CRC Press. DOI:10.1201/b19243-17.

Chowdhary, P., and Bharagava, A. (eds) (2019). *Emerging and Eco-Friendly Approaches for Waste Management*, Singapore: Springer.

Chowdhary, P., Raj, A., Verma, D., and Akhter, Y. (eds) (2020). *Contaminants and Clean Technologies*, Boca Raton, FL: CRC Press.

Das, S. K., Avasthe, R. K., and Gopi, R. (2014). Vermiwash: Use in organic agriculture for improved crop production. *Popular Kheti*, 2(4), 45–46.

Datta, S., Singh, J., Singh, S., and Singh, J. (2016). Earthworms, pesticides and sustainable agriculture: A review. *Environmental Science and Pollution Research*, 23(9), 8227–8243.

Edwards, C. A. (1997). *Earthworm Ecology*. Boca Raton, FL: CRC Press.

Elumalai, D., Kaleena, P. K., Fathima, M., and Hemavathi, M. (2013). Influence of vermiwash and plant growth regulators on the exomorphological characters of *Abelmoschus esculentus* (Linn.) Moench. *African Journal of Basic and Applied Sciences*, 5(2), 82–90.

Eo, J., and Park, K. C. (2019). Effect of vermicompost application on root growth and ginsenoside content of *Panax ginseng*. *Journal of Environmental Management*, 234, 458–463.

Esakkiammal, B., Lakshmibai, L., and Sornalatha, S. (2015). Studies on the combined effect of vermicompost and vermiwash prepared from organic wastes by earthworms on the growth and yield parameters of Dolichos lab. *Asian Journal of Pharmaceutical Science Technology*, 5(4), 246–252.

Gajalakshmi, S., and Abbasi, S. A. (2004). Neem leaves as a source of fertilizer-cum-pesticide vermicompost. *Bioresource Technology*, 92(3), 291–296.

Gajjela, S., and Chatterjee, R. (2019). Effect of foliar application of Panchagavya and vermiwash on yield and quality of bitter gourd (*Momordica charantia* L.). *International Journal of Chemical Studies*, 7(3), 218–224.

Gudeta, K., Julka, J. M., Kumar, A., Bhagat, A., and Kumari, A. (2021). Vermiwash: An agent of disease and pest control in soil, a review. *Heliyon*, 7(3), e06434.

Hanc, A., and Pliva, P. (2013). Vermicomposting technology as a tool for nutrient recovery from kitchen biowaste. *Journal of Material Cycles and Waste Management*, 15(4), 431–439.

Haralu, S., Karabhantanal, S. S., Jagginavar, S. B., and Naidu, G. K. (2018). Utilization of vermiwash as biopesticide in the management of pod borer, *helicoverpaarmigera* (hubner), in chickpea (*cicer arietinum* l.). *Applied Biological Research*, 20(1), 37–45.
Hatti, S. S., Londonkar, R. L., Patil, S. B., Gangawane, A. K., and Patil, C. S. (2010). Effect of *Eisenia fetida* vermiwash on the growth of plants. *Journal of Crop Science*, 1(1), 6.
Hau, J., Qiao, Y., Liu, G., and Dong, R. (2005). The influence of temperature, pH and C/N ratio on the growth and survival of earthworms in municipal solid waste. *Agricultural Engineering International: CIGR Journal*, vii.
Hiradeve, P. N., Deotale, R. D., Deogirkar, M. S., and Gaikwad, S. B. (2011). Effectivity of foliar sprays of vermicompost wash on chemical, biochemical, yield and yield contributing parameters of groundnut (*Arachishypogeae* L.). *Journal of Soils and Crops*, 21(2), 266–272.
Hussain, N., Abbasi, T., and Abbasi, S. A. (2020). Evaluating the fertilizer and pesticidal value of vermicompost generated from a toxic and allelopathic weed ipomoea. *Journal of the Saudi Society of Agricultural Sciences*, 19(1), 43–50.
Karmegam, N., and Daniel, T. (2008). Effect of vermicompost and chemical fertilizer on growth and yield of hyacinth bean, *Lablab purpureus* (L.) Sweet. *Dynamic Soil, Dynamic Plant*, 2(2), 77–81.
Karthikairaj, K., and Isaiarasu, L. (2013). Effect of vermiwash on the growth of mulberry cuttings. *World Journal of Agricultural Science*, 9(1), 69–72.
Karuna, K., Patil, C. R., Narayanswamy, P., and Kale, R. D. (1999). Stimulatory effect of earthworm body fluid (vermiwash) on crinkle red variety of *Anthurium andreanum*lind. *Crop Research*, 17(2), 253–257.
Kaur, P., Bhardwaj, M., and Babbar, I. (2015). Effect of vermicompost and vermiwash on growth of vegetables. *Research Journal of Animal, Veterinary and Fishery Sciences*, 3(4), 9–12.
Kaur, T. (2020). Vermicomposting: An effective option for recycling organic wastes. In Das, S. K. (ed.) *Organic Agriculture*. London: IntechOpen.
Khan, A., and Ishaq, F. (2011). Chemical nutrient analysis of different composts (Vermicompost and Pitcompost) and their effect on the growth of a vegetative crop *Pisum sativum*. *Asian Journal of Plant Science and Research*, 1(1), 116–130.
Khan, H. A. A. (2019). Characterization of permethrin resistance in a *Musca domestica* strain: Resistance development, cross-resistance potential and realized heritability. *Pest Management Science*, 75(11), 2969–2974.
Khapre, A., Khan, S. A., and Kumar, S. (2021). A laboratory-scale phytocover system for municipal solid waste landfills. *Environmental Technology*, 1–27. www.pubfacts.com/detail/33998978/A-Laboratory.
Khare, N. S. A., Bhargava, D., and Bhattacharya, S. (2005). Effect of initial substrate pH on vermicomposting using *Perionyx excavatus* (Perrier, 1872). *Applied Ecology and Environmental Research*, 4(1), 85–97.
Khater, H. F. (2012). Ecosmart biorational insecticides: Alternative insect control strategies. In Perveen. F. K. (ed.) *Insecticides: Advances in Integrated Pest Management*, In-Tech, pp. 17–61.
Kibatu, T., and Mamo, M. (2014). Vermicompost and vermiwash on growth, yield and yield components of Beetroot (*Beta vulgaris* L.). *World Applied Sciences Journal*, 32(2), 177–182.
Kumar, A., Prakash, C. H., Brar, N. S., and Kumar, B. (2018). Potential of vermicompost for sustainable crop production and soil health improvement in different cropping systems. *International Journal of Current Microbiology Applied Science*, 7(10), 1042–1055.
Lazcano, C., Arnold, J., Zaller, J. G., Martín, J. D., and Salgado, A. T. (2009). Compost and vermicompost as nursery pot components: Effects on tomato plant growth and morphology. *Spanish Journal of Agricultural Research*, 7(4), 944–951.
Lim, S. L., Wu, T. Y., Lim, P. N., and Shak, K. P. Y. (2015). The use of vermicompost in organic farming: Overview, effects on soil and economics. *Journal of the Science of Food and Agriculture*, 95(6), 1143–1156.
Mahanta, K., Rajkhowa, D. J., Kumar, M., Verma, H., Rao, K. K., and Saurabh, K. (2021). Vermicomposting for efficient recycling of biowastes. *Biotica Research Today*, *3*(6), 492–494.
Makkar, C., Singh, J., and Parkash, C. (2017). Vermicompost and vermiwash as supplement to improve seedling, plant growth and yield in *Linumusitassimum* L. for organic agriculture. *International Journal of Recycling of Organic Waste in Agriculture*, 6(3), 203–218.
Manyuchi, M. M., Phiri, A., Muredzi, P., and Chitambwe, T. (2013). Comparison of vermicompost and vermiwash bio-fertilizers from vermicomposting waste corn pulp. *World Academy of Science, Technology and Engineering International*, 7, 389–392.

Meena, A. L., Karwal, M., Raghavendra, K. J., and Kumari, P. (2021). Vermiwash: A potential tool for sustainable organic farming. *Food and Scientific Reports*, 2(5), 43–45.

Mishra, K., Singh, K., and Tripathi, C. P. M. (2015). Organic farming of rice crop and management of infestation of *Leptocoryzavaricornis* through combined effect of vermiwash with biopesticides. *Research Journal of Science and Technology*, 7(4), 205–211.

Nagavallemma, K. P., Wani, S. P., Lacroix, S., Padmaja, V. V., Vineela, C., Rao, M. B., and Sahrawat, K. L. (2004). Vermicomposting: Recycling wastes into valuable organic fertilizer. *Global Theme on Agroecosystems Report*, 8. Online.

Nandhini, D. U., and Venmathi, T. (2017). Vermiwash-A potential plant growth promoter. *Agric INTERNATIONAL*, 4(1), 27–30.

Nath, G., and Singh, K. (2011). Effect of foliar spray of biopesticides and vermiwash of animal, agro and kitchen wastes on soybean (*Glycine max* L.) crop. *Botany Research International*, 4(3), 52–57.

Nath, G., and Singh, K. (2012). Effect of vermiwash of different vermicomposts on the kharif crops. *Journal of Central European Agriculture*, 13(2), 379–402.

Nayak, H., Rai, S., Mahto, R., Rani, P., Yadav, S., Prasad, S. K., and Singh, R. K. (2019). Vermiwash: A potential tool for sustainable agriculture. *Journal of Pharmacognosy and Phytochemistry*, SP5, 308–312.

Parthasarathi, K., Balamurugan, M., and Ranganathan, L. S. (2008). Influence of vermicompost on the physicochemical and biological properties in different types of soil along with yield and quality of the pulse crop-blackgram. *Journal of Environmental Health Science and Engineering*, 5(1), 51–58.

Pathma, J., and Sakthivel, N. (2012). Microbial diversity of vermicompost bacteria that exhibit useful agricultural traits and waste management potential. *SpringerPlus*, 1(1), 1–19.

Patibanda, A. K., and Ranganathswamy, M. (2018). Effect of agrichemicals on biocontrol agents of plant disease control. In Panpatter, D.G., Jhala, Y.K., Sheat, H.N. and Vyas, R.V. (eds) *Microorganisms for Green Revolution* (pp. 1–21). Singapore: Springer.

Pattnaik, S., Parida, S., Mishra, S. P., Dash, J., and Samantray, S. M. (2015). Control of phytopathogens with application of vermiwash. *Journal of Pure and Applied Microbiology*, 9(2), 1697–1701.

Prasad, U., Sunkar, S., Gala, A. A., and Kumar, A. (2016). Formulation of vermiwash and humic acid and its application on *Allium cepa*. *Biosciences Biotechnology Research Asia*, 13(1), 523–529.

Rahman, K. M., and Zhang, D. (2018). Effects of fertilizer broadcasting on the excessive use of inorganic fertilizers and environmental sustainability. *Sustainability*, 10(3), 759.

Rajasooriya, A. S., and Karunarathna, B. (2020). Application of vermiwash on growth and yield of green gram (*Vigna radiata*) in sandy regosol. *AGRIEAST*, 14(2), 31–42.

Rana, S. S. (2018). Biological intensive nutrient management: Vermicompost. Available at: www.hillagric.ac.in/.../Lecture-10-BINM-Vermicompost.pdf

Rao, M. S., Kamalnath, M., Umamaheswari, R., Rajinikanth, R., Prabu, P., Priti, K., ... and Gopalakrishnan, C. (2017). *Bacillus subtilis* IIHR BS-2 enriched vermicompost controls root knot nematode and soft rot disease complex in carrot. *Scientia Horticulturae*, 218, 56–62.

Rao, M. S., Umamaheswari, R., Chakravarthy, A. K., Grace, G. N., Kamalnath, M., and Prabu, P. (2015). A frontier area of research on liquid biopesticides: The way forward for sustainable agriculture in India. *Current Science*, 108(9), 1590–1592.

Ravisankar, N., and Gangaiah, B. (2016). Organic farming: Scope and strategy for Andaman and Nicobar Islands. Unpublished.

Ray, S., and Ray, I. A. (2011). Impact of population growth on environmental degradation: Case of India. *Journal of Economics and Sustainable Development*, 2(8), 72–77.

Rengasamy, P. (2010). Soil processes affecting crop production in salt-affected soils. *Functional Plant Biology*, 37(7), 613–620.

Rena, Y. S., Patel, S., Killedar, D.J., Kumar, S., and Kumar, R. (2022). Eco-innovations and sustainability in solid waste management: An Indian upfront in technological, organizational, start-ups and financial framework. *Journal of Environmental Management*, 302, 113953. https://doi.org/10.1016/j.jenvman.2021.113953

Ruiz-Lau, N., Oliva-Llaven, M. A., Montes-Molina, J. A., and Gutiérrez-Miceli, F. A. (2020). Mitigation of salinity stress by using the vermicompost and vermiwash. In Bauddh, K. Kumar, S. Singh. R.P. and Korstad, J. (eds) *Ecological and Practical Applications for Sustainable Agriculture* (pp. 345–356). Singapore: Springer.

Sharma, D. P., Prajapati, J., and Tiwari, A. (2014). Effect of NPK, vermicompost and vermiwash on growth and yield of Okra. *International Journal of Basic and Applied Agricultural Research*, 12(1), 5–8.

Sharma, N., and Singhvi, R. (2017). Effects of chemical fertilizers and pesticides on human health and environment: A review. *International Journal of Agriculture, Environment and Biotechnology*, 10(6), 675–680.

Shivsubramanian, K., and Ganeshkumar, M. (2004). Influence of vermiwash on biological productivity of Marigold. *Madras Agricultural Journal*, 91(4–6), 221–225.

Singh, R. P., Singh, P., Araujo, A. S., Ibrahim, M. H., and Sulaiman, O. (2011). Management of urban solid waste: Vermicomposting a sustainable option. *Resources, Conservation and Recycling*, 55(7), 719–729.

Sreenivas, C., Muralidhar, S., and Rao, M. S. (2000). Vermicompost: A viable component of IPNSS in nitrogen nutrition of ridge gourd. *Annals of Agricultural Research*, 21(1), 108–113.

Srivastava, R. K., and Beohar, P. A. (2005). Vermicompost as a organic manure: A good substitute of fertilizers. *Verms and Vermitechnology*, 97.

Sulaiman, I. S. C., and Mohamad, A. (2020). The use of vermiwash and vermicompost extract in plant disease and pest control. In Egbuna, C.and Sawicka. B. (eds) *Natural Remedies for Pest, Disease and Weed Control* (pp. 187–201). New York: Academic Press.

Sundararasu, K., and Jeyasankar, A. (2014). Effect of vermiwash on growth and yield of brinjal, *Solanum melongena* (eggplant or aubergine). *Asian Journal of Science and Technology*, 5(3), 171–173.

Sundaravadivelan, C., Isaiarasu, L., Manimuthu, M., Kumar, P., Kuberan, T., and Anburaj, J. (2011). Impact analysis and confirmative study of physico-chemical, nutritional and biochemical parameters of vermiwash produced from different leaf litters by using two earthworm species. *Journal of Agricultural Technology*, 7(5), 1443–1457.

Suthar, S. (2007). Production of vermifertilizer from guar gum industrial wastes by using composting earthworm *Perionyx sansibaricus* (Perrier). *The Environmentalist*, 27(3), 329–335.

Suthar, S. (2010). Evidence of plant hormone like substances in vermiwash: An ecologically safe option of synthetic chemicals for sustainable farming. *Ecological Engineering*, 36(8), 1089–1092.

Tattini, M., Bertoni, P., Landi, A., and Traversi, M. L. (1990,). Effect of humic acids on growth and biomass partitioning of container-grown olive plants. In *II Symposium on Horticultural Substrates and Their Analysis, XXIII IHC 294* September (pp. 75–80).

Thakur, S., and Sood, A. K. (2019). Lethal and inhibitory activities of natural products and biopesticide formulations against *Tetranychusurticae* Koch (Acarina: Tetranychidae). *International Journal of Acarology*, 45(6–7), 381–390.

Tharmaraj, K., Ganesh, P., Kolanjinathan, K., Suresh Kumar, R., and Anandan, A. (2011). Influence of vermicompost and vermiwash on physico chemical properties of rice cultivated soil. *Current Botany*, 2(3).

Tudi, M., Daniel Ruan, H., Wang, L., Lyu, J., Sadler, R., Connell, D., … and Phung, D. T. (2021). Agriculture development, pesticide application and its impact on the environment. *International Journal of Environmental Research and Public Health*, 18(3), 1112.

Valdrighi, M. M., Pera, A., Agnolucci, M., Frassinetti, S., Lunardi, D., and Vallini, G. (1996). Effects of compost-derived humic acids on vegetable biomass production and microbial growth within a plant (*Cichoriumintybus*)—soil system: A comparative study. *Agriculture, Ecosystems and Environment*, 58(2–3), 133–144.

Varghese, S. M., and Prabha, M. L. (2014). Biochemical characterization of vermiwash and its effect on growth of capsicum frutescens. *Malaya Journal of Biosciences*, 1(2), 86–91.

Verma, S., Babu, A., Patel, A., Singh, S. K., Pradhan, S. S., Verma, S. K., … and Singh, R. K. (2018). Significance of vermiwash on crop production: A review. *Journal of Pharmacognosy and Phytochemistry*, 7(2), 297–301.

Yadav, K. D., Tare, V., and Ahammed, M. M. (2010). Vermicomposting of source-separated human faeces for nutrient recycling. *Waste Management*, 30(1), 50–56.

Zambare, V. P., Padul, M. V., Yadav, A. A., and Shete, T. B. (2008). Vermiwash: biochemical and microbiological approach as ecofriendly soil conditioner. *ARPN Journal of Agricultural and Biological Science*, 3(4), 1–5.

13 Phytoremediation of Mine Tailings

Biju P. Sahariah,[1,*] *Tanushree Chatterjee,*[2]
Jyoti K. Choudhari,[1,2] *Mukesh K. Verma,*[1,3]
Anandkumar J.,[3] *and Jyotsna Choubey*[2]
[1]Chhattisgarh Swami Vivekanand Technical University (CSVTU), Bhilai, Chattishgarh, India
[2]Raipur Institute of Technology, Raipur, Chattishgarh, India
[3]National Institute of Technology, Raipur, Chattishgarh, India
*Corresponding author Email: biju.sahariah@gmail.com

CONTENTS

13.1 Introduction 243
13.2 Composition and Treatment of Mine Tailings 244
- 13.2.1 Bioremediation Techniques 244
- 13.2.2 Phytoremediation 245
- 13.2.3 Phytoremediation Factors, Mechanisms and Approaches 245
- 13.2.4 Screening Criteria of the Potential Phytoremediation Agent 246
- 13.2.5 Phytoremediation Mechanisms and Strategies 246
- 13.2.6 Phytoremediation Approaches in View of MTs Treatment 248

13.3 Advantages and Limitations of Phytoremediation of Mine Tailings 249
13.4 Conclusion 252
References 252

13.1 INTRODUCTION

The fine waste material, generally from crushed rock, dumped soil, water or trace metal elements left after the target mineral extraction from ores in mining sites is termed mine tailings (MTs). The removal of the top soil horizon containing microbes, nutrients, seeds as well as humic substances initiates the mining operations. The presence of heavy metals in high amounts, low levels of micronutrients, organic/inorganic carbon-based material, and poor water-holding efficiency associated with fluctuating pH are general mine tailing conditions (Li et al., 2019b; Tardif et al., 2019; Wang et al., 2017). The impact of mine tailings contaminants and their disposal sites is responsible in terms of damage to environmental and population health, as unclaimed mining sites generally remain unvegetated for tens to hundreds of years, and exposed tailings can spread over tens of hectares via aeolian dispersion and water erosion intruding on human habitats. Acid mine drainage (AMD) generation from leachates of the MTs is another major concern related to mine tailings. AMD seriously affect the ecosystem, leading to a state of distinct absence of vegetation, which accelerates the migration of metals to the adjacent ecosystems (Gagnon et al., 2020b; Khoeurn et al., 2019). In three Norwegian mine tailings, fine particles revealed maximum toxicity during contact with sediments when released into sea water (Brooks et al., 2019). The presence and properties of particles greatly affect the soil and water composition of the areas near the mine sites whereas the same is also governed by the physicochemical property of the site (Miller et al., 2018).

DOI: 10.1201/9781003239956-15

13.2 COMPOSITION AND TREATMENT OF MINE TAILINGS

The extracted minerals from a mining site are treated for their high value, but the mine tailings are often ignored for natural activities or considered for very low treatment and valorization. The contaminant composition of MTs varies with the parent ore. However, toxic heavy metals and many organic/inorganic pollutants commonly result in heterogeneous pollutant complexes in MTs. The area is often characterized by heavy metals such as lead (Pb), cadmium (Cd), chromium (Cr), mercury (Hg), nickel (Ni) and arsenic (As), etc. in high amounts, the components used in processing the mining activity (Hg, As, etc.), low levels of micronutrients, carbon-based material, and poor water-holding efficiency associated with fluctuating pH (Li et al., 2019b; Tardif et al., 2019; Wang et al., 2017). Polyaromatic hydrocarbons (PAH), phenolics, nitrogen compounds (for example, ammonia), sulphur compounds, phosphorus, and thiocyanate, etc. are often traced in MTs (Falah et al., 2020; Gagnon et al., 2020b; Wang et al., 2017). The toxicity and its health/ environment impact of MTs components such as heavy metals, organic hydrocarbons, cyanide, etc. are well recognized. There is awareness of the toxicity and environmental risk raised by these pollutants, therefore, a reasonable, cost-effective and environmentally-friendly refinement system is required to treat them.

In the current scenario, due to the environmental consciousness of citizens and awareness regarding the negative impacts of MTs on populations and the environment, the practice of treating mine tailings has started and aqueous tailings dumping is no longer considered the solution. Also, the application of the 3R process (reducing, recycling, and re-using) of metals from the tailings instead of the unsafe practice of careless disposal is now important. Emphasis on the treatment through physical encapsulation, adsorption, metal recovery or other chemical and biological treatment processes for mine tailing reclamation is continually growing (Calderon et al., 2020; Demir and Derun, 2019; Falah et al., 2020; Kiventerä et al., 2019). The biological treatment process is used more than the other treatment options for mine tailings due to its natural phenomenon, its efficiency, being less prone to generate unwanted secondary pollutants, its complete mineralization of numerous organic and inorganic pollutants and its cost-effectiveness. A few significant studies of mine tailing are listed in Table 13.1 focussing on the treatment process and associated conditions.

13.2.1 Bioremediation Techniques

The bioremediation units are generally living or parts of dead biomass, for example, microbes (bacteria, fungi), algae and a number of plant species that possess the ability to bioremediate specific pollutants (Kumar et al., 2022, 2021) .The biological treatment process is significant due to its ability to achieve pollution abatement in a cost-effective way (Kumar, 2018; Kumar et al., 2020, 2018). High performance, eco-friendliness, its renewable nature, the low need for technical support, low maintenance and lack of chemicals, etc. are significant attractive factors offered by biological aspects during the bioremediation of various pollutants (Chandra et al., 2018; Chandra and Kumar, 2018; Kumar, 2021; Kumar and Chandra, 2020). The introduction of microbes or plant species in living or dead forms to the mine tailings can improve the quality of MTs through active interaction, can enhance the organic content of the impoverished area in terms of organic matter deposition and limit the migration of the metals into the environment by binding them in organic matters (Gagnon et al., 2020b; Munford et al., 2020). The bioremediation of pollutants from mine tailings follows biodegradation (for organics and inorganics, other than metal); and bioaccumulation, bioleaching and biosorption of the metal ions by various living and dead cells of microbes as well as the plant/ animal community.

During biodegradation, the agents derive the necessary energy and nutrients from the contaminants while transforming the latter into their simpler and elemental form. The bioaccumulation process is governed by either metabolism-independent extracellular adsorption or metabolism-dependent intracellular accumulation. Bioleaching or brine leaching of contaminants involves a direct (straight incorporation with metal) or an indirect mechanism (initiation of by-product generation that incorporates the metal) in suitable conditions to recover metals from low grade and concentration ore (Borja et al., 2019; Ngoma et al., 2018). Biosorption is either metabolically mediated or uses physico-chemical pathways in certain biomass (alive or dead), where heavy metals are accumulated through uptake and passively concentrated. The components of the cell and the spatial orientation of the cell wall greatly govern the biosorption of a metal. It is a passive phenomenon offered by the naturally present functional biomass where metal is confiscated by the chemical sites, even if the biomass is dead. These are mostly influenced by physical agents (electrostatic interaction or van der Waals forces) or chemical factors (metal ions displacement), binding, chelation, reduction, precipitation, and complexation, etc. Chemical and functional groups present in biosorbents, namely polysaccharides (alginate, chitin, cellulose, and glycan), amine, amide, imidazole, thioether, sulfonate, carbonyl, sulfhydryl, carboxyl, phosphodiester, phenolic, ester, and phosphate groups attract and impound metal ions (Vargas-García et al., n.d.; Velásquez and Dussan, 2009). Humic substances significantly influence the mobility and bioavailability of metals or metalloids by incorporating them via binding metal ions with various functional groups present in the humic substances (Ouni et al., 2014; Sheoran et al., 2016). Biochar is reported as highly effective in alleviating the toxicity risk of mine tailing though sometimes it shows contradictory behaviour through increasing or decreasing the bioavailable concentrations, depending on the presence of metal(loid)s in the mine tailings (Forján et al., 2016; Gu et al., 2020).

13.2.2 Phytoremediation

In a broad sense, the term "phytoremediation" means using plants and their associated microorganisms for remediation of contaminants and their toxicity from the environment (Greipsson, 2011). In general, any body part of the plant species may be useful for phytoremediation where roots act for uptake and other parts act as storage, detoxification of the contaminants or any necessary task required. The roots of the plants possess mechanisms to prevent toxicity and mainly play a role in uptake and accumulation of contaminants while extracting water and other essential nutrient elements from the soil. The root system with an enormous surface area balances the soil erosion, creates rhizospheres, secretes phytosiderophores (metal-solubilizing substances), drives the rhizosphere soil pH to be favourable for heavy metal desorption, enhances microbial diversity, so high microbial siderophores interactions increase the uptake of labile elements and hence limit contaminant migration. The shoots and other above-ground parts provide a storage place for the translocated elements and physically obstruct wind erosion, limiting the migration of contaminants into the atmosphere. Phytoremediation is recognized as a solar-driven process executed by plant biomass efficient in the stabilization of soil pH, and can enhance metalloid encapsulations, prevent soil erosion and the stimulation of microbial activity (Brevik and Sauer, 2014; Muñoz-Rojas, 2018). The revegetation of mine tailings for reclamation is based on the concept of phytoremediation. Planting robust plant species that can survive in poor soils can help in the deposition and enrichment of the organic matter and humic substances in the area, resulting in a boost of other organisms useful in that location. Direct revegetation of suitable species for phytoremediation of metalloids through reducing their mobility is reported to be feasible on sulfidic tailings (Xie and van Zyl, 2020). This chapter focuses on phytoremediation to treat MTs, describing its mechanisms, various approaches, the effective species and their favourable conditions to help researchers and mining professionals choose accurate methods for their respective MTs sites.

13.2.3 Phytoremediation Factors, Mechanisms and Approaches

The success of the phytoremediation process on the applied site depends on the selection of potential plants capable of remediation of certain pollutants existing in the site. According to the available conditions, plants apply their potential for growth and maintenance by providing phytoremediation of the site as a by-product, a highly beneficial aspect for the environment's health. Phytoremediation for a site starts with the selection of suitable efficient species and a better understanding of the basic mechanisms followed by plants for remediation.

13.2.4 Screening Criteria of the Potential Phytoremediation Agent

The halophytes (salt-tolerant plants) and metallophytes (metal-tolerant plants) are the plants most selected for phytoremediation. A few basic qualifications of plants species for the phytoremediation process are: (1) easy farming, fast-growing and high above ground biomass; (2) widely distributed root network; (3) tolerance of high toxic effects, translocation and accumulation of the target contaminants; (4) upright acclimation to existing immediate conditions; and (5) high resistance to pathogens/pests and repulsion of herbivores that would contaminate the food chain if they ingested the heavy metals..

The ultimate goal of phytoremediation is to remove the toxic contaminants and their effects on the mining sites. on the way to establishing an ecosystem with a suitable plant community complementing the soil quality improvement, microbial activity/diversity and reclamation of the ecosystem towards autarky. The two key measuring units of phytoremediation quantifications of metal/contaminants in plants and the selection of plants for the process are the bioconcentration or bioaccumulation factor (BF) and the translocation factor (TF) or shoot:root ratio (S:R). BF signifies the ratio of total concentration of contaminants in shoot tissue (x) to total contaminants concentrations in the soil (y) (Eid et al., 2020; Ladislas et al., 2012; Padmavathiamma and Li, 2007). TF is the ratio of total concentration of contaminants in shoot tissue (x) to total element concentration in the root tissue (z). In general, the value of TF or BF remains below one and exceeding value "1" recommends plants be only favourable for phytoaccumulation and not accepted for phytostabilization (Brooks et al., 1998). With reference to the specification of the MTs, suitable remediation agent selection is required where a high tolerance range of environmental stress, high adaptability, rapid growth and the requirement of less maintenance are highly desirable criteria, to a great extent, according to the available conditions for MTs' treatment.

13.2.5 Phytoremediation Mechanisms and Strategies

The species involved in phytoremediation employ a degradation/accumulation or sorption process, according to the exposed contaminants. The root is the principal contributor for the uptake of contaminants and follows the same mechanisms as that of essential micronutrient uptake from the environment. The plants generate various molecules such as organic acids, metal-mobilizing compounds (for example, phytosiderophores, carboxylates, etc.), and other chelating agents to bind foreign elements for uptake. These conditions are accompanied by plant-induced changes in pH and redox potential that enhance the solubilization of micronutrients in the soil that are not readily bioavailable. The uptake is regulated by various specialized proteins located in the plant cell plasma membrane; these act as a proton pump and co- or anti-transporter to generate electrochemical gradients for active ion uptake and also as channels for transporting ions to the cells (Tangahu et al., 2011). The evapotranspiration process, i.e., evaporation of water from the plant leaves, also facilitates the pumping of nutrients and other substances from the soil into the roots as well as translocation to the shoots. The uptake of contaminants or soil substances is influenced by a few factors, such as plant species with root zone and uptake capacity; and soil with environmental characteristics and the bioavailability of contaminants (Salido et al., 2010; Tangahu et al., 2011). Following the uptake of the contaminants, specialized transporters, mostly channel proteins, and H+ coupled carrier proteins located in the root cells, are responsible for influx-efflux translocation of contaminants from roots to shoots (DalCorso et al., 2019).

Plants follow two basic strategies as defence mechanisms primarily against metal contaminants: (1) avoidance/excluders; and (2) tolerance/accumulators while exposed to toxic

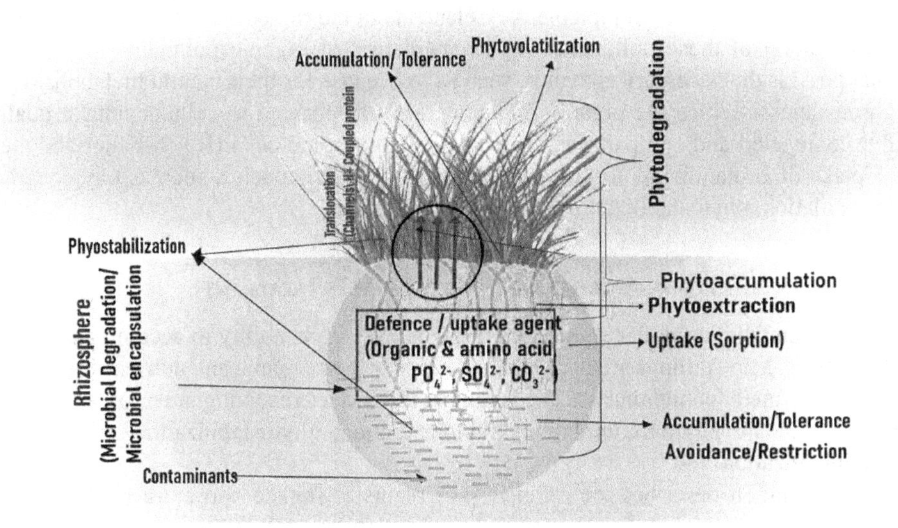

FIGURE 13.1 Phytoremediation approaches in polluted soil.

contaminates (Figure 13.1). These play a significant role in the phytoremediation mechanism. In the avoidance or the excluder process, plants restrict or limit the uptake of contaminants into their roots through the secretion of various root exudates in the form of organic acids or amino acids and cell wall pectins. These factors work as ligands for immobilization or chelation of the contaminant in the rhizosphere, modification or precipitation of metal ions, embedment in the cell wall and root sorption, etc. and restrict the entry of contaminants into the cells. Plants adopt a tolerance or accumulation strategy at the intracellular level for their survival against toxicity following the entry and accumulation of the contaminant into the cytosol. Plants accomplish minimization or detoxification of the accumulated ions by organic and inorganic ligands, such as organic acids, amino acids, phytochelatins (PCs), or metallothioneins (MTs), in the cytoplasm and cell wall, proteins/pectins/polyphenols capable of inactivation, chelation and compartmentalization, etc. of the contaminant (Dalvi et al., 2013; Gupta et al., 2013; Viehweger, 2014; Yan et al., 2020). Yan et al. (2020) in their thorough review of the phytoremediation process reported a few metal transporters families, such as ZIP (ZRT–IRT-like proteins), HMAs (heavy metal transporting ATPases), MTPs (metal transporter proteins), and NRAMPs (naturally resistant associated macrophage proteins) with sequence homology having a vital role in metal uptake, transport, homeostasis and tolerance. For example, the ZIP family is associated with heavy metal cations Fe, Mn, and Zn and P1B-type ATPases of HMAs are involved in Zn^{2+}, Cd^{2+}, Co^{2+}, and Pb transportation; the MTP family regulates the translocation of Zn^{2+} and Ni^{2+}, and NRAMPs facilitate the transportation of Cu^{2+}, Mn^{2+}, Co^{2+}, Fe^{2+}, and Cd^{2+}, etc.

In the case of metal and other inorganic contaminants, cells sequestrate the metal-chelator complex following the approaches of phytoaccumulation, phytostabilization, phytovolatilization or phytoextraction in the cellular parts of the plant and convert the contaminants to an inactivate form through compartmentalization aided by various diffusion facilitators (CDFs), cation exchangers (CAXs), and magnesium exchangers (He et al., 2012). Various plant species enhance the growth of microbes in their rhizospheres, which are efficient in bioremediation. Simultaneously, a number of microbes possess favourable characteristics such as stimulation of root proliferation, protection against pathogens, widening of the absorptive surface area of plant roots, increasing the plant tolerance to heavy metals, regulating the bio-availability of metal ions, improving plant nutrient or heavy metal uptake and translocation in the plant species.

In the case of organic contaminants, microbes living in the rhizosphere facilitate mineralization or extraction of the contaminants (rhizodegradation). The enhancement of microbial diversity by the roots through the secretion of complementary elements in the rhizosphere significantly contributes to the remediation of these pollutants. Phytoremediation of organic pollutants requires suitable plants that possess the necessary enzymes, such as oxygenase for their natural metabolism process. Through enzyme activities the organic pollutants are transformed to cellular uptake quality and thereafter conjugated and compartmentalized in the appropriate cells (Reichenauer and Germida, 2008). Uptake of contaminants to the plant is followed by approaches such as phytodegradation, phytoaccumulation, phytostabilization or phytovolatilization.

13.2.6 Phytoremediation Approaches in View of MTs Treatment

The key phytoremediation potential of a plant species lies in its ability to accumulate or eliminate the contaminants. Mine tailings are often characterized by low organic and nutrient contents and the presence of high metal contaminants. The foremost techniques receiving attention for contaminant phytoremediation are phytoaccumulation, phytoextraction, phytostabilization, phytodegradation and phytovolatilization.

Phytoaccumulation describes the efficiency of plants to uptake and extract the contaminants from soil via their roots, then translocate and accumulate them in other parts of the plant body. Various parts of the plant take part in the accumulation process to different extents. Many plant species are recognized to uptake and immobilize (phytostabilization) the contaminants, reducing their movement and bioavailability, depending on a number of factors, such as bioavailability and the condition of the metals, and the property of the soils. Phytostabilization is often carried out by the process of sorption by the roots, precipitation, complexations and metal valence reduction in the rhizosphere, accompanied by special redox enzymes. Phytostabilization is also recognized for its efficiency in converting toxic contaminants (heavy metals) to a comparatively less lethal state, decreasing possible metal stress, damage and leaching (Gagnon, Rodrigue-Morin, Tardif, et al., 2020; Gagnon, Rodrigue-Morin, Migneault et al., 2020; Xie and van Zyl, 2020). This prevents the migration of contaminants to pollute ground water and access the food chain.

Phytoextraction denotes the uptake of contaminants by plant roots followed by accumulation and translocation of the same at the shoots suitable for harvesting. A common difference between phytoextraction and phytostabilization is that the former is mostly associated with accumulation of the contaminants in its body parts, either shoots, leaves or roots whereas the latter principally works by confiscation of the contaminants within the rhizosphere instead of plant tissues (Yan et al., 2020; Zhang et al., 2012).

Phytovolatilization denotes the process of plant species transforming the contaminants, predominantly organic pollutants and a few heavy metals (for example, Se and Hg), into a volatile form released into the atmosphere following their uptake (Reichenauer and Germida, 2008). However, the plant can only subsequently transfer the contaminants from the soil to the atmosphere and deposition instead of complete remediation. Phytodegradation is the capacity of plants species to degrade contaminants where xenobiotics are accumulated and detoxified through metabolic activities with the help of various enzymes, such as dehalogenase and oxygenase. Also, researchers and mining officials have been continuously focussing on the phytomining approach to extract valuable elements, such as gold, platinum, etc. present in soil in trace concentrations (Brooks et al., 1998).

Considering the large areas covered in mining activities and MTs, their poor organic or nutrients availability and the presence of metal complexities, phytoremediation is a suitable, environmentally-friendly, solar-driven approach (Wang et al., 2017). The vegetation community pattern on the edge and in open mine tailings areas used to be significantly different, though not significantly correlated with metal concentrations, such as arsenic, but pH and nutrients (carbon and phosphorus) and organic matters were present (Munford et al., 2020). Tardif et al. (2019) reported substantial vegetation in vegetation surveys on a long-term basis (~50 years) in infertile and stressful substrates of gold mine tailings storage area. The slow succession is influenced by human disturbance rather than plant species and substrate composition and environmental elements. Bacterial diversity enhancement and community development in an acid mining area were made possible by long-term Lespedeza bicolour revegetation (Wu et al., 2018). *Pongamiapinnata* improved the soil condition in Vanadium-titanium magnetite tailings with an abundance of soil proteobacteria, enhancement of soil nitrogen fixation after legume remediation with keystone genera Rhizobium and Nordella during a soil community shift (Yu et al., 2019). Xie and van Zyl (2020), while applying multiple species, observed that direct revegetation for efficient phytostabolization of metal(loid)s can be established on sulfidic tailings if amended suitably with thorough geochemical knowledge and suitable species are selected. The phytoextraction potential of *Solanum nigrum* through cationic exchange or adsorption is influenced by the metal species and the temporal variation in a multi-metal contaminated mine tailings; however, phytoextraction improves with the addition of soil amendments (attapulgite and biochar) (Li et al., 2019b).

Researchers identified plant species, namely, *Sapiumsebiferum*, *Salix matsudana*, *Hibiscus cannabinus*, *Corchorus capsularis*, *Ricinus communis*, and *Populusnigra* as high Pb and Zn tolerant species, with notable growth characteristics and capacities to accumulate Pb and Zn from the mine tailings (Becerra-Castro et al., 2012; Laxman Singh et al., 2014; Tang et al., 2020). Due to the wide diversity of the plant community and its efficiency in metal and metalloid accumulation capacity, phytoremediation is intensively used for mine tailing treatment and a few studies are mentioned in Table 13.1.

13.3 ADVANTAGES AND LIMITATIONS OF PHYTOREMEDIATION OF MINE TAILINGS

Due to the complexity of the mine tailings pollutant components, the vastness of its area occupied, a single physical/chemical option or ex-situ treatment often fails to provide suitable results. Therefore, researchers are still working on various configurations of treatment options, such as individual or combined physico-chemical-biological processes to attend to toxic contaminants to reach a non-polluting harmless, dischargeable point as well as restoration and reclamation of MTs. A large number of plants species offer phytoremediation efficiency for handling multiple pollutants. The primary advantages of adopting the phytoremediation technique for MTs include: (1) it is suitable for cleaning up large-scale areas; (2) it is environmentally friendly as it preserves the top soils, improves the soil quality; and (3) it is applicable for a wide range of contaminants (mostly metals). Phytoremediation is an aesthetic, cost- effective, solar-driven process and requires low maintenance compared to many reactive soil remediation technologies, for example, incineration, thermal vaporization, solvent washing, etc. where damage to the biological components and alteration of the physical-chemical components of the soils are inevitable. Phytoremediation is highly dependent on MTs' components and the correct selection of suitable phytoremediation agents with high tolerance and biomass production property. Major limitations are: (1) it is a slow process not favourable for acute risk exposure; and (2) disposal of the plant species after phytoremediation of the sites needs to be considered due to the phytoremediation mechanisms of the plant species.

Scholars suggest that improving specific traits of phytoremediation plant species can help to overcome the limitations of slow growth in poor soil quality. Traditional breeding for somatic hybridization and genetic engineering to create transgenic plants through the fusion of high metal-accumulative and tolerant species with fast-growing plant species are highly effective and appreciated. Genetic engineering should be further used to improve the factors responsible for growth, the proteins involved in translocation/accumulation and the tolerance at the molecular level of the species. Also, modification of the rhizosphere by the addition of competent plant growth-promoting microbes can extensively improve the phytoremediation efficiency of the plant species.

TABLE 13.1
Contaminants in Mine Tailings and Plants Species with Their Phytoremediation Mechanisms

S. No.	Contaminants (MTs type)	Phytoremediation mechanism and plant species	Remark	Reference
1.	Al, Mn, Fe, U, V, Fe, Ni, Cu, Zn, Co and Se (Uranium MTs)	Phytoaccumulation: *Pterisvittata* *Physostegia digitalis, Cyperus compressus, Saccharum spontaneum,* *Crotonbun plandianus, Bacopamoneri,* *Calotropis procera* and *Azadira chtaindica*	*P. vittata* accumulates Al, V, Ni, Co, Se and U simultaneously followed by *P. digitalis, C. compressus* and *S. spantanium*. *C. bunplandianus* can accumulate Al, Mn, Fe, Ni, Co, Cu, Zn, Se and U simultaneously followed by *B. moneri, C. procera* and *A. indica*.	(Laxman Singh et al., 2014)
2.	Lead and zinc (Lead–zinc MTs)	*Chenopodium ambrosioides*	*C. ambrosioides* can support different metal-resistant bacterial communities *Paenibacillusjamilae* HTb8 and *Pseudomonas sp.* GTa5 in their rhizosphere.	(Zhang et al., 2012)
3.	Hu (Gold mine MTs)	Phytoextraction *Manihot esculenta* Crantz (cassava) to phytoextract Hg and Au	The best substrate contained 75% biosolids and 25% Au mine tailings. Fibrous roots accumulated up to 12.59 g kg−1 Hg and 18.99 mg kg−1 Au. And higher Hg concentrations suppressed phytoextraction of Au.	(Alcantara et al., 2017)
4.	Zn, Cr, Cu and Mn (Gold MTs)	Vetiver grass (*Vetiveria zizanioides* L. Nash).	Amending mine tailings with RHA together with NPK fertilizer application improved the chemical properties of mine tailings.The vetiver grass is recommended for stabilization of Zn and phytoextraction of Cr, Cu and Mn.	(Samsuri et al., 2019)
5.	Al and Pb (Metal MTs)	Phytoextraction *Brassica juncea* L. (Indian mustard) and plant growth-promoting bacteria such as Serratia K120, Enterobacter K125, Serratia MC107, Serratia MC119 and Enterobacter MC156	Roots of considered plant favour the transfer of metals to the plant, mainly Al and Pb from the eight analysed metals with bioaccumulation factors >1 for Al, Pb, Cd and Fe	(Mendoza-Hernández et al., 2019)
6.	Pb, Cd, Cu, Zn and Mn (Multi-metal MTs)	Phyto-stabilization Soybean, *Mucor circinelloides*, and A3 amendment (organic fertilizer: rice husk: biochar: ceramsite = 1:1:2:1)	Phyto-stabilization is enhanced with biochar and the removal rates of soil heavy metals are in the order of Pb>Cd>Cu>Zn>Mn	(Li et al., 2019a)
7.	Mn, Pb, Mn, and Zn (Copper MTs)	Phytoaccumulation *Prosopis tamarugo, Schinusmolle* and *Atriplexnum mularia* and arbuscular mycorrhizal fungi and organic and inorganic amendments	*P. tamarugo* and *A. nummularia* significantly accumulated Mn, Pb and Zn whereas *S. molle* showed accumulating characteristics of Cu, Mn, Pb, and Zn.	(Lam et al., 2017)

TABLE 13.1 (Continued)
Contaminants in Mine Tailings and Plants Species with Their Phytoremediation Mechanisms

S. No.	Contaminants (MTs type)	Phytoremediation mechanism and plant species	Remark	Reference
8.	As, Cu, Pb and Zn (Multi-metal contaminated MTs)	Phytostabilization *Penicillium aculeatum* PDR-4 and Trichoderma sp. PDR-16, isolated from the rhizosphere of *Pinus koraiensis* in mine tailing	Granted significant As, Cu, Pb and Zn tolerance and plant growth-promoting traits along with high phosphate and heavy metals mobilization and is better out during co-inoculation is more effective over single isolates for phytoremediation and biomass production	(Babu et al., 2014)
9.	Zn, Pb, Cd (Tailing pond)	Phytoaccumulation *Atriplexhalimus* and marble waste ($CaCO_3$; MW), pig slurry (PS) and their combination (MW + PS) as soil amendments	Grow of dominant species and accumulation of metals in leaf higher than in stem and root Significant increase of DTPA-extractable Zn in MW and MW + PS plots, Pb in all treatments except MW plot, Cd only in PS plot, and Cu only in MW + PS plot	(Acosta et al., 2018)
10.	Ni, Zn and Cd (Tailing heaps)	Phytostabilization *Ricinus communis* (Castor bean)	Castor bean plants Phyto-stabilize mine residues but do not accumulate high concentrations of Ni, Zn nor Cd	(Ruiz Olivares et al., 2013)
11.	Pb, Zn, Cd (Multi-metal contaminated MTs)	Phytoaccumulation Ectomycorrhizal (ECM) *species Pinus sylvestris (P. sylvestris)* *Pisolithus sp.1* and *Pisolithus sp.2;* *& Cenococcum geophilum (Cg)-. Laccaria sp.* (L1)- ECM, and non-ectomycorrhizal (NM) *P. sylvestris*	ECM *P. sylvestris* has high survival rate and better growth parameters and significantly accumulated much more Cd, Pb, and Zn with high soil enzyme activity than NM seedlings	(Liu et al., 2020)
12.	Pb, Cu and Cr (Copper MTs)	Phytostabilization Eleven Indigenous plants including *Solanum viarum* and *Baccharistrimera*	The plants found exhibited high tolerance to toxic metals and most of them possess phytoextraction potential of Pb or Cu. Among them *S. viarum* Dunaland B. trimera showed highest phytoextraction concentrations of Cr and Cu.	(Afonso et al., 2020)
13.	Pb, Zn and sulphide (Zn-Pb MTs)	Phytoaccumulation *Biscutella laevigata*	The *Biscutella laevigata* accumulated higher metal concentration in the tissue, especially Tl and, to a lesser extent, Zn, Pb, Fe and Mn, and it can influence metal mobility in the rhizo-soil	(Pavoni et al., 2017)

13.4 CONCLUSION

Mine tailings are a complex package of various pollutants of organic and inorganic origin, dominated by the major ore with a high potential to be an environmental health hazard. The presence of toxic heavy metals and complex organic compounds makes the treatment process even more challenging. Bioremediation is an efficient and cost-effective treatment process applied to mining area reclamation to deal with the mine tailings. The presence of heavy metals drastically influences the treatment of organics/inorganics and vice versa due to the inhibition mechanism. Phytoremediation, being a natural process, is certainly a suitable option to handle the widely distributed MTs sites with proper characterization of the MTs, selection of efficient native plant species and can provide the necessary conditions for acclimation of the plants in the initial stages. The symbiosis process, the selection of a suitable treating bio-agent and providing the optimum environmental and operational conditions can achieve an effective treatment result in a barren mine tailings site and restore a healthy ecosystem. However, compared to the vastness in areas, the loss of land and the negative impacts from MTs on the environment soil/water/atmosphere, phytoremediation is still the most suitable option for MTs, provided there is a detailed investigation of the efficient plant species in terms of metabolic and growth response and genetic level.

REFERENCES

Acosta, J.A., Abbaspour, A., Martínez, G.R., Martínez-Martínez, S., Zornoza, R., Gabarrón, M., and Faz, A., 2018. Phytoremediation of mine tailings with Atriplex halimus and organic/inorganic amendments: A five-year field case study. *Chemosphere* 204, 71–78. https://doi.org/10.1016/j.chemosphere.2018.04.027

Afonso, T.F., Demarco, C.F., Pieniz, S., Quadro, M.S., Camargo, F.A.O., and Andreazza, R., 2020. Bioprospection of indigenous flora grown in copper mining tailing area for phytoremediation of metals. *Journal of Environmental Management* 256, 109953. https://doi.org/10.1016/j.jenvman.2019.109953

Alcantara, H.J.P., Doronila, A.I., and Kolev, S.D., 2017. Phytoextraction potential of Manihot esculenta Crantz. (cassava) grown in mercury- and gold-containing biosolids and mine tailings. *Minerals Engineering* 114, 57–63. https://doi.org/10.1016/j.mineng.2017.09.010

Babu, A.G., Shim, J., Shea, P.J., and Oh, B.T., 2014. Penicillium aculeatum PDR-4 and Trichoderma sp. PDR-16 promote phytoremediation of mine tailing soil and bioenergy production with sorghum-sudangrass. *Ecoogical Engineering* 69, 186–191. https://doi.org/10.1016/j.ecoleng.2014.03.055

Becerra-Castro, C., Monterroso, C., Prieto-Fernández, A., Rodríguez-Lamas, L., Loureiro-Viñas, M., Acea, M.J., and Kidd, P.S., 2012. Pseudometallophytes colonising Pb/Zn mine tailings: A description of the plant-microorganism-rhizosphere soil system and isolation of metal-tolerant bacteria. *Journal of Hazardous Materials* 217–218, 350–359. https://doi.org/10.1016/j.jhazmat.2012.03.039

Borja, D., Nguyen, K.A., Silva, R.A., Ngoma, E., Petersen, J., Harrison, S.T.L., Park, J.H., and Kim, H., 2019. Continuous bioleaching of arsenopyrite from mine tailings using an adapted mesophilic microbial culture. *Hydrometallurgy* 187, 187–194. https://doi.org/10.1016/j.hydromet.2019.05.022

Brevik, E.C., and Sauer, T.J., 2014. The past, present, and future of soils and human health studies. *Soil Discussions* 1, 51–80. https://doi.org/10.5194/soild-1-51-2014

Brooks, R.R., Chambers, M.F., Nicks, L.J., and Robinson, B.H., 1998. Phytomining. *Trends in Plant Science* 3, 359–362. https://doi.org/10.1016/S1360-1385(98)01283-7

Brooks, S.J., Escudero-Oñate, C., and Lillicrap, A.D., 2019. An ecotoxicological assessment of mine tailings from three Norwegian mines. *Chemosphere* 233, 818–827.

Calderon, A.R.M., Alorro, R.D., Tadesse, B., Yoo, K., and Tabelin, C.B., 2020. Repurposing of nickeliferous pyrrhotite from mine tailings as magnetic adsorbent for the recovery of gold from chloride solution. *Resources, Conservation & Recycling* 161, 104971. https://doi.org/10.1016/j.resconrec.2020.104971

Chandra, R., Dubey, N.K., and Kumar, V., 2018. *Phytoremediation of Environmental Pollutants*. Boca Raton, FL: CRC Press.

Chandra, R., and Kumar, V., 2018. Phytoremediation: A green sustainable technology for industrial waste management. In Chandra, R., Dubey, N., and Kumar, V. (Eds), *Phytoremediation of Environmental Pollutants*. Boca Raton, FL: CRC Press. https://doi.org/10.1201/9781315161549-1

DalCorso, G., Fasani, E., Manara, A., Visioli, G., and Furini, A., 2019. Heavy metal pollutions: State of the art and innovation in phytoremediation. *International Journal of Molecular Science* 20, 3412. https://doi.org/10.3390/IJMS20143412

Dalvi, A., Dalvi, A.A., and Bhalerao, S.A., 2013. Response of plants towards heavy metal toxicity: An overview of avoidance, tolerance and uptake mechanism. *Annals of Plant Sciences* 2, 362–368.

Demir, F., and Derun, E.M., 2019. Modelling and optimization of gold mine tailings based geopolymer by using response surface method and its application in Pb^{2+} removal. *Journal of Cleaner Production* 237, 117766. https://doi.org/10.1016/j.jclepro.2019.117766

Eid, E.M., Galal, T.M., Sewelam, N.A., Talha, N.I., and Abdallah, S.M., 2020. Phytoremediation of heavy metals by four aquatic macrophytes and their potential use as contamination indicators: A comparative assessment. *Environmental Science and Pollution Research* 27, 12138–12151. https://doi.org/10.1007/s11356-020-07839-9

Falah, M., Ohenoja, K., Obenaus-Emler, R., Kinnunen, P., and Illikainen, M., 2020. Improvement of mechanical strength of alkali-activated materials using micro low-alumina mine tailings. *Construcion and Building Materials* 248, 118659. https://doi.org/10.1016/j.conbuildmat.2020.118659

Forján, R., Asensio, V., Rodríguez-Vila, A., and Covelo, E.F., 2016. Contribution of waste and biochar amendment to the sorption of metals in a copper mine tailing. *Catena* 137, 120–125. https://doi.org/10.1016/j.catena.2015.09.010

Gagnon, V., Rodrigue-Morin, M., Migneault, M., Tardif, A., Garneau, L., Lalonde, S., Shipley, B., ... Roy, S., 2020. Survival, growth and element translocation by 4 plant species growing on acidogenic gold mine tailings in Québec. *Ecological Engineering.* 151, 105855. https://doi.org/10.1016/j.ecoleng.2020.105855

Gagnon, V., Rodrigue-Morin, M., Tardif, A., Beaudin, J., Greer, C.W., Shipley, B., Bellenger, J.P., and Roy, S., 2020. Differences in elemental composition of tailings, soils, and plant tissues following five decades of native plant colonization on a gold mine site in Northwestern Québec. *Chemosphere* 250, 126243. https://doi.org/10.1016/j.chemosphere.2020.126243

Greipsson, S., 2011. Phytoremediation, *Nature Education Knowledge* 3(10), 7. http://sciepub.com/reference/278930

Gu, J., Yao, J., Duran, R., Sunahara, G., and Zhou, X., 2020. Alteration of mixture toxicity in nonferrous metal mine tailings treated by biochar. *Journal of Environmental Management* 265, 110511. https://doi.org/10.1016/j.jenvman.2020.110511

Gupta, D.K., Vandenhove, H., and Inouhe, M., 2013. Role of phytochelatins in heavy metal stress and detoxification mechanisms in plants. In Prasad, M.N.V. (Ed.), *Heavy Metal Stress in Plants* (pp. 73–94). Berlin: Springer.

He, S., He, Z., Yang, X., and Baligar, V.C., 2012. Mechanisms of nickel uptake and hyperaccumulation by plants and implications for soil remediation. *Advances in Agronomy* 117, 117–189. https://doi.org/10.1016/B978-0-12-394278-4.00003-9

Khoeurn, K., Sakaguchi, A., Tomiyama, S., and Igarashi, T., 2019. Long-term acid generation and heavy metal leaching from the tailings of Shimokawa mine, Hokkaido, Japan: Column study under natural condition. *Journal of Geochemical Exploration* 201, 1–12. https://doi.org/10.1016/j.gexplo.2019.03.003

Kiventerä, J., Piekkari, K., Isteri, V., Ohenoja, K., Tanskanen, P., and Illikainen, M., 2019. Solidification/stabilization of gold mine tailings using calcium sulfoaluminate-belite cement. *Journal of Cleaner Production* 239, 118008. https://doi.org/10.1016/j.jclepro.2019.118008

Kumar, V., 2018. Mechanism of microbial heavy metal accumulation from polluted environment and bioremediation. In Sharma, D., and Saharan, B.S. (Eds), *Microbial Fuel Factories.* Boca Raton, FL: CRC Press.

Kumar, V., 2021. Phytoremediation of distillery effluent: Current progress, challenges, and future opportunities. In Saxena, G., Kumar, V., and Shah, M.P. (Eds), *Bioremediation for Environmental Sustainability: Toxicity, Mechanisms of Contaminants Degradation, Detoxification and Challenges*. Oxford: Elsevier. https://doi.org/10.1016/B978-0-12-820524-2.00014-6

Kumar, V., Agrawal, S., Shahi, S.K., Singh, S., and Ramamurthy, P.C., 2022. Bioremediation potential of newly isolated *Bacillus albus* strain VKDS9 for decolourization and detoxification of biomethanated distillery effluent and its metabolites characterization for environmental sustainability. *Environmental Technology & Innovation* 26, 102260. https://doi.org/10.1016/j.eti.2021.102260

Kumar V., and Chandra R., 2020. Bioremediation of melanoidins containing distillery waste for environmental safety. In Bharagava, R., and Saxena, G. (Eds), *Bioremediation of Industrial Waste for Environmental Safety*. Singapore; Springer. https://doi.org/10.1007/978-981-13-3426-9_20

Kumar, V., Ferreira, L.F.R., Sonkar, M., and Singh, J. 2021. Phytoextraction of heavy metals and ultrastructural changes of *Ricinus communis* L. grown on complex organometallic sludge discharged from alcohol distillery. *Environmental Technology & Innovation* 22, 101382. https://doi.org/10.1016/j.eti.2021.101382

Kumar, V., Shahi, S.K., and Singh, S., 2018. Bioremediation: An eco-sustainable approach for restoration of contaminated sites. In Singh, J., Sharma, D., Kumar, G., and Sharma, N. (Eds), *Microbial Bioprospecting for Sustainable Development*. Singapore: Springer. https://doi.org/10.1007/978-981-13-0053-0_6

Kumar, V., Thakur, I.S., and Shah, M.P., 2020. Bioremediation approaches for pulp and paper industry wastewater treatment: Recent advances and challenges. In Shah, M.P. (Ed.), *Microbial Bioremediation & Biodegradation*. Singapore: Springer. https://doi.org/10.1007/978-981-15-1812-6_1

Ladislas, S., El-Mufleh, A., Gérente, C., Chazarenc, F., Andrès, Y., and Béchet, B., 2012. Potential of aquatic macrophytes as bioindicators of heavy metal pollution in urban stormwater runoff. *Water. Air.& Soil Pollution* 223, 877–888. https://doi.org/10.1007/s11270-011-0909-3

Lam, E.J., Cánovas, M., Gálvez, M.E., Montofré, Í.L., Keith, B.F., and Faz, Á., 2017. Evaluation of the phytoremediation potential of native plants growing on a copper mine tailing in northern Chile. *Journal of Geochemical Exploration* 182, 210–217. https://doi.org/10.1016/j.gexplo.2017.06.015

Laxman Singh, K., Sudhakar, G., Swaminathan, S.K., and Muralidhar Rao, C., 2014. Identification of elite native plants species for phytoaccumulation and remediation of major contaminants in uranium tailing ponds and its affected area. *Environment, Development and Sustainability* 17, 57–81. https://doi.org/10.1007/s10668-014-9536-7

Li, X., Wang, X., Chen, Y., Yang, X., and Cui, Z., 2019a. Optimization of combined phytoremediation for heavy metal contaminated mine tailings by a field-scale orthogonal experiment. *Ecotoxicology and Environmental Safety* 168, 1–8. https://doi.org/10.1016/j.ecoenv.2018.10.012

Li, X., Zhang, X., Wang, X., and Cui, Z., 2019b. Phytoremediation of multi-metal contaminated mine tailings with Solanum nigrum L. and biochar/attapulgite amendments. *Ecotoxicology and Environmental Safety* 180, 517–525. https://doi.org/10.1016/j.ecoenv.2019.05.033

Liu, B., Wang, S., Wang, J., Zhang, X., Shen, Z., Shi, L., and Chen, Y., 2020. The great potential for phytoremediation of abandoned tailings pond using ectomycorrhizal *Pinus sylvestris*. *Science of the Total Environment* 719, 137475. https://doi.org/10.1016/j.scitotenv.2020.137475

Mendoza-Hernández, J.C., Vázquez-Delgado, O.R., Castillo-Morales, M., Varela-Caselis, J.L., Santamaría-Juárez, J.D., Olivares-Xometl, O., Arriola Morales, J., and Pérez-Osorio, G., 2019. Phytoremediation of mine tailings by Brassica juncea inoculated with plant growth-promoting bacteria. *Microbiological Research.* 228, 126308. https://doi.org/10.1016/j.micres.2019.126308

Miller, K.A., Thompson, K.F., Johnston, P., and Santillo, D., 2018. An overview of seabed mining including the current state of development, environmental impacts, and knowledge gaps. *Frontiers in Marine Science*, online.. https://doi.org/10.3389/FMARS.2017.00418

Munford, K.E., Watmough, S.A., Rivest, M., Poulain, A., Basiliko, N., and Mykytczuk, N.C.S., 2020. Edaphic factors influencing vegetation colonization and encroachment on arsenical gold mine tailings near Sudbury, Ontario. *Environmental Pollution* 264, 114680. https://doi.org/10.1016/j.envpol.2020.114680

Muñoz-Rojas, M., 2018. Soil quality indicators: Critical tools in ecosystem restoration. *Current Opinions in Environmental Science and Health* 5, 47–52. https://doi.org/10.1016/j.coesh.2018.04.007

Ngoma, E., Borja, D., Smart, M., Shaik, K., Kim, H., Petersen, J., and Harrison, S.T.L., 2018. Bioleaching of arsenopyrite from Janggun mine tailings (South Korea) using an adapted mixed mesophilic culture. *Hydrometallurgy* 181, 21–28. https://doi.org/10.1016/j.hydromet.2018.08.010

Ouni, Y., Ghnaya, T., and Lakhdar, A., 2014. The role of humic substances in mitigating the harmful effects of soil salinity and improving plant productivity. *International Journal of Plant Production* 3, 353–374.

Padmavathiamma, P.K., and Li, L.Y., 2007. Phytoremediation technology: Hyper-accumulation metals in plants. *Water, Air, and Soil Pollution*, 184, 105–126. https://doi.org/10.1007/s11270-007-9401-5

Pavoni, E., Petranich, E., Adami, G., Baracchini, E., Crosera, M., Emili, A., Lenaz, D., Higueras, P., and Covelli, S., 2017. Bioaccumulation of thallium and other trace metals in Biscutella laevigata nearby a decommissioned zinc-lead mine (Northeastern Italian Alps). *Journal of Environmental Management*, 186, 214–224. https://doi.org/10.1016/j.jenvman.2016.07.022

Reichenauer, T.G., and Germida, J.J., 2008. Phytoremediation of organic contaminants in soil and groundwater. *ChemSusChem* 1(8–9), 708–717. https://doi.org/10.1002/cssc.200800125

Ruiz Olivares, A., Carrillo-González, R., González-Chávez, M. del C.A., and Soto Hernández, R.M., 2013. Potential of castor bean (*Ricinus communis* L.) for phytoremediation of mine tailings and oil production. *Journal of Environmental Management* 114, 316–323. https://doi.org/10.1016/j.jenvman.2012.10.023

Salido, A.L., Hasty, K.L., Lim, J.-M., and Butcher, D.J., 2010. Phytoremediation of arsenic and lead in contaminated soil using Chinese brake ferns (*Pteris vittata*) and Indian mustard (*Brassica juncea*). *International Journal of Phytoremediation*, 5, 89–103. https://doi.org/10.1080/713610173

Samsuri, A.W., Tariq, F.S., Karam, D.S., Aris, A.Z., and Jamilu, G., 2019. The effects of rice husk ashes and inorganic fertilizers application rates on the phytoremediation of gold mine tailings by vetiver grass. *Applied Geochemistry* 108, 104366. https://doi.org/10.1016/j.apgeochem.2019.104366

Sheoran, V., Sheoran, A.S., and Poonia, P., 2016. Factors affecting phytoextraction: A review. *Pedosphere* 26, 148–166. https://doi.org/10.1016/S1002-0160(15)60032-7

Tang, Y., Kang, H., Qin, Z., Zhang, K., Zhong, Y., Li, H., and Mo, L., 2020. Significance of manganese resistant bacillus cereus strain WSE01 as a bioinoculant for promotion of plant growth and manganese accumulation in Myriophyllum verticillatum. *Science of the Total Environment* 707, 135867. https://doi.org/10.1016/j.scitotenv.2019.135867

Tangahu, B.V., Sheikh Abdullah, S.R., Basri, H., Idris, M., Anuar, N., and Mukhlisin, M., 2011. A review on heavy metals (As, Pb, and Hg) uptake by plants through phytoremediation. *International Journal of Chemical Engineering* 2011. Online. https://doi.org/10.1155/2011/939161

Tardif, A., Rodrigue-Morin, M., Gagnon, V., Shipley, B., Roy, S., and Bellenger, J.P., 2019. The relative importance of abiotic conditions and subsequent land use on the boreal primary succession of acidogenic mine tailings. *Ecological Engineering* 127, 66–74. https://doi.org/10.1016/j.ecoleng.2018.11.003

Vargas-García, M.C., Suárez-Estrella, F., López, M.J., Guisado, G., and Moreno, J., 2010. *Bioremediation of Heavy Metals with Microbial Isolates*. Madrid: Ministerio de Ciencia y Technología.

Velásquez, L., and Dussan, J., 2009. Biosorption and bioaccumulation of heavy metals on dead and living biomass of *Bacillus sphaericus*. *Journal of Hazardoous Materials* 167, 713–716. https://doi.org/10.1016/j.jhazmat.2009.01.044

Viehweger, K., 2014. How plants cope with heavy metals. *Botanical Studies* 55, 1–12. https://doi.org/10.1186/1999-3110-55-35

Wang, L., Ji, B., Hu, Y., Liu, R., and Sun, W., 2017. A review on in situ phytoremediation of mine tailings. *Chemosphere* 184, 594–600. https://doi.org/10.1016/j.chemosphere.2017.06.025

Wu, Z., Yu, F., Sun, X., Wu, S., Li, X., Liu, T., and Li, Y., 2018. Long-term effects of Lespedeza bicolor revegetation on soil bacterial communities in Dexing copper mine tailings in Jiangxi Province, China. *Applied Soil Ecology* 125, 192–201. https://doi.org/10.1016/j.apsoil.2018.01.011

Xie, L., and van Zyl, D., 2020. Distinguishing reclamation, revegetation and phytoremediation, and the importance of geochemical processes in the reclamation of sulfidic mine tailings: A review. *Chemosphere* 252, 126446. https://doi.org/10.1016/j.chemosphere.2020.126446

Yan, A., Wang, Y., Tan, S.N., Mohd Yusof, M.L., Ghosh, S., and Chen, Z., 2020. Phytoremediation: A promising approach for revegetation of heavy metal-polluted land. *Frontiers in Plant Science* 359. Online. https://doi.org/10.3389/FPLS.2020.00359

Yu, X., Kang, X., Li, Y., Cui, Y., Tu, W., Shen, T., Yan, M., Gu, Y., Zou, L., … Chen, Q., 2019. Rhizobia population was favoured during in situ phytoremediation of vanadium-titanium magnetite mine tailings dam using Pongamia pinnata. *Environmental Pollution* 255, 113167. https://doi.org/10.1016/j.envpol.2019.113167

Zhang, W.H., Huang, Z., He, L.Y., and Sheng, X.F., 2012. Assessment of bacterial communities and characterization of lead-resistant bacteria in the rhizosphere soils of metal-tolerant Chenopodium ambrosioides grown on lead-zinc mine tailings. *Chemosphere* 87, 1171–1178. https://doi.org/10.1016/j.chemosphere.2012.02.036

14 Advanced Functional Approaches of Nanotechnology in Food and Nutrition

Sonia Morya,[1,*] *Chinaza Godswill Awuchi,*[2] *and Farid Menaa*[3]
[1]Department of Food Technology and Nutrition, School of Agriculture, Lovely Professional University, Phagwara, India
[2]School of Natural and Applied Sciences, Kampala International University, Kampala, Uganda
[3]Departments of Internal Medicine, Nanomedicine, and Advanced Technologies, Fluorotronics-CIC, La Jolla, San Diego, CA, USA
*Corresponding author Email: sonia.morya8911@gmail.com

CONTENTS

14.1 Introduction....257
14.2 Nanomaterials Applied in Developing Food Nanocomposite Polymers....259
14.2.1 Nanofibers (NFs)....259
14.2.2 Nanolayers (NLs)....261
14.2.3 Nanoparticles (NPs)....261
14.3 Nanocarriers for Food Ingredients....262
14.4 Nanocomposites' Applications in Food Packaging....263
14.4.1 Applications in Active Food Packaging....263
14.4.1.1 Active Packaging for Oxygen Scavenging....263
14.4.1.2 Anti-Microbial Food Packaging Applications....264
14.4.2 Improvement of Mechanical, Thermal, and Barrier Properties....266
14.4.3 Nanoparticles as Sensors in Smart Packaging....267
14.5 Nanosolutions Against Food Fraud....268
14.6 Conclusion....268
Acknowledgments....269
References....269

14.1 INTRODUCTION

In recent years, nanotechnology has found significant applications in food science, nutrition, and nutraceuticals (Menaa, 2015b). Nanotechnology can be used in a wide variety of food applications, such as in food safety for the detection of pathogens, in food security acting as smart, intelligent, and active packaging, in food processing for the encapsulation of bioactive, flavor enhancers, probiotics, etc. (Morya et al., 2016). In general, nanotechnology involves the design, production, characterization, and application systems at the nanoscale. It mostly embraces innovation and advances in engineered

DOI: 10.1201/9781003239956-16

materials, biological systems, electronic components, and instrumentation. The desired sizes range from the atomic level to 100 nm, as the properties of the materials at this level often vary remarkably in comparison to materials at larger scales (e.g., microscale, macroscale) (Batool et al., 2020). The ratio of surface area-to-volume of nanomaterials increases dramatically, resulting in an increase in the reactivity of chemical surfaces; the quantum mechanical effects also have relevance (Assadpour and Mahidi Jafari, 2019; Awuchi et al., 2022; Menaa, 2013; Hernández-Muñoz et al., 2019).

The functional advantages of nanoparticles (NPs) in food and nutrition systems are extensive. Food scientists and nutrition experts have explored their advantages and applications in different areas, including food packaging, nutrient delivery, and bioavailability, medicines, food formulation, catalysis, active packaging, optical improvement, sensory improvement, nutraceutical formulations, or delivery of bioactive compounds (Assadpour and Mahidi Jafari, 2019; Batool et al., 2020; Khan et al., 2018; Menaa, 2015b; Menaa et al., 2013; Menaa et al., 2014; Pyo et al., 2017).

Improvements in moisture and gas barriers in materials used for food packaging using nanofillers, nanosensors, intelligent packaging, and novel nanocomposite polymer films for anti-bacteria and anti-fungal applications are a few examples of the advantages of nanotechnology in improving the quality and safety of foods (Hernández-Muñoz et al., 2019). The nanoscale is essential when at least 50% of particles within a system have sizes below 100 nm (Pyo et al., 2017). According to IUPAC and other studies, the uppermost limit of NPs' size is believed to be 500 nm to obtain a stable emulsion and other desirable features (Khan et al., 2015; Menaa, 2014; Sharif et al., 2015; Solans et al., 2005).

Due to nanotechnology advances in various fields and areas of specialization, including food science, medicine, and pharmacy, scientists have established advanced techniques for nanoencapsulation to protect, control, and target the release of several bioactive ingredients and components, especially nutraceuticals, food components, and pharmaceuticals, consequently enhancing their safety, bioavailability, and utilization in the body (Assadpour and Mahidi Jafari, 2019; Esfanjani et al., 2018; Hernández-Muñoz et al., 2019; Menaa, 2013, 2015a, 2015b; Menaa et al., 2013). Microencapsulation techniques were mostly used in scientific and industrial activities, although these techniques were restricted to a few chemical and physical processes including conventional emulsions, liposomes, coacervation, extrusion, spray drying, etc. (Assadpour and Mahidi Jafari, 2019). However, the recent significant improvements in nanoencapsulation and nanocarriers address many challenges formerly encountered with nanotechnology in these areas. Many think that nanocarriers may give more bioavailability by increasing the ratios of surface-to-volume and, as a result, a greater likelihood of muco-adhesiveness in the small intestine as well as greater interaction with metabolic factors, including enzymes and other biomolecules in the body (Assadpour and Mahidi Jafari, 2019; McClement and Jafari, 2018); additionally, NPs may simply pass through walls of cells and find their way into targeted cells, to eventually release their components (Jafari and McClements, 2017; Katouzian et al., 2017; Menaa et al., 2013). Similarly, and from the nanotechnical perspective, nanocarriers enhance hydrophobic nutraceuticals' solubility more efficiently, with negligible impact on the appearance of the final food product, such as beverages and drinks (McClement and Jafari, 2018).

Currently, several nanocarriers have been well developed with the aim of controlling, protecting, and targeting the release of nutraceuticals and bioactive ingredients in foods; all of which have been applied in food formulations to improve nutrient bioavailability. Some are made using simple equipment including sonication devices and high-pressure homogenizers (Assadpour and Mahdi Jafari, 2019). Several nanocarriers are only prepared using sophisticated novel equipment such as spraying and electrospinning (Iqbal et al., 2020; Razzaq et al., 2021). Due to the predicted functionality and solubility of bioactive constituents, there is the possibility of selecting a suitable system for delivery at the nano range (Assadpour and Mahidi Jafari, 2019; Esfanjani et al., 2018; Hernández-Muñoz et al., 2019).

This chapter focuses on the advanced functional approaches of nanotechnology in foods and nutrition, including (active, smart) food packaging, food, and nutrients nanoencapsulation, nutraceuticals, and food fraud, among others. Researchers in the food and pharmaceutical industries should promptly work toward enhancing the safety, sensory properties, shelf-life, and quality of foods through the application of nanotechnology, such as nanoencapsulation of functional bioactive constituents (e.g., polyphenols) while enhancing the sensory properties (texture, taste, flavor, etc.) of the final food products, preventing the proliferation of pathogenic microbes in foods, enhancing the oxidation stability of oil-rich foods and edible oils, etc. (Hernández-Muñoz et al., 2019). These are new problems that may be solved efficiently through nanoencapsulation of the bioactive ingredients in foods. Additionally, in-depth studies can be carried out to explore ways to improve the safety and quality of nanocarriers, nanocomposites, and NPs in the body as they are a concern to food manufacturers, public authorities, and consumers.

14.2 NANOMATERIALS APPLIED IN DEVELOPING FOOD NANOCOMPOSITE POLYMERS

Nanocomposites are composites with one or more phases having at least one dimension on the nanoscale. Polymer nanocomposites have polymer or copolymer with nanofillers spread inside the nanocomposite matrix (Hernández-Muñoz et al., 2019). Nanotechnology applications in the packaging of foods are based on using materials with at least one particle in the dimensions within the nanoscale, typically within 1–100 nm (Kumar et al., 2011). As the particle sizes cross into this range, the physical and chemical properties of NPs usually interrelate, offering enhanced physicochemical properties (Naffakh et al., 2013). Additionally, as the polymer chain sizes are usually within 2–20 nm, the contacts between the nanomaterials in nanocomposite polymers also become vital as they have comparable dimensions (ibid.). The nanomaterials are generally grouped into three categories (Table 14.1), according to the structure of their dimensions (Barba et al., 2002; Hernández-Muñoz et al., 2019; Naffakh et al., 2013):

1. Nanofibers (NFs): materials that are one-dimensional but show two dimensions on the nanoscale, e.g., dendrimers, nanorings, nanowires, nanorods, nanotubes, nanowhiskers, etc.
2. Nanolayers (NLs): bidimensional materials with one dimension on the nanoscale, e.g., nanoshells, nanoplatelets, nanodisks, nanoflakes, nanosheets, etc.
3. NP: equiaxed or iso-dimensional materials with three dimensions on the nanoscale, e.g., fullerenes, nanoclusters, nanospheres, nanocrystals, nanogranules, etc.

Due to their unique large surface-to-volume ratio, NFs and NLs are commonly applied as fillers for reinforcement in many polymer nanocomposites, while NPs are suitable for developing intelligent and active nanocomposites, including biosensors, oxygen scavengers, and antimicrobial polymers, because of their smaller sizes (Hernández-Muñoz et al., 2019).

14.2.1 NANOFIBERS (NFs)

NFs are nanomaterials which display various particles with a nanometric width and thickness, but a millimetric or micrometric length. As a result, such structural fibers commonly have great aspect ratios, making them ideal fillers for reinforcement in polymer nanocomposites; they are mostly used to improve the mechanical properties, and some physicochemical properties, including thermal resistance or a gas barrier (Cushen et al., 2012; Hernández-Muñoz et al., 2019; Iqbal et al., 2020; Razzaq et al., 2021). Undeniably, fibers in general have many mechanisms for energy dissipation during the process of fracturing, such as fiber pullout, matrix deformation/fracture, crack bridging or matrix/fiber debonding, and fiber fracture, which contribute to improving the tensile performance of the resulting materials, based on their matrix, spatial arrangement, interphase, and geometric form (Cushen et al., 2012; Hernández-Muñoz et al., 2019). The NFs are mostly integrated with the aim of

TABLE 14.1
Classification of Nanomaterials and Their Common Application in Food Systems

Nanomaterials' classification	Examples	Properties	Uses and applications	References
NFs	Dendrimers, nanorings, nanowires, nanorods, nanotubes, nano-whiskers, etc.	They are materials that are one-dimensional but showing two dimensions on the nanoscale	NFs commonly have great aspect ratios, making them ideal fillers for reinforcement in polymer nanocomposites; they are mostly used to improve mechanical properties, and some physicochemical properties,including thermal resistance or gas barrier	(Hernández-Muñoz et al., 2019; Naffakh et al., 2013; Barba et al., 2002)
NLs	Nanoshells, nanoplatelets, nanodisks, nanoflakes, nanosheets, etc.	They are bidimensional materials with one dimension on the nanoscale	Currently, NLs have gained more relevance in polymer nanocomposites manufacturing, attracting huge research interest and commercial applications for food packaging and other related uses	(Naffakh et al., 2013; Paul and Robeson, 2008; Barba et al., 2002).
NPs	Fullerenes, nanoclusters, nanospheres, nanocrystals, nanogranules, etc.	They are equiaxed or iso-dimensional materials with three dimensions on the nanoscale	NPs are divided into: (a) artificially made NPs, which are mostly made in geometrical and regular shapes, therefore intentionally bioengineered to achieve specific and targeted purpose, and (b) NPs of natural origin (minerals) mostly found in crystalline or irregular shapes. Natural and artificial NPs are widely applied in manufacturing polymer nanocomposites for applications in food packaging, especially in developing novel nanomaterials for intelligent and active food packages. Along with the antimicrobial properties, metal and metallic oxide NPs are generally excellent reducing agents, sensors, catalysts, and sorbents.	(Hernández-Muñoz et al., 2019;Kumar et al., 2011;Ray and Okamoto, 2003).

reinforcing the polymer nanocomposites, including carbon NFs, glass NFs, and some fibrous clays, such as boehmite nanorods, halloysite nanotubes, and sepiolite NFs (Barba et al., 2002; Tjong, 2006). Recently, novel synthetic materials, including carbon nanotubes (CNTs) and their analogues such as nanotubes of boron nitride (BN), nanotubes of silicon nitride (Si_3N_4), and nanowires of gallium nitride (GaN), have attracted researchers' interest, due to their unique chemical, biological, and physical properties (Kumar et al., 2011; Naffakh et al., 2013). Nano-enabled materials (NEM) such as carbon nanotubes are used to increase the mechanical strength of packaging. Research shows that *Escherichia coli* is inhibited when it comes into contact with carbon nanotubes (Kang et al., 2007; Morya et al., 2016).

CNTs, are single-walled (SWCNTs) or multiple-walled (MWCNTs) single-atomic thick graphite sheets infused in tubular structures, with outstanding reinforcement properties as packaging nanopolymers, eliciting excellent thermal, mechanical, and barrier performances (Batool et al., 2020; Naffakh et al., 2013; Paul and Robeson, 2008; Tjong, 2006), because of their admirable strength, tensile modulus, conductivity, and high aspect ratios (Rhim et al., 2013). Additionally, recent studies have shown antimicrobial properties of CNTs, which has resulted in their increased application in active food packaging (Silvestre et al., 2011). The thin and long morphological features of CNTs are believed to puncture the cells of microorganisms, resulting in irreversible destruction of their cytoplasmic structures; however, similar studies suggested that they may also be cytotoxic to human cells due to this mechanism, especially CNTs in contact with the skin/tissues and lungs (Choudalakis and Gotsis, 2009).

Bio-based NFs, cellulose-based NFs, and chitin-based NFs have recently been studied and tested as organic nanofillers for reinforcing natural biological polymers, including polyhydroxyalkanoates, polylactic acid (PLA), natural rubber, soy protein isolate, chitosan (CS), and starch; biodegradable synthetic polymers have also attracted attention, including polycaprolactone (PCL) and polyvinyl alcohol (PVOH) (Paul and Robeson, 2008). Chitin and cellulose are among the most abundant biopolymers in nature, and they give long NFs which easily yield several stiff nano whiskers under partial acid hydrolysis (Naffakh et al., 2013).

14.2.2 Nanolayers (NLs)

As stated earlier, NLs are bidimensional materials with one dimension on the nanoscale. When particles of certain materials contain one dimension on the nanoscale, with two other larger ones, they commonly form platelet-like disk, flake, or sheet-like structures. Nevertheless, these laminar structures seldom occur alone in nature, because of the strong hydrogen bond, electrostatic, or van der Waals interactions which surround their surfaces (Kumar et al., 2011). Consequently, they stack in dozens instead, leading to the formation of great layered lumps like card decks (tactoids), as the materials are called (Kumar et al., 2011; Naffakh et al., 2013; Pavlidou and Papaspyrides, 2008). Typical examples include some metallic oxides, chalcogenides, most micas, halides, graphene and its derived products, and most clays (Menaa et al., 2015; Menaa et al., 2021; Naffakh et al., 2013).

Currently, the latter nanomaterials have gained more relevance in polymer nanocomposites manufacturing, attracting huge research interest and commercial applications for food packaging and other related uses. They are generally recognized as safe (GRAS) (Pavlidou and Papaspyrides, 2008). Clays are excellent NLs, and can be categorized based on: (1) the proportion and type of aluminum substituents; (2) the type of counterions present; and (3) the platelets' inner structure (Kumar et al., 2011).

14.2.3 Nanoparticles (NPs)

NPs are nanomaterials with iso-dimensional nanometric particles which have several shapes. NPs are divided into: (1) artificially made NPs, which are mostly made in geometrical and regular shapes, therefore intentionally bioengineered to achieve specific and targeted purpose, and (2) NPs of natural origin (minerals) mostly found in crystalline or irregular shapes (Naffakh et al., 2013).

NPs are usually nanocrystals, rod-shaped, and have a high tensile strength and elastic modulus; their incorporation into matrices of the polymers results in significant enhancement and control over several physical and chemical properties of the resulting nanobiocomposites, including biocompatibility, surface wettability, water resistance, thermal stability, or mechanical reinforcement, among others (Naffakh et al., 2013; Paul and Robeson, 2008).

Natural and artificial NPs are widely applied in manufacturing polymer nanocomposites for use in food packaging, especially in developing novel nanomaterials for intelligent and active food packages (Kumar et al., 2011). Typical examples are metallic NPs which have anti-microbial properties, such as zinc (Zn), copper (Cu), gold (Au), and silver (Ag) NPs (Batool et al., 2020; Ray and Okamoto, 2003). Although these materials are inert at the macroscale, their atoms show significant activity and catalytic behavior at small sizes, which usually results in significant anti-fungal and anti-bacterial activities at their outermost surface parts (Barba et al., 2002; Naffakh et al., 2013). Additional related examples are metallic oxides NPs, including zinc oxide (ZnO), copper oxide (CuO), magnesium oxide (MgO), and titanium dioxide (TiO_2), with photocatalytic behavior which confers on them substantial anti-microbial properties when irradiated using ultraviolet (UV) or visible (or UV-vis) radiation (Ahmad et al., 2021; Kumar et al., 2011; Wijesinghe et al., 2021).

Along with the anti-microbial properties, metallic oxides and metal NPs are generally excellent reducing agents, sensors, catalysts, and sorbents. AgNPs have the ability to absorb and degrade ethene (Rhim and Ng, 2007), while titania (TiO_2) nanocrystals act as photocatalytic oxygen scavengers and UV-blockers in acceptable working condition; conversely, NPs of antimony-tin oxide (Sb_2O_5/SnO2), zirconia (ZrO_2), alumina (Al_2O_3), silica (SiO_2), silicon carbide (SiC), and calcium carbonate ($CaCO_3$) are known to show no considerable active behavior, and as a result, they are only used as reinforcement fillers in matrices of polymers to enhance their surface, thermal, barrier, and mechanical properties (Barba et al., 2002; Hernández-Muñoz et al., 2019; Rhim et al., 2013; Tjong, 2006). This reinforcement can also be accomplished with natural biopolymers through the incorporation of organic nanofillers with excellent compatibility, such as CSNPs or nanocrystals of starch into their structural matrix, which subsequently lead to improvement of their chemical and physical properties with no significant effect on their respective biodegradability and edibility (Naffakh et al., 2013). However, as CS is among the polysaccharides that have natural/inherent anti-microbial activities, it is likely to be applied for active and reinforcement purposes in the production of polymeric nanobiocomposites, and as such plays the same role as the inorganic TiO_2NPs and AgNPs (Hernández-Muñoz et al., 2019; Paul and Robeson, 2008).

14.3 NANOCARRIERS FOR FOOD INGREDIENTS

Food nanocarriers can generally be classified into five classes according to the main equipment/ ingredients applied for formulation during the nanoencapsulation (Jafari, 2017):

1. *Biopolymer NPs* represent a key category of nanocarriers for food components. They are either formulated by single biopolymer NPs, usually produced from carbohydrates precipitation or protein desolvation, or by complexing two biopolymers with different surface-charges (Ghasemietal., 2018; Hosseini et al., 2017); examples include nanofibrils or nanotubes of whey protein (Jafari, 2017), nano gels of some bio polymers such as CS, whey protein, soy protein, and alginates (Mokhtari et al., 2017).
2. *Lipid-based nanocarriers*, which are among the most common nanocarriers in food systems. They are formulated with fat and oil. Nanoemulsions are significant nanocarriers in this class that are produced by water, oil, and biopolymers/surfactants in various forms of single oil-in-water (O/W) or water-in-oil (W/O) nanoemulsions, double nanoemulsions

(water-in-oil-in-water (W/O/W) or oil-in-water-in-oil (O/W/O)), structural nanoemulsions stabilized with double/single biopolymer coatings layer, and pickering nanoemulsions stabilized with NPs biopolymer (Akhavan et al., 2018; Jafari et al., 2017; Khan et al., 2015; Menaa, 2014; Sharif et al., 2015). Based on the bioactive ingredients' solubility, they could be trapped or dissolved in the aqueous phases (hydrophilic constituents) or oil phases (hydrophobic constituents) (Khan et al., 2015; Menaa, 2014; Sharif et al., 2015). Nanoliposomes, another set of lipid-based nanocarriers, are usually formulated using oils, phospholipids, and various solvents (Menaa, 2014). Nanoliposomes' central aqueous cavities are suitable for nanoencapsulation of hydrophobic constituents and entrapping hydrophilic constituents; they have to be placed in nanoliposome membranes, which may be double layers or a single bilayer of phospholipids (Demirci et al., 2017). There are other subcategories of nanoliposomes. One subcategory of nanovehicles is generally called nanolipid carriers and they are usually formulated using a solid lipid and oils (NLCs) mixture or solid lipids (SLNs) alone (Menaa and Menaa, 2012). These are novel lipid-based nanocarriers believed to have benefits over conventional nanoemulsions because of less leakage from entrapped bioactive ingredients, improved size control, and enhanced release control process (Katouzianetal., 2017; Menaa and Menaa, 2012; Pyo et al., 2017).

3. *Nature-driven nnanocarriers* are another class of nanocarriers used in foods. They are nanovehicles that exist naturally, and originally act as systems for nanoencapsulation of bioactive ingredients such as various forms of caseins, cyclodextrins, and amylosenano structures (Gharibzahedi and Jafari, 2017; Haratifar and Guri,2017; Jafari,2017; Katouzianetal., 2017; Pyo et al., 2017).
4. *Nanocarriers* in food systems that need to be produced using special equipment. These include electro-spraying/spinning, a nanospray dryer, as well as nanofluidic systems (Ghorani et al., 2017). It is most unlikely that one would be able to formulate relevant nanocarriers without such special equipment, while for the other nanocarriers classes, common and available equipment is used, such as rotor-stator devices, sonication devices, high-pressure homogenizers, etc. (Hernández-Muñoz et al., 2019).
5. *Miscellaneous food nanocarriers* are NPs formulated from chemical polymeric substances. These include nanocrystals, inorganic NPs, and nanostructured surfactants (Assadpour and Mahidi Jafari, 2019; Hernández-Muñoz et al., 2019). Although these nanocarriers are applied in pharmaceutical industries, there is not sufficient information about their application in food systems and nutraceuticals. This of course opens more avenues and opportunities for further research and evaluations.

14.4 NANOCOMPOSITES' APPLICATIONS IN FOOD PACKAGING

14.4.1 Applications in Active Food Packaging

Nanocomposites have been widely applied in active food packaging. Active packaging is used to attain active functions for the food safety and quality. Active food packaging is referred to as the extension of foods' shelf-life, safety improvement, and quality enhancement through the removal or release of materials from or into the packed foods or the headspace. Active food packaging, which usually makes use of polymer nanocomposites, has attracted interest for its potential applications mainly in the formulation of oxygen scavengers and anti-microbial films.

14.4.1.1 Active Packaging for Oxygen Scavenging

The presence of oxygen (O_2) affects food quality, including its sensory properties, appearance, and freshness, especiallyO_2-sensitive foods. Direct oxidation reaction induces rancidity in oils and browning in fruits, while indirect oxidation causes deterioration of foods by aerobic microbes

(Hatzigrigoriou and Papaspyrides, 2011). To preserve the foods prone to O_2-mediated negative effects, O_2 should not be allowed into the food package, including the headspace. Vacuuming can be extensively applied for O_2 elimination just before sealing the food and/or nutraceutical packages. However, this does not fully eliminate the O_2 presence which may permeate the package from the external environments and find its way into packages, depending on the materials used to formulate the package. One alternative used to reduce concentrations of O_2 in packaging of food is by using modified atmosphere packaging (MAP). Nevertheless, both techniques do not completely prevent residual O_2 which may be sufficient to reduce the quality of foods (Hernández-Muñoz et al., 2019). At the nanoscale, this is usually resolved using packaging films, O_2 scavengers, or nano-enabled devices, which have the ability to start a chemical interaction with O_2 as well as the capacity to remove traces of O_2 from the package, subsequently protecting the contents against the undesirable and unwanted effects of O_2.

An optimum O_2 scavenger has been built to focus more on the permeability of the packaging material, the level of O_2 in the headspace, and the O_2 trapped inside foods. Along with the ability to absorb O_2 in large amounts, the O_2 scavenger is designed not to pose any risk to the human body and at the same time be cost-effective (Jafari, 2017; Naffakh et al., 2013). O_2 scavengers can be included in sachets, bottle crowns, plastic trays, labels, and plastic films (Realini and Marcos, 2014). Most of O_2 scavengers depend on the oxidation mechanism of metallic iron (Fe) explained by the following reactions:

$$Fe \rightarrow Fe^{2+} + 2e^{-} \text{ (initial reaction, yielding ferrous iron)}$$

$$1/2O_2 + H_2O + 2e^{-} \rightarrow 2OH^{-} \text{ (liberated electrons interact with oxygen and moisture)}$$

$$Fe^{2+} + 2OH^{-} \rightarrow Fe(OH)_2 \text{ (the formed hydroxyl ion interacts with ferrous iron)}$$

$$Fe(OH)_2 + 1/4O_2 + 1/2\,H_2O \rightarrow Fe(OH)_3 \text{ (ferric hydroxide interact oxygen and moisture)}$$

With respect to the mechanism of Fe oxidation, systems of Fe-based scavengers are required to be activated by moisture. In the case of dry (or low water activity) foods, molecules which are photo-sensitive are usually integrated into films to increase the rate of reaction of the Fe-based scavengers of O_2 (Hernández-Muñoz et al., 2019). The mechanism activation has film irradiation at the wavelength for suitable photo-sensitizer excitation by light, which sensitizes molecules of O_2 diffused into films, transforming them to singlet O_2 (O_2 in a quantum state with all electrons spin paired). The excited singlet O_2 has high reactivity and rapidly reacts with the scavengers (Childress et al., 2019).

Fe-based scavengers are also formulated on the nanoscale. Fe can perform the oxidation reaction efficiently, behaving as optimum O_2 scavengers. Nanometric-scale O_2 scavengers formulated with NPs of Fe have been made and tried on roasted sunflower seeds and walnuts, which led to promising outcomes for potential applications in active food packaging (Hernández-Muñoz et al., 2019). Fe-based O_2 scavengers are attached, absorbed, or embedded in clay NPs for applications in plastic food packages. Additional instances of the application of nanotechnology in active food packaging rely on O_2 scavengers activated by heat, which consist of a-tocopherol, moisture, and Fe (II) chloride ($FeCl_2$) nanoencapsulated in PCL.

14.4.1.2 Anti-Microbial Food Packaging Applications

In addition to nanotechnology applications for improving the functional properties of materials used for food packaging, nanopolymer matrices are applied in active packaging designing which depend on developing films with the ability to release anti-microbial substances into foods or the headspace (Hernández-Muñoz et al., 2019). Morya et al. discussed in their review article how nano-enabled materials are known for their anti-microbial characteristics. Research revealed that more than 650 pathogenic microbes can be destroyed by using nano-silver particles in packaging materials (Morya et al., 2016). A downside in designing active food packaging aimed at releasing active substances is the integration of the active agent into the matrix of the polymer. It is often added during polymer processing, which usually has many possible inconveniences. As the polymer processing takes place under humid or dry conditions, based on the active molecule's physical and chemical properties, it can be degraded/inactivated or evaporate in the process (Zhang et al., 2014). Nanotechniques usually overcome this challenge by loading the nanofillers together with the active molecules prior to incorporation into polymer films (Xiong et al., 2018). Based on this, the active substances can be physically and/or chemically adsorbed in the nanofillers. Additionally, nano-formulated polymers may modulate the release of active substances to improve their efficiency when the foods are used (Hernández-Muñoz et al., 2019). More research is urgently required to explore ways of improving the incorporation and release of active compounds in food packages.

Polymer nanocomposite films with anti-microbial properties based on metal ions released from Cu NPs and AgNPs are commercially applied and well studied. Ag ions (Ag^+) are currently integrated into inorganic materials including zirconium phosphate, glass, and zeolites, to be applied as active antimicrobial agents (Hosseini and Gómez-Guillén, 2018). AgNPs have been studied for their anti-microbial characteristics for food applications (Hernández-Muñoz et al., 2019). The high surface-to-volume ratio of such NPs offers a greater contact area with microbes, and as a result increases the efficacy of the anti-microbial properties. Studies indicate that the anti-bacterial effects of AgNPs are based on the NPs' size; NPs ranging from 1–10 nm present sufficient direct interactions with bacterial cells, thereby increasing their effectiveness. Anti-microbial activities of nanopolymers combined with AgNPs have been associated with Ag^+ release from a matrix of polymers (ibid.). AgNPs then function as a nano-reservoir for Ag^+ delivery from nanostructured films to foods. Also, the combination of AgNPs with AuNPs or zeolites NPs is known to have higher anti-microbial activities (Duncan, 2011; Lopez-Carballo et al., 2013). In the same way as AgNPs, CuNPs function as a nanoreservoir for sustainable Cu ions (Cu^{2+}) release from Cu oxidation in the surface of the NPs. Cu, a Cu^{2+} impregnated nanofiller, and copper oxide (CuO) NPs have been integrated into many natural and synthetic polymeric substances (Ravishankar and Jamuna, 2011; Silvestre et al., 2011). ZnO NPs have anti-microbial properties which rely on generating hydrogen peroxide (H_2O_2) once stimulated through visible light along with zinc ions (Zn^{2+}) release (Ravishankar and Jamuna, 2011; Silvestre et al., 2011). They are incorporated into many polymers for active packaging. Nanocrystals of Zn have been incorporated into polymers for anti-microbial active packaging (Cioffi et al., 2005; Duncan, 2011).

Nanoparticles of titanium oxide have anti-microbial properties similar to reactive oxygen species (ROS) photocatalytic generation under UV light. Nanoparticles of titanium oxide have self-disinfecting photocatalytic properties which are applied in the food industries to disinfect contact surfaces of foods and to prevent the formation of biofilms. Applications of titanium oxide NPs in active packaging of foods are currently being explored further. They are integrated into polymers or used for coating the surface of polymeric films to prevent or decrease the growth of microorganisms on solid food surfaces (Ravishankar and Jamuna, 2011). Photolytic microbial inactivations and visible light absorbance of titanium oxide are enhanced through metallic doping. Noble metals (silver (Ag), gold (Au), platinum (Pt), osmium (Os), iridium (Ir), palladium (Pd), ruthenium (Ru), rhodium (Rh)), especially silver ions and metallic silver are applied to achieve this purpose. Photocatalysts of titanium oxide, capable of operating not just under ultraviolet radiation but also visible light radiation, are developed through ions of transition metals implanted in the catalyst bulk (Duncan, 2011; Silvestre et al., 2011). There are other nanoparticles of metals and metal oxides with possible application in developing anti-microbial active packaging for foods, such as gold, magnesium oxide, cadmium telluride/selenide, aluminum oxide, etc. (Duncan, 2011; Ravishankar and Jamuna, 2011).

14.4.2 Improvement of Mechanical, Thermal, and Barrier Properties

Polymers are conventionally filled using several inorganic substances in high concentrations, usually above 20%, to reduce the cost of production and recycle waste (Pavlidou and Papaspyrides, 2008; Rhim and Ng, 2007). Nevertheless, the practice has considerable disadvantages for the resulting composites, including opacity, brittleness, and weight increase, while inducing higher wear and poorer processability rates in the process facilities (Pavlidou and Papaspyrides, 2008; Tjong, 2006). However, there was an improvement in 1986, when researchers at Toyota succeeded in synthesizing nanocomposites of nylon-6/montmorillonite through polymerization *in situ*, which reported a significant improvement in barrier, mechanical, and thermal properties, in relation to neat polymers, with only minute quantities of clay loadings less than 5% (Pavlidou and Papaspyrides, 2008; Rhim et al., 2013).This led to the development of novel similar hybrid materials, and the study of their possible characteristics and application (Rhim and Ng, 2007). These materials are known to have improved in glass transition, deformation, tensile strength, water permeability, gas (carbon dioxide (CO_2), O_2) permeability, dimensional stability, thermal stability, heat distortion temperature, etc., without impairment in their optical characteristics, melt flow, and density (Cushen et al., 2012; Pavlidou and Papaspyrides, 2008; Rhim et al., 2013).

The main aim of incorporating NPs into a matrix of polymers is to increase the polymers' stiffness or modulus through mechanisms of reinforcement explained by conventional composites theories (Kumar et al., 2011; Tjong, 2006). The theories predict a strong enhancement in the resulting nanocomposites' Young modulus, with only minute additions of appropriately aligned and dispersed inorganic NPs that have natural resistance to straining because of their high modulus (Paul and Robeson, 2008; Pavlidou and Papaspyrides, 2008). These theoretical enhancements are broadly confirmed through experimental trends, and the results are attributed to additional reinforcement mechanically to platelet-related structural components rather than to fibrous structures, because of their physical effects and two-dimensional properties (Paul and Robeson, 2008). As a result, platelets have the capability to deflect microcracks' propagation into tortuous pathways emerging from all sides in sample planes when aligned inside, while providing a huge surface area which enables concentration and stress transfer in the reinforcement phase (Paul and Robeson, 2008; Pavlidou and Papaspyrides, 2008; Tjong, 2006). The tensile strength can deteriorate or improve in regard to neat polymers, based on the extent of the interfacial adhesiveness facing the inorganic fillers to the polymeric matrices (Paul and Robeson, 2008). Nanocomposites of high strength are accomplished if both phases of materials are bonded strongly, while their chemical incompatibilities often translate into early failure and brittleness (Paul and Robeson, 2008; Pavlidou and Papaspyrides, 2008).

Thermal behavior is very relevant in nanocomposites of polymers made for packaging foods with features, such as sterilizable packages, microwaveable, cook-in-the-bag, hot-fill packaging. During the cooling of the molten polymer, NPs function as heterogeneous nucleating agents to polymer spherulites and crystals, thereby significantly decreasing their sizes while raising their irregularity and number (Pavlidou and Papaspyrides, 2008). At a low loading level (less than 5%), NPs increase the crystallization of the polymer and at the same time reduce its melting point, which decreases the temperature and cycle time of the machine, thereby improving the efficiency of the production process (Pavlidou and Papaspyrides, 2008; Silvestre et al., 2011; Tjong, 2006). Additionally, NPs that have larger surface areas interact with chains of polymers to stabilize the metastable phases, therefore encouraging crystal polymorphism that contributes to improving the mechanical characteristics of the nanocomposites' polymers (Hernández-Muñoz et al., 2019). NPs often increase the glass transition temperature when packages are formed and increase the heat distortion temperature of the polymer matrix, due to the formation of an intercalated structure that restricts the segmental movement of the polymer chains (Pavlidou and Papaspyrides, 2008; Ray and Okamoto, 2003; Rhim and Ng, 2007), hence increased mechanical stability.

Another main benefit of the application of nanocomposites polymers in food packaging is their improved barrier properties to aromatic compounds, water vapor, and permanent gases such as CO_2, O_2, and nitrogen (N) (Choudalakis and Gotsis, 2009). These improvements are due to the decrease in diffusivity and solubility, and consequently, to the permeability of the gas molecules when migrating inside the matrix of nanocomposites, as initiated through incorporating non-permeable fillers to neat polymers (Tan and Thomas, 2016). Such NPs can have direct effects on the tortuous path once penetrants are made to move around them; it can be indirect once they initiate alignment of the polymer chains, or polymer crystallites' modification and alignment. It is worth noting that the diffusion resistance results from an increase in the distance of the path of tortuosity round the particles of the fillers (Falla et al., 1996).

14.4.3 Nanoparticles as Sensors in Smart Packaging

The exceptional electro-optical and chemical properties of NPs make them appropriate for applications in sensors' development in food packaging (Ranjan et al., 2014; Sozer and Kokina, 2012). The properties of NPs could be improved in accordance with their shape, size, and chemical surrounding. The contact with the required molecule can change the physico-chemical properties including the redox potentials, conductivity, and surface plasmon resonance, and produce detectable signals (Mills and Hazafy, 2009). The great interest in applications of nanotechnology in smart packaging of foods has focused on nanosensors' evolution in detecting pathogens, spoilage, product tampering, and contamination (Hernández-Muñoz et al., 2019; Mills and Hazafy, 2009). The nanosensors' application in food packaging mainly involves the detection of gases. Consequently, developing sensors capable of detecting the presence of O_2, CO_2, gaseous amines, ethanol, acetone, ethylene, and moisture have attracted the attention of many researchers and food processors.

For O_2 sensors, considerable progress has been made with the use of SnO_2 NPs and TiO_2 NPs. In the presence of methylene blue (redox-active dye), SnO_2 NPs and TiO_2 NPs function as photo-activated indicators, through a change of color when the O_2 concentration is low (Lee et al., 2005; Mills, 2005).

CO_2 monitoring in MAP can be done with nanobeads fluorophore-encapsulated polymers as an on-pack sensor strip showing reversible and rapid CO_2 response at different concentrations (von Bultzingslowen et al., 2002).

Gaseous amines are detected from changes of conductance in nanocomposites of TiO_2 microrods and SnO_2 NPs. The lowest detection limits can be detected with fluorescence quenching of perylene-based fluorophores nanofibrils (Che and Zang, 2009). The presence of gaseous amine indicates meat and fish spoilage. Optical sensors of the calorimeter, based on the uniqueness of the surface plasmon resonance (SPR) of AgNPs and AuNPs, are known for their notable activities in detecting trace amines in gaseous phase and in dissolution (Abargues et al.,2014; Pandey et al., 2013). Similarly, nanocomposites of ZnO and SnO, which are semiconducting oxides (Ahmad et al., 2021; Wijesinghe et al., 2021), are applied to detect volatile organics, such as carbon monoxide (CO), ethanol, and acetone (Barreca et al., 2007; Nayak et al., 2015). Also, the presence of ethylene gas, a hormone responsible for ripening in fruits, can be detected with WO_3-SnO_2 nanocomposites (El-Sayed et al., 2015; Nayak et al., 2015; Pimtong-Ngam et al., 2007).

To sense humidity, carbon-coated CuNPs of 20–50 nm in size have been developed, to prepare porous, ten-sided, and wetting agent-loaded thin films (Luechinger et al., 2007). When these films are placed in controlled atmospheres with a growing volume of ethanol/water vapors, there are significant changes in color, and spectral shifts of at least 50 nm per high level of relative humidity (70–80%) (van Dommelen et al., 2018). When exposed to a humid environment, the polymer matrix swelling caused greater inter-NP separations (Zhang et al., 2018).

14.5 NANOSOLUTIONS AGAINST FOOD FRAUD

In the European Union, food fraud has been defined as: "Food fraud is committed when food is deliberately placed on the market, for financial gain, with the intention of deceiving the consumer" whereas in the United States food fraud is defined as "Food fraud is a collective term used to encompass the deliberate and intentional substitution, addition, tampering, or misrepresentation of food, food ingredients, or food packaging; or false or misleading statements made about a product, for economic gain" (Ryan, 2015). Food fraud encompasses the trade of ill-suited and probable harmful foodstuff, mislabeling, item or element substitution, etc. to fool the customers. It offers chances of contaminating high value food ingredients that are far too easy to mask by vision and most tests (Spink, 2012). As the solution, novel tests and the solicitation of former technological tests have been established in new forms and methods of nanotechnology to combat customer exposure to duplicitous food activities. Nanotechnological methods of bioactive encapsulation, nanosensors, etc. are effective ways of preventing food fraud (Wheatley et al., 2013). Food processing, agricultural production, food preservation, food packaging and pathogen detection are such food sectors where technology on the nanoscale is important. This technology includes functions like nanodelivery of nutraceuticals in systems, for example, polymer and lipid-based monoglyceride self-amassed arrangements or nanosystems as supply vehicles (Garcia et al., 2010; Sozer and Kokini, 2009). Moreover, these nanosystems can be used for nutrient or flavor delivery in food areas. One such sample is the application of liposome as a nutrient, flavor and food anti-microbial in the nutrient delivery system (Taylor et al., 2005). Liposome-encapsulated anti-microbial peptides showed advantages when compared to bacteriocins (Nisin) and its food shelf-life is proving to be a promising delivery system in food areas (da Silva Malheiros et al., 2010). Nanosensors are a vital defense against food fraud, as these devices are able to sense and enumerate even very tiny concentrations of organic composites, pathogens, and other chemicals. Besides, they demonstrate high sensitivity, a rapid response and repossession, which offers good potential for examining susceptible contaminants on a massive scale, among others (Otles and Yalcin, 2010). Nano-biosensors have also been used for the detection of mycotoxins and pathogens. The rapid detection potential of nanobiosensors is easy in comparison to conventional microbiological controls against toxins as they are complicated but reliable. *E. coli*, Salmonella and Listeria are the most common microorganisms studied under the light of nanobiosensor technology as they are the usual food fraud ingredients. It was also stated that a nanobiosensor through microgravimetric quartz crystal microbalance and DNA was brought in to detect *E. coli*. In this sensor, streptavidin-coated ferrofluid, which was a Fe_3O4 nanoparticle, was used to augment the signal (Valdés et al., 2009).

14.6 CONCLUSION

Risk is regularly assessed as the likelihood of something bad occurring and that likelihood is frequently hard to evaluate for many reasons. Creating preventive systems concentrating on removing danger is the duty of all food providers. However, keeping fraudulent items from entering the natural food chain requires accreditation programs and arduous testing innovations and systems to keep harmful food items out of the inventory network. Nanotechnology has an amazingly high potential to benefit society through the food industry. Advances in microfluids and inorganic nanoparticles have empowered the planning of proficient sensors to quickly distinguish microorganisms or pesticides in food on the farm or in the item available in the market. The nanobiosensor can likewise be applied for ecological contamination control in the food business. Functionalized food with nanoparticles as flavors and supplements transporters can improve food quality and security. The significant issue in the short term will be the toxicity of these nanomaterials in plants, creatures and humans and afterwards this should be examined case by case. It is also important to assess the accumulation of these nanostructures in plants and their biotransformations. These examinations will deliver better and more secure items and hence will prevent fraudulent food activities.

ACKNOWLEDGMENTS

The authors are very obliged to Lovely Professional University for its support in bringing this chapter to its final form.

REFERENCES

Abargues, R., Rodriguez-Canto, P. J., Albert, S., Suarez, I., and Martínez-Pastor, J. P. (2014). Plasmonic optical sensors printed from Ag–PVA nanoinks. *Journal of Materials Chemistry C*, 2(5), 908–915.

Ahmad, M., Rehman, W., Khan, M. M., Qureshi, M. T., Gul, A., Haq, S., Ullah, R., Rab, A., and Menaa, F. (2021). Phytogenic fabrication of ZnO and gold decorated ZnO nanoparticles for photocatalytic degradation of Rhodamine B. *Journal of Environmental Chemical Engineering*, 9(1), 104725.

Akhavan, S., Assadpour, E., Katouzian, I., and Jafari, S. M. (2018). Lipid nano scale cargos for the protection and delivery of food bioactive ingredients and nutraceuticals. *Trends in Food Science and Technology*, 74, 132–146.

Assadpour, E., and Mahdi Jafari, S. (2019). A systematic review on nanoencapsulation of food bioactive ingredients and nutraceuticals by various nanocarriers. *Critical Reviews in Food Science and Nutrition*, 59(19), 3129–3151.

Awuchi, C. G., Morya, S., Dendegh, T. A., Okpala, C. O. R., and Korzeniowska, M. (2022). Nanoencapsulation of food bioactive constituents and its associated processes: A revisit. *Bioresource Technology Reports*, 101088.

Barba, A., Beltrán, V., Feliu, C., Garcia, J., Ginés, F., Sánchez, E., and Sanz, V. (2002). *Materiasprimas para la fabricación de suportes de baldosascerâmicas*, 2nd edn. Castellón, Spain: Instituto de Tecnologia Cerámica.

Barreca, D., Comini, E., Ferrucci, A. P., Gasparotto, A., Maccato, C., Maragno, C., ... and Tondello, E. (2007). First example of ZnO–TiO_2 nanocomposites by chemical vapor deposition: structure, morphology, composition, and gas sensing performances. *Chemistry of Materials*, *19*(23), 5642–5649.

Batool, A., Menaa, F., Uzair, B., Khan, B. A., and Menaa, B. (2020). Progress and prospects in translating nanobiotechnology in medical theranostics. *Current Nanoscience*, 16(5), 685–707.

Che, Y., and Zang, L. (2009). Enhanced fluorescence sensing of amine vapor based on ultrathin nanofibers. *Chemical Communications*, 34, 5106–5108.

Childress, K. K., Kim, K., Glugla, D. J., Musgrave, C. B., Bowman, C. N., and Stansbury, J. W. (2019). Independent control of singlet oxygen and radical generation via irradiation of a two-color photosensitive molecule. *Macromolecules*, 52(13), 4968–4978.

Choudalakis, G., and Gotsis, A.D., 2009. Permeability of polymer/clay nanocomposites: A review. *European Polymer Journal*, 45(4), 967e984.

Cioffi, N., Torsi, L., Ditaranto, N., Tantillo, G., Ghibelli, L., Sabbatini, L., Bleve-Zacheo, T., ... Traversa, E. (2005). Copper nanoparticle/polymer composites with antifungal and bacteriostatic properties. *Chemistry of Materials*, 17(21), 5255–5262.

Cushen, M., Kerry, J., Morris, M., Cruz-Romero, M., and Cummins, E. (2012). Nanotechnologies in the food industry: Recent developments, risks and regulation. *Trends in Food Science and Technology*, 24(1), 30–46.

da Silva Malheiros, P., Daroit, D. J., and Brandelli, A. (2010). Food applications of liposome-encapsulated antimicrobial peptides. *Trends in Food Science and Technology*, 21(6), 284–292.

Demirci, M., Caglar, M. Y., Cakir, B., and Gülseren, İ. (2017). Encapsulation by nanoliposomes. In Jafari, S. M. (Ed.) Nanoencapsulation Technologies for the Food and Nutraceutical Industries, London: Academic Press, pp. 74–113.

Duncan, T. V. (2011). Applications of nanotechnology in food packaging and food safety: Barrier materials, antimicrobials and sensors. *Journal of Colloid and Interface Science*, 363(1), 1–24.

El-Sayed, M., Hassan, Z., Awad, M. I. F. R. A., and Salama, H. (2015). Chitosan-whey protein complex (CS-WP) as delivery systems to improve bioavailability of iron. *International Journal of Applied Pure Science in Agriculture*, 1, 34–46.

Esfanjani, A. F., Assadpour, E., and Jafari, S. M. (2018). Improving the bioavailability of phenolic compounds by loading them within lipid-based nanocarriers. *Trends in Food Science and Technology*, 76, 56–66.

Falla, W. R., Mulski, M., and Cussler, E. L. (1996). Estimating diffusion through flake-filled membranes. *Journal of Membrane Science*, 119(1), 129–138.

García, M., Forbe, T., and Gonzalez, E. (2010). Potential applications of nanotechnology in the agro-food sector. *Food Science and Technology*, 30(3), 573–581.

Gharibzahedi, S. M., and Jafari, S. M. (2017). Nanocapsule formation by cyclodextrins. In Jafari, S. M. (Ed.) *Nanoencapsulation Technologies for the Food and Nutraceutical Industries*. London: Academic Press, pp. 187–261.

Ghasemi, S., Jafari, S. M., Assadpour, E., and Khomeiri, M. (2018). Nanoencapsulation of d-limonene within nanocarriers produced by pectin-whey protein complexes. *Food Hydrocolloids*, 77, 152–162.

Ghorani, B., Alehosseini, A., and Tucker, N. (2017). Nanocapsule formation by electrospinning. In Jafari, S. M. (Ed.) *Nanoencapsulation Technologies for the Food and Nutraceutical Industries*. London: Academic Press, pp. 264–319.

Haratifar, S., and Guri, A. (2017). Nanocapsule formation by caseins. In Jafari, S. M. (Ed.) *Nanoencapsulation Technologies for the Food and Nutraceutical Industries*. Academic Press, pp. 140–164.

Hatzigrigoriou, N. B., and Papaspyrides, C. D. (2011). Nanotechnology in plastic food-contact materials. *Journal of Applied Polymer Science*, 122(6), 3719–3738.

Hernández-Muñoz, P., Cerisuelo, J. P., Domínguez, I., López-Carballo, G., Catalá, R., and Gavara, R. (2019). Nanotechnology in food packaging. In Amparo, L. R., Rovira, M., Sanz, M. and Gomez Gomez-Mascaraque, L. (Eds) *Nanomaterials for Food Applications*. Oxford: Elsevier. pp. 205–232.

Hosseini, S. F., and Gómez-Guillén, M. C. (2018). A state-of-the-art review on the elaboration of fish gelatin as bioactive packaging: Special emphasis on nanotechnology-based approaches. *Trends in Food Science and Technology*, 79, 125–135.

Hosseini, S. M., Ghiasi, F., and Jahromi, M. (2017). Nanocapsule formation by complexation of biopolymers. In Jafari, S. M. (Ed.) *Nanoencapsulation Technologies for the Food and Nutraceutical Industries*. London: Academic Press, pp. 447–492.

Iqbal, H., Khan, B. A., Khan, Z. U., Razzaq, A., Khan, N. U., Menaa, B., and Menaa, F. (2020). Fabrication, physical characterizations and in vitro antibacterial activity of cefadroxil-loaded chitosan/poly (vinyl alcohol) nanofibers against Staphylococcus aureus clinical isolates. *International Journal of Biological Macromolecules*, 144, 921–931.

Jafari, S. M. (2017). An introduction to nano encapsulation techniques for the food bioactive ingredients. In Jafari, S. M. (Ed.) *Nanoencapsulation of Food Bioactive Ingredients*. London: Academic Press, pp. 1–62.

Jafari, S. M., and McClements, D. J. (2017). Nanotechnology approaches for increasing nutrient bioavailability. *Advances in Food and Nutrition Research*, 81, 1–30.

Jafari, S. M., Paximada, P., Mandala, I., Assadpour, E., and Mehrnia, M. A. (2017). Encapsulation by nanoemulsions. In Jafari, S. M. (Ed.) *Nanoencapsulation Technologies for the Food and Nutraceutical Industries*. : Academic Press, pp. 36–73.

Kang, S., Pinault, M., Pfefferle, L. D., and Elimelech, M. (2007). Single-walled carbon nanotubes exhibit strong antimicrobial activity. *Langmuir*, 23(17), 8670–8673.

Katouzian, I., Esfanjani, A. F., Jafari, S. M., and Akhavan, S. (2017). Formulation and application of a new generation of lipid nano-carriers for the food bioactive ingredients. *Trends in Food Science and Technology*, 68, 14–25.

Khan, B. A., Akhtar, N., Menaa, A., and Menaa, F. (2015). A novel Cassia fistula (L.)-based emulsion elicits skin anti-aging benefits in humans. *Cosmetics*, 2(4), 368–383.

Khan, B. A., Mahmood, T., Menaa, F., Shahzad, Y., Yousaf, A. M., Hussain, T., and Ray, S. D. (2018). New perspectives on the efficacy of gallic acid in cosmetics and nanocosmeceuticals. *Current Pharmaceutical Design*, 24(43), 5181–5187.

Kumar, P., Sandeep, K. P., Alavi, S., and Truong, V. D. (2011). A review of experimental and modeling techniques to determine properties of biopolymer-based nanocomposites. *Journal of Food Science*, 76(1), E2–E14.

Lee, S. K., Sheridan, M., and Mills, A. (2005). Novel UV-activated colorimetric oxygen indicator. *Chemistry of Materials*, 17(10), 2744–2751.

López-Carballo, G., Higueras, L., Gavara, R., and Hernández-Muñoz, P. (2013). Silver ions release from antibacterial chitosan films containing in situ generated silver nanoparticles. *Journal of Agricultural and Food Chemistry*, 61(1), 260–267.

Luechinger, N. A., Loher, S., Athanassiou, E. K., Grass, R. N., and Stark, W. J. (2007). Highly sensitive optical detection of humidity on polymer/metal nanoparticle hybrid films. *Langmuir*, 23(6), 3473–3477.

McClements, D. J., and Jafari, S. M. (2018). General aspects of nanoemulsions and their formulation. In Jafari, S. M. (Ed.), *Nanoemulsions*. London: Academic Press, pp. 3–20.

Menaa, F. (2013). When pharma meets nano or the emerging era of nano-pharmaceuticals. *Pharmaceutica Analytica Acta*, 4, 223.

Menaa, F. (2014). Emulsions systems for skin care: From macro to nano-formulations. *Journal of Pharma Care Health Systems*, 1, e104.

Menaa, F. (2015a). Curcumin nano-sized delivery systems against cancers: From bench to clinics. *Journal of Pharma Care Health Systems*, 2, e135.

Menaa, F. (2015b). Food nanotechnology: A safe innovation for production and competition. *BAOJ Nutrition*, 1(003).

Menaa, F., Abdelghani, A., and Menaa, B. (2015). Graphene nanomaterials as biocompatible and conductive scaffolds for stem cells: Impact for tissue engineering and regenerative medicine. *Journal of Tissue Engineering and Regenerative Medicine*, 9(12), 1321–1338.

Menaa, F., Fatemeh, Y., Vashist, S. K., Iqbal, H., Sharts, O. N., and Menaa, B. (2021). Graphene, an interesting nanocarbon allotrope for biosensing applications: Advances, insights, and prospects. *Biomedical Engineering and Computational Biology*, 12, 1179597220983821.

Menaa, F., Menaa, A., and Menaa, B. (2014). Polyphenols nano-formulations for topical delivery and skin tissue engineering. In Watson, R.R., Preedy, V. R. and Zibadi, S. (Eds.) *Polyphenols in Human Health and Disease*. London: Academic Press, pp. 839–848.

Menaa, F., Menaa, A., Treton, J., and Menaa, B. (2013). Nanoencapsulations of dietary polyphenols for oncology and gerontology: Resveratrol as a good example—resveratrol nano-formulations: Suitable for cancer patients and the elderly. *Introduction to Functional Food Science*, 1, 383–404.

Menaa, F., and Menaa, B. (2012). Development of mitotane lipid nanocarriers and enantiomers: Two-in-one solution to efficiently treat adreno-cortical carcinoma. *Current Medicinal Chemistry*, 19(34), 5854 –5862.

Mills, A. (2005). Oxygen indicators and intelligent inks for packaging food. *Chemical Society Reviews*, 34(12), 1003–1011.

Mills, A., and Hazafy, D. (2009). Nanocrystalline SnO_2-based, UVB-activated, colourimetric oxygen indicator. *Sensors and Actuators B: Chemical*, 136(2), 344–349.

Mokhtari, S., Jafari, S. M., and Assadpour, E. (2017). Development of a nutraceutical nano-delivery system through emulsification/internal gelation of alginate. *Food Chemistry*, 229, 286–295.

Morya, S., Amoah, A. D., Thompkinson, D. K., and Rout, S. (2016). Nano-enabled materials in food industry and their potential hazards to human health and environment. *International Journal of Pharma and Bio Sciences*, 7(2), 589–593.

Naffakh, M., Diez-Pascual, A. M., Marco, C., Ellis, G. J., and Gomez-Fatou, M. A. (2013). Opportunities and challenges in the use of inorganic fullerene-like nanoparticles to produce advanced polymer nanocomposites. *Progress in Polymer Science*, 38(8), 1163–1231.

Nayak, A. K., Ghosh, R., Santra, S., Guha, P. K., and Pradhan, D. (2015). Hierarchical nanostructured WO_3–SnO_2 for selective sensing of volatile organic compounds. *Nanoscale*, 7(29), 12460–12473.

Otles, S., and Yalcin, B. (2010). Nano-biosensors as new tool for detection of food quality and safety. *LogForum*, 6(4), 7.

Pandey, S., Goswami, G. K., and Nanda, K. K. (2013). Green synthesis of polysaccharide/gold nanoparticle nanocomposite: An efficient ammonia Sensor. *Carbohydrate Polymers*, 94(1), 229–234.

Paul, D. R., and Robeson, L. M. (2008). Polymer nanotechnology: Nanocomposites. *Polymer*, 49(15), 3187–3204.

Pavlidou, S., and Papaspyrides, C. D. (2008). A review on polymer-layered silicate nanocomposites. *Progress in Polymer Science*, 33(12), 1119–1198.

Pimtong-Ngam, Y., Jiemsirilers, S., and Supothina, S. (2007). Preparation of tungsten oxide–tin oxide nanocomposites and their ethylene sensing characteristics. *Sensors and Actuators A: Physical*, 139(1–2), 7–11.

Pyo, S. M., Müller, R. H., and Keck, C. M. (2017). Encapsulation by nanostructured lipid carriers. In Jafari, S. M. (Ed.) *Nanoencapsulation Technologies for the Food and Nutraceutical Industries* London: Academic Press, pp. 114–137.

Ranjan, S., Dasgupta, N., Chakraborty, A. R., Samuel, S. M., Ramalingam, C., Shanker, R., and Kumar, A. (2014). Nanoscience and nanotechnologies in food industries: Opportunities and research trends. *Journal of Nanoparticle Research*, 16(6), 1–23.

Ravishankar, R. V. and Jamuna, B. A. (2011). Nanoparticles and their potential application as antimicrobials, *Science Against Microbial Pathogens: Communicating Current Research and Technological Advances*, 1(1), 10.

Ray, S. S., and Okamoto, M. (2003). Polymer/layered silicate nanocomposites: A review from preparation to processing. *Progress in Polymer Science*, 28(11), 1539–1641.

Razzaq, A., Khan, Z. U., Saeed, A., Shah, K. A., Khan, N. U., Menaa, B., Iqbal, H. and Menaa, F. (2021). Development of cephradine-loaded gelatin/polyvinyl alcohol electrospun nanofibers for effective diabetic wound healing: In-vitro and in-vivo assessments. *Pharmaceutics*, 13, 349.

Realini, C. E., and Marcos, B. (2014). Active and intelligent packaging systems for a modern society. *Meat Science*, 98(3), 404–419.

Rhim, J. W., and Ng, P. K. (2007). Natural biopolymer-based nanocomposite films for packaging applications. *Critical Reviews in Food Science and Nutrition*, 47(4), 411–433.

Rhim, J. W., Park, H. M., and Ha, C. S. (2013). Bio-nanocomposites for food packaging applications. *Progress in Polymer Science*, 38(10–11), 1629–1652.

Ryan, J. M. (2015). *Food Fraud*. London: Academic Press.

Sharif, A., Akhtar, N., Khan, M. S., Menaa, A., Menaa, B., Khan, B. A., and Menaa, F. (2015). Formulation and evaluation on human skin of a water-in-oil emulsion containing Muscathamburg black grape seed extract. *International Journal of Cosmetic Science*, 37(2), 253–258.

Silvestre, C., Duraccio, D., and Cimmino, S. (2011). Food packaging based on polymer nanomaterials. *Progress in Polymer Science*, 36(12), 1766–1782.

Solans, C., Izquierdo, P., Nolla, J., Azemar, N., and Garcia-Celma, M. J. (2005). Nano-emulsions. *Current Opinion in Colloid and Interface Science*, 10(3–4), 102–110.

Sozer, N., and Kokini, J. L. (2009). Nanotechnology and its applications in the food sector. *Trends in Biotechnology*, 27(2), 82–89.

Sozer, N., and Kokina, J. L., 2012. The applications of nanotechnology. In Pico, Y. (Ed.), *Chemical Analysis of Food: Techniques and Applications*. Oxford: Elsevier, pp. 145–176.

Spink, J. (2012). Defining food fraud and the chemistry of the crime. In Ellefson, W., Zach, L. and Sullivan, D. (Eds.) Improving Import Food Safety, Hoboken. NJ: Wiley, pp. 195–216.

Tan, B., and Thomas, N. L. (2016). A review of the water barrier properties of polymer/clay and polymer/graphene nanocomposites. *Journal of Membrane Science*, 514, 595–612.

Taylor, T. M., Weiss, J., Davidson, P. M., and Bruce, B. D. (2005). Liposomal nanocapsules in food science and agriculture. *Critical Reviews in Food Science and Nutrition*, 45(7–8), 587–605.

Tjong, S. C. (2006). Structural and mechanical properties of polymer nanocomposites. *Materials Science and Engineering: R: Reports*, 53(3–4), 73–197.

Valdés, M. G., González, A. C. V., Calzón, J. A. G., and Díaz-García, M. E. (2009). Analytical nanotechnology for food analysis. *Microchimica Acta*, 166(1–2), 1–19.

van Dommelen, R., Fanzio, P., and Sasso, L. (2018). Surface self-assembly of colloidal crystals for micro-and nano-patterning. *Advances in Colloid and Interface Science*, 251, 97–114.

von Bültzingslöwen, C., McEvoy, A. K., McDonagh, C., MacCraith, B. D., Klimant, I., Krause, C., and Wolfbeis, O. S. (2002). Sol–gel based optical carbon dioxide sensor employing dual luminophore referencing for application in food packaging technology. *Analyst*, 127(11), 1478–1483.

Wheatley, V. M., and Spink, J. (2013). Defining the public health threat of dietary supplement fraud. *Comprehensive Reviews in Food Science and Food Safety*, 12(6), 599–613.

Wijesinghe, U., Thiripuranathar, G., Iqbal, H., and Menaa, F. (2021). Biomimetic synthesis, characterization, and evaluation of fluorescence resonance energy transfer, photoluminescence, and photocatalytic activity of zinc oxide nanoparticles. *Sustainability*, 13(4), 2004.

Xiong, R., Grant, A. M., Ma, R., Zhang, S., and Tsukruk, V. V. (2018). Naturally-derived biopolymer nanocomposites: Interfacial design, properties and emerging applications. *Materials Science and Engineering: R: Reports*, 125, 1–41.

Zhang, D., Tong, J., and Xia, B. (2014). Humidity-sensing properties of chemically reduced graphene oxide/polymer nanocomposite film sensor based on layer-by-layer nano self-assembly. *Sensors and Actuators B: Chemical*, 197, 66–72.

Zhang, P., von Plessen, G., and Offenhäusser, A. (2018). Nanoparticle based strain and chemical sensors. Doctoral dissertation, Universitätsbibliothek der RWTH Aachen.

15 The Role of Nanotechnology in Insect Pest Management

S.A. Dwivedi[*,1] *and Lelika Nameirakpam*[2]
[1]Department of Entomology, School of Agriculture, Lovely Professional University Punjab, India
[2]FPC Manager, Clover Organic Private Limited, Dehradun Uttara Khand (India)
*Corresponding author Email: sunil.21186@lpu.co.in

CONTENTS

15.1 Introduction 273
15.2 Nanoscale Components and Their Use 275
15.3 Nanomaterials 276
15.3.1 Synthesis of Nanoparticles 276
15.3.2 Natural Availability of Nanomaterials in an Insect Pest 277
15.3.3 Application of Nanoparticles Against Stored Grain Insect Pests 278
15.3.4 Efficacy of Nanoparticles Against Polyphagous Pests 279
15.3.5 Chemical and Biological Nanoparticles Against Insect Pests 280
15.4 Methods for the Preparation of Nanomaterials Based on Controlled Release Formulations (CRF) for Biocides Application 280
15.4.1 Nano-Based Formulations of Pesticides 282
15.4.2 Favourable Development of Nanopesticides Formulation 283
15.4.3 Nano-Encapsulation Pesticides Formulation Assists in Pest Management 284
15.4.4 Nano-Emulsions (NEs) 284
15.4.5 Polymers 285
15.5 Novel Nano-Insecticides Are Potential Tools Against Insect Pests in Crops 286
15.6 Applications of Nanoformulations of Pesticide 287
15.7 Conclusion 287
References 288

15.1 INTRODUCTION

Nanotechnology is a recently discovered process adopted by various related disciplines, such as medicine, biological science, biotechnology, analytical chemistry, engineering, physics and agriculture. Nanotechnology is the study and use of microscopic particles as various sizes of molecules. Around 2.5 mt of regular synthetic chemicals are used every year in the management of insect pests, as assessed in nanotechnology. This amount is expected to increase and cause overall serious harm, due to the long duration in use of pesticides, the shortage of logical plans, the pesticide permeation into the soil, into water bodies and the environment that causes contamination or build-up on the crop surfaces, thus affecting nature and general well-being. However, the current integrated pest management (IPM) system is inadequate to limit the creepy crawly populace. In this way, finding new and current methodologies for the eradication and control of bugs is urgently required. The solution is nanotechnology in farming. Subsequently, nanomaterials can be used against bug invasions

DOI: 10.1201/9781003239956-17

with this methodology. Nanotechnology has great potential, including to expand bug spray plans using nanoparticles (Ragaei and Sabry, 2014). One nanometre is one-billionth of a metre or one-billionth of a millimetre or one-thousandth of a micrometre. Nanotechnology is a newly discovered field that can be used in clinical sectors, natural science, biotechnology, logical science, designing, material science and horticulture. By and large, nano refers to a decrease in scale encased by 1 and 100 nm. One-billionth is nano; it is identified with particles included in a billionth of a metre. A nanoscale is one for every 80,000th width of a human hair or pretty much 10 'H' molecules wide (Manjunatha et al., 2019). The anticipated populace in 2050 will be 9.2 billion and overall food production will have to be based on sustainable intensification of crop production and flexibility (Friedrich, 2015). Nanotechnology can possibly upset the agrarian industry and food manufacture with its new methods, such as recent instruments for sub-atomic administration of medicines or quick infection location by upgrading the capacity of plants to retain supplements. These techniques can improve science at different harvests as well as possibly upgrade yields or dietary benefits, such as creating an improved framework for checking ecological parameters and improving the capacity of plants to retain supplements or insecticides (Tarafdar et al., 2013). New risks in developing countries include a lack of sanitation, an increased risk of illness and risks to horticulture from changing climate patterns (Biswal et al., 2012; Chowdhary et al., 2020a). The present horticultural business is unable to battle infections and other harvest microbes and thus nanotechnology can help. Some time before the infestations start, the incorporated detecting, checking and controlling framework of nanotechnology could identify the plant sickness and advise the farmer to enact bioactive frameworks, for example, pesticides, supplements, drugs, aerobatics, naturalistic arts and impalpable cell bio-reactors. Sooner rather than later, nanostructured forces will be available which build up the productivity of synthetic pesticides, and require low amounts of application. Nanotechniques can protect the soil by implication using elective (sustainable power source) supplies, and channels to decrease contamination and tidy up extant toxins present in water as well as in the soil. In rural areas, the innovative work of nanotactics is probably going to help by improving the condition of hereditarily changed harvests, animal creation inputs, replacing synthetic pesticides and offering exactness in cultivating methods (Prasad, Bagde et al., 2012). Nanotechnology is especially effective and can improve the soil fertility in developing nations (Campbell et al., 2014). As of late, food safety and nutritional safety are now part of recent wisdom. Farming progress additionally relies upon social incorporation, well-being, atmospheric variation, efficiency, the ecological community, raw materials, and must likewise have specific target situated objectives. Consequently, improving agrarian livelihoods is a useful route to avoid poverty. Agriculture is thus directed towards recuperation, in this way, the ecological presentation is required and simultaneously the natural pecking order of the biological systems is maintained, corresponding to farming food production (Thornhill et al., 2016). Nanotechniques have been created in the last ten years and have produced numerous novel nanoparticles to be applied in all areas of life. These fields incorporate nanoinsecticides, which includes those used against pests. After the Second World War, excess use of chemical pesticides, such as chlorinated hydrocarbon, organophosphorus, or carbamates occurred. Because of the broad use of every single traditional pesticide against pests, the pests developed resistance to traditional pesticides. Researchers worked on new strategies for pest control, for example, new and offbeat bug sprays to defeat the creepy crawlies. Sadly, these new strategies were unable to defeat the creepy crawlies. Over the last ten years, researchers have been working on novel bug spray items dependent on nanotechnology. Nanopesticides were first used in the agrochemical labs; however, these sprays have not yet achieved public acceptance or are accessible on the market. After all, those nanoinsecticides with upgraded properties will be modified sooner rather than later and will definitely bring about both new risks as well as advantages for mankind and ecological well-being (Prasad, et al., 2014; Prasad, Bhattacharyya et al., 2017). At present, there is great interest in a quick, solid as well as low cost ecosystem to identify, observe, and determine natural host particles in farming areas (Vidotti et al., 2011; Sagadevan and Periasamy, 2014).

15.2 NANOSCALE COMPONENTS AND THEIR USE

Nowadays, nanoscale particles and other attractive components of comparative worth can be used in nanotechniques. In the environment, the way of lifeforms from microbes to creepy crawlies depends on nanometre moulded protein that works on each thing. Nanometre categories include carbon dark, that improves the mechanical attributes of tyres, nanometre silver materials advance films, nanometre materials are the premise of impetuses basic to the carboniferous business (Fiorito, 2008; Huck, 2008). Nanotechnology has recently offered an incredible prospect for use in natural security (Nowack, 2009; Ying, 2009;). Nanoparticles are being applied in gadgets for contamination detecting, therapeutics and antiseptics. Use of nanomaterials for contamination avoidance in undisturbed earth is additionally being investigated (Duebendorf, 2008). What's more, nanostructured particles are being used in different fabricated items, mainly in composites particles and stuffing. The efficiency of nanotechnology to reform social welfare, in using materials, is recorded as well as its corresponding innovation, and vitality divisions have been very much proved. A few items empowered by nanotechnology are now available on the market, including prophylactic condiment, straightforward sunscreen creams, safe herbs and nanomaterials in eco-accommodating toxicants (Barik et al., 2008; Gha-Youthful et al., 2008; Bhattacharyya, 2009).

The range of nanomaterials used to manage plant infections includes nanoforms of silica, alumino-silicates carbon and silver. Nanotechniques have amazed the academic world in light of the fact that at the nanometre scale, particles present various attributes. The use of nanocategories of silver material as antimicrobic operators has become progressively more regular as innovation expands, making production increasingly practical. After all, silver presents various methods of deactivating microbes, so it might be used to manage different disease-causing organisms in a generally more secure manner, as opposed to the application of fungicides. Silver influences numerous bio-chemical procedures in the microbes, remembering the progress in procedure capacities and plasma film. Such particles additionally forestall the outflow of adenosine triphosphate (ATP) creation-related protein (Yamanka et al., 2005). The exact instrument of bio-particles restraint is not yet fully understood. Accordingly, use of nanoparticles has been viewed as a substitute and a compelling methodology which is environmentally accommodating as well as wise in the management of disease-causing organisms (Kumar and Yadav, 2009; Prasad et al., 2011; Swamy and Prasad, 2012; Prasad and Swamy, 2013). Nanomaterials have an extraordinary strength in the administration of plant diseases compared to engineered fungicides (Park et al., 2006). Magnesium oxide (MgO), zinc oxide (ZnO) nanomaterials are powerful aseptic as well as anti-odour operators (Shah and Towkeer, 2010). Simplicity and observable straightforwardness and perfection create ZnO and MgO nano-arrangements as an appealing antiseptic. Both additionally have been proposed as suitable repellents of microorganism additives for timber or foods (Aruoja et al., 2009; Huang et al., 2005; Sharma et al., 2009). Appropriate functional nanocapsules give better infiltration via insect fingernail skin as well as permit moderate, managed arrival of dynamic fixing on reaching the target weed. Use of nanobiopesticides is increasing since they are safe for plants and offer less ecological contamination compared to customary substance toxicants (Barik et al., 2008). Nanosilver particles are used in the bio-framework. For some time a solid repressive and bactericidal impact has been sought as well as a wide range of antiseptic exercises (Swamy and Prasad, 2012; Prasad, Swamy et al., 2012; Prasad and Swamy, 2013). Silver (Ag) nanomaterials, with exterior territory as well as a good portion of surface molecules, give high antimicrobial impact compared to their mass (Suman et al., 2010). The fungicidal attributes of a nanocategory Ag colloidal arrangement is used as a specialist for antimycotic treatment of different organisms that cause plant diseases; most noteworthy, the restraint of plant pathogenic parasites has been6 observed on Potato Dextrose Agar (PDA) and 100 ppm of AgNPs (Kim et al., 2012).

15.3 NANOMATERIALS

To counteract the *Pediculus humanus capitis* adult and nymphal activity, blended Ag nanomaterials were applied to a leaf extract concentrate for *Tinospora cordifolia* and indicated a very effective extinction of head mite *Pediculus humanu* as well as fourth instar nymphs of *Anopheles subpictus* and *Culex-quinque fasciatus* (Jayaseelan et al., 2011). Nanomaterials stacked with garlic basic oil are effective against red flour beetle, *Tribolium castaneum* (Herbst, 1797) (Yang et al., 2009). Nanotubes with aluminosilicate can adhere to plants externally, while elements of the nanotube can adhere to the outer threads of the creepy crawly bugs and finally enter the body and attack specific physiological capacities (Patil, 2009). Nanomaterials present opportunities to progressively viably manage nuisances, yet our related deficiency of data on how they react and how they can be accommodated are making officials pause before permitting their discharge into the earth (Khot et al., 2012). Nanopesticides were praised for reducing the ecological impression left by ordinary toxic components.

15.3.1 Synthesis of Nanoparticles

The concoction strategies depend upon a synthetic bond shaped between the dynamic compound and a wrapping grid, for example, a polymer. This bond is in two unique locales: principal polymeric bracelets or inner side-chain. First, are novel "macromolecules", otherwise known as ace equivalents, in light of the fact that the compound will retain its attributes in full at the time of discharge. Next, the bug spray atoms will connect to the side-chains of one monomer as well as afterwards the polymerization response happens or if the polymerization happens only from that point onward, the biocide connects to the side-chain (Wilkins, 2004).

Perspectives available for nanoparticles synthesis are bottom-up and top-down (Arole and Munde, 2014). The latter approach is excessive as well as stable, whereas the former necessitates self-building of atomic size materials to enlarge the nanosized particles that are attained by corporeal as well as synthesis means (Thakkar et al., 2010) as shown in Table 15.1. Nevertheless, eco-friendly green syntheses are economical, and proliferate and trigger stable NP formation, as shown in Figure 15.1.

TABLE 15.1
Chemical and Physical Synthesis of AgNPs

Type	Reducing agent	Characterization	Biological activities	Reference
Chitosan-loaded AgNPs	Polysaccharide chitosan	TEM, FTIR, XRD, DSC, TGA	Antibacterial	Ali et al., 2011
PVP-coated AgNPs	Sodium borohydrine	UV-vis, TEM, EDS, DLS, FIFFF	NANA	Tejamaya et al., 2012
AgNPs	Ascorbic acid	UV-vis, EFTEM	Antibacterial	Pal et al., 2007
AgNPs	Hydrazine, D-glucose	UV-vis, TEM	Antibacterial	Shrivastava et al., 2007
Polydiallyldimethylammonium chloride and polymethacrylic acid–caped AgNPs	Methacrylic acid polymers	UV-vis, reflectance spectrophotometery	Antimicrobial	Dubas et al., 2006

Notes: NPs, nanoparticles; TEM, transmission electron microscopy; FTIR, Fourier-transform infrared; XRD, X-ray diffraction; DSC, differential scanning calorimetry; TGA, thermogravimetric analysis; UV-vis, ultraviolet-visible (spectroscopy); EDS, energy-dispersive spectroscopy; DLS, dynamic light scattering; Fl-FFF, flow field-flow fractionation; EFTEM, energy-filtered TEM; NA, not applicable.

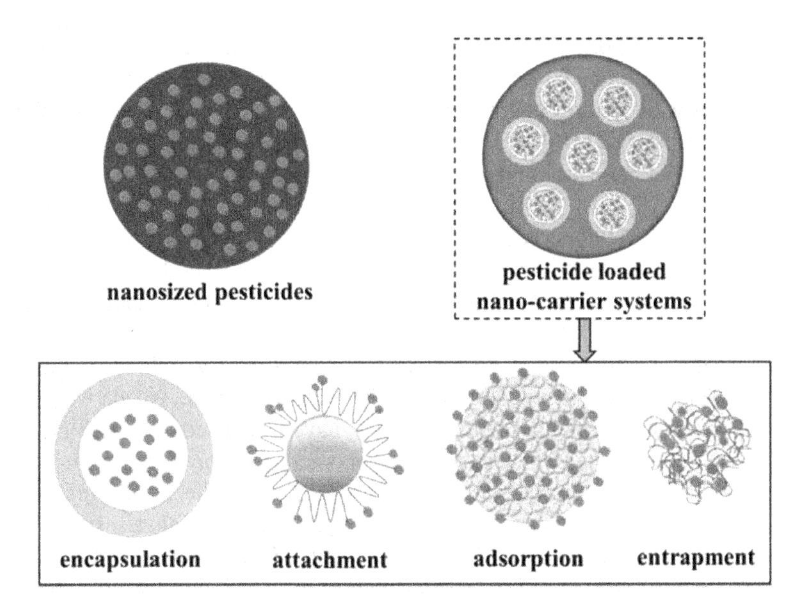

FIGURE 15.1 Different approaches to synthesis of Ag nanoparticles (NPs).

Synthetically orchestrated nanoparticles nowadays are treated as harmful to the ecosystem, so because of this, nanocomponents may connect to the plant framework and are studied as green nanotechniques (Prasad, 2014). This is a protected procedure, it decreases waste and reduces ozone-depleting substance emanations. Use of inexhaustible materials underway in such items is effective, these procedures have less effect on the ground surface (Prasad et al., 2014; Prasad et al., 2016). Nanoparticles are eco-friendly feasible as well as effective and belong to the field of green nanotechnology. How will the ecological sustainability of this technology be accomplished in future? These dangers are surely relieved by promoting green nanotechniques (Kandasamy and Prema, 2015). The improvement of the cutting-edge farming framework through designed clever nanotools could be a brilliant procedure to cause a revolution in rural practices, and along these lines lessen and additionally remove the influence of present-day agriculture on the earth, by upgrading both the predation systems and the amount produced (Sekhon, 2014; Liu and Lal, 2015). Bio-sensors are likewise something offering numerous qualities of nanotactics, along these lines, this tool can do a basic job. Because of the exceptional attributes of nanoparticles, bio-sensors could effectively advance progress (Fraceto et al., 2016). Numerous novel possibilities are presented by bio-sensors (Sertova, 2015).

15.3.2 Natural Availability of Nanomaterials in an Insect Pest

Unfortunately, nanoformations are being ignored, despite being a conceivably potential source of items that meet certain particulars (Watson and Watson, 2004). Their increasing incorporation depends upon nanotechniques being used so far as a free innovation accessible in the environment (Ehrlich et al., 2008). A genuine model is the arranged hexadic stuffed exhibit from the wings of a cicada, for example, *Psaltoda claripennis* Ashton, as well as the white ants family Rhinotermitidae (Zhang and Liu, 2006). Likewise, the ferromagnetic component distinguished in the Italian Honey bee's *Apis mellifera* Linnaeus mid-sections is also recognized as reasonable use (Desoil et al., 2005). Attractive nanoparticles in the *A. mellifera* mid-regions are acknowledged as associated with their magneto reception component (El-Jaick et al., 2001). Bionic attractive attributes of bumble bees and *A. mellifera* are examined with the end goal of learning the honey bee's metabolic reaction to an attractive field Billen, 2006; Hsu et al., 2007; Chambarelli et al., 2008). Also the polarizations of the honey bee's mid-section, forelegs, head and thorax are estimated independently as elements of temperature as well as farm conditions (Esquivel et al., 2002; Desoil et al., 2005). Both the anti-ferromagnetic reactions of ferrihydrite centres of the iron stockpiling ferritin protein as well as ferromagnetic reactions of nanoscale magnetite ($Fe3O_4$) materials are recorded (Dawson, 2008; Tai Kai, 2008).

Moderately huge magnetite materials (ca. 30 nm or progressively), equipped for maintaining permanent polarization at ambient temperatures are found in the mid-regions, however, but missing in the heads and thorax. Subterranean red imported fire ant (*Solenopsis invicta*) labourers, sovereigns and winged, are broken down by attractive reverberation imagery (X-ray) to identify regular attraction. Every single ferromagnetic material is attractive to nanoparticles, which are exclusively able to restrict explicit heading for food in a host of creepy crawlies. As of late, cicada wings examined by nuclear power microscopy (Atomic Force Microscopy) have been used for watching nanoparticles. Experiments demonstrate that the dimensions of the nanomaterials may range from 200 to 1000 nm. The frameworks will in general have adjusted the proportions at the zenith and it is only some 150–350 nm out from the external surface, as such winged nanomaterials assist in the streamlined proficiency of the insects. Disconnected nanomaterials of the insects have widths of around 12 and 11 nm in the mid-section with petiole and head with radio wires, individually. Nanocomponent parts are additionally present in creepy crawlies. Wings of butterflies have brilliant shading parts and these shading segments are only nanoparticles. As of late, a novel photodegradable bug spray including nanoparticles has been prepared (Guan et al., 2008). These metal nanomaterials can be used in defining bug sprays, toxic substances and creepy crawly-repulsing synthetic compounds (Esteban-Tejedaet al., 2010; Zahir et al., 2012). Nanosilica that is set up from silica can be used as a nanopesticide. Creepy crawlies use a variety of lipids on their fingernail skin to ensure water on their bodies along these lines to keep them from dying of dryness. This system of creepy crawly deterrent is used by the nanosilica that is consumed in the lipids of fingernail skin by physiosorption, consequently causing bug death exclusively by physical means, when this pesticide is sprayed on external areas of the leaves and the stem (Barik et al., 2008). Nanomaterials will bring about an improvement in productive and potential methodologies in the extinction of pests. Some specialists globally are working on this project, and henceforth, these nanotechniques will show definite progress in this field (De et al., 2013). Nanotubes loaded with aluminosilicate can adhere to the external part of a plant, while a constituent of a nanotube can adhere to the surface hair of creepy crawlies and enter their body and affect its physiological capacities (Patil et al., 2009). There is additionally a report of the adequacy of Diatomaceous Earths (DEs), made chiefly out of shapeless silica and obtained from fossilized phytoplankton, working against bugs on the off chance that it has high silica content with uniform size dispersion. Normally, unstable phyto-chemicals and nanoparticles of bugs are exclusively responsible for the association between plant and creepy crawly (Gorb and Gorb, 2009). In addition, electro-spun silk fibroin-based strands with a normal distance across of 700 nm, made by a silk-producing moth, are set up from watery recovered silkworm silk arrangements (Jin et al., 2004; Zhang et al., 2007). The electro-spun nanocomposite of silkworm silk helps in creating single divider carbon nanotubes for a sedate conveyance framework (Ayutsede et al., 2008).

15.3.3 Application of Nanoparticles Against Stored Grain Insect Pests

Nanomaterials stacked with garlic, *Allium sativum*, basic oil are significant in the case of red flour beetle, *Tribolium castaneum* Herbst (Yang et al., 2009). The danger of the imidacloprid increased when covered with nanoparticles, the poison of the half nano SDS/Ag/TiO2-IMI was higher in the adult stage of *Martiansder mestoides* Chevrolat contrasted with the 95% IMI, as demonstrated by low LC_{50} estimated (Guan et al., 2008). SiO_2 and Ag nano particles could be potential pesticides on larval and adults pulse beetle, *Callosobruchus maculatus* (Rouhani et al., 2012). The LC_{50}

consideration for SiO_2 and Agnanoparticles were determined as 0.68 and 2.06 g/kg cowpeas on adults and 1.03 and 1.00 g/kg on grubs, separately. Results indicated that both silica and silver nanomaterial are exceptionally successful on adults and grubs with 100% and 83% death rate, respectively. According to Debnath et al. (2011), the surface-functionalized silica nanoparticle (SNP) may be a practical option compared to ordinary pesticides. The entomotoxicity of SNP was tried against rice weevil *Sitophilus oryzae* and its efficiency was compared to mass estimated silica (singular particles bigger than 1 smaller scale m). Formless SNP was seen as exceptionally compelling against this creepy crawly, causing over 90% mortality, showing the efficiency of SNP to control bugs. As indicated by Vani and Brindhaa (2013), silica nanoparticles were incorporated by altered Stober's sol-gel technique and described by filtering electron magnifying lens, x-beam diffraction and vitality dispersive x-beam spectroscopy, and had an entomotoxic impact against the grain bug *Corcyra cephalonica*. Nebulous silica nanoparticles were seen as profoundly viable against this creepy crawly bug, causing 100% mortality. There is likewise a report of insecticidal efficiency of indistinct lipophilicsilica nanoparticles on red flour beetle (*Tribolium castaneum*) (Debnath et al., 2012; Zahir et al., 2012). Also considered are the entomotoxic impacts of silver nanoparticles (AgNPs) (orchestrated by using watery leaves concentrates of *Euphorbia prostrate*) on *Sitophilus oryzae.* Pesticidal bioassay tests are performed at different fixations for 14 days. The LD_{50} estimations of fluid concentrate, the $AgNO_3$ arrangement and the incorporated AgNPs are 213.32, 247.90, 44.69 mg/kg; LD_{90} = 1648.08, 2675.13, 168.28 mg/kg, separately such outcomes propose that the leaves' moist extracts of *Eclipta prostrata* and integrated Ag NPs can possibly be used as a perfect eco-accommodating base for the control of the rice weevil, *Sitophilus oryzae.* Stadler et al. (2010) effectively applied nano-aluminium against two grain pests, *S. oryzae* and the lesser grain borer, *Rhyzopertha dominica.*

Consumers' objections regarding foreign matter in foodstuffs involve finding pests like insects; mould and toxic fungi which adulterate the wrapping in a short time in a stockpile. Cereals are a valuable stored food grain. For the assessment of effectiveness, silver nanocomponents were assimilated into wrapping polymer on developing fungus as well as insect pests. *Sitotroga cerealella* is checked due to its financial value in the storage industry of protected grains. Use of AgNPs is via excellent wrapping polymers. Results showed that not only AgNPs favourably release poisonous microbes but also have a toxic effect on pests. Thus, the attributes of antimicrobic and toxicity of AgNPs as a nonchemical pesticide should be examined in storage entomology for the safety of food grains (Allahvaisi, 2016).

15.3.4 Efficacy of Nanoparticles Against Polyphagous Pests

Different nanoparticles like Cds, nano-Ag and Nano-TiO_2 have demonstrated a critical impact on the tobacco caterpillar, *Spodoptera litura*. The portion reaction information of the second instar *S. litura* larvae showed the destructive power of nanoparticles, the LC_{50} of compact discs was 508.84, of nano-TiO_2 and nano-Ag the LC_{50} was 791.10 and 1403.14 ppm, individually. Compact discs nanoparticles had a higher larval mortality of 21.41, 93.79% at 150 and 2400 ppm, separately (Chakravarthy et al., 2012). Use of nanosilica on tomato plants may limit *Spodoptera littoralis* pervasion. It gives a moderate level of protection, however, it presents the upside of being plausibly incorporated together with the broad strategies in controlling this nuisance. Nanosilica splashes influence the inclination of the *Spodoptera littoralis*, along these lines expanding the protection of tomato plants. Likewise it influences the natural parameters of the creepy crawly, such as its lifespan and procreation, subsequently decreasing the conceptive potential (El-Bendary and El-Helaly, 2013). Rouhani et al. (2012) announced the entomotoxic impacts of Ag and Zn nanoparticles (combined via a solo thermal technique) on sweet paper aphid, *Aphis nerii.* For examination purposes, imidacloprid was likewise used as a customary bug spray. In the trials, the LC_{50} consideration for imidacloprid, Ag and Ag-Zn nanoparticles was determined to be 0.13m/L, 424.67 mg m/L, and 539.46 mg m/L, individually.

The result indicated that Ag nanoparticles can be used as a significant instrument in bug projects of *A. nerii*. A nanogel has been set up from pheromone, methyl eugenol using a low-sub-atomic weight gelator. This is very stable in unsecured surrounding situations and hindered the dissipation of pheromones appreciably. This enabled a decrease in the recurrence of pheromone energizing in the plantation. Eminently the association of the nanogelled pheromone presented a powerful deterrent to the oriental fruit fly, *Bactrocera dorsalis* (Hendel), a common pest destructive of various horticultural products including guava, mango, etc. (Bhagat et al., 2013). Insect pests are creature populaces, which occur in each conceivable condition with a differed number of animal groups. Among them, a few pests act as vectors of various diseases and cause damage to plants, thus influencing the production of yields in national and global markets. Harvested outcomes show the misfortunes brought about by insect pests in farming, a few synthetic substances have been applied to control them (Ragaei, 2014). Larval nuisances are one of the main sources that influence farming efficiency, causing billions of dollars of misfortune every year (Ramya, 2008). The significant harvest plants that are influenced by the pod borer of Bengal gram are maize, cotton (Malik, 2002), nut, sorghum, pigeon pea and tomato. This pest has been affecting more than 180 types of hosts and 45 groups of wild cultivars (Sullivan, 2007). Production of cotton is low because of 150 distinctive attacks by pests at different life stages (Gandhi, 2009). This forces an analysis to assemble the customary and advanced strategies to beat different pests around the globe. Collaboration prompts the actuation of a safe framework for plants and there is cross-discussion between different ways of creating both essential as well as auxiliary mortalities. A cell divider is a hindrance to a caterpillar but is crossable later so the deadly pests can progress to conclusively kill the plant. The auxiliary metabolites created are an explicit protective framework, as in that way metabolites have a broad role in the development of different metabolites with a clear trademark. The plant auxiliary metabolites help the plant in several different ways, for example, counteraction of phytophagous and causal organism attack, fascination of pollination, symbionts, crops correspondence (Panghal et al., 2014).

15.3.5 Chemical and Biological Nanoparticles Against Insect Pests

Nanoparticles show entomotoxicity against insect pests. Nanosilica splashes influence the extinction of the cotton leaf worm, *Spodoptera litura*, expanding the protection of tomato plants. It reduces the lifespan, fertility and regenerative capability of females on the tomato plants and therefore decreases the creepy crawly populace, reducing the harm and yield misfortunes to the harvest (El-Bendary and El-Helaly, 2013). Silica and silver nanoparticles are very powerful against adult and hatchlings of cowpea seed insect, *Callosobruchus maculates* with a mortality rate of 100% and 83%, respectively (Rouhani et al., 2012). Chemical and biological materials are listed in Table 15.2.

15.4 METHODS FOR THE PREPARATION OF NANOMATERIALS BASED ON CONTROLLED RELEASE FORMULATIONS (CRF) FOR BIOCIDES APPLICATION

As indicated by (Wilkins, 2004), the strategies for CRF (controlled release formulations) planning can be isolated in compounds or in physical ones. The concoction techniques depend on a synthetic bond (normally a covalent one) between the dynamic compound and the covering framework, for example, a polymer. This bond can be placed in two distinct locales: in the fundamental polymeric chain or in a side-chain. In the former, the new "macromolecule" is additionally called a professional biocide, on the grounds that the compound will gain its properties in reality when it is discharged. In the subsequent one, the bug spray atom can connect at first to the side-chain of one monomer and afterwards the polymerization response happens or the polymerization happens only from that point onward, the biocide connects to the side-chain. There is yet a third way, in light

TABLE 15.2
List of Chemical and Biological Nanoparticles Against Pests

Metals	Biological agents	Insect pests	Reference
Chemical nanomaterials	-		
Ca	-	*Bactrocera dorsalis*	Christenson and Foote, 1960
CdS, Ag and TiO_2	-	*Spodoptera litura*	Chakravarthy et al., 2012
Ag and Zn	-	*Aphis nerii*	Rouhani et al., 2012
AgNPs	-	*S. litura and Achaea janata*	Yasur and Rani, 2015
SNPs	-	*S. litura*	Debath et al., 2012
AgNps	Bifenthrin	*Lygus hesperus*	Louder, 2015
Bio nano materials			
Gold, CdS, TiO_2	DNA	*S. litura*	Chakravarthy et al., 2012
Gold	DNA	*S. litura*	Chandrashekharaiah et al., 2015
AgNPs	*Aristoolochia indica* and *Cassia occidentalis*	*Helicoverpa armigera*	Siva and Kumar, 2015

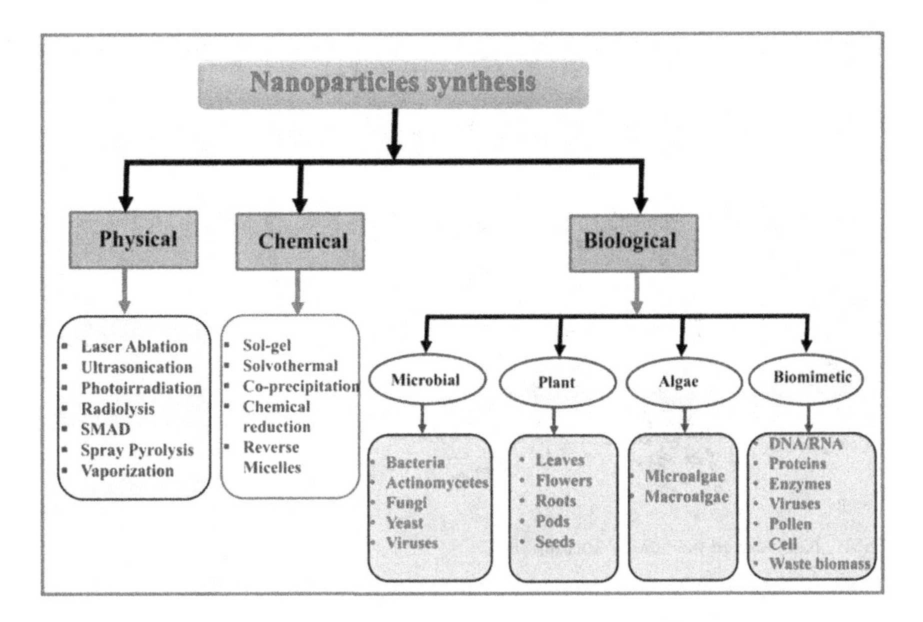

FIGURE 15.2 Physical methods of CRF preparation.

of the intermolecular communications. For this situation, the biocide is "immobilized" in the net delivered by the cross-linkages in the polymer. The physical strategies can likewise be part of two unmistakable classes. In the initial one, a blend of biocide and polymer is made. In the other one, the polymeric chain frames a "film", detaching the bioactive compound from the outer condition. This is simply the technique which will create the nanocapsules themselves. In spite of the fact that there are various types of nanomaterials that can be used in CR plans, the small scale and nanocapsules are by a long shot the most generally used method for the controlled arrival of biocides. Physical methods for CRF are presented in Figures 15.2 and 15.3.

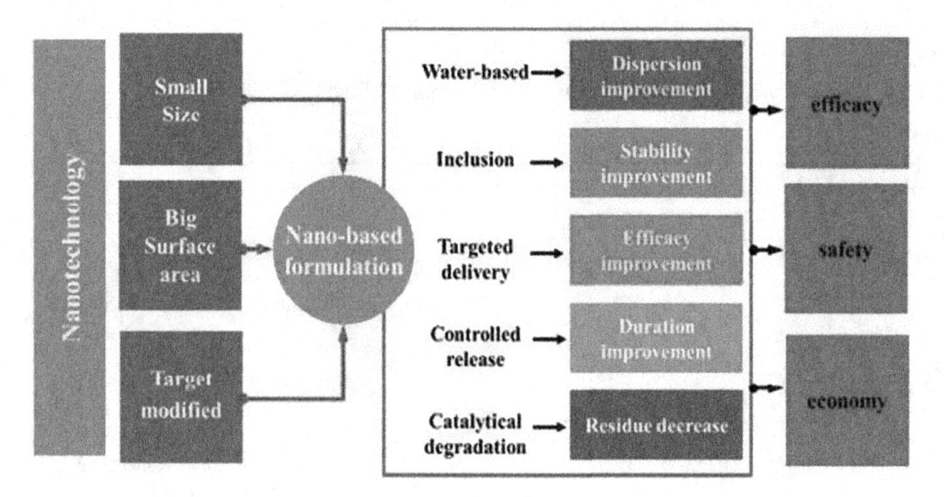

FIGURE 15.3 Chemical methods of CRF preparation.

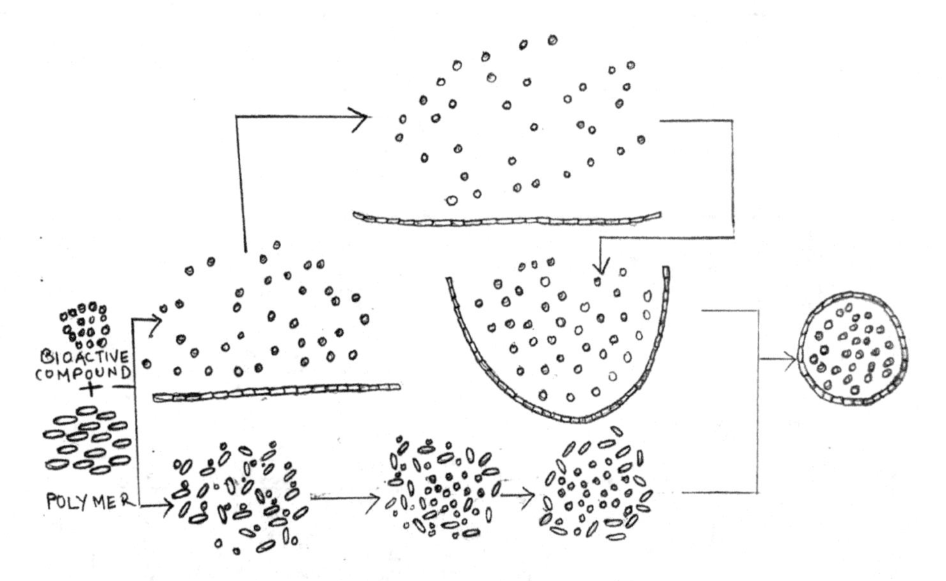

FIGURE 15.4 Nano-based pesticides' formulation.

15.4.1 Nano-Based Formulations of Pesticides

Nanopesticides can be applied to ceramics, proteins, emulsion, polymer, lipids and metals. The general procedure for the manufacture of nanotoxic substances is to use an oil in water emulsion system (micro as well as nano-emulsion) conversion of emulsion to organic nanomaterial by grinding, solvent dissipation, coacervation and condensation methods (Elek et al., 2010). Nanopesticides can be modified by straight processing within the nanomaterials or by loading a toxic material with a nanotransporter in a conveyance arrangement. In nanocarrier systems, toxic substances are filled by encapsulation inside the nanomaterial polymeric shell, absorption onto the nanoparticle surface, attachment on the particle core via ligands or entrapment within the polymeric matrix. A variety of nanoformulation types have been developed, covering nano-emulsions, nanocapsules, nanospheres, nanosuspensions, solid lipid nanoparticles, mesoporous nanoparticles and nanoclays shown in Figure 15.4 (Ao et al., 2013).

Nanopesticides can be comprised of natural fixings of metal oxides in different structures, i.e., particles and micellular. Nano-articulations follow the common regular paths of other toxic substances and are comprised of: expanding the clear solvency of an ineffective solvent dynamic fixing, discharging the dynamic fixing in a moderate/directed way and additionally securing the dynamic fixing against untimely debasement. The nanoparticles used in biopesticides control are described below:

- The best-known nanomaterials that have been used in controlled discharge formulations for biocides conveyance are *nanospheres*: the total in dynamic composite is homogeneously disseminated in the polymeric structure.
- *Nanocapsules*: the total in which the dynamic compound is thought to be close to the inside centre, connected by the lattice polymer.
- *Nanogels*: hydrophilic (for the most part cross-connected) polymers which can assimilate high volumes of water.
- *Micelles*: the total framed in an aqueous arrangement by particles containing hydrophilic and hydrophobic moieties (Ragaei and Sabry, 2014).

Numerous attempts have been made to overcome the plague of creepy crawlies, for instance, using natural control, which is very tedious. Controlled discharge frameworks appear in this situation as an extremely alluring option in this combat zone. Controlled discharge details partner the dynamic composite with a dormant substance. The latter are liable for ensuring as well as dealing with the pace of compound discharge into the objective site in a characterized timeframe. The principal motivation behind controlled discharge frameworks is administering the bio-accessibility of a dynamic composite after its utilization (Wilkins, 2004). Nanocapsules have been generally used as medication bearers in the therapy of different ailments (Radhika et al., 2011), from tropical diseases (Kuntworbe et al., 2012) to malignant growths (Joshi et al., 2012). Micro-encapsulation is a flexible apparatus for a hydrophobic toxicant, improving its scattering in a watery admixture and permitting its arrival at the dynamic composite. As a keen conveyance framework, they present greater differentiation, and lack of blocking in the bioactive mixes with regards to objective microbes (Peteu et al., 2010). Nanomaterial-based bug sprays offer different favourable conditions, such as being able to plan using a process that is more promptly scattered in arrangement. It lessens the issues related with floating and filtering, because of its strong ecosystem as well as prompting an increasingly compelling collaboration with the target creepy crawly. These sprays enable the use of less amounts of dynamic compound per zone, and the plan may provide an ideal fixation conveyance to the objective bug spray for longer periods because there is no need for respraying; they likewise reduce the cost and the problem of the human sludge-film, offering plant toxicity as well as reducing natural harm to other non-targeted living beings and even the yields themselves (Margulis-Goshen and Magdassi, 2012). In a couple of words, nanotechniques can be used in several different ways to upgrade the viability of bug sprays in crop protection. Nowadays research and demonstrations on nanopesticides provide strength to decrease the adverse effect of synthetic pesticides as well as give excellent results in the management of target pests. Improved nanoframeworks can reduce problems in agribusiness, such as natural diversity, food security and sustenance efficiency (Dwivedi and Singh, 2020).

15.4.2 Favourable Development of Nanopesticides Formulation

Nanomaterials are used as agrochemicals for pest management, compared to traditional pesticide formulations, in that traditional condition more than 90% of chemicals are lost in the crop ecosystem and available in harvested products as residues. Pest management systems simplify the uninterrupted release of active ingredients onto the target pest over a widespread range in a short period

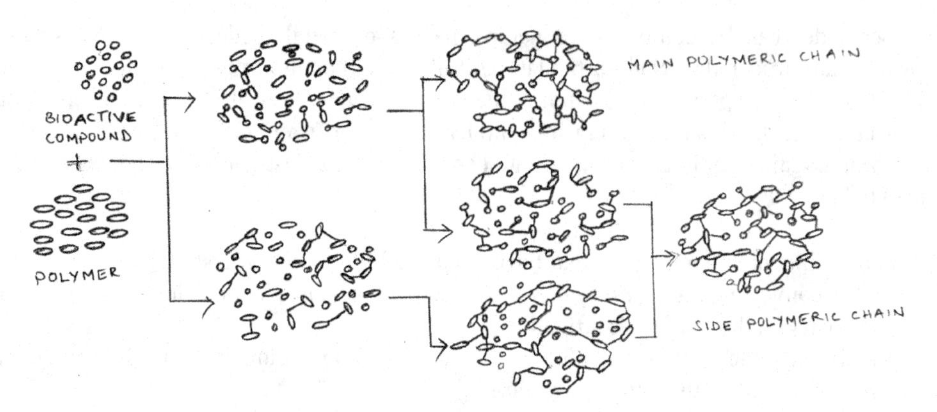

FIGURE 15.5 Favourable development of nanopesticides' formulation.

of time, due to widening the targeted release effectiveness of the chemical on insects, pests and pathogens (John et al., 2017), increasing the solubility as well as scattering as fat-dissolvable toxic substances in fusible explication (Anjali et al., 2012). Nanopesticides decrease the frequency of pesticidal spraying and treatment abundance by increasing the stable efficacy of their duration (Yu et al., 2017), improving their bio-potency, decreasing the pesticidal input in plants by avoiding the problem of non-target toxicity, as well as enhancing its workability as light-sensitive components using contract photo-degradation (Anjali et al., 2012). Constant discharge of pesticide harms the biodiversity of the environment (John et al., 2017). The favourable development of nanopesticides formulation is represented in Figure 15.5.

15.4.3 Nano-Encapsulation Pesticides Formulation Assists in Pest Management

Nano-encapsulation is the method by which a composite is slowly but productively released to the targeted host for creepy crawly pest management. Release techniques assimilate decomposing biodegradation, dispersal and absorbent weight with an explicit pH (Vidyalakshmi et al., 2009). Epitomized citronella oil nano-emulsion is set up by an elegant synthesis of 2.5% wetting agents and 100% glycerol, to create secure beads for the expansion of the oil as well as tolerate discharge. The rate of discharge is based on the assurance time, thus a reduction in discharge can extend the mosquito protection time (Sakulk et al., 2009). Nanopesticides, nanofungicides and nanoherbicides are being applied successfully in fruit and ornamental crops (Owolade et al., 2008). The extermination of fruit flies, including sex pheromones, is valuable in reducing unfortunate pest populaces responsible for destroying the production harvest. A nanogel was set up from a methyl eugenol using a low-sub-atomic weight gelator. It is really steady in open surroundings and prevented the dissipation of the pheromone altogether (Bhagat et al., 2013). The classification of the most effective nanoparticles is being applied in the management of discharge formulations for pesticides (Ragaei and Sabry, 2014).

15.4.4 Nano-Emulsions (NEs)

NEs are self-emulsifying colloidal frameworks, having the interior stage normally smaller than 100 nm, scattered in a continuous medium (Anton, 2011). This attribute improves some physiochemical properties, for example, security and bioavailability. Truth be told, the small size of the inner stage permits the framework to sidestep the issues identified with gravity, maintaining a strategic distance as dusting or accumulation. Besides, this low surface and assemblage strains advance the appropriate spread and entrance of the dynamic mixes (Tadros, 2004). NEs are commonly composed of an aqueous stage, a slick stage, a surfactant specialist and a potential co-surfactant. Thus, they can fuse both hydrofoil and lipophilic mixes. The selection of NEs' parts is carefully identified with their application. For instance, it is conceivable to choose a few sleek stages between manufactured oils (triglycerides ethyl oleate and squalene), or vegetable and mineral oil (soybean, olive and sunflower). For the most part, the slick stage is used to solubilize and convey lipophilic atoms, yet at times the sleek portion, on account of plant fundamental oils (EOs), can likewise be the dynamic fixing. EOs have been broadly used in conventional medication around the globe since medieval times, essentially for their antimicrobial and cancer prevention agent properties. An essential point on the detailing of EOs is the choice of reasonable surfactant specialists. The amphibological attributes of a surfactant are matched by the hydrophilic–lipophilic equalization (HLB) consideration. The decision on the appropriate HLB consideration relies upon the idea of consistent stages. Notwithstanding, it ought to be easy to choose a surfactant with a moderate worth since it will separate between the watery and the sleek stage, bringing down the interfacial strain and giving the ideal bend of the layer, to ensure the arrangement and adjustment of the beads. Contingent upon the synthetic equity, ions are isolated into various classes: non-ionic, cationic, zwitterionic and anionic. Mostly discussed are polisorbates (anionic, for example, Tween 80 (HLB 16.7) and Range 80(HLB8.6). Lately, enthusiasm has been developing for abusing the surfactant properties of regular items, for example, proteins (lectin) sugar esters and polysaccharides for the improvement of natural formulations. NEs profoundly checked out, since they have some down-to-earth points of interest: ease of plan, modern scale-up and high strength for use in some applications. To start with, it is essential to state that, in spite of the prefixes "nano", they are characterized by two unique requests of size, i.e., 10–6 and 10–9, individually, and the size of the scattered stage (for the most part slick beads) for both of these two frameworks falls into the nanometric extent. It is difficult to precisely characterize the scope of molecule size appropriation, since different writers report various outcomes within the nanometric request of magnitude (Gasco, 1997; Talegaonkar, 2008). Regardless, it has been stated that MEs are defined as a smaller size of the scattered stage with respect to NEs (Rao, 2011; Clements, 2012).

15.4.5 Polymers

Amphiphilic copolymers, blended from poly (ethylene glycols) and different aliphatic di acids, which self-gather into nanomaterial totals in fluid media, are used to create controlled discharge (CR) definitions of imidacloprid [1-(6 chloro-3-pyridinyl methyl)-N-nitro imidazolidin-2-ylideneamine] using an embodiment strategy. High solubilization forces and low basic micelle fixation (CMC) of these amphiphilic polymers may build the efficiency of plans. Definitions were portrayed by infrared (IR) spectroscopy, dynamic light dissipating (DLS) and a transmission electron magnifying instrument (TEM). Exemplification proficiency, loading limit and dependability after a quickened stockpiling trial of the created plans were contemplated. Discharge from the business detailing was quicker than the CR definitions. The dispersion type (estimation) of imidacloprid in water ran from 0.22 to 0.37 in the testicle while the time taken for arrival of half of imidacloprid extended from 2.32 to 9.31 days for the CR details. The created CR definitions can be used for productive irritation in various yields. At the point when a business definition for a commonsense field application is wanted, it is imperative to use materials that are perfect for the proposed applications: condition cordial, promptly biodegradable, not creating poisonous corruption by items, and minimal effort. The use of a few biopolymers, i.e., polymers that are delivered by common sources, which simultaneously have great physical and substance properties and still present mellow biodegradation conditions, is a fascinating way to maintain a strategic distance from the use of petrochemical subsidiaries that may be another source of ecological damage (Adak et al., 2012). The basic polymers (manufactured and normal ones) used in CRFs for bug sprays application are recorded in Table 15.3.

TABLE 15.3
Several Examples of Polymers Often Used in the Nanoparticles' Production

Polymer	Active compound	Nanomaterial	Reference
Lignin-polyethylene Glycol-ethyl cellulose	Imidacloprid	Capsule	Flores-Cespedes et al., 2012
Polyethylene glycol	B-Cyfluthrin	Capsule	Loha et al., 2012
Chitosan	Etofenprox Piperonyl	Capsule	Hwang et al., 2011
Polyethylene	Butoxide and Deltamethrin	Capsule	Frandsen et al., 2010
Polyethylene glycol	Garlic essential oil	Capsule	Yang et al., 2009
Acrylic acid-Buacrylate	Itraconazole	Capsule	Goldshtein et al., 2005
Carboy methyl cellulose	Carbaryl	Capsule	Isiklan, 2004
Polyamide	Pheromones	Fibre	Hellmann et al., 2011
Starch-based polyethylene	Endosulfan	Film	Jana et al., 2001
Methyl methacrylate and methacrylic acid with and without 2-hydoxy ethyl methacrylate cross linkage	Cypermethrin	Gel	Rudzinski et al., 2003
N-(octadecanol-1-glycidyl ether)-O-sulphate chitosan-octadecanol glycidyl ether	Rotenone	Micelle	Lao et al., 2010
Polyethyleneglycol-dimethyl esters	Carbofuran	Micelle	Shakil et al., 2010
Carboxymethyl chitosan-ricinoleic acid	Azadirachtin	Particle	Feng and Peng, 2012
Chitosan-poly (lactide)	Imidacloprid	Particle	Li et al., 2011
Polyvinylchloride	Chlorpyrifos	Particle	Liu et al., 2002
Glyceryl ester of fatty acids	Carbaryl	Spheres	Quaglia et al., 2001
Anionic surfactants (sodium linear alkyl benzene sulfonate, naphthalene sulfonate condensate sodium salt and sodium dodecyl sulphate)	Novaluron	Powder	Elek et al., 2010

15.5 NOVEL NANO-INSECTICIDES ARE POTENTIAL TOOLS AGAINST INSECT PESTS IN CROPS

Nanoparticles, such as silver, silica, carbon, titanium dioxide, zeolites, or copper alumina, are proposed as prospective components in the management of insect pests. Nanomaterials proceed in various dimensions (unpredictable frequently >100 nm), from rigid disabled particles to a polymer and oil–water-based formation, apart from as a preservative as well as an active ingredients (a.i.). Botanical nanomaterials change the function or risk side of a.i. They have many advantages over ordinary insecticides, including an effective formulation, safe and quick utilization, effectiveness of targeting specific pests, low dose of application and a safe ecological point of view (Walker et al., 2017). Actuated opposition in plants, because of the use of silica, against numerous herbivorous arthropods has been well recorded (Keeping and Kvedaras, 2008). Various forms of nanopesticides are successful in bug sprays, fungicides, herbicides (Matsumoto et al., 2009; Peteu, 2010). Nanotechnology is rising as an exceptionally appealing device for detailing and conveyance of pesticide dynamic fixings. For example, nanocapsules dependent on polymers are intended for the controlled arrival of dynamic fixing and upgraded conveyance through improved infiltration through leaves. Nanotechniques have successfully been used in nanoparticle-interceded gene transfer. They very well may be used to convey DNA and other wanted synthetic substances into tissues culture for the security of the targeted host against creepy crawlies (Torney, 2009). Nanomaterials will bring about the advance in productive and potential methodologies in the extermination of bugs. Research is being carried out globally as there is a pressing need to apply nanotechnology and this warrants nitty-gritty investigation in this field (De et al., 2013). Nanotubes loaded with aluminosilicate can adhere to plant surfaces, while elements of the nanotube can adhere to the surface hair of the bugs and finally enter the bug body and impact certain physiological capacities (Patil et al., 2009). There is additionally a report of the viability of Diatomaceous Earths (DEs), made primarily out of formless silica and obtained from fossilized phytoplankton working against bugs, on the off chance that it has high silica content with uniform size dissemination. Hence optional metabolites-based bio-pesticides are a regular source that could be focused on to dispose of pest nuisances and are eco-friendly. These bio-pesticides are natural, generally safe, environmentally benevolent, compared to manufactured pesticides. Along these lines, plants are investigated that have strength against insect pests and control their populace and spare crops, for example, weeds having no significance in farming that contain numerous helpful optional metabolites that can be successful against pests. Plants have an innate ability to bargain; concentrations represent diverse pests-safe mixes such as limonene and b-caryophyllene from Ocimum kilimandscharicum (Singh et al., 2014).

15.6 APPLICATIONS OF NANOFORMULATIONS OF PESTICIDE

Actuated opposition in plants, because of silica, against numerous herbivorous arthropods has been recorded (Keeping and Kvedaras, 2008; Chowdhary et al., 2020b). Various nanopesticides are successful in bug sprays, fungicides, herbicides (Matsumoto et al., 2009; Peteu 2010; Bharagava and Chowdhary, 2019). Nanotoxic substances are arranged in fungicides and bug sprays and their efficiency can be contrasted to the customary items. Nanohexaconazole is portrayed by SEM, TEM as well as FT-IR and so forth, and it was below 100 mm in dimensions. It is manifold increasingly powerful in managing microbes and nanosulphur is manifold progressively viable for the eradication of vermin when contrasted with its wettable powder (WP) details (Gopal et al., 2011). To guarantee the material, it was assessed before propelling it. They assessed the creation of a nicotine carboxylate nano-emulsion using a progression of fatty acids (C10-C18) as well as surface oily water nano-emulsion and indicated a mono-modal conveyance size of molecule sizes of 100 nm. The bio-action of the bug spray doses are appraised to deal with adult common fruit fly, *D. melanogaster* by evaluating the parameters of the lethal time 50 (LT_{50}). The embodiment effectiveness was reduced with the expanding size of the unsaturated fats tried. The bio-actions follow a similar pattern, with better activity when the chain length is reduced (Casanova et al., 2005).

15.7 CONCLUSION

Nanoparticles of various shapes commonly are applied for effective control of target insect pests via the formulations of pesticides, resulting in a supply of green and ecologically sound substitutes regarding crop security. With the progress of nanotechniques, the use of green chemistry to combine nanoparticles in different biotic assets has decreased the application of poisonous pesticides. The supervised discharged attributes of nano-encapsulation particles liberate the active ingredients to target the regions, and apply the auto-detect capability necessary for additional exploration. Even if absolute quality associated with nanoparticles has not yet been established, as well as encouraging nano-encapsulated insecticides, these are the primary stages of improvement, and it is expected that such technology will reduce, first of all, the amount of toxic chemicals required for food safety, as well as the use of poison. The main outcomes considered to appear from favourable repercussions of nanopesticides are the use of nanomaterial to encapsulate and balance bio-pesticides that will stop threats to the crop ecosystem. Although the noxiousness of nanoparticles has still not definitely been recognized, they perform a key role in crop safety due to their distinct physical as well as chemical resources. Using nanoparticles is the latest study in entomology for food safety, therefore it requires additional fieldwork analysis.

REFERENCES

Adak, T., Kumar, J., Shakil, N.A., and Walia, S. 2012. Development of controlled release formulations of imidacloprid employing novel nano-ranged amphiphilic polymers, *Journal of Environmental Science and Health, Part B: Pesticides, Food Contaminants, and Agricultural Wastes*, 47(3): 217–225.

Ali, S.W., Rajendran, S., and Joshi, M. 2011. Synthesis and characterization of chitosan and silver loaded chitosan nanoparticles for bioactive polyester. *Carbohydrate Polymers*, 83(2): 438–446. doi: 10.1016/j.carbpol.2010.08.004

Allahvaisi, S. 2016. Effects of silver nanoparticles (AgNPs) and polymers on stored pests for improving the industry of packaging foodstuffs. *Journal of Entomology and Zoology Studies*, 4(4): 633–640

Anjali, C.H., Sharma, Y., Mukherjee, A., and Chandrasekaran, N. 2012. Neem oil (*Azadirachta indica*) nanoemulsion—a potent larvicidal agent against *Culex quinquefasciatus*. *Pest Management Science*, 68: 158–163.

Anton, N., and Vandamme, T.F. 2011. Nano-emulsions and micro-emulsions: Clarifications of the critical differences. *Pharmaceutical Research*, 28(5): 978–985.

Ao, M., Zhu, Y., He, S., Li, D., Li, P., Li, J., and Cao, Y. 2013. Preparation and characterization of 1-naphthylacetic acid-silica conjugated nanospheres for enhancement of controlled-release performance. *Nanotechnology*, 24: 035601.

Arole, V., and Munde, S. 2014. Fabrication of nanomaterials by top-down and bottom-up approaches: An overview. *Journal of Material Science*, 1: 89–93.

Aruoja, V., Dubourguier, H., Kasamets, C., and Kahru, K.A. 2009. Toxicity of nanoparticles of CuO, ZnO and TiO_2 to microalgae, *Pseudokirchneriella subcapitata*. *Science of the Total Environment*, 407(4): 1461–1468.

Ayutsede, J., Gandhi, M., Sukigara, S., and Ko, F. 2008. Carbon nanotube reinforced *Bombyx mori* nano fibers by the electro spinning process. Warrendale, PA, Materials Research Society.

Barik, T.K., Sahu, B., and Swain, V. 2008. Nanosilica: From medicine to pest control. *Parasitology Research*, 103(2): 253–258.

Bhagat, D., Samantha, S.K., and Bhattacharya, S. 2013. Efficient management of fruit pests by pheromone. *Nanogels Science Reports*, 3: 1–8.

Bhattacharyya, A., Gosh, M., Chinnaswamy, K.P., Sen, P., Barik, B., Kundu, P., and Mandal, S. 2008. Nanoparticle (allelochemicals) and silkworm physiology. In Chinnaswamy, K.P. and Vijaya Bhaskar Rao, A. (Eds.) *Recent Trends in Seribiotechnology*. Bangalore: Times of India Press, pp. 58–63.

Billen, J. 2006. Signal variety and communication in social insects. *Proceedings of the Section Experimental and Applied Entomology—Netherlands Entomological Society*, 17: 9.

Biswal, S.K., Nayak, A.K., Parida, U.K., and Nayak, P.L. 2012. Applications of nanotechnology in agriculture and food sciences. *International Journal of Science and Innovative Discovery*, 2(1): 21–36.

Campbell, F., Conti, G., Heckman, J. J., Moon, S. H., Pinto, R., Pungello, E., and Pan, Y. 2014. Early childhood investments substantially boost adult health. *Science*, 343(6178), 1478–1485.

Casanova, H., Araque, P., and Ortiz, C. 2005. Nicotine carboxylate insecticide emulsions: Effect of the fatty acid chain length. *Journal of Agricultural and Food Chemistry*, 53(26): 9949–9953.

Chakravarthy, A.K., Chandrashekharaiah, S.B., Kandakoor, A., Bhattacharyya, K., Dhanabala, K., Gurunatha, K., and Ramesh, P. 2012. Bio efficacy of inorganic nanoparticles CdS, Nano-Ag and Nano TiO_2 against *Spodoptera litura* (Fabricius) (Lepidoptera: Noctuidae). *Current Biotica*, 6: 271–281.

Chambarelli, L.L., Pinho, M.A., Abracado, L.G., Esquivel, D.M.S., and Wajnberg, E. 2008. Temporal and preparation effects in the magnetic nanoparticles of *Apis mellifera* body parts. *Journal of Magnetism and Magnetic Materials*, 320(14): e207–e210.

Chandrashekharaiah, M., Kandakoor, S.B., Gowda, G.B., Kammar, V., and Chakravarthy, A.K. 2015. Nanomaterials: A review of their action and application in pest management and evaluation of DNA-tagged particles. In Chakravarthy, A.K. (Ed.), *New Horizons in Insect Science*. Cham: Springer, pp. 113–126.

Chowdhary, P., and Bharagava, A. (Eds.) 2019. *Emerging and Eco-Friendly Approaches for Waste Management*. Singapore: Springer:

Chowdhary, P., Raj, A., Verma, D., and Akhter, Y. (Eds.) 2020a. *Contaminants and Clean Technologies*. Boca Raton, FL: CRC Press.

Chowdhary, P., Raj, A., Verma, D., and Akhter, Y. (Eds.) 2020b. *Microorganisms for Sustainable Environment and Health*. Oxford: Elsevier.

Christenson, L.D., and Foote, R.H. 1960. Biology of fruit flies. *Annual Review of Entomology*, 5: 171–192.

Clements, A., Young, J.C., Constantinou, N. and Frankel, G. 2012. Infection strategies of enteric pathogenic *Escherichia coli. Gut Microbes*, 3(2): 71–87.

Dawson, K. 2008. Bio nano interactions—a rational approach to the interaction between nanoscale materials and living matter? *Nanobioeurope*, 8: 1–2.

De, A., Bose, R., Kumar, A., and Mozumdar, S. 2014. Management of insect pests using nanotechnology: As modern approaches. In *Targeted Delivery of Pesticides Using Biodegradable Polymeric Nanoparticles*. New Delhi: Springer, pp. 29–33.

Debnath, N.S., Mitra, S., Das, S., and Goswami, A. 2012. Synthesis of surface functionalized silica nanoparticles and their use as entomotoxic nanocides. *Power Technology*, 221: 252–256.

Debnath, N.S., Sumistha, D., Dipankar, S., Ramesh, C., Somesh, C.H. and Bhattacharya, A.G. 2011. Entomotoxic effect of silica nanoparticles against *Sitophilus oryzae* (L.). *Journal of Pesticide Science*, 84: 99–105.

Desoil, M., Gillis, P., Gossuin, Y., Pankhurst, Q.A., and Hautot, D. 2005. Definitive identification of magnetite nanoparticles in the abdomen of the honeybee *Apis mellifera. Journal of Physics: Conference Series*, 17: 45–49.

Dubas, S.T., Kumlangdudsana, P., and Potiyaraj, P. 2006. Layer-by-layer deposition of antimicrobial silver nanoparticles on textile fibers. *Colloids and Surfaces A*, 289(1): 105–109. doi: 10.1016/j.colsurfa.2006.04.012

Duebendorf, St. G.T. 2008. How do nanoparticles behave in the environment? Paper presented at NanoEco-Empa international meeting on nanoparticles in the environment. Published by Bernd Nowack and Nicole Mueller, in *Environmental Science and Technology*. http://pubs.acs.org/cgibin/ asap.cgi/esthag/asap/html/es7029637.html

Dwivedi, S.K., and Tomer, A. 2020. Applications for nanotechnology in the polyphagous destructive insect pest management of agricultural crops, nanotechnology for food, agriculture, and environment, nanotechnology in the life sciences. *Springer Nature*. https://doi.org/10.1007/978-3-030-31938-0_9

Ehrlich, H., Janussen, D., Simon, P., Bazhenov, V.V., Shapkin, N.P., Erler, C., Mertig, M., ... Vournakis, J.N. 2008. Nano structural organization of naturally occurring composites—Part II: Silica-chitin-based biocomposites. *Journal of Nanomaterials*, 3*:* 1–8.

El-Bendary, H.M. and El-Helaly, A.A. 2013. First record nanotechnology in agricultural: Silica nano-particles a potential new insecticide for pest control. *Applied Science Report*, 4(3): 241–246.

Elek, N., Hoffman, R., Raviv, U., Resh, R., Ishaaya, I., and Magdassi, S. 2010. Novaluron nanoparticles: Formation and potential use in controlling agricultural insect pests. *Colloids and Surfaces A: Physicochemical and Engineering Aspects*, 372(1–3): 66–72.

El-Jaick, L.J., Acosta-Avalos, D., Esquivel, D.M., De, S., Wajnberg, E., Linhares, and M.P. 2001. Electron paramagnetic resonance study of honeybee *Apis mellifera* abdomens. *European Biophysics Journal*, 29(8): 579–586.

Esquivel, D.M.S., Wajnberg, E., Cernicchiaro, G.R., Acosta-Avalos, D., and Garcia, B.E. 2002. Magnetic material arrangement in *apis mellifera* abdomen. *MRS Online Proceedings Library Archive*, 724.

Esteban-Tejeda, L., Malpartida, F., Pecharroman, C., and Moya, J.S. 2010. High antibacterial and antifungal activity of silver mono dispersed nanoparticles embedded in a glassy matrix. *Advanced Engineering Materials*, 12(7): 292–297.

Feng, B.H., and Peng, L.F. 2012. Synthesis and characterization of carboxy methyl chitosan carrying ricinoleic functions as an emulsifier for azadirachtin. *Carbohydrate Polymers*, 88: 576–582.

Fiorito, S. 2008. Carbon nanotubes: Angels or demons? Rome: University La Sapienza Institute of Neurobiology and Molecular Medicine, CNR Rome, Italy.

Flores-Cespedes, F., Figueredo-Flores, C.I., Daza-Fernandez, I., Vidal-Pena, F., Villafranca Sanchez, M., and Fernandez-Perez, M. 2012. Preparation and characterization of imidacloprid lignin-polyethylene glycol matrices coated with ethyl cellulose. *Journal of Agricultural Food Chemistry*, 60: 1042–1051.

Fraceto, L.F., de Grillo, R., Medeiros, G.A., Scognamiglio, V., Rea, G., and Bartolucci, C. 2016. Nanotechnology in agriculture: Which innovation potential does it have?. *Frontiers in Environmental Science*, 4: 20.

Frandsen, M.V., Pedersen, M.S., Zellweger, M., Gouin, S., Roorda, S.D. and Phan, T.Q.C. 2010. Piperonyl Butoxide and Deltamethrin containing insecticidal polymer matrix comprising HDPE and LDPE. Patent number WO 2010015256 A2 20100211.

Friedrich, T. 2015. A new paradigm for feeding the world in 2050: The sustainable intensification of crop production. *Resource Management*, 22: 18.

Gandhi, V.P., and Namboodiri, N.V. 2009. Economics of BT cotton versus non-BT cotton in India: A study across four major cotton-growing states. Ahmedabad, India: Centre for Management in Agriculture, Indian Institute of Management, pp. 1–127.

Gasco, M.R. 1997. Micro emulsions in the pharmaceutical field: Perspectives and applications. *Surfactant Science Series*, 6: 7–122.

Gha-Young, K., Joonmok, S., Min-Su, K., and Seung-Hyeon, M. 2008. Preparation of a highly sensitive enzyme electrode using gold nanoparticles for measurement of pesticides at the ppt level. *Journal of Environmental Monitoring,* 10: 632–637.

Goldshtein, R., Jaffe, I. and Tulbovich, B. 2005. Hydrophilic dispersions of nanoparticles of inclusion complexes of amorphous compounds. Patent number US 20050249786 A120051110.

Gopal, M., Chaudhary, S.R., Ghose, M., Dasgupta, R., Devakumar, C., Subrahmanyam. B., Srivastava, C., … Goswami, A. 2011. Samfungin: A novel fungicide and the process for making the same. Indian Patent Application No. 205.

Gorb, E.V., and Gorb, S.N. 2009. Contact mechanics at the insect-plant interface: How do insects stick and how do plants prevent this? In Borodich, F.M. (Ed.), *IUTAM Symposium on Scaling in Solid Mechanics.* Dordrecht: Springer, pp. 243–252.

Guan, H., Chi, D., Yu, J., and Li, X. 2008. A novel photodegradable insecticide: Preparation, characterization and properties evaluation of nano Imidacloprid. *Pesticide Biochemistry and Physiology*, 92(2): 83–91.

Hellmann, C., Greiner, A., and Wendorff, J.H. 2011. Design of pheromone-releasing nano fibers for plant protection. *Polymers for Advanced Technologies*, 22(4): 407–413.

Hsu, C.Y., Ko, F.Y., Li, C.W., Fann, K., and Lue, J.T. 2007. Magneto reception system in honeybees (*Apis mellifera*). *PLoS ONE*, 2(4): 395.

Huang, L., Dian-Qing, L., Yan-Jun, W., Min David, G., and Xue, E.D. 2005. Controllable preparation of nano-MgO and investigation of its bactericidal properties. *Journal of Inorganic Biochemistry*, 99(5): 986–993.

Huck, W.T.S. 2008. Responsive polymers for nano scale actuation. *Materials Today*, 11(7–8): 24–32.

Hwang, I.C., Kim, T.H., Bang, S.H., Kim, K.S., Kwon, H.R., Seo, M.J., Youn, Y.N., … Yu, Y.M. 2011. Insecticidal effect of controlled release formulations of etofenprox based on nano-bio technique. *Journal of the Faculty of Agriculture, Kyushu University*, 56(1): 33–40.

Isiklan, N. 2004. Controlled release of insecticide carbaryl from cross linked carboxy methyl cellulose beads. *Free Environmental Bulletin*, 13: 537–544.

Jana, T., Roy, B.C., and Maiti, S. 2001. Biodegradable film 6 modification of the film for control release of insecticides, *European Polymer Journal*, 37(4): 861–864.

Jayaseelan, C., Rahuman, A.A., Rajakumar, G., Vishnu, A.K., Santhosh Kumar T., Marimuthu, S., Bagavan, A., … Elango, G. 2011. Synthesis of pediculocidal and larvicidal silver nanoparticles by leaf extract from heartleaf moonseed plant, *Tinospora cordifolia Miers. Parasitology Research*, 109: 185–194.

Jin, H.J., Chen, J., Karageorgiou, V., Altman, G.H., and Kaplan, D.L. 2004. Human bone marrow stromal cell responses on electro spun silk fibroin mats, *Biomaterials*, 25(6): 1039–1047.

John, H., Lucas, J. Clare, W., and Dusan, L. 2017. Nanopesticides: A review of current research and perspectives. In Grumezescu, A.M. (Ed.), *New Pesticides and Soil Sensors*. London: Academic Press, pp. 193–225.

Joshi, M.D., Unger, W.J., Storm, G., van Kooyk, Y. and Mastrobattista, E. 2012. Targeting tumor antigens to dendritic cells using particulate carriers. *Journal of Controlled Release*, 161: 25–37.

Kandasamy, S., and Prema, R.S. 2015. Methods of synthesis of nanoparticles and its applications. *Journal of Chemical and Pharmaceutical Research*, 7: 278–285.

Keeping, M.G. and Kvedaras, O.L. 2008. Silicon as a plant defence against insect herbivory: Response to Massey, Ennos and Hartley. *Journal of Animal Ecology*, 77(3): 631–633.

Khot, L.R., Sankaran, S.J.M., Ehsani, R. and Schuster, E.W. 2012. Applications of nano materials in agricultural production and crop protection: A review. *Crop Protection*, 35: 64–70.

Kim, S.W., Jung, J.H., Lamsal, K., Kim, Y.S., Min, J.S., and Lee, Y.S. 2012. Antifungal effects of silver nanoparticles (AgNPs) against various plant pathogenic fungi. *Mycobiology*, 40: 53–58.

Kumar, V., and Yadav, S.K. 2009. Plant-mediated synthesis of silver and gold nanoparticles and their applications *Journal of Chemical Technology & Biotechnology: International Research in Process, Environmental & Clean Technology*, 84(2): 151–157.

Kuntworbe, N., Martini, N., Shaw, J., and Al-Kassas, R. 2012. Malaria intervention policies and pharmaceutical nanotechnology as a potential tool for malaria management. *Drug Development Research*, 73(4): 167–184.

Lao, S.B., Zhang, Z.X., Xu, H.H., and Jiang, G.B. 2010. Novel amphiphilic chitosan derivatives: Synthesis, characterization and micellar solubilization of rotenone. *Carbohydrate Polymers*, 82: 1136–1142.

Li, M., Huang, Q., and Wu, Y. 2011. A novel chitosan-poly (Lactide) copolymer and its submicron particles as imidacloprid carriers. *Pest Management Science*, 67: 831–836.

Liu, R., and Lal, R. 2015. Potentials of engineered nanoparticles as fertilizers for increasing agronomic productions. *Science of the Total Environment*, 514: 131–139.

Liu, Y., Laks, P., and Heiden, P. 2002. Controlled release of biocides in solid wood. ii. efficacy against *Trametes versicolor* and *Gloeophyllum trabeum* wood decay fungi. *Political Science*, 86: 608–614.

Loha, K.M., Shakil, N.A., Kumar, J., Singh, M.K. and Srivastava, C. 2012. Bio-efficacy evaluation of nano formulations of β-cyfluthrin against *Callosobruchus maculatus* (Coleoptera: Bruchidae). *Journal of Environmental Science and Health, Part B*, 47(7): 687–691.

Louder, J.K. 2015. Nanotechnology in agriculture: Interactions between nanomaterials and cotton agrochemicals. PhD thesis. Texas Tech University.

Malik, M.F., Khan, A.G., Hussiany, S.W., Rehman, D.U., and Amin, M. 2002. Scouting and control of *Helicoverpa armigera* by synthetic pheromones technology in apple. *Asian Journal of Plant Sciences*, 3: 663–664.5.

Manjunatha, R.L., Naik, D., and Usharani, K.V. 2019. Nanotechnology application in agriculture: A review. *Journal of Pharmacognosy and Phytochemistry*, 8(3): 1073–1083.

Margulis-Goshen, K., and Magdassi, S. 2012. Nanotechnology: An advanced approach to the development of potent insecticides. In Ishaaya, I., Palli, S.R., and Horwitz, A.R. (Eds.), *Advanced Technologies for Managing Insect Pests*, Dordrecht: Springer, pp. *295–314*.

Matsumoto, S., Christie, R.J., Nishiyama, N., Miyata, K., and Ishii, A. 2009. Environment-responsive block copolymer micelles with a disulfide cross-linked core for enhanced siRNA delivery, *Biomacromology*, 10: 119–127.

Nowack, B. 2009. Is anything out there? What life cycle perspectives of nano-products can tell us about nanoparticles in the environment, *Nano Today*, 4: 11–12.

Owolade, O.F., Ogunleti, D.O., and Adenekan, M.O. 2008. Titanium dioxide affects disease development and yield of edible cowpea. *Journal of Agricultural and Food Chemistry*, 7(50): 2942–2947.

Pal, S., Tak, Y.K., and Song, J.M., 2007. Does the antibacterial activity of silver nanoparticles depend on the shape of the nanoparticles? A study of Gram negative bacterium *Escherichia coli*. *Applied Environmental Microbiology*, 73: 1712–1720.

Panghal, D., Nagpal, M., Thakur, G.S. and Arora, S. 2014. Dissolution improvement of atorvastatin calcium using modified locust bean gum by the solid dispersion technique. *Scientia Pharmaceutica*, 82(1): 177–192.

Park, H.J., Kim, S.H., Kim, H.J., and Choi, S.H. 2006. A new composition of nano sized silica-silver for control of various plant diseases. *Journal of Plant Pathology*, 22 : 25–34.

Patil, S.A. 2009. *Economics of Agri Poverty: Nano-bio Solutions*. New Delhi: Indian Agricultural Research Institute.

Peteu, S.F., Oancea, F., Sicuia, O.A., Constantinescu, F., and Dinu, S. 2010. Responsive polymers for crop protection. *Polymers*, 2: 229–251.

Prasad, K.S., Pathak, D., Patel, A., Dalwadi, P., Prasad, R., Patel, P., and Kaliaperumal, S.K. 2011. Biogenic synthesis of silver nanoparticles using *Nicotiana tobaccum* leaf extract and study of their antibacterial effect. *African Journal of Biotechnology*, 10(41): 8122–8130.

Prasad, R. 2014. Synthesis of silver nanoparticles in photosynthetic plants. *Journal of Nanoparticles*() 8. doi: 10.1155/2014/963961

Prasad, R., Bagde, U.S., and Varma, A. 2012. Intellectual property rights and agricultural biotechnology: An overview. *African Journal of Biotechnology*, 11(73): 13746–13752.

Prasad, R., Bhattacharyya, A., and Nguyen, Q.D. 2017. Nanotechnology in sustainable agriculture: Recent developments, challenges, and perspectives. *Frontiers of Microbiology*, 8: 1014. https://doi.org/10.3389/fmicb.2017.01014

Prasad, R., Kumar, V., and Prasad, K.S. 2014. Nanotechnology in sustainable agriculture: Present concerns and future aspects. *African Journal of Biotechnology*. 13(6): 705–713. doi:10.5897/AJBX2013.13554.

Prasad, R., Pandey, R., and Barman, I. 2016. Engineering tailored nanoparticles with microbes: Quo vadis? *WIREs Nanomedicine and Nanobiotechnology*, 8: 316–330. doi: 10.1002/wnan.1363.

Prasad, R., and Swamy, V.S. 2013. Antibacterial activity of silver nanoparticles synthesized by bark extract of *Syzygium cumini*. *Journal of Nanoparticles*, 2013: 1–6 http://dx.doi.org/10.1155/2013/431218.

Prasad, R., Swamy, V.S., and Varma, A. 2012. Biogenic synthesis of silver nanoparticles from the leaf extract of *Syzygium cumini* (L.) and its antibacterial activity. *International Journal of Pharmacy and Biological Science*, 3(4): 745–752.

Quaglia, M., Chenon, K., Hall, A. J., De Lorenzi, E., and Sellergren, B. 2001. Target analogue imprinted polymers with affinity for folic acid and related compounds. *Journal of the American Chemical Society*, 123(10), 2146–2154.

Radhika, P.R., Sasikanth, A., and Sivakumar, T. 2011. Nanocapsules: A new approach in drug delivery. *International Journal of Pharmaceutical Sciences and Research*, 2(6): 1426–1429.

Ragaei, M., and Sabry, A.H. 2014. Nanotechnology for insect pest control. *International Journal of Science, Environment and Technology*, 3(2): 528–554.

Ramya, S., Rajasekaran, C., Sundararajan, G., Alaguchamy, N., and Jayakumararaj, R. 2008. Antifeedant activity of leaf aqueous extracts of selected medicinal plants on VI instar larva of *Helicoverpa armigera*, *Ethnobotanical Leaflets*, 12: 938–943.

Rao, J., and McClements, D.J. 2011. Formation of flavor oil microemulsions, nanoemulsions and emulsions: Influence of composition and preparation method. *Journal of Agricultural and Food Chemistry*, 59, 5026–5035.

Rouhani, M., Samih, M. A., and Kalantari, S. 2012. Insecticidal effect of silica and silver nanoparticles on the cowpea seed beetle, *Callosobruchus maculatus F. (Col.: Bruchidae)*. *Journal of Entomology Research*, 4: 297–305.

Rudzinski, W.E., Chipuk, T., Dave, A.M., Kumbar, S.G., and Aminabhavi, T.M. 2003. pH sensitive acrylic-based copolymeric hydrogels for the controlled release of a pesticide and a micronutrient. *Journal of Applied Polymer Science*, 87: 394–403.

Sagadevan, S., and Periasamy, M. 2014. Recent trends in nano biosensors and their applications: A review. *Reviews on Advanced Materials Science*, 36: 62–69.

Sakulk, N.U., Uawongyart, P., Soottitantawat, N., and Ruktanonchai, U. 2009. Characterization and mosquito repellent activity of citronella oil nano emulsion. *International Journal of Pharmaceutics*, 372: 105–111.

Sekhon, B.S. 2014. Nanotechnology in agri-food production: An overview. *Nanotechnology, Science and Applications,* 7: 31–53. doi: 10.2147/NSA.S39406

Sertova, N.M. 2015. Application of nanotechnology in detection of mycotoxins and in agricultural sector. *Journal of Central European Agriculture*, 16: 117–130. doi: 10.5513/JCEA01/ 16.2.1597

Shah, M.A., and Towkeer, A. 2010. *Principles of Nanosciences and Nanotechnology*. New Delhi: Narosa Publishing House.

Shakil, N.A., Singh, M.K., Pandey, A., Kumar, J., Parmar, V.S., Singh, M.K., Pandey, R.P., and Watterson, A.C. 2010. Development of poly (*Ethylene glycol*) based amphiphilic co polymers for controlled release delivery of carbofuran. *Journal of Macromolecular Science, Part A: Pure and Applied Chemistry*, 47: 241–247.

Sharma, V.K., Yngard, R.A., and Lin, Y. 2009. Silver nanoparticles: Green synthesis and their antimicrobial activities. *Advances in Colloid and Interface Science*, 145: 83–96.

Shrivastava, S., Bera, T., Roy, A., Singh, G., Ramachandrarao, P., and Dash, D. 2007. Characterization of enhanced antibacterial effects of novel silver nanoparticles. *Nanotechnology*, 18(22): 225103. doi: 10.1088/0957-4484/18/49/495102

Singh, P., Jayaramaiah, R.H., Sarate, P., Thulasiram, H.V., Kulkarni, M.J., and Giri, A.P. 2014. Insecticidal potential of defense metabolites from *Ocimum kilimandscharium* against *Helicoverpa armigera*. *PLoS One*, 9(8): e104377.

Siva, C., and Kumar, M.S. 2015. Pesticidal activity of eco-friendly synthesized silver nanoparticles using *Aristolachia indica* extract against *Helicoverpa armigera* Hubner (Lepidoptera: Noctuidae). *International Journal on Advanced Science Technology Research*, 2: 197–226.

Stadler, T., Buteler, M. and Weaver, D.K. 2010. Novel use of nanostructured alumina as an insecticide. *Pest Management Science*, 66: 577–579.

Sullivan, M., and Molet, T. 2007. CPHST pest datasheet for *Helicoverpa armigera*. USDA-APHIS-PPQ-CPHST. Washington, DC: USDA.

Suman, P.R., Jain, V.K., and Varma, A. 2010. Role of nanomaterials in symbiotic fungus growth enhancement. *Current Science*, 99: 1189–1191.

Swamy, V.S., and Prasad, R. 2012. Green synthesis of silver nanoparticles from the leaf extract of *Santalum album* and its antimicrobial activity. *Journal of Optoelectronic and Biomedical Materials*, 4(3): 53–59.

Tadros, T., Izquierdo, P., Esquena, J., and Solans, C. 2004. Formation and stability of nano-emulsions. *Advances in Colloid and Interface Science*, 108: 303–318.

Tai, K.N.G. 2008. Structural analysis of crystalline and nanocrystalline materials. In *Proceedings of the Department of Physics*. Hong Kong: The Hong Kong University of Science and Technology.

Talegaonkar, S., Azeem, A., Ahmad, F., Khar, R., Pathan, S., and Khan, Z. 2008. Microemulsions: A novel approach to enhanced drug delivery. *Recent Patents on Drug Delivery & Formulation*, 2(3): 238–257.

Tarafdar, J.C., Sharma, S., and Raliya, R. 2013. Nanotechnology: Interdisciplinary science of applications. *African Journal of Biotechnology*, 12(3): 219–226.

Tejamaya, M., Romer, I., Merrifield, R.C., and Lead, J.R. 2012. Stability of citrate, PVP, and PEG coated silver nanoparticles in ecotoxicology media. *Environmental Science & Technology*, 46(13): 7011–7017.

Thakkar, K.N., Mhatre, S.S., and Parikh, R.Y. 2010. Biological synthesis of metallic nanoparticles. *Nanomedicine*, 6(2): 257–262. doi: 10.1016/j.nano.2009.07.002

Thornhill, A.H., Mishler, B.D., Knerr, N.J., González-Orozco, C.E., Costion, C.M., Crayn, D. M. and Miller, J.T. 2016.. Continental-scale spatial phylogenetics of Australian angiosperms provides insights into ecology, evolution and conservation. *Journal of Biogeography*, 43(11): 2085–2098.

Torney, A. 2009. Nanoparticle mediated plant transformation. Emerging technologies in plant science research. Inter-Departmental Plant Physiology Major Fall Seminar Physics. p. 696.

Vani, C. and Brindhaa, U. 2013. Silica nanoparticles as nanocides against *Corcyra cephalonica* (S.), the stored grain pest. *International Journal of Pharma and Bio Sciences*, 4(3): 1108–1118.

Vidotti, M., Carvalhal, R.F., Mendes, R.K., Ferreira, D.C.M., and Kubota, L.T. 2011. Biosensors based on gold nanostructures. *Journal of the Brazilian Chemical Society*, 22(1): 3–20.

Vidyalakshmi, R., Bhakyaraj, R. and Subhasree, R.S. 2009. Encapsulation, the future of probiotics: A review. *Advances in Biological Research*, 3(3–4): 96–103.

Walker, G.W., Kookana, R.S., Smith, N.E., Kah, M., Doolette, C. L., Reeves, P.T., and Navarro, D.A. 2017. Ecological risk assessment of nano-enabled pesticides: A perspective on problem formulation. *Journal of Agricultural and Food Chemistry*, 66(26): 6480–6486.

Watson, G.S., and Watson, J.A. 2004. Natural nano-structures in insects: Possible functions of ordered arrays characterized by atomic force microscopy. *Applied Surface Science*, 235(1–2): 139–144.Wilkins, R.M. 2004. Controlled release technology, agricultural. In Seidel, A. (Ed.), *Kirk Othmer Encyclopedia of Chemical Technology,* 5th edn. Hoboken, NJ: John Wiley & Sons.

Yamanka, M., Hara, K., and Kudo, J. 2005. Bactericidal actions of silver ions solution on *Escherichia coli*: Studying by energy filtering transmission electron microscopy and proteomic analysis. *Journal of Applied Environment and Microbiology*, 71: 7589–7593.

Yang, F.L., Li, X.G., Zhu, F., and Lei, C.L., 2009. Structural characterization of nanoparticles loaded with garlic essential oil and their insecticidal activity against *Tribolium castaneum* (Herbst) (Coleoptera: Tenebrionidae). *Journal of Agricultural Food Chemistry*, 57(21): 10156–10162.

Yasur, J., and Rani, P.U. 2015. Lepidopteran insect susceptibility to silver nanoparticles and measurement of changes in their growth, development and physiology. *Chemosphere*, 124: 92–102.

Ying, J.Y. 2009. Nanobiomaterials. *Nano Today*, 4: 1–2.

Yu, M., Yao, J., Liang, J., Zeng, Z, Cui, B., Zhao, X., Sun, C., … Cui, H. 2017. Development of functionalized abamectin poly(lactic acid) nanoparticles with regulatable adhesion to enhance foliar retention. *RSC Advances*, 7: 11271–11280.

Zahir, A.A., Bagavan, A., Kamaraj, C., Elango, G., and Rahuman, A.A. 2012. Efficacy of plant-mediated synthesized silver nanoparticles against *Sitophilus oryzae*. *Journal of Biopesticides*, 288(Suppl. 5): 95–102.

Zhang, J., and Liu, Z. 2006. Cicada wings: A stamp from nature for nanoimprint. *Lithography*, 2(12): 1440–1443.

Zhang, Y.Q., Shen, W.D., Xiang, R. L., Lan-Jian Zhuge, L.J., Gao, W.J., and Wang, W.B. 2007. Formation of silk fibroin nanoparticles in water-miscible organic solvent and their characterization. *Journal of Nanoparticle Research*, 9(5): 885–900.

16 Polymer Nanocomposites for Wastewater Treatment

Adnan Khan,[1] *Sumeet Malik,*[1] *Nisar Ali,*[2,*]
Muhammad Bilal,[3,*] *Yong Yang,*[2] *Mohammed Salim Akhter,*[4]
Vineet Kumar,[5] *and Hafiz M.N. Iqbal*[6]
[1]Institute of Chemical Sciences, University of Peshawar, Khyber Pakhtunkhwa, Pakistan
[2]Key Laboratory of Regional Resource Exploitation and Medicinal Research, Faculty of Chemical Engineering, Huaiyin Institute of Technology, Huaian, Jiangsu Province, China
[3]Institute of Chemical Technology and Engineering, Faculty of Chemical Technology, Poznan University of Technology, Berdychowo 4, PL-60695 Poznan, Poland
[4]Department of Chemistry, College of Science, University of Bahrain, Bahrain
[5]Department of Basic and Applied Sciences, School of Engineering and Sciences, GD Goenka University, Haryana, India
[6]Tecnológico de Monterrey, School of Engineering and Sciences, Monterrey, Mexico
Corresponding authors Email: nisarali@hyit.edu.cn; muhammad.bilal@put.poznan.pl

CONTENTS

16.1 Introduction 295
16.2 Sources of Natural Polymers 296
16.3 Classification of Nanocomposites 297
16.3.1 Polymeric Matrix Nanocomposite 297
16.3.2 Metal Matrix Nanocomposites 298
16.3.3 Polymer/Ceramic Nanocomposites 298
16.3.4 Polymer/Layer Silicate (PLS) Nanocomposites 298
16.4 Applications of Polymer-Based Nanocomposites in Wastewater Treatment 298
16.4.1 Metal Ion Removal 298
16.4.2 Removal of Organic Contaminants 299
16.4.3 Oil and Water Separation 300
16.4.4 Removal of Other Pollutants 301
16.5 Conclusion 302
Acknowledgments 302
References 302

16.1 INTRODUCTION

Water pollution is caused by different toxic organic and inorganic materials that are dangerous to all life forms: human beings, aquatic life, and animals (Unuabonah and Taubert, 2014; Khan, M., Khan,

DOI: 10.1201/9781003239956-18

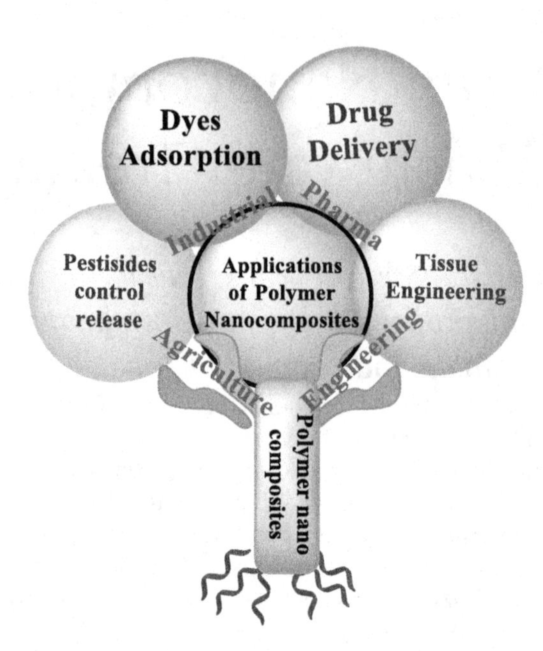

FIGURE 16.1 Possible applications of polymer nanocomposites.

et al., 2021). Water is very important in daily life, but, lately, industrialization and other processes have been contaminating the potable water and making it unsuitable for drinking and other purposes as well (Khodakarami and Bagheri, 2021; Khan, S., Kahn, et al., 2021). These pollutants are a threat to the entire biosphere.

For some time, we have been witnessing the situation that nanotechnology has become very important in various fields such as engineering, transport, beverages, pharmaceuticals, foods (Figure 16.1), and many other things (Patel et al., 2006; Ali et al., 2021). This technology is widely used in all sectors, whether they are electrical or mechanical. So, first, we need to know what a polymer is and what a nanocomposite is. Polymers are materials formed by combining different materials, as the word poly means many (Huang et al., 2019; Yang et al., 2021). Generally composite materials that are formed by integrating two or more materials possess different chemical and physical properties. So a composite form has unique properties comprising the properties of both materials combined. A composite usually comprises two phases, i.e., the matrix phase and the dispersed phase. Now we need to know what a nanocomposite is (Rytwo, 2012; Saleh et al., 2019; Nawaz et al., 2020). A nanocomposite is a form of composite in which one phase has morphology on the nanoscale, i.e. nanotubes or nanoparticles, etc. They are formed by combining two materials with different phases, and the property of the materials formed depends upon the individual property of the materials incorporated in the first place (Ali, Uddin, et al., 2020; Mukhopadhyay et al., 2020). They have high multi-functional stability, mechanical stability, chemical stability, physical stability, and a huge inter-phase zone. So, having learned the basics of composites and nanoparticles, now we will study how we can use nanocomposites to treat waste and polluted water, focusing on their importance for the present day (Badruddoza et al., 2013; Ali, Ahmed, et al., 2020) (Figure 16.2).

16.2 SOURCES OF NATURAL POLYMERS

Polymers are found in nature in many different forms. Most of them are water-based. Some examples of polymers that occur naturally are DNA, silk, wool, etc. (Ahmad et al., 2021; Ali, Bilal, Khan,

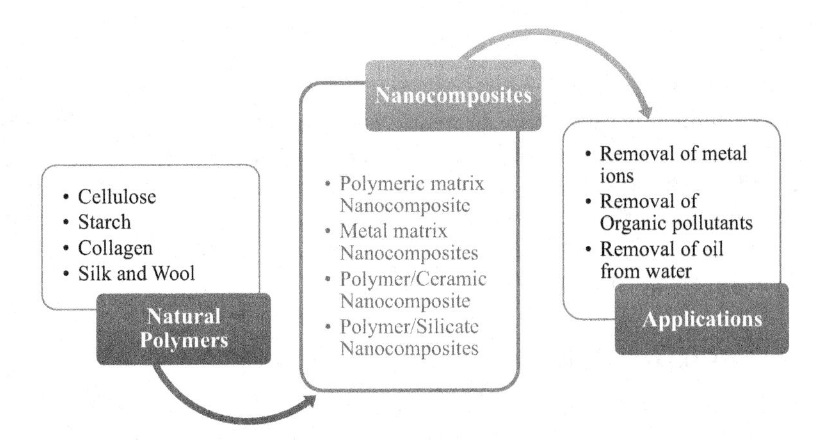

FIGURE 16.2 Sources of natural polymer, their composites, and applications.

et al., 2020).Vulcanized rubber is also a natural source of polymer (Söderqvist Lindblad et al., 2001; Khan, H., et al., 2020). The naturally occurring polymers may differ with respect to the origin of their presence. One of the most abundant natural polymers is cellulose. Cellulose is most commonly found in the cell wall of plants (Ali, Khan, Malik, et al., 2020; Khan, S.U., et al., 2016). Another naturally occurring polymer is collagen, which is mostly present as collective tissue in the skin of human beings (Ali, Khan, Bilal, et al., 2020). Latex is another natural polymer that is found in a variety of plants and rubber trees. Starch is more commonly found in grains, cereals, and potatoes. Pectin is found in the cell walls of terrestrial plants (Ali, Khan, Nawaz, et al., 2020). Another naturally occurring polymer, chitin, is abundantly present in the waste of the fishing industry, while carrageenan is present in red edible seaweeds (Aziz et al., 2020). Gelatin is obtained from the hydrolysis of collagen extracted from the skin, connective tissues, and bones of animals, while alginate and xyloglucan are present in brown seaweed and plant cell walls, respectively (Sartaj et al., 2020; Khan, 2019a).

16.3 CLASSIFICATION OF NANOCOMPOSITES

Nanocomposites are classified into two types: polymer-based, and non-polymer-based materials. Nanomaterials are also classified into nanophase nanoparticle materials and nanostructured materials. Nanophase nanoparticles are nano-sized particles that are totally dispersed in different media (Lochab et al., 2014; Ali, Naz, et al., 2020). Nanostructured particles are defined as bulk material particles made from grains that are of nanometer size. Polymer nanocomposites are in turn divided into different types, which are discussed below.

16.3.1 Polymeric Matrix Nanocomposite

By using fillers on high surface areas, scientists are trying to find ways to improve and enhance the properties of these materials. Scientists have developed nanocomposites using this method (Yin and Dend, 2015; Ali, Bilal, Nazir, Khan, et al., 2020). We can enhance the properties of these nanocomposites by adding some particles under the following conditions:

- There should be a strong interaction between the matrix and the nanoparticles.
- Matrix particles should be well dispersed. A high aspect ratio should be $L/h>300$ and 5% of enhanced materials added (Naskar et al., 2016; Ali, Azeem, et al., 2020).

16.3.2 Metal Matrix Nanocomposites

When a large volume of an alloy or metal is added as a nanomaterial in the matrix of the composite, the product formed is called a mater matrix nanocomposite. They have the characteristics of both metals and ceramics (Casati and Vedani, 2014; Ali, Bilal, Khan, Ali, et al., 2020). They are tough, ductile, and have high thermal stability. Metal matrixes are combined with carbon nanotubes (CNTs) with a substantial surface area and are mechanically and thermally stable. By this process, we can achieve wide applications. Their applications are used in the aerospace and automotive industries (Ahn et al., 2015; Ali, Khan, Malik, et al., 2020).

16.3.3 Polymer/Ceramic Nanocomposites

Ceramic materials are brittle, and have high chemical and thermal stability. However, the low toughness of ceramic materials is still a challenge in a wide range of industries (Kumar et al., 2018; Ali, Zada et al., 2019). To overcome this flaw, ceramic nanocomposites are becoming known because of their enhanced mechanical properties and their wide applications (Graz et al., 2009). For instance, the fracture toughness of whiskers, fibers, and other materials can be increased by this method. Summing up, the addition of high-strength fibers to ceramic materials can help in the formation of materials with high thermal stability and mechanical properties much higher than that of average ceramic materials (Hua et al., 2004; Khan, A., Shah, et al., 2019).

16.3.4 Polymer/Layer Silicate (PLS) Nanocomposites

PLS polymers are also available these days. The process of adding a layer of silicates to polymers is almost 50 years old. They have attracted great interest in academia as well as for industrial applications (Yeh and Change, 2008; Khan, H., Khalil, et al., 2019). Montmorillonite and Hactorite are the most commonly used types of layer silicate. This is a Smectite-type layer silicate (Krikorian and Pochan, 2003; Ali, Kamal, et al., 2018). The method of preparation depends upon the organization, the polymer matrix, and the layered silicate.

16.4 APPLICATIONS OF POLYMER-BASED NANOCOMPOSITES IN WASTEWATER TREATMENT

Due to their unique properties, polymer-based nanocomposites are often used to purify water. Different technologies of polymer-based nanocomposites include adsorption, oxidation, ion exchange, coagulation and flocculation, membrane separations, biodegradation, microbial treatment, and an advanced oxidation process.

16.4.1 Metal Ion Removal

Different toxic metals, such as lead, cadmium, arsenic, selenium, mercury, and lead, when present in water, are very harmful (Sohni et al., 2018). These metals can penetrate the human body through the food chain. These metals' intake can cause serious problems, such as nausea, neurological disorders, loss of appetite, cancer, muscular weakness, diabetes, etc. Zinc is also present, but it is harmful in a high dose, a small dose is not that worrying (Ali, Ismail, et al., 2018). Masoumi et al. (2014) studied the adsorptive removal of different metals through a magnetic poly (MMA-MA)@ Fe_3O_4 nanocomposite. The polymer was prepared through the radical polymerization technique. The results showed a sorption capacity of 90.09, 90.91, 109.89, and 111.11 mg g−1 for Co^{2+},Cr^{3+}, Zn^{2+},Cd^{2+}, respectively. Different operating conditions, such as pH, time, sorbent mass, etc., were also optimized. The sorption process followed the Langmuir sorption isotherms and pseudo-second-order rate law (Figure 16.3).

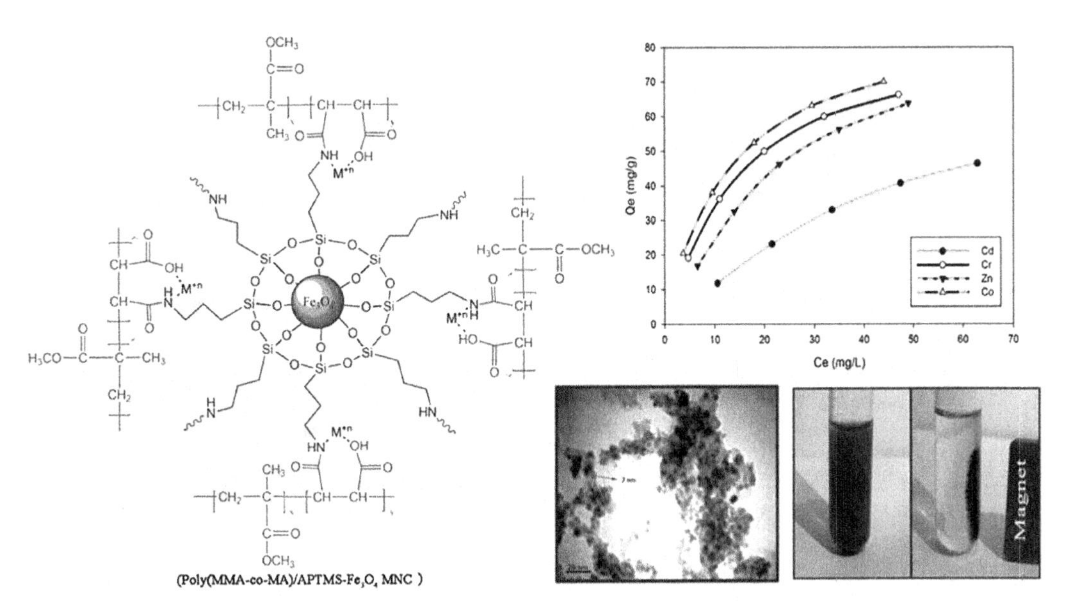

FIGURE 16.3 Metal ion removal applications of poly(MMA-co-MA)/Fe_3O_4 Magnetic material from wastewater.

Source: Reproduced from Masoumi et al. (2014) with permission from ACS.

Mahapatra et al. (2013) used the electrospinning method to synthesize an Fe_3O_4/Al_2O_3iron oxide-alumina fiber nanocomposite. First, the boehmite nanoparticle was constructed by the sol-gel process following impregnation in an PVP-iron acetyl acetonate solution. The solution was subjected to electrospin to prepare the polymer fiber nanocomposite. The obtained polymer nanocomposite was used to remove Cu^{2+}, Ni^{2+},Pb^{2+}, and Hg^{2+} ions. The results obtained showed a sorption capacity of 4.98 mg/g, 32.36 mg/g, 23.75 mg/g, and 63.69 mg/g for the respective ions. Asghari et al. (2018) prepared three types of nanosorbents, namely PANI-surfactant@Fe3O4, PTh-surfactant@Fe3O4, and Ppy-surfactant@Fe3O4, which have both magnetic and synergically conductive properties. The nanosorbents were then compared for the ligandless pre-concentration of Ni^{2+} and Pb^{2+} in the samples of some vegetables simultaneously. The magnetic nanoparticles were prepared using the co-precipitation method following the in-situ sonochemical oxidative polymerization process. The results showed a high extraction percentage of 85–95% for the microextraction of traces of Pb^{2+} and Ni^{2+} ions in complex plant matrices.

16.4.2 Removal of Organic Contaminants

Organic pollutants also contaminate water. Hydrophilic and hydrophobic are both compounds present in organic compounds (Khan, S.U., et al., 2016; Shah et al., 2018). The most common organic pollutants include dyes, pigments, pharmaceutical drugs, metabolites, etc. The organic pollutants also have a harmful impact on the surrounding environment (Rahim et al., 2016; Saeed et al., 2018). The various types of nanosorbents, such as zeolites, CNT, or polymeric materials have high adsorbent tendencies. γ-Fe_2O_3/CSCs, γ-Fe_2O_3/SiO_2/chitosan composite, L-Cht/γ-Fe_2O_3, chitosan/Fe_2O_3/MWCNTs, banana peel, bagasse pith, hydrolyzed nanosilica incorporated polyacrylamide grafted xanthan gum, methyl violet onto poly(acrylic acid-co-acrylamide)/kaolin hydrogel are also used to remove dyes (Hussain et al., 2016; Wahid et al., 2017). Guerritore et al. (2020) developed the cross-linked nanocomposite of styrene (HCLN) containing mesoporous silica nanoparticles (MSN). The resultant material possessed a 1600 m^2/g (high surface area) and enhanced hydrophilic

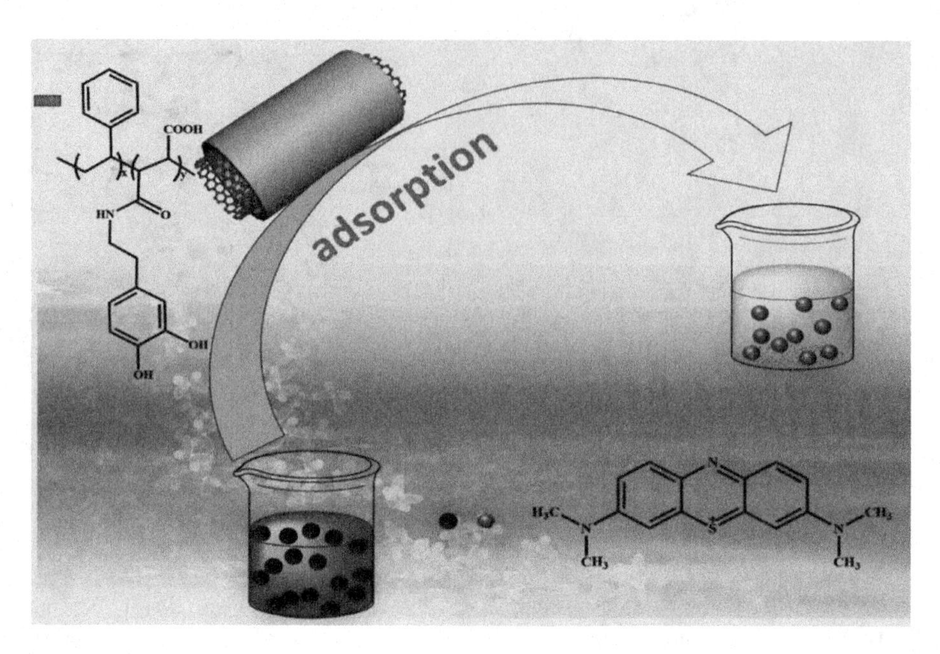

FIGURE 16.4 Methylene blue removal applications of carbon nanotubes-based polymer nanocomposites.

Source: Reproduced from Gan et al. (2020) with permission from Elsevier.

character. The prepared material was used for the sorption of Remazol Brilliant Blue R (RB). The results showed a satisfactory performance for the effective adsorption of the said dye. Bhat et al. (2020) also studied the synthesis of NiO nanoparticle doped-PVA-MF polymer nanocomposites. The prepared polymer nanocomposite (K1-K5) was successfully used for the sorption of Congo red dye. An efficient CR dye removal of about 75% was achieved confirming the good sorption ability of the polymer nanocomposite film. Gan et al. (2020) used a ring opening reaction for the preparation of the composite of CNTs (CNTs@poly(S-MA-DA)). The prepared nanocomposite was applied for the successful elimination of methylene blue dye (Figure 16.4). The different operating parameters, such as pH, temperature, etc., were also optimized. The results revealed that the removal of MB dye was a spontaneous and endothermic process.

16.4.3 Oil and Water Separation

In the petrochemical, pharmaceutical and food industries, oil is also a pollutant that needs to be cleaned (Neelofar et al., 2017). Wettability of the surface of the material is very important for the oil and water separation. Wettability depends on the geometric microstructure and chemical composition. The materials are used in two ways:

- membrane fibers coated with hydrophobic materials;
- hydrophobic materials directly (Khan, A., et al., 2017).

Li et al. (2017) prepared an eco-friendly material based on lignin grafted carbon nanotubes (L-CNTs) by a three-step method. The prepared material was used to remove oil droplets and Pb^{2+} ions. The nanocomposite showed good efficiency in the removal of contaminants. The enhanced removal efficacy of the water droplets from the water bodies was ascribed to the tridimensional

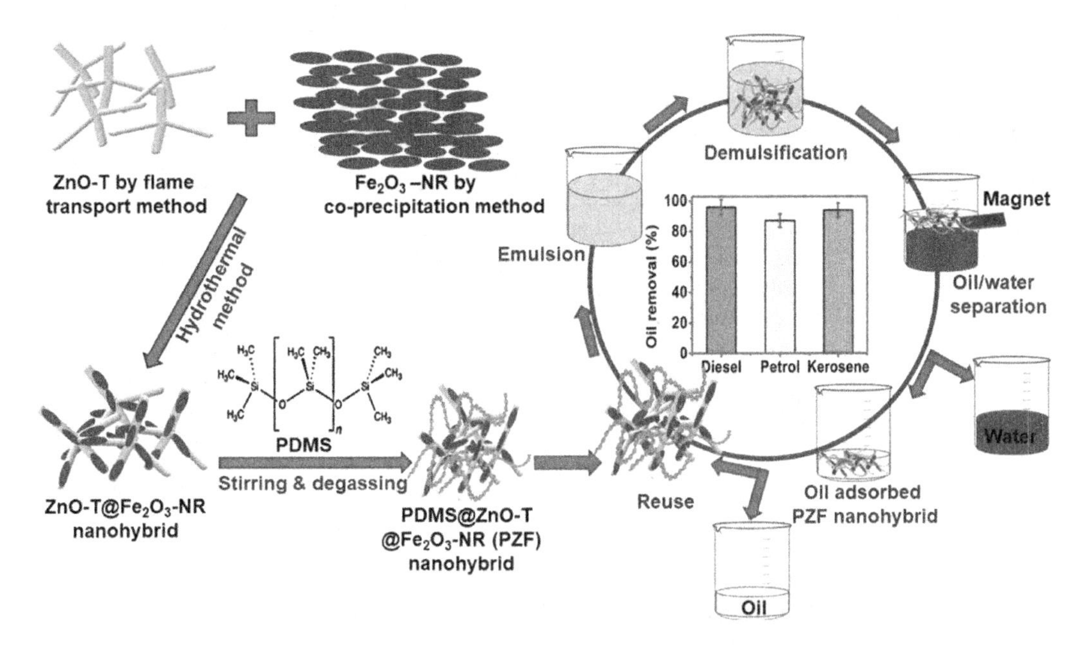

FIGURE 16.5 Illustration of the synthesis of superhydrophobic polymer-coated tetrapodal magnetic nanocomposite as efficient sorbent for the removal of oil from wastewater.

Source: Reproduced from Sharma et al. (2019) with permission from Elsevier.

structure and large surface area of the L-CNTs. The incorporation of the low-cost natural polymer, i.e., lignin, also reduces the high costs of the nanocomposite. Sharma et al. (2019) used a hydrothermal method for the preparation of rod-shaped polydimethylsiloxane@zinc oxide tetrapod@ iron oxide nanohybrid material. The prepared nanohybrid was used to remove oil from wastewater bodies. The results showed a percentage oil removal of 96% and a sorption capacity of 1135mg/g for diesel oil removal. The kinetics showed a rapid equilibrium for the removal of oil following the pseudo-first-order kinetics model. The high sorption capability of the nanohybrid was attributed to the magnetic, superhydrophobic-superoleophobic properties of the nanohybrid (Figure 16.5). Thanikaivelan et al. (2012) studied the development of collagen and iron oxide as stable magnetic nanocomposite (SPIONs) following a simple method using leather industry wastes. Different techniques such as spectroscopic, microscopic, and calorimetric processes were applied to study the molecular interactions between the spherical SPIONs and helical collagen fibers. The results showed a satisfactory selective oil absorption and magnetic tracking ability, making it a necessary component to be exploited for oil-removing purposes.

16.4.4 Removal of Other Pollutants

Along with the aforementioned pollutants, some other pollutants need to be removed. These pollutants are pathogens, microorganisms, pesticides, and other organic materials. They can also form disinfection by-products, so the risk is very high. Polymer nanocomposites remove pathogens by removing the toxic chemicals present in them. This method has proved quite effective over the years, but it also has some side effects related to health. It can react with DNA to disrupt hydrogen bonding. Some metals are effective and not injurious to health while others are harmful to the human body so we must choose them according to our requirements (Table 16.1).

TABLE 16.1
Removal of Various Pollutants from Wastewater Bodies Using Polymer-Based Nanocomposites

Polymer	Nanocomposite	Contaminants	Reference
Alginate	Sericin-derived activated carbon	Methylene blue dye	Kwak et al. (2018)
Cellulose	Carboxymethyl cellulose-*g*-poly (acrylic acid)/attapulgite (CMC-*g*-PAA/APT) hydrogel composites	Pb^{2+}	Liu et al. (2010)
Cellulose	Silica sol	Oil separation	Liu et al. (2011)
Chitosan	Chitosan nanofibers cross-linked with glutaraldehyde	Cr^{+6}	Li et al. (2015)
	Modified polymer nanocomposite	Sunset yellow, Rhodamine B	Ptaszkowska-Koniarz et al. (2017)
Cyclodextran	Fe_3O_4 NPs functionalized by Carboxymethyl-β-cyclodextrin (CM-β-CD) (CDpoly-MNPs)	Pb^{2+}, Cd^{2+}, Ni^{2+}	Badruddoza et al. (2013)
	Polyaniline (PANI) nanoparticles	Methylene blue	Ayad et al. (2013)
	Uniform polyaniline (PANI) microspheres	Methyl orange	Ai et al. (2010)
Poly(ortho-phenylenediamine)		Hg(II) and Ag(I),	Li et al. (2009)
	Poly(*m*-phenylenediamine) (P*m*PD) microspheres	Hg(II)	Li et al. (2020)
Poly(aniline-5-sulfo-2-anisidine)	Aniline–sulfoanisidine copolymer nanosorbents	Ag^{1}	Li et al. (2010)
Polydopamine	Graphene oxide polymer nanocomposites	Methylene blue	Wan et al. (2017)

16.5 CONCLUSION

To conclude, as we have seen above, nanocomposites provide new opportunities and technologies for all industrial sectors. They have effective properties and are environmentally friendly as well. Nanocomposites are well suited to meet the demands of the industrial sector. Some limitations do pose challenges, but these are common in almost every process and can be dealt with. They are also very effective in water treatment, as discussed above. Considering the unique chemical, thermal, magnetic, optical, and other properties of nanocomposites, we should consider their applications in various processes.

ACKNOWLEDGMENTS

The authors acknowledge support from the key laboratories for special resource development and medicinal research in Jiangsu Province, project number LPRK202101.

REFERENCES

Ahmad, N., Anae, J., Khan, M. Z., Sabir, S., Yang, X. J., Thakur, V. K., … and Coulon, F. (2021). Visible light-conducting polymer nanocomposites as efficient photocatalysts for the treatment of organic pollutants in wastewater. *Journal of Environmental Management*, 295, 113362.

Ahn, B., Zhilyaev, A. P., Lee, H. J., Kawasaki, M., and Langdon, T. G. (2015). Rapid synthesis of an extra hard metal matrix nanocomposite at ambient temperature. *Materials Science and Engineering: A*, 635, 109–117.

Ai, L., Jiang, J., and Zhang, R. (2010). Uniform polyaniline microspheres: A novel adsorbent for dye removal from aqueous solution. *Synthetic Metals*, 160(7–8), 762–767.

Ali, N., Ahmad, S., Khan, A., Khan, S., Bilal, M., Ud Din, S., … and Khan, H. (2020). Selenide-chitosan as high-performance nanophotocatalyst for accelerated degradation of pollutants. *Chemistry—An Asian Journal*, 15(17), 2660–2673.

Ali, N., Azeem, S., Khan, A., Khan, H., Kamal, T., and Asiri, A. M. (2020). Experimental studies on removal of arsenites from industrial effluents using tridodecylamine supported liquid membrane. *Environmental Science and Pollution Research*, 27(11), 11932–11943.

Ali, N., Bilal, M., Khan, A., Ali, F., and Iqbal, H. M. (2020). Design, engineering and analytical perspectives of membrane materials with smart surfaces for efficient oil/water separation. *TrAC Trends in Analytical Chemistry*, 127, 115902.

Ali, N., Bilal, M., Khan, A., Ali, F., Yang, Y., Khan, M., … and Iqbal, H. M. (2020). Dynamics of oil-water interface demulsification using multifunctional magnetic hybrid and assembly materials. *Journal of Molecular Liquids*, 312, 113434.

Ali, N., Bilal, M., Khan, A., Ali, F., Yang, Y., Malik, S., … and Iqbal, H. M. (2021). Deployment of metal-organic frameworks as robust materials for sustainable catalysis and remediation of pollutants in environmental settings. *Chemosphere*, 272, 129605.

Ali, N., Bilal, M., Nazir, M. S., Khan, A., Ali, F., and Iqbal, H. M. (2020). Thermochemical and electrochemical aspects of carbon dioxide methanation: A sustainable approach to generate fuel via waste to energy theme. *Science of the Total Environment*, 712, 136482.

Ali, N., Ismail, M., Khan, A., Khan, H., Haider, S., and Kamal, T. (2018). Spectrophotometric methods for the determination of urea in real samples using silver nanoparticles by standard addition and 2nd order derivative methods. *Spectrochimica Acta Part A: Molecular and Biomolecular Spectroscopy*, 189, 110–115.

Ali, N., Kamal, T., Ul-Islam, M., Khan, A., Shah, S. J., and Zada, A. (2018). Chitosan-coated cotton cloth supported copper nanoparticles for toxic dye reduction. *International Journal of Biological Macromolecules*, 111, 832–838.

Ali, N., Khan, A., Bilal, M., Malik, S., Badshah, S., and Iqbal, H. (2020). Chitosan-based bio-composite modified with thiocarbamate moiety for decontamination of cations from the aqueous media. *Molecules*, 25(1), 226.

Ali, N., Khan, A., Malik, S., Badshah, S., Bilal, M., and Iqbal, H. M. (2020). Chitosan-based green sorbent material for cations removal from an aqueous environment. *Journal of Environmental Chemical Engineering*, 8, 104064.

Ali, N., Khan, A., Nawaz, S., Bilal, M., Malik, S., Badshah, S., and Iqbal, H. M. (2020). Characterization and deployment of surface-engineered chitosan-triethylenetetramine nanocomposite hybrid nano-adsorbent for divalent cations decontamination. *International Journal of Biological Macromolecules*, 152, 663–671.

Ali, N., Naz, N., Shah, Z., Khan, A., and Nawaz, R. (2020). Selective transportation of molybdenum from model and ore through poly inclusion membrane. *Bulletin of the Chemical Society of Ethiopia*, 34(1), 93–104.

Ali, N., Uddin, S., Khan, A., Khan, S., Khan, S., Ali, N., … and Bilal, M. (2020). Regenerable chitosan-bismuth cobalt selenide hybrid microspheres for mitigation of organic pollutants in an aqueous environment. *International Journal of Biological Macromolecules*, 161, 1305–1317.

Ali, N., Zada, A., Zahid, M., Ismail, A., Rafiq, M., Riaz, A., and Khan, A. (2019). Enhanced photodegradation of methylene blue with alkaline and transition-metal ferrite nanophotocatalysts under direct sun light irradiation. *Journal of the Chinese Chemical Society*, 66(4), 402–408.

Asghari, A., Parvari, S. M., Hemmati, M., and Rajabi, M. (2018). Statistical evaluation of three kinds of sonochemically-prepared magnetic conductive polymer nanocomposites for ultrasound-assisted ligandless uptake of some deleterious metal ions in vegetable samples. *Journal of Molecular Liquids*, 268, 867–874.

Ayad, M., El-Hefnawy, G., and Zaghlol, S. (2013). Facile synthesis of polyaniline nanoparticles; its adsorption behavior. *Chemical Engineering Journal*, 217, 460–465.

Aziz, A., Ali, N., Khan, A., Bilal, M., Malik, S., Ali, N., and Khan, H. (2020). Chitosan-zinc sulfide nanoparticles, characterization and their photocatalytic degradation efficiency for azo dyes. *International Journal of Biological Macromolecules*, 153, 502–512.

Badruddoza, A. Z. M., Shawon, Z. B. Z., Tay, W. J. D., Hidajat, K., and Uddin, M. S. (2013). Fe_3O_4/cyclodextrin polymer nanocomposites for selective heavy metals removal from industrial wastewater. *Carbohydrate Polymers*, 91(1), 322–332.

Bhat, S. A., Zafar, F., Mirza, A. U., Mondal, A. H., Kareem, A., Haq, Q. M. R., and Nishat, N. (2020). NiO nanoparticle doped-PVA-MF polymer nanocomposites: Preparation, Congo red dye adsorption and anti-bacterial activity. *Arabian Journal of Chemistry*, 13(6), 5724–5739.

Casati, R., and Vedani, M. (2014). Metal matrix composites reinforced by nano-particles—A review. *Metals*, 4(1), 65–83.

Gan, D., Dou, J., Huang, Q., Huang, H., Chen, J., Liu, M., ... and Wei, Y. (2020). Carbon nanotubes-based polymer nanocomposites: Bio-mimic preparation and methylene blue adsorption. *Journal of Environmental Chemical Engineering*, 8(2), 103525.

Graz, I., Krause, M., Bauer-Gogonea, S., Bauer, S., Lacour, S. P., Ploss, B., ... and Wagner, S. (2009). Flexible active-matrix cells with selectively poled bifunctional polymer-ceramic nanocomposite for pressure and temperature sensing skin. *Journal of Applied Physics*, 106(3), 034503.

Guerritore, M., Castaldo, R., Silvestri, B., Avolio, R., Cocca, M., Errico, M. E., ... and Ambrogi, V. (2020). Hyper-crosslinked polymer nanocomposites containing mesoporous silica nanoparticles with enhanced adsorption towards polar dyes. *Polymers*, 12(6), 1388.

Hua, F., Cui, T., and Lvov, Y. M. (2004). Ultrathin cantilevers based on polymer−ceramic nanocomposite assembled through layer-by-layer adsorption. *Nano Letters*, 4(5), 823–825.

Huang, X., Wang, R., Jiao, T., Zou, G., Zhan, F., Yin, J., ... and Peng, Q. (2019). Facile preparation of hierarchical AgNP-loaded MXene/Fe_3O_4/polymer nanocomposites by electrospinning with enhanced catalytic performance for wastewater treatment. *ACS Omega*, 4(1), 1897–1906.

Hussain, S., Ullah, Z., Gul, S., Khattak, R., Kazmi, N., Rehman, F., ... and Khan, A. (2016). Adsorption characteristics of magnesium-modified bentonite clay with respect to acid blue 129 in aqueous media. *Polish Journal of Environmental Studies*, 25(5), 1947–1953.

Khan, A., Ali, N., Bilal, M., Malik, S., Badshah, S., and Iqbal, H. (2019). Engineering functionalized chitosan-based sorbent material: Characterization and sorption of toxic elements. *Applied Sciences*, 9(23), 5138.

Khan, A., Begum, S., Ali, N., Khan, S., Hussain, S., and Sotomayor, M. D. P. T. (2017). Preparation of crosslinked chitosan magnetic membrane for cations sorption from aqueous solution. *Water Science and Technology*, 75(9), 2034–2046.

Khan, A., Shah, S. J., Mehmood, K., Ali, N., and Khan, H. (2019). Synthesis of potent chitosan beads a suitable alternative for textile dye reduction in sunlight. *Journal of Materials Science: Materials in Electronics*, 30(1), 406–414.

Khan, H., Gul, K., Ara, B., Khan, A., Ali, N., Ali, N., and Bilal, M. (2020). Adsorptive removal of acrylic acid from the aqueous environment using raw and chemically modified alumina: Batch adsorption, kinetic, equilibrium and thermodynamic studies. *Journal of Environmental Chemical Engineering*, 8: 103927.

Khan, H., Khalil, A. K., and Khan, A. (2019). Photocatalytic degradation of alizarin yellow in aqueous medium and real samples using chitosan conjugated tin magnetic nanocomposites. *Journal of Materials Science: Materials in Electronics*, 30(24), 21332–21342.

Khan, H., Khalil, A. K., Khan, A., Saeed, K., and Ali, N. (2016). Photocatalytic degradation of bromophenol blue in aqueous medium using chitosan conjugated magnetic nanoparticles. *Korean Journal of Chemical Engineering*, 33(10), 2802–2807.

Khan, M., Khan, A., Khan, H., Ali, N., Sartaj, S., Malik, S., ... and Bilal, M. (2021). Development and characterization of regenerable chitosan-coated nickel selenide nano-photocatalytic system for decontamination of toxic azo dyes. *International Journal of Biological Macromolecules*, 182, 866–878.

Khan, S., Khan, A., Ali, N., Ahmad, S., Ahmad, W., Malik, S., ... and Bilal, M. (2021). Degradation of carcinogenic Congo red dye using ternary metal selenide-chitosan microspheres as robust and reusable catalysts. *Environmental Technology and Innovation*, 22, 101402.

Khan, S. U., Khan, F. U., Khan, I. U., Muhammad, N., Badshah, S., Khan, A., ... and Nasrullah, A. (2016). Biosorption of nickel (II) and copper (II) ions from aqueous solution using novel biomass derived from Nannorrhops ritchiana (Mazri Palm). *Desalination and Water Treatment*, 57(9), 3964–3974.

Khodakarami, M., and Bagheri, M. (2021). Recent advances in synthesis and application of polymer nanocomposites for water and wastewater treatment. *Journal of Cleaner Production*, 296, 126404.

Krikorian, V., and Pochan, D. J. (2003). Poly (L-lactic acid)/layered silicate nanocomposite: fabrication, characterization, and properties. *Chemistry of Materials*, 15(22), 4317–4324.

Kumar, S., Supriya, S., and Kar, M. (2018). Enhancement of dielectric constant in polymer-ceramic nanocomposite for flexible electronics and energy storage applications. *Composites Science and Technology*, 157, 48–56.

Kwak, H. W., Hong, Y., Lee, M. E., and Jin, H. J. (2018). Sericin-derived activated carbon-loaded alginate bead: An effective and recyclable natural polymer-based adsorbent for methylene blue removal. *International Journal of Biological Macromolecules*, 120, 906–914.

Li, L., Li, Y., Cao, L., and Yang, C. (2015). Enhanced chromium (VI) adsorption using nanosized chitosan fibers tailored by electrospinning. *Carbohydrate Polymers*, 125, 206–213.

Li, X. G., Feng, H., and Huang, M. R. (2010). Redox sorption and recovery of silver ions as silver nanocrystals on poly (aniline-co-5-sulfo-2-anisidine) nanosorbents. *Chemistry—A European Journal*, 16(33), 10113–10123.

Li, X. G., Huang, M. R., Tao, T., Ren, Z., Zeng, J., Yu, J., ... and Imahori, H. (2020). Highly cost-efficient sorption and desorption of mercury ions onto regenerable poly (m-phenylenediamine) microspheres with many active groups. *Chemical Engineering Journal*, 391, 123515.

Li, X. G., Ma, X. L., Sun, J., and Huang, M. R. (2009). Powerful reactive sorption of silver (I) and mercury (II) onto poly (o-phenylenediamine) microparticles. *Langmuir*, 25(3), 1675–1684.

Li, Z., Chen, J., and Ge, Y. (2017). Removal of lead ion and oil droplet from aqueous solution by lignin-grafted carbon nanotubes. *Chemical Engineering Journal*, 308, 809–817.

Liu, J., Huang, W., Xing, Y., Li, R., and Dai, J. (2011). Preparation of durable superhydrophobic surface by sol-gel method with water glass and citric acid. *Journal of Sol-Gel Science and Technology*, 58(1), 18–23.

Liu, Y., Wang, W., and Wang, A. (2010). Adsorption of lead ions from aqueous solution by using carboxymethyl cellulose-g-poly (acrylic acid)/attapulgite hydrogel composites. *Desalination*, 259(1–3), 258–264.

Lochab, B., Shukla, S., and Varma, I. K. (2014). Naturally occurring phenolic sources: Monomers and polymers. *RSC Advances*, 4(42), 21712–21752.

Mahapatra, A., Mishra, B. G., and Hota, G. (2013). Electrospun Fe_2O_3–Al_2O_3 nanocomposite fibers as efficient adsorbent for removal of heavy metal ions from aqueous solution. *Journal of Hazardous Materials*, 258, 116–123.

Masoumi, A., Ghaemy, M., and Bakht, A. N. (2014). Removal of metal ions from water using poly (MMA-co-MA)/modified-Fe_3O_4 magnetic nanocomposite: Isotherm and kinetic study. *Industrial and Engineering Chemistry Research*, 53(19), 8188–8197.

Mukhopadhyay, R., Bhaduri, D., Sarkar, B., Rusmin, R., Hou, D., Khanam, R., ... and Ok, Y. S. (2020). Clay–polymer nanocomposites: Progress and challenges for use in sustainable water treatment. *Journal of Hazardous Materials*, 383, 121125.

Naskar, A. K., Keum, J. K., and Boeman, R. G. (2016). Polymer matrix nanocomposites for automotive structural components. *Nature Nanotechnology*, 11(12), 1026–1030.

Nawaz, A., Khan, A., Ali, N., Ali, N., and Bilal, M. (2020). Fabrication and characterization of new ternary ferrites-chitosan nanocomposite for solar-light driven photocatalytic degradation of a model textile dye. *Environmental Technology and Innovation*, 20, 101079.

Neelofar, N., Ali, N., Khan, A., Amir, S., Khan, N. A., and Bilal, M. (2017). Synthesis of Schiff bases derived from 2-hydroxy-1-naphth-aldehyde and their tin (II) complexes for antimicrobial and antioxidant activities. *Bulletin of the Chemical Society of Ethiopia*, 31(3), 445–456.

Patel, H. A., Somani, R. S., Bajaj, H. C., and Jasra, R. V. (2006). Nanoclays for polymer nanocomposites, paints, inks, greases and cosmetics formulations, drug delivery vehicle and wastewater treatment. *Bulletin of Materials Science*, 29(2), 133–145.

Ptaszkowska-Koniarz, M., Goscianska, J., and Pietrzak, R. (2017). Adsorption of dyes on the surface of polymer nanocomposites modified with methylamine and copper (II) chloride. *Journal of Colloid and Interface Science*, 504, 549–560.

Rahim, M., Ullah, I., Khan, A., and Haris, M. R. H. M. (2016). Health risk from heavy metals via consumption of food crops in the vicinity of District Shangla. *Journal of the Chemical Society of Pakistan*, 38(1).

Rytwo, G. (2012). The use of clay-polymer nanocomposites in wastewater pretreatment. *The Scientific World Journal*, 2012, n.p.

Saeed, K., Sadiq, M., Khan, I., Ullah, S., Ali, N., and Khan, A. (2018). Synthesis, characterization, and photocatalytic application of Pd/ZrO 2 and Pt/ZrO 2. *Applied Water Science*, 8(2), 60.

Saleh, T. A., Parthasarathy, P., and Irfan, M. (2019). Advanced functional polymer nanocomposites and their use in water ultra-purification. *Trends in Environmental Analytical Chemistry*, 24, e00067.

Sartaj, S., Ali, N., Khan, A., Malik, S., Bilal, M., Khan, M., … and Khan, S. (2020). Performance evaluation of photolytic and electrochemical oxidation processes for enhanced degradation of food dyes laden wastewater. *Water Science and Technology*, 81(5), 971–984.

Shah, S., ud Din, S., Khan, A., and Shah, S. A. (2018). Green synthesis and antioxidant study of silver nanoparticles of root extract of Sageretia thea and its role in oxidation protection technology. *Journal of Polymers and the Environment*, 26(6), 2323–2332.

Sharma, M., Joshi, M., Nigam, S., Avasthi, D. K., Adelung, R., Srivastava, S. K., and Mishra, Y. K. (2019). Efficient oil removal from wastewater based on polymer coated superhydrophobic tetrapodal magnetic nanocomposite adsorbent. *Applied Materials Today*, 17, 130–141.

Söderqvist Lindblad, M., Ranucci, E., and Albertsson, A. C. (2001). Biodegradable polymers from renewable sources. New hemicellulose-based hydrogels. *Macromolecular Rapid Communications*, 22(12), 962–967.

Sohni, S., Gul, K., Ahmad, F., Ahmad, I., Khan, A., Khan, N., and Bahadar Khan, S. (2018). Highly efficient removal of acid red-17 and bromophenol blue dyes from industrial wastewater using graphene oxide functionalized magnetic chitosan composite. *Polymer Composites*, 39(9), 3317–3328.

Thanikaivelan, P., Narayanan, N. T., Pradhan, B. K., and Ajayan, P. M. (2012). Collagen based magnetic nanocomposites for oil removal applications. *Scientific Reports*, 2(1), 1–7.

Unuabonah, E. I., and Taubert, A. (2014). Clay–polymer nanocomposites (CPNs): Adsorbents of the future for water treatment. *Applied Clay Science*, 99, 83–92.

Wahid, F., Mohammadzai, I. U., Khan, A., Shah, Z., Hassan, W., and Ali, N. (2017). Removal of toxic metals with activated carbon prepared from Salvadora persica. *Arabian Journal of Chemistry*, 10, S2205–S2212.

Wan, Q., Liu, M., Xie, Y., Tian, J., Huang, Q., Deng, F., … and Wei, Y. (2017). Facile and highly efficient fabrication of graphene oxide-based polymer nanocomposites through mussel-inspired chemistry and their environmental pollutant removal application. *Journal of Materials Science*, 52(1), 504–518.

Yang, Y., Ali, N., Khan, A., Khan, S., Khan, S., Khan, H., … and Bilal, M. (2021). Chitosan-capped ternary metal selenide nanocatalysts for efficient degradation of Congo red dye in sunlight irradiation. *International Journal of Biological Macromolecules*, 167, 169–181.

Yeh, J. M., and Chang, K. C. (2008). Polymer/layered silicate nanocomposite anticorrosive coatings. *Journal of Industrial and Engineering Chemistry*, 14(3), 275–291.

Yin, J., and Deng, B. (2015). Polymer-matrix nanocomposite membranes for water treatment. *Journal of Membrane Science*, 479, 256–275.

17 Ohmic Heating as an Advantageous Technology for the Food Industry

Prospects and Applications

Sonia Morya,[1,*] *Chinaza Godswill Awuchi,*[2]
Pankaj Chowdhary,[3] *Suneel Kumar Goyal,*[4] *and Farid Menaa*[5]
[1]Department of Food Technology and Nutrition, School of Agriculture, Lovely Professional University, Punjab, India
[2]Department of Physical Sciences, Kampala International University, Kampala, Uganda
[3]Department of Microbiology, Babasaheb Bhimrao Ambedkar University, Lucknow, Uttar Pradesh, India
[4]Department of Farm Engineering, Institute of Agricultural Sciences, Banaras Hindu University, Varanasi, Uttar Pradesh, India
[5]Departments of Internal Medicine, Nanomedicine, and Advanced Technologies, Fluorotronics-CIC, San Diego, USA
[*]Corresponding author Email: sonia.morya8911@gmail.com, and menaateam@gmail.com

CONTENTS

17.1 Introduction 308
17.2 History of Ohmic Heating 309
17.3 Principles of Ohmic Heating 310
17.4 Process and Components of the Ohmic Heating System 310
17.4.1 The Set-Up of an Ohmic Heating System 311
17.5 Ideal Food Products 312
17.6 Benefits of Ohmic Heating 313
17.6.1 Nutritional Benefits 313
17.6.2 Microbial Inactivation Benefits 313
17.7 Process Parameters and Factors That Influence the Ohmic Heating Process 314
17.7.1 Electrical Conductivity (EC) 314
17.7.2 Field Strength 314
17.7.3 Frequency and Waveform 314
17.7.4 Electrodes 314
17.7.5 Particle Size 315
17.7.6 Ionic Concentration 315
17.7.7 Particle Concentration 315
17.7.8 Particle Location 316
17.8 Applications of Ohmic Heating 316
17.8.1 Pasteurization and Sterilization 316

DOI: 10.1201/9781003239956-19

17.8.2 Inactivation of Microorganisms....317
17.8.3 Enzyme Stabilization....317
17.8.4 Extraction....317
17.8.5 Blanching....318
17.8.6 Thawing....318
17.8.7 Flours and Starch....318
17.8.8 Fermentation....319
17.8.9 Electroporation Effects....319
17.8.10 Inactivation of Enzymes....319
17.9 Advantages of Ohmic Heating....320
17.10 Limitations and Disadvantages....321
17.11 Mathematical Modelling....321
17.12 Ohmic Heating Effects on Quality Characteristics of Foods....322
17.12.1 Water Absorption Capacity (WAC) and Water Solubility Index (WSI)....322
17.12.2 Pasting Properties....323
17.12.3 Thermal Properties....323
17.13 Current Status of Ohmic Heating Globally....324
17.14 Conclusion....324
References....325

17.1 INTRODUCTION

Ohmic heating, also referred to as "resistance heating", "joule heating", or "electroconductive heating", is an electrical heating technique that generates heat through passing electrical currents through food which resists the electric flow (Kumar et al., 2014; Varghese et al., 2014; Fellows, 2017). Heat is quickly and often uniformly generated in the food, producing a high quality sterilized product, which makes it appropriate for aseptic processing (Varzakas and Tzia, 2015; Fellows, 2017). Energy in the form of electricity is linearly transformed to thermal energy with an increase in electrical conductivity; it is a key process parameter which influences the heating rate and uniformity (Fellows, 2017). Ohmic heating is suitable for foods with suspended particulates in a weak medium containing salt because of their higher resistance characteristics (Varghese et al., 2014). The ohmic heating technique is also valuable because of its capacity to inactivate microbes via thermal and non-thermal cell destruction (Ramaswamy et al., 2014; Varghese et al., 2014; Fellows, 2017). It can inactivate anti-nutrients, thus maintaining nutrients' availability and sensory characteristics (Ramaswamy et al., 2014). Nevertheless, ohmic heating has been limited by fouling, electric conductivity, and viscosity (Varghese et al., 2014; Fellows, 2017). Though ohmic heating is yet to be approved by some regulatory authorities, such as the FDA for commercial applications, it has numerous potential uses, including cooking, fermentation, and sterilization (Fellows, 2017). Ohmic heating has been applied as a sterilization technique in some countries of the developing world. However, more robust research needs to be done to establish its effectiveness.

Ohmic heating is gradually causing a revolution in food industries due to its broad applications and cleanliness. The advantages of ohmic heating over conventional thermal processing techniques, such as induction heating, "dielectric heating", radio-frequency heat treatment, and microwave heat treatment have gained more traction from researchers and industrial operators (Kumar et al., 2014; Kaur and Singh, 2016; Kumar, 2018). Less cooking time, better quality of product, better energy efficiency, cost effectiveness, and environmental friendliness are the major advantages of ohmic heating. It is mainly commercially industrialized for sterilization and pasteurization; however, attempts have been made to apply ohmic heating in other areas (Kumar, 2018; Kumar et al., 2014), such as fermentation, cooking, etc. Most early challenges involving the use of ohmic heating have been addressed, with more studies currently looking into ways to optimize and improve the process. The future of ohmic heating is guaranteed in food applications. The rate of non-uniform generation and distribution of heat is a result of particles of electrical heterogeneity, and the difference in particle orientation and shape is recommended for further studies (Kumar, 2018). For performance evaluation and reduction in cost of ohmic heating for commercial use in developed, developing, and underdeveloped nations, several technical and commercial studies are required. Numerous potential applications of ohmic heating exist, including sterilization, heating, fermentation, pasteurization, blanching, evaporation, and dehydration. Other than heating, the application of the electric field in ohmic heating leads to cell membrane electroporation, which decreases the gelatinization temperature, increases extraction, and decreases enthalpy. Ohmic treatment causes quicker food heating in addition to maintaining nutritional value and sensory properties.

The foods placed in-between the electrodes play a resistance role in the electric circuit. The process involves the passage of an electric current through foods aimed at heating them. The optimal design of ohmic heating requires a knowledge of food electrical conductivity (EC) and the required mathematical models (Kumar et al., 2014). The EC of foods often increases with increased moisture and temperature. A positive linear correlation between EC and temperature (T) exists. Ohmic heating offers many potential and actual applications in the future, including its use in fermentation, extraction, dehydration, blanching, and evaporation (USFDA, 2000), pasteurization, heating, and sterilization of foods in military or prolonged space missions (Kumar et al., 2014). The energy in ohmic heating is directly dissipated into the foods. Ohmic heating is a direct heating method. EC is an essential tool in designing ohmic heating equipment.

Ohmic heating can also be considered an emerging technique which offers numerous industrial applications in all parts of the world. The main ohmic heating principle is electrical energy dissipation as heat which makes use of an electrical conductor. The quantity of heat released is directly proportional to the current flow induced by the gradients of the voltage in foods and the EC of the foods (Kumar et al., 2014; Kaur and Singh, 2016; Kumar, 2018). The water absorption capacity (WAC), thermal properties, the water solubility index (WSI), and pasting properties are affected by ohmic heating. Ohmic heating has been reported to cause pre-gelatinization of starch granules, thereby reducing the energy needed during processing. However, its high cost, lack of use in fatty foods, and lack of adequate knowledge limit its use.

The mathematical model predictions using EC equations were accurate (Kumar, 2018). There is a need to develop ohmic heating for commercial applications in industrial food processing and research is also needed for the development of an ohmic heating system for domestic purposes.

17.2 HISTORY OF OHMIC HEATING

The idea of ohmic heating started before 1897 (Jones, 1897). In 1841, James P. Joule made the discovery that an electric current passage released heat, which explains why the process is also called Joule heating. James P. Joule discovered that the heat generated was directly proportional to the wire resistance multiplied by the square of the electricity current. Resistance of electricity is measured in Ohms (Ω). As the heat is caused by the electrical resistance, it was called "ohmic" heating. The ohmic heating application as a successful commercial processing technique was introduced in the 1920s for the first time, and was called the "Electro-Pure" process (Anderson and Finkelstein, 1919). About 50 sets of electric milk sterilizing equipment were in operation in the 1930s, however, they had all disappeared by the 1950s. The reason for the discontinuity was the high amount of electricity required and the high cost of processing (Fryer and Li, 1993). Other uses also stopped due to the complications of inappropriate contact of the electrodes and foods, electrolysis and contamination of food because of unsuitable electrode materials (Mizrahi et al., 1975). However, efforts were directed towards reusing ohmic heating in food processing. APV Baker Limited established the first ohmic heating system for the commercial sterilization of foods (Skudder, 1992). Recently, more studies have been done to optimize the process of ohmic heating, reduce the cost, and improve the efficiency of the electricity usage.

17.3 PRINCIPLES OF OHMIC HEATING

Many foods have high dissolved salts and moisture, which are electric conductors through electrolytic conduction as the dissolved salts and moisture are electrolytes. When foods are placed in an ohmic heating system and the electric field is connected, the ions in the electrolytes are attracted towards the opposite charged electrodes (-ve to +ve and +ve to -ve). The movement of ions in the electrolytes generates heat. The food creates electric resistance to the ions' movement, leading to increased kinetic energy within the system, resulting in heating of the food (Singh and Heldman, 2014; Kumar, 2018). In the same way, heat is directly produced in the food through Joule heating as an alternating current (AC) passes through the food, resulting in the internal generation of heat, which causes an increase in temperature (Kumar, 2018). To put it simply, ohmic heating makes use of the electrical resistance of the food to generate the heat required for the food processing (Singh and Heldman, 2014; Kumar, 2018). The heat is immediately and volumetrically generated in the foods because of the ionic motion; this is known as the "Joule effect". The quantity of generated heat is directly proportional to the electric current induced by the gradient of the voltage in the electrical field and the EC of the food under processing (Kumar, 2018).

The rate of heat generation in ohmic heating is described as (Samprovalaki et al., 2007):

$$Q = \sigma E^2 \qquad \text{(Eq. 17.1)}$$

This is equivalent to:

$$I^2R.\ E = \text{electric field strength (V m}^{-1}\text{)}$$

where

σ = local electrical conductivity (S m^{-1}),
Q = rate of internal energy generation (W m^{-3}).

The voltage distribution is:

$$\nabla(\sigma \nabla E) = 0 \qquad \text{(Eq 17.2)}$$

The voltage distribution depends on the EC distribution in the ohmic heating system and the system geometry. Eqn (17.2) differs from the normal Laplace's equation ($\nabla 2\ E = 0$) as it has to do with the medium wherein the EC is a function of temperature and position.

17.4 PROCESS AND COMPONENTS OF THE OHMIC HEATING SYSTEM

There are various arrangements for a "continuous ohmic heating system". Figure 17.1 shows the basic process of an ohmic heating system. A power source is required to generate an electrical current (Varghese et al., 2014). Electrodes pass electricity via the food matrix (ibid.). The distance between the two electrodes is adjustable to achieve the optimum field strength. The supplied electric current moves to the first electrode, then goes via the food sample in the space between the electrodes (ibid.). The food resistance to the current flow causes internal heat generation (Fellows, 2017). The current keeps flowing from the second electrode back to the source of power for circuit closure (Varghese et al., 2014). Insulator caps round the electrodes control the space in the ohmic heating system (ibid.). The EC, field strength and time of residence are the key to heat generation

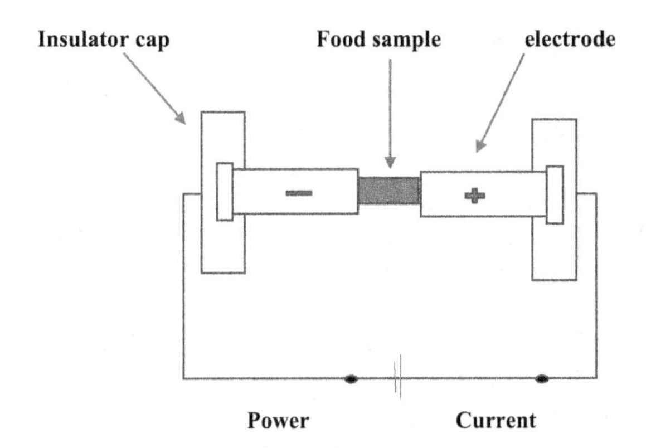

FIGURE 17.1 The basic ohmic heating process.

(Fellows, 2017). Ohmic heating equipment basically contains a heating chamber, an alternating power source, and two electrodes.

17.4.1 The Set-Up of an Ohmic Heating System

The set-up of an ohmic heating system is mainly made up of an AC power source, a data logger system, a heating cell, a voltage control compartment, electrodes, and thermocouple. Many researchers use various construction materials for the ohmic heater. Marra and other researchers analyzed the transfer of heat in ohmic heating of a solid diet. They used cells made of stainless steel of 7.6 cm outer diameter, 7.2 cm inner diameter, internal length of 11.5 cm, and length of 14.5 cm. The inner surface of the cell was lined using Teflon, with three thermocouples placed at the cell top (Kaur and Singh, 2016). Thermocouples were made using T thermocouple wires in a 2 mm diameter sheath of stainless steel. Titanium electrodes coated with platinum of 6.9 cm diameter were fixed to the cell ends. Also, the system contains a voltage control compartment (VCU). According to Marra et al. (2009): "The control panel received 230 V, 50 Hz AC current; an integrated transformer was applied to regulate voltage to 100 V." Kaur and Singh (2016) reported that "an experiment was done on starch gelatinization temperature using ohmic heating, in which the scientists used data logger system and T type thermocouple which was covered using silicon". The voltage applied heating was measured using an HP data logger. The system contained flat electrodes manufactured with titanium, a transformer, and a heating box enclosed using lids. The "power supply" was alternating current at 50 Hz with "100 V".

In another research study on ohmic heating of fruit and meat, researchers made use of a set-up made of 10 cylindrical chambers fitted with platinized titanium electrodes. The EC of the ten samples was measured at 140°C. The foods were fastened at the ends using electrodes within every cell, and "a T-type copper-constantan, Teflon-coated thermocouple" (ibid.). The ohmic heating system was linked to relay the electric switch that determined the order of which cell was heating. Current and voltage "transducers" measured the current flowing through the samples and the voltage across the samples. A data logger recorded the data at fixed intervals. In 2009, Icier used ohmic heating equipment made up of "variable transformers" (0 to 240 V), a power source, a microprocessor board, and a transformer (50 Hz) in an experiment on "influence of ohmic heating on rheological and electrical properties of reconstituted whey solutions" (Icier, 2009). An electronic temperature sensor coated in Teflon using compression fitting was applied in order to estimate the temperature at various segments of the samples in testing cells. The microprocessor board recorded the voltage, temperatures, and current applied. The information was transmitted to a microcomputer at fixed intervals.

17.5 IDEAL FOOD PRODUCTS

The most important aspect of ohmic heating is the food EC. Most food materials with dissolved ionic salts and at least 30% moisture have been reported to be sufficiently suitable for ohmic heating application. Foods suitable in ohmic heating application are "viscous" and have sufficient particles (Fellows, 2017); for example, egg, soymilk, whey, salsa, fruit in syrup, stews, milk, thick soups, sauces, heat-sensitive liquids, and ice cream mix. The efficiency of converting the electric flow to heat relies on moisture, the dissolved salt content, and the globules of fat because of their resistance and thermal conductivity (Ramaswamy et al., 2014). In foods with particles, the particles heat quicker than the liquid matrix as a higher resistance to the electric flow and conductivity contributes to thorough heating in an ohmic heating system (Fellows, 2017). It avoids overheating the liquid matrix and at the same time the particles attain adequate heat. High EC values show a higher number of ionic components suspended in the food, and are proportional to the heating rate (Varghese et al., 2014). The EC values increase in the presence of polar compounds, including salts and acids, but decrease in the presence of nonpolar compounds, such as fats (ibid.). Also, the EC of foods in general increases as the temperature increases, and usually changes when changes in structure occur while heating, e.g. starch gelatinization (Fellows, 2017). The pH, bulk density, and "specific heat" of different constituents in the food affect the degree of heating (Ramaswamy et al., 2014). The EC of selected foods is shown in Table 17.1.

1. *Meat*: Ohmic heating can be applied as a quicker-alternative way for cooking and thawing meat. Ohmic heating provides potential for safer meat through effectively preventing the growth of microorganisms via the uniform distribution of heat in the food and instant cooking inside of the food (Mitelut et al., 2011). Turkey meat cooked with ohmic heating yielded high quality products and an 8- to 15-fold decrease in duration of cooking (Zell et al., 2010).
2. *Milk*: Ohmic heating of milk was first suggested by Anderson and Finkelstein (Kumar, 2018). Ohmic heating has been proved to be superior for milk pasteurization, which can reduce fouling. Ohmic heating not only has thermally-induced lethal effects on microbes, it also has non-thermally-induced lethal effects on microbes (Sun et al., 2008).

TABLE 17.1
Electrical Conductivity (EC) of Common Foods

Food	T (°C)	EC (S/m)
Apple juice	20	0.239
Beer	22	0.143
Carrot juice	22	1.147
Carrot	19	0.041
Coffee (black)	22	0.182
Chicken meat	20	0.19
Beef	19	0.42
Starch solution (5.5%) with		
• 2% salt	19	4.3
• 0.55% salt	19	1.3
• 0.2% salt	19	0.34

Source: Fellows (2017).

Notes: T = temperature. EC = electrical conductivity. Concentration of salt is directly proportional to EC. Moisture is directly proportional to EC.

3. *Fruits*: Several commercial, perishable, local, and heat-sensitive fruit products have been processed with ohmic heating, e.g. apple, orange, sour cherry juice, strawberry pulps, pineapple juices, orange juice, pomegranate juice, and lemon juice (Darvishi et al., 2011; Kumar, 2018).
4. *Seafoods*: Ohmic heating improved the efficacy of cooking seafood, including surimi, shrimps, etc. Ohmic heating was reported to be quicker and uniform, giving the same yield, texture, and colour even more than conventional cooking methods (Lascorz et al., 2016). Ohmically heated gel indicated at least two times shear strain and shear stress more than gel heated with a water bath (Kumar, 2018).
5. *Vegetables*: The effects of cooking with ohmic heating on cauliflower firmness have been evaluated (Kumar, 2018). The data showed that the combination of pre-cooking in low-temperature saline solutions and ohmic heating offers a real result to sterilization using HTST for cauliflower florets (ibid.). Ohmic heating avoids the lack of firmness in vegetables in comparison to conventional methods.

17.6 BENEFITS OF OHMIC HEATING

The benefits of using the ohmic heating method include reduced cooking time, less capital cost, uniform and quick heating (>1°C/s), better energy efficiency, and volumetric heating in comparison to aseptic processes, PEF, and canning (Varghese et al., 2014). The volumetric heating ensures proper internal heating as opposed to heat transfer from other mediums. It leads to the production of high quality, safe foods with very little or no change in structure, nutrition, and sensory properties. In ohmic heating, the transfer of heat is uniform and always reaches the interior parts of foods which otherwise would have been difficult to heat to the core (Fellows, 2017). It has an excellent heat penetration rate, with "less fouling on electrodes compared to other conventional methods" (Varghese et al., 2014). Also, ohmic heating requires less maintenance and cleaning, leading to an environmentally-friendly heating process (Fellows, 2017).

17.6.1 NUTRITIONAL BENEFITS

The reduced processing duration in ohmic heating preserves the nutrients and organoleptic characteristics of food. Ohmic heating inactivates anti-nutrients such as pectinase, polyphenol oxidase (PPO), and lipoxygenase (LOX) because of the removal of the enzymes' active metallic groups by the electrical field during the process (Ramaswamy et al., 2014). Just like other methods of heating, ohmic heating causes protein agglutination, fat melting, and starch gelatinization (Fellows, 2017). The nutrients which are soluble in water are preserved in the liquid suspension, resulting in no loss of nutrients when the food is eaten (Kaur et al., 2016).

17.6.2 MICROBIAL INACTIVATION BENEFITS

Ohmic heating is used for microbial inactivation through thermal cellular damage as well as by non-thermal cellular damage caused by the ohmic heating electrical field (Varghese et al., 2014). Ohmic heating destroys microbes through membrane rupture, cell membranes' electroporation, and cell lysis (Ramaswamy et al., 2014; Fellows, 2017). In electroporation, leakage of excess ions as well as intramolecular components leakage result in the death of the cells (Ramaswamy et al., 2014). In membrane rupturing, the cells swell because of increased diffusion of moisture across the membranes of the cell (Varghese et al., 2014). In cell lysis, considerable disruption and breakdown of cytoplasmic membranes and cell walls result in cell lysis (Fellows, 2017; Ramaswamy et al., 2014; Varghese et al., 2014).

17.7 PROCESS PARAMETERS AND FACTORS THAT INFLUENCE THE OHMIC HEATING PROCESS

Several factors including the EC, concentration, size of particles, electrodes, and field strength influence the ohmic heating of food.

17.7.1 Electrical Conductivity (EC)

Electrical conductivity (EC) is the most crucial food property which influences ohmic heating. The EC of food depends on the applied field strength, the temperature, the microstructural set-up, and the ionic break-up of the food. The increase in EC is directly proportional to the temperature increase during the process of ohmic heating; the voltage gradient can be constant (Castro et al., 2004). The design of efficient ohmic heating equipment depends on the food's EC. "The ohmic heating rate is proportional to the square of the strength of the electric field and the EC" (Sastry and Palaniappan, 1992). Studies have been done on the EC of fruit products such as purees and juices (Castro et al., 2004; Icier, 2012). The EC of meat has been studied recently. The electrical conductivities of various porks at 20°C indicated that lean portions have higher EC than fatty portions. The EC of selected foods can be seen in Table 17.1. In 2011, Darvishi et al. reported that the EC reduced with an increase in temperature when bubbling commenced. The EC decrease may be due to increased solid concentration because of moisture evaporation, resulting in drag in the movement of the ions. An increase in moisture leads to an increase in the EC of foods.

17.7.2 Field Strength

High field strength leads to effective microbial inactivation. The heating rate in an ohmic heating system increases with the voltage gradient increase (Kumar, 2018). The stronger the intensity of the electric field, the higher the EC and the quicker is the rate of heating. Effects of the strength of the electric field and several thermal processes on strawberry EC were studied (Castro et al., 2004; Kumar, 2018). "Increase of EC with strength of electric field was reported for strawberry filling and strawberry pulps" (Kumar, 2018). However, it was not reported for strawberry-apple sauce and strawberry topping. The heating caused the destruction of the membrane and a consequent increase in free water. The application of field strength leads to increased fluid motion via capillaries that are proportional to the EC. High field strength of 104–105 Vcm^{-1} led to effective microbial inactivation (Barbosa-Cánovas et al., 1998). Expressing as well as the extraction process improved in various foods with "field strength" below 100 Vcm^{-1} (Icier, 2012).

17.7.3 Frequency and Waveform

Frequency and waveform alteration of the applied AC current also affect the rate of heating. Reducing the frequency of the AC current significantly increases the yield in oil extraction (Kumar, 2018). Reducing the frequency of the AC current improves the stabilization of enzymes and enhances other processes.

17.7.4 Electrodes

Electrodes are the main influence on heat loss, thereby affecting the effectiveness of ohmic heating (De Alwis and Fryer, 1990). Thicker electrodes reduce the temperature increase rate which shows its larger mass and reduced resistance to electric flow (Zell et al., 2010). Zell and other scientists evaluated numerous metals, including platinized titanium, titanium, aluminium, and stainless steel, for suitability as materials for electrodes. Thinner aluminium was found to provide the highest temperatures at the interface of the electrodes. Loss of heat from the exterior parts of the foods depends mostly on the kind of product cell content as well as the electrode, resulting in unacceptable "high temperature gradients" in the food. Electrodes are possible contributors to loss of heat in ohmic heating. "The thicker the electrode, the lesser the rate of increase in temperature, that is reflecting larger mass and lesser resistance to electric flow" (Zell et al., 2011). Reduced temperature at the surface of the electrodes was observed. A huge difference in temperature was reported between the thermocouples located near the electrodes and the thermocouples placed at the ohmic heating cell centre. The product contamination risks from reactions in the electrode can be reduced with bipolar pulses of 10 kHz and also by using electrodes made with platinized titanium (Kaur and Singh, 2016; Kumar, 2018). Electrodes made with platinized titanium were confirmed as most appropriate as they showed more resistance to electrolysis, with a suitable rate of heating, and were more robust, unlike the electrodes made with thinner aluminium (Kaur and Singh, 2016).

17.7.5 Particle Size

The ohmic heating rate is influenced by the particle size. An increase in particle size leads to a corresponding decrease in the EC. An increase in the duration of heating was reported as proportional to an increase in carrot cubes concentration in the ohmic heating system. The heating rate has been reported to decrease as the particle size of the carrot increases (Kaur and Singh, 2016; Kumar, 2018). This may be due to the fact that the larger the particle size, the longer the time required for heating. It has been reported that "heating duration required to attain the same temperature increased as particle size increases" (Kaur and Singh, 2016; Kumar, 2018). The heating rate reduces as the particle size increases; showing lower efficiency in heating of the liquid with increased particle diameter.

17.7.6 Ionic Concentration

Electrical conductivity (EC) increase during ohmic heating involving biological tissues takes place because of the ionic mobility increase due to changes in the structure of the tissue, including a decrease in the aqueous phase viscosity, tissue softening, non-conductive gas bubbles expulsion, and the breakdown of the cell wall protopectin (Kaur and Singh, 2016). The stronger the ionic concentration, the quicker the rate of heating has been noticed. Icier and other researchers resolved that puree from apricot heated faster than peach puree because it is more acidic. Greater ionic concentration has a significant impact on the heating characteristics of ohmically heated food. A greater EC could be associated with softer tissues and as a result greater ionic mobility compared to firmer tissues of pear, pineapple, and apples (ibid.). In 2009, Zell et al. found minced beef and salted meats had the highest EC in comparison with the whole meat heated in perpendicular or parallel path (Zell et al., 2009). Shirsat et al. (2004) proposed that a mincing effect was to enable moisture release and also the release of inorganic components from the tissues of myofibrillar during the mincing and chopping processes. Salting seemed to improve the EC difference due to the orientation of the fibre. Tumbling and injection have to be done carefully to attain uniform distribution of salt and for ohmic heating optimization (Zell et al., 2009).

17.7.7 Particle Concentration

Particle concentration affects the rate of heating in the two phases. In 2004, Castro et al. reported a decrease in strawberry pulp EC with an increase in concentration of strawberry in the pulp. The pulp with initial Brix 14.5°, pH 4.0, and starch content of 2.5%, had more EC than the pulp with Brix 26.5°, pH 4.0, and without starch. Prolonged heating time (2x) is needed for a 20–80°C temperature increase compared to the less solid concentration because with the increase in particle size, the bulk density (BD) decreases. They also reported that the heating time was seen to increase with carrot cubes concentration in the food matrix, and their results agree with previous studies (Castro et al., 2004; Kaur and Singh, 2016).

> The ohmic heating rate improved with reduction in concentration of juices from 60–20 percent, as solid contents of purees in their study were less than that of juices in the reference, the rates of ohmic heating of the purees were estimated to be higher than concentrated juices at every voltage gradient, 20 to 70 Vcm^{-1}.
>
> (Kaur and Singh, 2016)

17.7.8 Particle Location

The location and orientation of the particles play a considerable role in the ohmic heating system. The heating characteristics of the particle mass as affected by the particle location in relation to the electrodes have been evaluated. The surface of the electrode was partially or fully touching the particle mass in the system during the heating operation (Kaur and Singh, 2016). "For parallel connection, solid phase heated slower than liquid phase, whereas in the series connection, solid phase heated faster than liquid phase" (ibid.).

17.8 APPLICATIONS OF OHMIC HEATING

The ohmic heating system has huge potential applications in many food processing operations, such as enzyme stabilization, thawing, gelatinization of starch, fermentation, extraction, blanching, inactivation of microorganisms, pasteurization, and sterilization (Fellows, 2017) (Table 17.2). Ohmic heating can pasteurize foods with particulates for hot filling, aseptically process refrigerated foods, process ready-to-eat foods, and pre-heat foods before the canning operation (Varghese et al., 2014). As the data on the EC of solid foods is not sufficient, it is hard to demonstrate the quality and safety of the ohmic heating system; the US Food and Drug Administration (FDA) and some regulatory authorities are yet to approve ohmic heating for food applications (USFDA, 2000). In addition, successful 12D decrease for prevention of *Clostridium botulinum*is is yet to be confirmed (ibid.).

17.8.1 Pasteurization and Sterilization

Ohmic heating is used for juice sterilization. Ohmically heated orange juices contains more flavouring compound concentrations and has longer sensory properties shelf-life than juice conventionally pasteurized (Kumar, 2018).

TABLE 17.2
Ohmic Heating in Food Applications

Foods	Food applications
Milk, soups, cauliflower florets, stews, fruit slices and purees, and juices	Heating liquid food with big particulates, aseptic processing, heating heat-sensitive liquid foods, and sterilization
Meat patties, pork, hamburger patties, vegetable, chicken, and minced beef	Cooking
Stew type food	Military rations; Space food
Shrimp blocks	Thawing
Orange juice, juices, and process fish cake	Microbial spore inactivation, enzymes inactivation
Vegetable purees, soymilk extraction, sucrose extraction from sugar beets, and potato slices	Extraction; blanching

17.8.2 Inactivation of Microorganisms

Inactivation of microorganisms in ohmic heating is mostly thermally induced, just like conventional methods of heating. The electric field in ohmic heating could result in slight non-thermally induced cellular destruction (Sun et al., 2008; Kumar, 2018). Ohmic heating at low frequencies, often 50–60 Hz, makes the cell wall form pores as well as accumulate charges, and it is for this reason that the extra impact of ohmic heating is needed. This effect leads to the decreased D value necessary for the inactivation of microorganisms in comparison with traditional methods of heating. This decrease was reported for *Bacillus subtilis*, *Streptococcus thermophilus*, *Bacillus licheniformis*, *Escherichia coli*, and *Byssochlamysfulva* (Kumar, 2018). By subjecting foods to ohmic heating pre-treatment, the strength of further thermal/heat applications for additional inactivation of microorganisms can be lessened. The curves of inactivation of microorganisms for ohmic heating are related to the curves for other heating methods apart from the slope difference that could possibly be explained using the electric field presence and strength (USFDA, 2000). Ohmic heating has been known to destroy nearly every microbe, thus thermal death kinetic evaluations for spoilage microbes and pathogens could be followed during the design of systems and experiments which make use of the ohmic heating system. The heat treatment necessary for the inactivation of microorganisms in foods and other biological materials may be decreased if there is sublethal damage or extra lethal effects caused by the electric current.

17.8.3 Enzyme Stabilization

In 2013, Moreno et al. reported that ohmic heating totally inactivated the enzyme polyphenol oxidase (PPO) in products of apple cubes (Moreno et al., 2013). Ohmic heating is effective for the inhibition of lipases for rice bran stabilization, similar to steaming.

17.8.4 Extraction

Ohmic heat of apple tissues before the extraction of juice mechanically significantly increased the juice yields more than non-ohmically heated apple tissue (Kaur and Singh, 2016). The use of ohmic heating improved the total percentage lipid extracts from the bran of rice. Electric fields of strength below 100Vcm^{-1} improve the processes of expressing and extraction in various foods (Wang and Sastry, 2002). In ohmic heating, it is essential to apply a field strength of 100Vcm^{-1} for 1–100 seconds. The electroporation mechanism or electrical breakdown was thought to dominate in the studies done on ohmic heating application in expression and extraction (ibid.). Praporscic in 2006 studied the yield of juice from apple and potato, as influenced by the ohmic heating system. "Two types of studies involving various treatment compartments were done", including conductivity and a textural study of juice yield and cylindrical samples evaluations of slices of tissues (Praporscic et al., 2006). The study showed that the degree of tissue disintegration and yield of juice depend on the plant tissue type, the treatment duration, the intensity of the electric field, and the temperature. Extraction of juice was at an optimum when the tissues of the plants were ohmically heated at 50°C. This can be explained using the combination effects of thermal tissue softening and cell membranes' electroporation. Ohmically heated potatoes and apples at a strength of electric field below 100Vcm^{-1} provide mechanical tissues softening and high membrane damage even at 50°C. Extraction of juice using pressing was reported to give the highest yield when tissues were ohmically heated at 50°C. Ohmic heating has been applied to improve soy milk extraction from soybeans and to increase the yield of extraction of sucrose from sugar beet (Kaur and Singh, 2016). The advantages include the maintenance of nutrients and colour of foods, higher yields in extraction, and quicker processing times (Castro et al., 2004; Wang and Sastry, 2002).

17.8.5 Blanching

The use of ohmic heating for blanching may significantly decrease the leaching of solutes, in comparison to hot water processing in quick blanching time, irrespective of the size and shape of the food. Electric fields improved the loss of moisture in blanching of potato using ohmic heating (Kumar, 2018).

17.8.6 Thawing

Ohmic heating application in thawing has the benefits of better quality of food and less microbial growth in comparison to other methods of thawing (Icier, 2012). Ohmic heating had significant effects on the energy utilization ratio (EUR) and the rate of thawing of frozen beef. The use of ohmic heating in thawing frozen meat offers reduced loss of weight (Duygu and Umit, 2015).

17.8.7 Flours and Starch

Ohmic-treated starch had the lowest reduction in enthalpy, thus lower energy need. The EC of suspensions of native starch is proportional to the temperature. Starch is used in foods as a stabilizer, a thickener, and for moisture retention. Extrusion processes and drum driers used in industry for the preparation of pre-gelatinized starch resulted in the severe breakdown of the starch granules, thereby leading to an increase in the soluble solid levels. In 2007, An and King studied the heat characteristics of ohmically treated and conventional heated starch and flour from rice at numerous voltages and frequencies (An and King, 2007). Since the rice samples were subjected to pre-gelatinization because of conventional and ohmic heat treatment, there was an increase in the gelatinization temperature (GT) and they were more rigid because of the interactions of the starch chain. Additionally, there was a decrease in enthalpy for ohmic heated and conventional heated starch during the gelatinization. Thus, lower gelatinization energy was needed in the differential scanning calorimetry (DSC) analysis. Ohmic heated starch had the lowest enthalpy reduction possibly due to the highest pre-gelatinization extent via ohmic heating. Flour from brown rice had the highest GT (Kaur and Singh, 2016).

The enthalpy of ohmic heated starch at 20 Vcm^{-1} was the least, as the lower voltage led to total pre-gelatinization from a much longer time of heating needed to attain a temperature of 100°C. Ohmic heating at 70 Vcm^{-1}reduced the onset of GT of white flour, and faster swelling produced by rice flour, while the sample heated conventionally had better thermal resistance. The results of one research study indicated that peaks of endothermic gelatinization were seen on EC curves and thermo grams of DSC, with comparable temperature ranges and comparable shapes. The percentage SG data generated from the EC curves and thermograms of DSC agree in the mid and low range of gelatinization but differ when the percentage SG increased, because of the high rate of ohmic heating. The EC increased as the temperature increases, but the EC decreased as the gelatinization degree decreases, obviously resulting from changes in structure and bound water increase. Morya et al. (2016) studied various millets (sorghum, barnyard, pearl millet, oats) for their dough rheological characteristics and found gelatinization of starch takes place at peak torque (heating stage). Another study examined ohmic heating application wherein an electric current was passed via the food to produce heat, attain satisfactory gelatinization, and minimize solid losses in cassava and jicama starch (Kaur and Singh, 2016). The heating resulted in higher gelatinization with shorter processing times and high voltage; 123V for 10 minutes. For starch of jicama, the WAC and the WSI ranged from 3.2–4.7% and 1.2–2.2%, respectively, at 50°C, while for starch of cassava, 3.7–5.4% and 3.2–6.1% respectively. From thermal properties and viscosity analysis, it was reported that 39.1% of starch of cassava and 70.0% of starch of native jicama were gelatinized with lengthier processing times and maximum voltage. The starches obtained had various functional properties, including numerous gelatinization rates which can be applied industrially.

17.8.8 Fermentation

Ohmic heating may reduce the lag phase of fermenting microbes and thus reduce the time of fermentation in the production of wine, beer, cheese, and yogurt. Ohmic heating application was evaluated on *Lactobacillus acidophilus* fermentation. It was reported that the electrical field application in the processes could result in electroporation of the cell membrane, causing faster as well as more efficient transport of nutrient to the cell interior, thereby decreasing the duration of the lag phase in fermentation. But productivity decreased in subsequent stages of fermentation due to ohmic heating. This could be because of mild electroporation presence, which enhances substrates transport at early fermentation stages, thereby hastening the fermentation process. In subsequent stages, electroporation will enhance the transport of metabolites into cells and inhibit fermentation (USFDA, 2000). The mechanisms of pore formation on cell tissues have remained supported by the latest studies.

17.8.9 Electroporation Effects

Cell electroporation is the "formation of pores in the membranes" of cells as a result of the presence of an electric field and as a consequence the membrane permeability is improved and diffusion of material throughout the membrane is accomplished through electro-osmosis (An and King, 2007). It is believed the electroporation mechanism or electric breakdown is central for non-thermal ohmic heating effects (ibid.). Electroporation takes place as the membrane of the cell has definite dielectric strength that could be surpassed by an electrical field within the ohmic heating application, the field results in cell membranes' electroporation. The dielectric strength of the membrane of the cell is associated with the quantity of lipid insulator in the membrane. The formed pores can differ in size which depends on the electric field strength, and can seal again after a while. Kaur and Singh in 2016 stated that too many exposures result in cell death because of intracellular components' leakage via the pores. Consequently, electroporation is very cell-destructive and results in increased likelihood of microbial death. Another study done in almost identical conditions of temperature shows the kinetics of "*Bacillus subtilis* spores' inactivation" could definitely be increased through ohmic heating. In 2002, Yoon et al. reported, in ohmic heating, the electrical field seemed to have an effect on the cell wall directly and indirectly, and the intracellular materials, which have coenzymes, nucleic acids, protein, amino acids, and related food components, were leached to the medium of the culture (Yoon et al., 2002). They found that at less than 50°C, exudates concentration in the supernatant of yeast was comparable in both conditions, while beyond 50°C, the concentrations were greater in ohmic heating than in "conventional" heat treatment. They discovered higher rates of exudation depend on yeast-destructive rates and the heating method used. The electrical field in the ohmic heating system may have had increasing effects on the electroporation rate, and result in excessive exudation as well as the death of the cells. They also found that levels of protein exudates significantly increased as the electric field was increased (10–20 Vcm^{-1}). Absorbance at 260 nanometers credited to the nucleic acids was double at 20 Vcm^{-1} and the total content of the protein was three times higher compared to the one at 15 Vcm^{-1} (ibid.).

17.8.10 Inactivation of Enzymes

In ohmic heating, applied electrical fields resulted in faster polyphenol oxidase and lipoxygenase inactivation than in the conventional method of heating (Castro et al., 2004). In the same way, ohmic heating was more efficient for the needed inactivation of microorganisms and pectin esterase because of the lower residence time as the flavor substances released did not break down as fast as in pasteurization that used the conventional method. Icier and others studied the inactivation of peroxidase and changes in colour in pea puree blanching with ohmic heating, wherein samples of pea puree were subjected to blanching using conventional heating and ohmic heating system. Blanching with ohmic heating was done at four various gradients of voltage from 20–50 Vcm^{-1}.

The samples of puree were heated starting from 30–100°C and then kept at 100°C for sufficient blanching. Blanching with conventional methods was done using a water bath at 100°C. The ohmic heating applied with at least 30 Vcm^{-1} voltage gradients inactivated the peroxidase in a shorter time compared to blanching with a water bath. Blanching with ohmic heating at 50 Vcm^{-1} resulted in the shortest inactivation time (54 seconds) and the best quality of colour. Colour changes when blanching with ohmic heating can be explained using first-order reaction. The hue angle is the greatest suitable combination with R2= 0.95 (approximated), and it closely describes the reaction kinetics of the overall changes in colour of the pea puree for ohmic heating at 20 Vcm^{-1}.

The deactivation kinetics of polyphenol oxidase in ohmically heated grape juice has been evaluated (Icier et al., 2008). The temperatures of critical deactivation were reported at 60°C or less for 40 Vcm^{-1}, while the critical deactivation temperature for 20 Vcm^{-1} and 30 Vcm^{-1} was reported at 70°C. Numerous kinetic models for polyphenol oxidase deactivation in ohmic heating at 30 Vcm^{-1} were used in data of the experiment. The simplest kinetic model with "one step first-order deactivation" was reported as being better than the additional complex models. The energy of activation for deactivation of polyphenol oxidase within temperatures of 70°C to 90°C was seen to be at 83.5 kilojoule per mole. The enzyme pectin methyl esterase (PME) exists in almost all plant tissues, various bacteria, and fungi. Pectin methyl esterase does not have a prosthetic group. PEM de-esterifies "galactosyluronate methyl esters of pectins", releasing methanol and protons into the medium. Exogenous PEM inactivation in cloudberry jams and apple juice is reported to obey first-order kinetics (Wilinska et al., 2008). The same was reported in ohmic heating, after studies involving the same foods were done. Ohmic heating in a continuous electric field of AC was used on orange juice with spores of *Bacillus subtilis* to evaluate its inactivation effects. Effective inactivation of the spores of *Bacillus subtilis* was attained with electric sterilization under a pressurized system, using a combination of a high electric field and a high temperature, in a shorter time. Similarly, peculiar smell development and ascorbic acid losses were reduced using the ohmic heating system (Uemura and Isobe, 2003).

17.9 ADVANTAGES OF OHMIC HEATING

Ohmic heating has many advantages which include lower capital cost, less cooking time, better energy efficiency, better product quality, an environmentally-friendly process, among others. Others include shorter processing times, colour maintenance, retention of food nutritional value, and higher yield in comparison to heating using conventional methods (Darvishi et al., 2011). In addition, the cleaning of ohmic heating equipment is relatively less compared to traditional heat exchangers as a result of the less fouling on the contact surface of the food in ohmic heating. Ohmic heating application in industries allows energy efficiencies above 90% and a decrease in consumption of energy around 82–97% in comparison to other conventional methods such as smoke-house heating. Ohmic heating is a green technological process (ibid.). Generally, ohmic heating has many advantages including:

1. Low risk of food damage due to burning due to no hot surface for transfer of heat.
2. Simplified control of the process with low cost of maintenance.
3. In ohmic heating, a higher temperature is quickly attained, e.g., temperatures for UHT processing.
4. Continuous process with no surface for heat transfer.
5. Improved efficiency and cost effectiveness.
6. Environmentally friendly.
7. Heating of liquid–particle and particulate foods mixtures.
8. Increased efficiency of energy conversion.
9. It is suitable for shear-sensitive foods due to its low flow velocity.

10. Immediate system shutdown.
11. Required less cleaning frequency.
12. Reduced risk of food burning.
13. Colour maintenance,
14. Retention of food nutritional value.
15. Capital investment and product safety optimization due to high solid loading.
16. Fast and uniform heating of solid and liquid phases with little or no heat damage and loss of nutrients.
17. Less fouling compared to other heating methods used conventionally.
18. Less cost of maintenance as there are no mechanically moving parts.

17.10 LIMITATIONS AND DISADVANTAGES

Some of the limitations of ohmic heating include viscosity and change in the EC (Varghese et al., 2014; Fellows, 2017). The particles' density in the liquid suspension can reduce the extent of processing. Fluid with a higher viscosity will give more heating resistance, making the food mix heat faster than food with low viscosity (Fellows, 2017). The EC of food is dependent on the product composition, temperature, and frequency (Fellows, 2017; Varghese et al., 2014). It can be increased through the addition of ionic compounds, or reduced through the addition of non-polar components. Changes in the EC limit ohmic heating because it makes it hard to model thermal processes after a temperature increase in foods with multi-constituents (Varghese et al., 2014).

Heating and freezing in "conventional tubular heat exchanger" and ohmic heating are initially more capital-intensive than processing using conventional methods. Another known disadvantage has been reported that foods with fats are ineffectively heated in an ohmic heating system, as they are non-conductive because there is no presence of salt and water, which are critical in ohmic heating. If the fat particles occur in a region with high EC where the electric currents circumvent them, they could heat more slowly because of the lack of the EC. The pathogens which might be in the fat globules might be exposed to insufficient heat compared to the substances in other regions. There is also the likelihood of "runaway" heating (US FDA, 2000). As the temperature of the system increases, the EC increases because of the quicker electron movements.

17.11 MATHEMATICAL MODELLING

Mathematical modelling is applied as valuable assistance in developing, understanding and validating emerging heating technologies. Ohmic heating has been improved with mathematical modelling. The mathematical model of the ohmic heating operation should have the capacity to quantify loss of heat, detect likely cold and hot spots, and evaluate the effects of important variables, e.g. the sample EC and field strength (Kumar et al., 2014). Ohmic heating mathematical modelling and the characterization of EC of some foods have been carried out by many researchers. Conventional estimation of the cold-spot temperature of mixtures was shown by a "mixed fluid model" rather than a static model after the cold-spot takes place in the particle, occurring as the medium EC increases more than that of the solid EC (Kaur and Singh, 2016). Nevertheless, the static fluid model offers better conservative predictions of temperatures of the mixture cold-spot when the cold-spot is in fluid, especially as the EC of solids increases more than the EC of the medium. To examine the role of the coefficient of the heat transfer of the fluid-to-particle, parametric simulations were done, and it can be applied to circumstances with mixed fluids and high concentration of solids. It was established that both particle centre heating rates and the heating rate of the average liquid increased with a decrease in the coefficient of the "fluid-to-particle" heat transfer. In 2009, Marra et al. established a mathematical model for a solid food under heating in a "cylindrical batch ohmic heating cell" (Marra et al., 2009). Temperature distribution and temperature profiles of the process of ohmic heating were simulated and evaluated through mathematical and experimental modelling. The simulation offered a good relationship between the mathematical and experimental modelling. Cell shape optimization and the optimization of electrode configurations, as well as validation and ascertaining of safe pasteurization for other solid foods, may be carried out using the designed experimental and mathematical model (Kaur and Singh, 2016). A simplified qualitative-based tool has been made from the data obtained with the simulation model, to evaluate the losses via leaching during the process of blanching. The model depends on the process of unidirectional transfer of mass and heat in an infinite plate, and was designed to permit changes in the product and medium volume, adjusting the initial concentration of the solute in the medium and also in the food, modifying the partition factors, changing the medium and the plate temperature, adjusting the coefficients of mass transfer and external heat, as well as the plate thickness (ibid.).

Mathematical modelling of the EC of heterogeneous foods was done using series, parallel, "effective medium theory (EMT)", "two forms of Maxwell–Eucken (ME-1 and ME-2)" (Rocha and Cruz, 2001; Kaur and Singh, 2016). Researchers carried out experimental investigations and mathematical modeling of a case with fewer conductive particles bounded by a medium of high EC, which was subjected to ohmic heating under static conditions. They reported that the electric current passes more via the medium with a high EC and avoided the particle with a lower EC. The absence or presence of alternative conducting paths via the surrounding medium influence the rates of energy generation and voltage drops in the two media. Tumpanuvate and Jittanitin (2012) developed mathematical models to predict the changes in temperature of botanical beverages, concentrated orange and pineapple purees and juices in ohmic heating. The estimation of temperature of the concentrated juices and botanical beverages were more accurate if and only if evaporated moisture and the loss of heat to surroundings were captured in the mathematical models.

17.12 OHMIC HEATING EFFECTS ON QUALITY CHARACTERISTICS OF FOODS

17.12.1 Water Absorption Capacity (WAC) and Water Solubility Index (WSI)

The WSI and WAC of foods show how foods interact with water. They are essential for the prediction of how foods will behave under ohmic heating if further processed; WAC and WSI are among the key functional properties of foods (Godswill, 2019; Nnennaya et al., 2020; Twinomuhwezi et al., 2020). The rate of starch conversion from its granule during food processing can be examined through WAC and WSI (Godswill, 2019). In 2005, Martínez-Bustos et al. stated that starch and flours that were subjected to ohmic heating had more WSI and WAC compared to starch and flours not treated with ohmic heating (Martínez-Bustos et al., 2005). This can be associated with structural changes and the rearrangement which occurred in the process, giving these starch granules more access to the absorption of moisture. Molecular weight, amylopectin chain length, and the ratio of short to long chains of the starch granules also have effects (Godswill, 2019; Twinomuhwezi et al., 2020). The starch obtained using ohmic heating had satisfactory WAC and was found to be easily dispersed in cold water, and has a low water solubility index in comparison to the one obtained through drum driers and extrusion. Morya et al. (2016) reported that for starch that was pre-gelatinized, without any applied cutting force to the swollen granules, small incomplete amylose leaching occurred, the starch components slightly broke down, and possibly continued connection in a continuous matrix. Consequently, the foods are readily accessed by water as a result of their softened structures and only have limited cold water solubility as a result of the high molecular weight components. The foods easily disperse in cold water, forming moderate and stable suspensions, and can be applied mostly in baby food, fruit purées, as adhesives in textile, potato purées, and dry soup mixes.

17.12.2 Pasting Properties

Viscosity can be achieved by heating starch and this depends on the starch quality as well as the amylase activity (Morya et al., 2016). Viscosity determination is applied in the measurement of treatment severity and fragmentation degree of starch granules. There is a possibility that ohmic treatment causes the changes which decrease the re-accommodation of the branched chains of amylopectin and/or linear chains of amylose. In 2000, Karapantsios et al. studied the EC of gelatinization of starch in conventional heating process. They reported that EC had a linear correlation with time till the gelatinization temperature was attained and then showed a decreasing trend (Karapantsios et al., 2000). Li et al. (2004) reported from their study that the EC and the starch suspension temperature had a linear correlation after and before gelatinization. The cause of the decrease in the EC in a range of gelatinization was as a result of the increase in viscosity and starch granule swelling, and led to a decrease in the area for movement of the starch particles and an increase in the swollen particles' resistance motion. Wang and Sastry (2002) suggested that gelatinization takes place any time starch is heated using excessive moisture. Owing to the resulting changes in the chemical and physical properties of foods, gelatinization is among the most significant quality parameters in design and analysis of food processes (Awuchi et al., 2020). As said earlier, viscosity determination is applied in the measurement of the treatment severity and fragmentation degree of the starch granules; also, there is a possibility that the ohmic treatment causes the changes which decrease the re-accommodation of the branched chains of amylopectin and/or linear chains of amylose.

Greater maximum viscosity for starch from jicama than the one from maize, though less than starch cassava and the one from potato, has been reported. In 1990, Halden and other researchers reported that the changes in structure during starch gelatinization, i.e., the changes between first viscosity and final viscosity and increase in bound water in the ohmic heating process, were caused by the gelatinization of starch induced by the EC which increases as the temperature increases and decreases as degree of gelatinization decreases (Halden et al., 1990). Martínez-Bustos et al. (2005) reported that 107–115V for 10–20 minutes in ohmic heating caused a greater decline in the viscosity as well as a total retrogradation of starch from cassava and jicama. All the treatments had less total retrogradation and decreased final viscosity, relative to native starch. The starch retrogradation intensity could be as a result of the fact that, when pressurized, gelatinization limits amylase exudation from starch granules (Douzals et al., 1998). Voltage was the determinant for pre-gelatinization.

17.12.3 Thermal Properties

Starch thermal properties can be studied with differential scanning calorimetry (DSC). Starch gelatinization is significant in determining the general cooking characteristics and functional properties of foods and flours (Godswill, 2019). During the heating of starch, changes in some properties, the gelatinization temperature (GT), the peak viscosity, the specific heat capacity, and enthalpy occur, indicating the degree of gelatinization of starch (Awuchi et al., 2020). The molecular order losses can be measured using gelatinization endothermic enthalpy. Several studies have been done to evaluate the starches' thermal properties for various reasons, such as "the effect of amylose-lipid complexes", "the effect of lipid and protein on starch gelatinization", "the effect of heat-moisture treatment on starch properties", and "the effect of annealing on starch properties" (Kaur and Singh, 2016). An and King (2007) concluded that the conditions of ohmic heating may alter the thermal properties of starch from rice and rice flours with different levels of fats and proteins, in comparison with samples heated conventionally. when compared with native CRS, the onset and peak GT of ohmically and conventionally heated foods were higher statistically, although a significant reduction in enthalpies of all treated foods is as a result of the total pre-gelatinization from ohmic heating or conventional heating so they needed lower energy as they pass via DSC heating. In comparison to conventionally heated food for 20 minutes and 3.8 minutes, the enthalpies of rice starch subjected to ohmic heating were lower at 20V and 40V, respectively. They added that a faster rate of heating appeared to gelatinize foods less due to the shorter time of cooking, leading to additional energy requirements for the gelatinization of starch. Wang and Sastry (2002) reported that less enthalpies through DSC show that more gelatinization of starch occurs in advance, so less energy was required for total starch gelatinization than was required for raw foods. Also, they stated that changes in EC synchronize with the gelatinization of starch. The percentage SG data obtained from DSC thermograms and curves of EC fitted well in the mid and low ranges of gelatinization. Martínez-Bustos et al. (2005) stated that higher energy was needed for gelatinization of starch from the jicama and cassava compared to starches subjected to ohmic heating. The greatest gelatinization degree for jicama starch was achieved at a voltage gradient of 123 V for 10 minutes, and for cassava, 123 V for 10 minutes too. It corresponds to 39.1% and 70.03% gelatinization of the starch from cassava and jicama respectively.

17.13 CURRENT STATUS OF OHMIC HEATING GLOBALLY

Although many studies on ohmic heating have been carried out to investigate the potential applications, numerous designs and industrial operations, equipment improvement, and process parameters optimization are still ongoing. Even though ohmic heating has been explored mostly in some parts of the world including the USA, Canada, the United Kingdom, Japan, Italy, Sweden (Yildiz and Guven 2014), Nigeria, India, and Switzerland, currently, many countries are exploring the possibility of ohmic heating application in food industries.Some of the companies producing ohmic heating equipment are shown in Table 17.3. Currently, there are at least 18 plants making ohmic heating systems for commercial operations throughout the globe and the equipment is used for sterilization and pasteurization of pumpable foods, including milk, egg, ice-cream mix, fruit and vegetable products (such as purees, juices, pulps), whey, soymilk, soups, etc., and also for aseptic packaging (Icier, 2012).

17.14 CONCLUSION

The energy in ohmic heating is directly dissipated into foods. Ohmic heating is a direct heating method. Electrical conductivity (EC) is an essential tool in designing ohmic heating equipment. Numerous potential applications of ohmic heating exist, including sterilization, heating, fermentation, pasteurization, blanching, evaporation, and dehydration. Other than heating, the application of an electric field in ohmic heating results in cell membrane electroporation, which reduces gelatinization temperature, increases extraction rates, and reduces enthalpy. Ohmic heating causes quicker food heating in addition to maintaining nutritional value and sensory properties. The water absorption capacity (WAC), thermal properties, the water solubility index (WSI), and the pasting properties are affected by ohmic heating. Ohmic heating has been reported to cause pre-gelatinization of

TABLE 17.3
Some Companies Producing an Ohmic Heating System

Country	Company	References
UK	APV Baker Ltd., C-Tech Innovation	(Yildiz and Guven, 2014; Kumar, 2018)
US	Raztek	(Icier, 2012; Kumar, 2018)
Sweden	Alfa Laval	(Kumar, 2018)
Canada	Agro process, IAI Group	(Kumar, 2018; Yildiz and Guven, 2014)
Japan	Yanagiya Machinery Co. Ltd.	(Kumar, 2018; Icier, 2012)
Switzerland	Kasag	(Kumar, 2018)
Italy	Emmepiemme SRL	(Kumar, 2018)

starch granules, thereby reducing the energy needed during processing. However, its high cost, lack of use in fatty foods, and lack of adequate knowledge limit the use of ohmic heating.

REFERENCES

An, H.J., and King, J.M. (2007). Thermal characteristics of ohmically heated rice starch and rice flours. *Journal of Food Science*, 72(1), C084–C088.

Anderson, A.K., and Finkelstein, R. (1919). A study of the electropure process of treating milk. *Journal of Dairy Science*, 2(5), 374–406.

Awuchi, C.G., Owuamanam, I.C., Ogueke, C.C., and Hannington, T. (2020). The assessment of the physical and pasting properties of grains. *European Academic Research*, 8(2): 1072–1080.

Barbosa-Cánovas, G.V., Gongora-Nieto, M.M., and Swanson, B.G. (1998). Nonthermal electrical methods in food preservation/Métodos eléctricos no térmicos para la conservación de alimentos. *Food Science and Technology International*, 4(5), 363–370.

Castro, I., Teixeira, J.A., Salengke, S., Sastry, S.K., and Vicente, A.A. (2004). Ohmic heating of strawberry products: Electrical conductivity measurements and ascorbic acid degradation kinetics. *Innovative Food Science and Emerging Technologies*, 5(1), 27–36.

Darvishi, H., Hosainpour, A., Nargesi, F., Khoshtaghaza, M.H., and Torang, H. (2011). Ohmic processing: Temperature dependent electrical conductivities of lemon juice. *Modern Applied Science*, 5(1), 210–216.

De Alwis, A.A.P., and Fryer, P.J. (1990). A finite-element analysis of heat generation and transfer during ohmic heating of food. *Chemical Engineering Science*, 45(6), 1547–1559.

Douzals, J P., Perrier Cornet, J.M., Gervais, P., and Coquille, J.C. (1998). High-pressure gelatinization of wheat starch and properties of pressure-induced gels. *Journal of Agricultural and Food Chemistry*, 46(12), 4824-4829.

Duygu, B., and Ümit, G. (2015). Application of ohmic heating system in meat thawing. *Procedia: Social and Behavioral Sciences*, 195, 2822–2828.

Fellows, P.J. (2017). Food processing technology. 4th edn. Kidlington: Woodhead Publishing, pp. 831–838.

Fryer, P.J., and Li, Z. (1993). Electrical resistance heating of foods. *Trends in Food Science Technology*, 4: 364–369.

Godswill, A.C. (2019). Proximate composition and functional properties of different grain flour composites for industrial applications. *International Journal of Food Sciences*, 2(1), 43–64.

Halden, K., De Alwis, A.A.P., and Fryer, P.J. (1990). Changes in the electrical conductivity of foods during ohmic heating. *International Journal of Food Science and Technology*, 25(1), 9–25.

Icier, F. (2009). Influence of ohmic heating on rheological and electrical properties of reconstituted whey solutions. *Food and Bioproducts Processing*, 87(4), 308–316.

Icier, F. (2012). Ohmic heating of fluid foods. In: Cullen, P.J., Tiwari, B.K., and Valdramidis, V.P. (Eds.), *Novel Thermal and Non-thermal Technologies for Fluid Foods*. London: Academic Press, pp. 305–367.

Icier, F., Yildiz, H., and Baysal, T. (2008). Polyphenoloxidase deactivation kinetics during ohmic heating of grape juice. *Journal of Food Engineering*, 85(3), 410–417.

Jones, F. (1897). Apparatus for electrically treating liquids. United States Patent and Trademark Office. US Patent (592), 735.

Karapantsios, T.D., Sakonidou, E.P., and Raphaelides, S.N. (2000). Electrical conductance study of fluid motion and heat transport during starch gelatinization. *Journal of Food Science*, 65(1), 144–150.

Kaur, N., and Singh, A.K. (2016). Ohmic heating: Concept and applications: A review. *Critical Reviews in Food Science and Nutrition*, 56(14), 2338–2351.

Kaur, R., Gul, K., and Singh, A.K. (2016). Nutritional impact of ohmic heating on fruits and vegetables: A review. *Cogent Food and Agriculture*, 2(1), 1159000.

Kumar, J.P., Ramanathan, M., and Ranganathan, T.V. (2014). Ohmic heating technology in food processing: A review. *International Journal of Food Engineering Research and Technology*, 3(2), 1236–1241.

Kumar, T. (2018). A review on ohmic heating technology: Principle, applications and scope. *International Journal of Agriculture, Environment and Biotechnology*, 11(4), 679–687.

Lascorz, D., Torella, E., Lyng, J.G., and Arroyo, C. (2016). The potential of ohmic heating as an alternative to steam for heat processing shrimps. *Innovative Food Science and Emerging Technologies*, 37, 329–335.

Li, L. T., Li, Z., and Tatsumi, E. (2004). Determination of starch gelatinization temperature by ohmic heating. *Journal of Food Engineering*, 62(2), 113–120.

Marra, F., Zell, M., Lyng, J.G., Morgan, D.J., and Cronin, D.A. (2009). Analysis of heat transfer during ohmic processing of a solid food. *Journal of Food Engineering*, 91(1), 56–63.

Martínez-Bustos, F., López-Soto, M., Zazueta-Morales, J.J., and Morales-Sánchez, E. (2005). Preparation and properties of pre-gelatinized cassava (Manihotesculenta.Crantz) and jicama (Pachyrhizuserosus) starches using ohmic heating. *Agrociencia*, 39(3), 275–283.

Mitelut, A., Popa, M., Geicu, M., Niculita, P.E.T.R.U., Vatuiu, D., Vatuiu, I., … and Cramariuc, R. (2011). Ohmic treatment for microbial inhibition in meat and meat products. *Romanian Biotechnological Letters*, 16(1), 149–152.

Mizrahi, S., Kopelman, I.J., and Perlman, J. (1975). Blanching by electro-conductive heating. *International Journal of Food Science and Technology*, 10(3), 281–288.

Moreno, J., Simpson, R., Pizarro, N., Pavez, C., Dorvil, F., Petzold, G., and Bugueño, G. (2013). Influence of ohmic heating/osmotic dehydration treatments on polyphenoloxidase inactivation, physical properties and microbial stability of apples (cv. Granny Smith). *Innovative Food Science and Emerging Technologies*, 20, 198–207.

Morya, S., Amoah, A.E.D., Thompkinson, D.K., and Charan, A.A. (2016). Studies on the dough characteristics of millets under the action of mechanical stress. *Journal of Applied Thoughts*, 5, 48–78.

Nnennaya, A.N., Kate, E.C., Evelyn, B.N., Godswill, A.C., Linda, A.C., Julian, I.C., and Moses, O. (2020). Study on the nutritional and chemical composition of "Ogiri" condiment made from sandbox seed (Huracrepitans) as affected by fermentation time. *GSC Biological and Pharmaceutical Sciences*, 11(2), 105–113.

Praporscic, I., Lebovka, N.I., Ghnimi, S., and Vorobiev, E. (2006). Ohmically heated, enhanced expression of juice from apple and potato tissues. *Biosystems Engineering*, 93(2), 199–204.

Ramaswamy, H.S., Marcotte, M., Sastry, S., and Abdelrahim, K. (Eds.) (2014). *Ohmic Heating in Food Processing*. Boca Raton, FL: CRC Press, pp. 93–102.

Rocha, R.P., and Cruz, M.A.E. (2001). Computation of the effective conductivity of unidirectional fibrous composites with an interfacial thermal resistance. *Numerical Heat Transfer: Part A: Applications*, 39(2), 179–203.

Samprovalaki, K., Bakalis, S., and Fryer, P.J. (2007). Ohmic heating: Models and measurements. *Heat Transfer in Food Processing*. Southampton: WIT Press, pp. 159–186.

Sastry, S. K., and Palaniappan, S. (1992). Mathematical modeling and experimental studies on ohmic heating of liquid-particle mixtures in a static heater 1. *Journal of Food Process Engineering*, 15(4), 241–261.

Shirsat, N., Lyng, J.G., Brunton, N.P., and McKenna, B.M. (2004). Ohmic processing: Electrical conductivities of pork cuts. *Meat Science*, 67(3), 507–514.

Singh, R.P., and Heldman, D.R. (2014). *Introduction to Food Engineering.* 5th edn. Orlando, FL: Academic Press Incorporation.

Skudder, P. (1992). Long life products by ohmic heating. *International Food Ingredients*, 4, 36–41.

Sun, H., Kawamura, S., Himoto, J.I., Itoh, K., Wada, T., and Kimura, T. (2008). Effects of ohmic heating on microbial counts and denaturation of proteins in milk. *Food Science and Technology Research*, 14(2), 117–123.

Tumpanuvatr, T., and Jittanit, W. (2012). The temperature prediction of some botanical beverages, concentrated juices and purees of orange and pineapple during ohmic heating. *Journal of Food Engineering*, 113(2), 226–233.

Twinomuhwezi, H., Awuchi, C.G., and Rachael, M. (2020). Comparative study of the proximate composition and functional properties of composite flours of amaranth, rice, millet, and soybean. *American Journal of Food Science and Nutrition*, 6(1), 6–19.

Uemura, K., and Isobe, S. (2003). Developing a new apparatus for inactivating *Bacillus subtilis* spore in orange juice with a high electric field AC under pressurized conditions. *Journal of Food Engineering*, 56(4), 325–329.

USFDA (U.S. Food and Drug Administration) (2000). Kinetics of microbial inactivation for alternative food processing technologies. FDA Center for Food Safety and Applied Nutrition report—A report of the IFT for the FDA of the US Department of Health and Human Services. June, 2, Washington, DC: USFDA.

Varghese, K.S., Pandey, M.C., Radhakrishna, K., and Bawa, A.S. (2014). Technology, applications and modelling of ohmic heating: A review. *Journal of Food Science and Technology*, 51(10), 2304–2317.

Varzakas, T., and Tzia, C. (2015). *Handbook of Food Processing: Food Preservation*. Boca Raton, FL: CRC Press.

Wang, W.C., and Sastry, S.K. (2002). Effects of moderate electrothermal treatments on juice yield from cellular tissue. *Innovative Food Science and Emerging Technologies*, 3(4), 371–377.

Wilińska, A., de Figueiredo Rodrigues, A.S., Bryjak, J., and Polakovič, M. (2008). Thermal inactivation of exogenous pectin methylesterase in apple and cloudberry juices. *Journal of Food Engineering*, 85(3), 459–465.

Yildiz, H., and Guven, E. (2014). Industrial application and potential use of ohmic heating for fluid foods. *Bulgarian Chemical Communications*, 46, 98–102.

Yoon, S.W., Lee, C.Y.J., Kim, K.M., and Lee, C.H. (2002). Leakage of cellular materials from Saccharomyces cerevisiae by ohmic heating. *Journal of Microbiology and Biotechnology*, 12(2), 183–188.

Zell, M., Lyng, J.G., Cronin, D.A., and Morgan, D.J. (2009). Ohmic heating of meats: Electrical conductivities of whole meats and processed meat ingredients. *Meat Science*, 83(3), 563–570.

Zell, M., Lyng, J.G., Cronin, D.A., and Morgan, D.J. (2010). Ohmic cooking of whole turkey meat: Effect of rapid ohmic heating on selected product parameters. *Food Chemistry*, 120(3), 724–729.

Zell, M., Lyng, J.G., Morgan, D.J., and Cronin, D.A. (2011). Minimising heat losses during batch ohmic heating of solid food. *Food and Bioproducts Processing*, 89(2), 128–134.

18 Nanoparticles Synthesis from Kitchen Waste

Opportunities, Challenges, and Future Prospects

*Deepa and Raunak Dhanker**
Department of Basic and Applied Sciences, School of Engineering and Sciences, GD Goenka University, Gurugram, Haryana, India
*Corresponding author Email: raunakbiotech@gmail.com

CONTENTS

18.1 Introduction 329
18.2 Green Synthesis of Metal Nanoparticles Using Kitchen Food Waste 330
 18.2.1 Metal Ion Solution 330
 18.2.2 Biological Reducing Agents 330
 18.2.3 Separation of MNPs 331
 18.2.4 Monitoring of MNPs 332
 18.2.5 Mechanism of MNPs' Synthesis 332
 18.2.6 Factors Affecting MNPs Synthesis 332
18.3 Silver and Gold Nanoparticles Produced by Kitchen Food Waste 333
 18.3.1 Silver (Ag) Nanoparticles 333
 18.3.2 Gold (Au) Nanoparticles 335
 18.3.3 Other Types of Nanoparticles 335
18.4 Applications of MNPs 335
18.5 Conclusion 336
References 337

18.1 INTRODUCTION

Nanotechnology is one of the most advanced and promising fields in present-day science. Nanotechnology is the study of materials (size range, 1–100 nm) at an atomic and molecule scale. Nowadays drug delivery, biosensing, cosmetics, optical energy, agricultural areas, imaging, and others are in one way or another way related to nanoparticles (Deepa et al., 2022). Nanotechnology has established the new horizon in the field of drug delivery sciences (Patra et al., 2019). The effective recycling of wastes through innovative techniques can produce useful materials from untapped natural resources, which are of global concern to support a community of 12.3 billion people until 2100 (Dhanker et al., 2021). A large amount of domestic waste is generated in India whose only 30% is recycled. Therefore, the rest of this waste can be used for the preparation of different kinds of nanoparticles for various applications (Dhanker et al., 2021). Nanoparticles (NPs) possess exceptional properties, attributable to their small size (1–100 nm), expanded reactivity, and larger surface area. Nanoparticles can be generated from various metals such as platinum (Pt), gold (Au), and silver (Ag), however, other metals such as zinc (Zn), iron (Fe) and palladium (Pd) have also been used

DOI: 10.1201/9781003239956-20

because of their broad application in various areas, for example, medicine, cosmetics, energy, agricultural business, and medical services. (Garg et al., 2022) Two decades ago, kitchen-generated organic waste was not considered a major resource loss (Baiano, 2014). However, current alarms about malnutrition, preservation, environmental deprivation, have raised many socio-economic concerns about kitchen food waste, which has prompted more studies to focus on advancing techniques for reducing wastage and the procurement of garbage disposal (Kantor et al., 1997; Gustavsson et al., 2011; Ghosh et al., 2015). The aim of such strategies is to maximize the benefits and utility value of kitchen waste, while minimizing the amount of garbage that ends up in landfills (Loehr, 2012) (Dhanker et al., 2013). The present chapter highlights the green synthesis and different fabrication techniques of silver and gold nanoparticles. Kitchen waste and nanoparticle synthesis are a unique combination of the fabrication process (Kim et al., 2012; Madhumitha et al., 2012; Yang et al., 2014). Due to the presence of antioxidant compounds and various proteins, the aqueous plant extract works as a stabilizing agent. Kitchen waste contains a large number of bioactive substances (Prasad et al., 2014). Nanoparticles have unique qualities due to their small dimensions and great surface-to-volume ratio when compared to the metal substrate (Perez, 2005). For example, gold (Au) nanoparticles are widely applied in healthcare for clinical diagnosis (Torres-Chavolla et al., 2010; Puvanakrishnan et al., 2012). Silver (Ag) nanostructures have antibacterial action against a wide range of human and animal infections (Asha Rani et al., 2009; Li et al., 2011; Shah et al., 2016). As a catalyst, platinum (Pt) and palladium (Pd) nanomaterials have already been used (Cheong et al., 2010; Lin et al., 2011; Coccia et al., 2012). Likewise, metallic oxide nanoparticles, such as oxides of copper (Cu_2O, CuO) and oxide of zinc (ZnO), have also been found to exhibit antibacterial properties. ZnO has thus been used in a wide range of active packaging (Espitia et al., 2012) and many more.

The use of kitchen food waste material in the green fabrication of profitable products like metal and metal oxide nanoparticles is a recent field of study. The chapter highlights the latest research in this field and analyzes several experimental conditions that control nanoparticle fabrication. Besides this, the chapter also discusses the potential applicability of silver and gold nanoparticles generated from kitchen wastage in different areas.

18.2 GREEN SYNTHESIS OF METAL NANOPARTICLES USING KITCHEN FOOD WASTE

The green fabrication of metallic nanoparticles and oxides of metal nanoparticles from kitchen food waste is a new method, which is eco-friendly, cost-effective, and a unique alternative to the traditional methods. Green synthesis of metal nanoparticles has prompted scientists to explore more effective resource utilization and management strategies. Various applications of nanotechnology for the effective recycling of kitchen waste have enormous potential in India, where only ~30% of domestic waste is recycled.

18.2.1 Metal Ion Solution

The most important criterion for the production of green metal nanoparticles (MNPs) is a metal ion solution, such as the $AgNO_3$ solution in the case of silver NPs (nanoparticles) and the $HAuCl_4$ solution in the case of gold NPs, as well as a reductive biological agent. Because reducing agents as well as other compounds that act as regulating and/or capping agents are commonly found in plant cells, there is no need to introduce them independently (Sahu et al., 2012; Arunachalam et al., 2013).

18.2.2 Biological Reducing Agents

In biological systems, several reducing agents are widely distributed. Green synthesis of MNPs has been possible in the laboratory from vegetation extracts. For the green synthesis of nanoparticles, the plant extract can be more advantageous compared to other sources as it acts as a reducing as well as a capping agent. The plants used to make MNPs range from microalgae to angiosperms; unfortunately, there are few reports available on kitchen food waste, which is the perfect approach because it is freely available and discarded as trash everywhere. Plant parts such as leaves, pulp, bark, juices, roots, and stems have been used to fabricate MNPs (Aromal and Philip, 2012). Different kinds of food waste materials, such as fruit and vegetable peel, seeds, leaves, roots, and rinds, have been studied so far for the green synthesis of metallic nanoparticles in different sizes and shapes (Dhanker et al., 2021). Kitchen waste also contains an enormous collection of flavonoids, amino acids, enzymes, phenolic compounds, polypeptides, polysaccharides, and alkaloids. Micronutrients are biomolecules that can be used for nanoparticle formation. Green synthesis of metallic nanoparticles is a bottom-up process in which particles of matter interact to generate initial key components, which is further self-assembled (Al Momani et al., 2008). Gold, silver, platinum, and palladium are examples of noble metal nanoparticles that can be successfully produced via plant-based synthesis from the following resources:

1. Papaya fruit extract to synthesize the plant-mediated silver nanoparticles and evaluate their antimicrobial activities (Jain et al., 2009; Shah et al., 2015).
2. Banana peel extract to synthesize nanoparticles studied for their antimicrobial activity against microbes. The numerous factors that can have a huge impact in defining the features of the nanoparticles have also been investigated (Bankar et al., 2010a; Narayanamma et al., 2016).
3. Cavendish banana peel extract has been used to synthesize silver nanoparticles and for its antibacterial properties (Kokila et al., 2015).
4. Waste banana stem extract used for biogenic synthesis of silver nanoparticles and their anti-microbial activity in contradiction of *E. coli* and *Staphylococcus epidermis* (Dang et al., 2017).
5. The peel of *Punica granatum* for the synthesis of Ag and Au nanoparticles was studied by Ahmad et al. (2012). For the same purpose, AgNPs (silver nanoparticles) from seed and peel extracts of *Punica granatum* were studied for the antimicrobial activity.
6. The synthesized silver nanoparticles using pomegranate seed and peel extract do not show antibacterial activity in favor of *Pseudomonas*, *Bacillus cereus*, and *Staphylococcus albus* (Nisha et al., 2015a).
7. Although plant extracts have been used for minimization and stability of NPs, more research is needed involving separation and characterization of the components present in a sample (Ahmad et al., 2012).
8. The manufacturing, characterization, and bio-effectiveness of silver nanoparticles that have been synthesized from vegetable waste have been performed by following the green synthesis approach (Kumar et al., 2015).
9. Also, it has been established that high-value nanoparticles can be generated by biogenic processes by using intensive aquaculture and horticultural foodstuff (Ghosh et al., 2017).
10. The green manufacture of silver nanoparticles from plant origin has been proven, as well as analyzed for their antibacterial efficacy.
11. *Pisumsativum* peel aqueous extract used to elucidate superficial green biosynthesis of silver nanoparticles (Patra et al., 2019).
12. Similarly, from watermelon rind extract palladium nanoparticles of 96 nm dimension were synthesized (Lakshmipathy et al., 2015).

18.2.3 Separation of MNPs

For the separation of MNPs, different methods are used.

1. Thermo-gravimetric analysis (TGA) has been used to determine the impact of $AgNO_3$ on the number of organic materials in manufactured MNPs (Morales-Sanchez et al., 2011) and to determine the thermostability of MNPs (Kora et al., 2010).
2. Using inductive coupled plasma (ICP), the proportion and transformation of MNPs were investigated (Song et al., 2009a).

18.2.4 Monitoring of MNPs

The effect of pH and ionic strength on the synthesis of MNPs, as well as the size stability of generated MNPs, were studied using UV-visible spectroscopic analysis. In most of the studies, the scanning electron microscopy (SEM) morphological analysis revealed spherical MNPs, however, a few researchers reported other shapes as well. Most of the researchers determined that the emergence of face-centered cubic (FCC) crystallographic structural MNPs was analyzed by x-ray diffraction (XRD) studies (Mourdikoudis, Pallares and Thanh, 2018).

18.2.5 Mechanism of MNPs' Synthesis

The presence of a significant amount of biological molecules such as proteins, enzymes, phenolic compounds, alkaloids, and other organic chemicals capable of producing electrons for the reduction of M^+ ions to M^0, results in the creation of NPs by the biological system. The processes involved in the formation of MNPs are depicted in Figure 18.1.

18.2.6 Factors Affecting MNPs Synthesis

Major parameters that can affect the formation of nanoparticles include: (1) the type of food waste; (2) the concentration of food waste; (3) the temperature; and (4) the metal ion concentration. These factors can alter the mechanism of NPs' formation and their properties (Haytham, 2015). Factors affecting the bio-fabrication of nanoparticles are given in Table 18.1. The formation of nano-sized

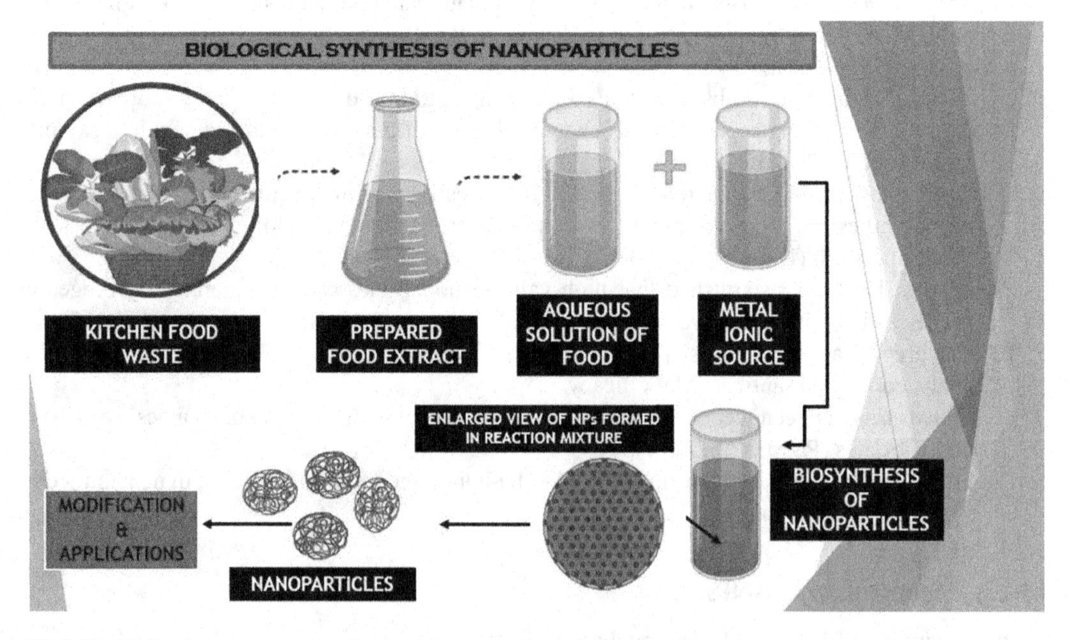

FIGURE 18.1 Processes involved in metal nanoparticles formation.

TABLE 18.1
Factors Affecting Biofabrication of Nanoparticles

	Factors	Effect on fabrication of nanoparticles
1	Regulatory factors	Effect on biological synthesis of MNPs
2	pH	Change in size and shape
3	Reactants concentration	Change in shape
4	Time	Time ∞ size of MNPs
5	Temperature	Change in size, shape, yield and stability

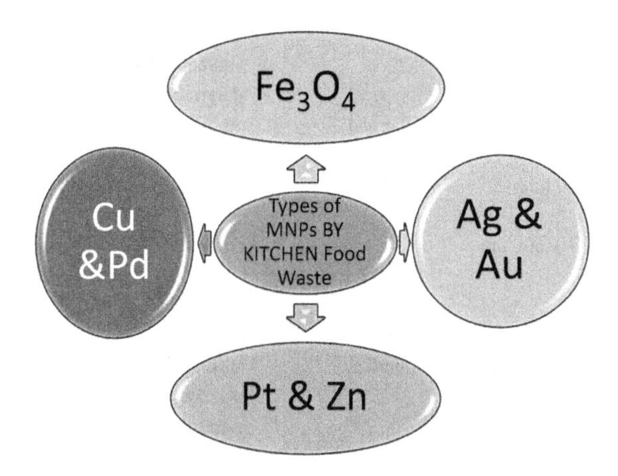

FIGURE 18.2 Types of nanoparticles generated by kitchen food waste.

metal nanoparticles can be confirmed by its characteristic surface plasmon absorption at various peak ranges in UV–visible spectra (Kumar et al., 2015). Metal ion elimination is triggered by green waste extract, which then stimulates nucleation. Smaller neighboring particles gradually accumulate on respective low-energy edges, resulting in the development of stable nanostructures. Food waste organic substances serve as biological capping agents throughout this phase, influencing the order and synthesis of smaller particles as they proceed (Jha et al., 2009).

18.3 SILVER AND GOLD NANOPARTICLES PRODUCED BY KITCHEN FOOD WASTE

A large number of NPs that can be synthesized from kitchen waste are grouped under various types, depending on their morphology, size, and chemical properties. Table 18.2 presents a list of recent work done on the green production of metal and metal oxide nanoparticles using various kitchen food waste products. The main types of nanoparticles that may be synthesized from the kitchen food waste are Fe_3O_4, Cu, Pd, Ag, Au, Pt, and Zn, as presented in Figure 18.2.

18.3.1 Silver (Ag) Nanoparticles

The variety of kitchen waste has motivated many researchers to explore its prospective use in the production of metal nanoparticles. Due to the sheer extraordinary antibacterial activities of Ag (silver) compounds, Ag nanoparticles have created an opportunity to explore this new area. Ag

TABLE 18.2
List of Nanoparticles, Their Morphology and Sources

	Source	Nanoparticle size	Nanoparticle shape	Application	References
1	Papaya fruit extract	15 nm	Cubic	Antimicrobial activities	Jain et al., 2009
2	Banana peel	23.7nm	Spherical	Antimicrobial activities	Haytham, 2015
3	Bottle gourd peel	5–40nm	Spherical	Antimicrobial activities	Kumar, 2015
4	Punica granatum seed and peel	10–30nm	Spherical	Antimicrobial activities	Nisha et al. 2015a
5	Punica granatum leaf	10–30nmm	Nanorods	Antimicrobial activities	Nisha et al., 2015a, 2015b
6	Cavendish banana peel	34nm	Grain size	Antibacterial and free radical scavenging assay	Kokila, 2015
7	Watermelon rind	96nm	Spherical	Catalytic application	Lakshmipathy et al., 2015;
8	Waste vegetable peel	20nm	Spherical	Antimicrobial activities	Sharma et al., 2016
9	Banana peel	10 nm	Spherical	Antimicrobial activities	Narayanamma et al., 2016
10	Banana stem	70–600 nm	Cube, triangular, hexagonal	Antimicrobial activities	Huu Dang et al., 2017
11	Neem, tea, kitchen waste	–	–	Antimicrobial activities	Yadav S. et al., 2018
12	Pea peel	10–25 nm	Spherical	Antidiabetic, cytotoxicity, antioxidant, antibacterial activity	Patra et al., 2019

nanoparticles have been used in a range of healthcare applications and medications due to their unique physical, chemical, and enhanced antimicrobial characteristics (Chen et al., 2008; Sotiriou, 2011). For centuries, nano-sized Ag particles have been used to cure certain diseases in Ayurveda practice. Recently, several researchers have demonstrated a reduction of silver ions in an aqueous medium including kitchen waste (Jain et al., 2009; Ahmad et al., 2012). Ahmad and Sharma (2012) used an extract from the *Ananas comosus* plant to create sphere-shaped Ag nanoparticles with sizes ranging from 5 to 35 nm. Similarly, from the extract of *Citrus sinensis* (Orange) peel, spherical-shaped crystalline Ag nanoparticles ranging in size from 3 to 12 nm were produced. Papaya fruit extract was also used (Jain et al., 2009; Konwarha et al., 2011). Researchers created plant-mediated silver nanoparticles and tested their antibacterial properties. Moreover, a study was conducted to manufacture nanoparticles from banana peel extract (Bankar et al., 2010a; Haytham, 2015; Narayanamma et al., 2016). Furthermore, Dang et al. (2017) studied the biologically active production of silver nanoparticles from waste banana stem extract and its antibacterial efficacy against *E. coli* and *S. epidermis*. An investigation was carried out by Ahmad et al. (2012) in which silver and gold nanoparticles were synthesized using *Punica granatum* peel. Additionally, the antimicrobial properties of silver nanoparticles generated from pomegranate leaves was investigated (Nisha et al., 2015b). An important study (Kumar et al., 2015) reported identification and bio-efficacy of biocompatible silver nanoparticles using vegetables waste. Silver nanoparticles may have new applications as a result of research into their antioxidant properties. Similarly, *Pisum sativum* L. peel aqueous leaf extract has been used (Patra et al., 2019) for the biosynthesis of silver nanoparticles through facile green synthesis and further investigated for its anti-diabetic, cytotoxicity, antioxidant, and antibacterial activities.

18.3.2 Gold (Au) Nanoparticles

Due to a broad array of applications in different areas such as catalysis, biology, biosensing, medications, image analysis, and diagnosis, gold nanocomposites are a potential candidate (Cai et al., 2008; Ghosh et al., 2008; Liu et al., 2008; Torres-Chavolla et al., 2010; Mikami et al., 2013). Nanoparticles of varying size (20–25 nm), spherical in shape, can be synthesized from the waste of grape skins, stalks, and seeds (Krishnaswamy et al., 2014). Similarly, using an extract from *Pyrus* sp., triangular and hexagon-shaped polycrystalline gold nanoparticles with sizes ranging from 200 to 500 nm have been produced (Ghodake et al., 2010). Moreover, an extract of *Mangifera indica* (mango) peel was used to synthesize the gold nanocomposite ranging in size from 6.03 ±2.77 nm to 18.01 ±3.67 nm (Yang et al., 2014). Similarly, to control nanoparticle size, research with different concentrations of *Aloe vera* leaf extracts was done with chloroauric ions (Chandra et al., 2006). The nanoparticles' dimensions ranging between 50 and 350 nm were controlled by altering the concentrations of the extract, which in turn influenced the shape of the nanoparticles from spherical to triangular.

18.3.3 Other Types of Nanoparticles

Palladium (Pd) nanoparticles were synthesized from *Musa paradisiac* (banana) peel resulting in crystalline and irregular shapes, with a size range of 50 nm (Bankar et al., 2010a, 2010b). Abdel-Shafy and Mansour (2018) reported the eco-friendly manufacture of nanoparticles using organic food waste and their use for the environment. Furthermore, Lakshmipathy et al. (2015) employed a watermelon rind extract to produce Pd nanoparticles of a 96 nm particle size. Furthermore, tea and coffee residue extracts were capable of producing both Ag and Pd nanoparticles. Nanomaterials generated from them were spherical and ranged in size from 5 nm to 100 nm, with the majority of the particles falling between 20 and 60 nm (Nadagouda et al., 2008). Furthermore, a current study found that from tea extracts wastes magnetic ferric oxide (Fe_3O_4) NPs can be generated. The nanoparticles were cubes and pyramids in shape with sizes ranging from 5 to 25 nm (Lunge et al., 2014). Magnesium oxide (MgO) and manganese (II, III) oxide (Mn_3O_4) are other kinds of metallic oxide nanoparticles which have been synthesized from the extracts of the *Citrus sinensis* (orange) peel and *Musa paradisiac* (banana) peels, respectively. MgO nanoparticles with a mean size of 29 nm and a spherical shape have been synthesized from the extract of *Citrus sinensis* (Ganapathi et al., 2015).

18.4 APPLICATIONS OF MNPs

MNPs have enormous potential to be used in different fields such as antibacterial, biomedical, electromechanical, biosensing, cosmetology, electro-optic energy, agriculture, image analysis, water treatments, and so on, due to their unique properties. To verify and evaluate the antimicrobial activity of MNPs, several novel approaches have been used. Lots of research is going on, and an extensive amount of literature is available on the antibacterial action of NPs. As an effective antibacterial and excellent biocompatibility component, AgNPs have received considerable amount of attention from researchers. AgNPs with higher concentrations have faster membrane permeability than those with lower concentrations, causing bacteria to breach their cell walls (Kasthuri et al., 2009). *Citrus sinensis* peel extract has been used as a reducing and capping agent to produce biologically synthesized silver nanoparticles in a single process. Silver nanoparticles were effectively reduced by *C. sinensis* aqueous peel extracts, and efficiently showed antimicrobial activity against *E. coli, Pseudomonas aeruginosa*, and *Streptococcus aureus* (Kaviya et al., 2011).

Nowadays nanoproducts are used in our daily life, because they are eco-friendly with high competence value. One of the profitable examples of this is a water purifier. Silica nanoparticles

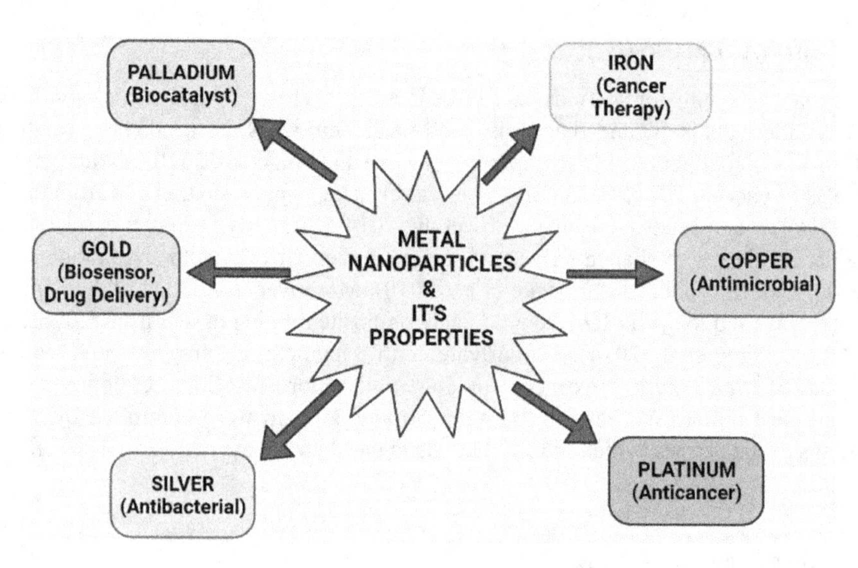

FIGURE 18.3 Different types of nanoparticles and applications.

are used in a wide range of commercial goods. Enhanced silica-based nanomaterials are effective pesticide control agents with a wide range of commercial purposes. Purification of wastewater with nanoparticles is the best route as it possesses a high surface area (Amin et al., 2014). AgNPs are applied in a variety of robotic processes as silver metal is an exceptional heat-conducting substance. It is mostly found in heat-sensitive devices such as PCR lids and UV spectrophotometers. Nanosilver, which is used as a surface-coating material, can create instrument components. It is exceptionally heat-resistant and devoid of sample distortion (Weiss et al., 2006). Food items become infected with microbes in the food sector as a result of different operations, such as raw material production, packaging and transportation. As a result, there is a need to enhance cost-effective sensors to calculate product quality. MNPs have been commercialized as nanosensors, and they identified hazardous bacteria and checked for multiple levels of contamination at a minimal cost.

In the food and cosmetic industries, metal nanoparticles are used as preservatives. The novel aspects of MNPs are found in a variety of commercial applications, including cosmetic products, medications, and food preservatives (Song and Kim, 2009b; Kokura et al., 2010) (Figure 18.3). Sustainable MNPs are used in the food and healthcare industries to replace preservatives. Bone and tooth adhesives, face cream, and homemade items are just a few of the additional uses of nanoparticles (Kouvaris et al., 2012). Ag and Si nanoparticles are found in a variety of items, including sunscreens, anti-aging lotions, toothpaste, mouthwash, hair products, and fragrances (Kumar and Yadav, 2009).

18.5 CONCLUSION

Recycling and utilization of the food waste produced by the kitchen have several benefits. Large quantities of waste produced globally in our own kitchen provide a better platform for renewable sources for active compounds. Green synthesis processing strategies are environmentally friendly, able to reduce the waste volume in landfills, and can be reworked for the formation of many profitable products. The significant characteristic of green synthesis is to produce NPs free from toxic chemicals, solvents, and reductants that are commonly used in traditional methods. Despite the numerous advantages of using food waste extracts to produce nanomaterials, there are still a number of unaddressed issues, such as nanoparticles' size and shape uniformity. At present, only a few studies have used food waste to generate nanoparticles. Due to the limited research, there is still sufficient space for researchers to perform research in this area.

Metal and metal oxide nanoparticles have been used in antibacterial, biosensing, catalyst supports, pharmacology, cancer treatment, and packaging applications and can be produced through traditional physical and chemical processes. There are many possibilities to produce innovative metal-based nanomaterials with exceptional features, given the wide variety of food waste availability and the ability to control the effects of various parameters during fabrication. Despite the high level of interest in the green fabrication of nanoparticles by using kitchen wastes, there are presently only a few practical applications accessible. Scientific comparison of nanoparticles' properties between traditionally manufactured nanoparticles and those produced from the discarded kitchen materials must be responsible for the recommendation of property variances and possible uses. This method is eco-friendly, and cost-effective with stable NPs. The existing literature promises advanced potential benefits for future perspectives in the research field. Only a few researchers have described the use of food waste to synthesize nanoparticles so far. So there is a lot of scope and potential in this field for further investigation.

REFERENCES

Abdel-Shafy, H.I., and Mansour, M.S.M. (2018). Green synthesis of metallic nanoparticles from natural resources and food waste and their environmental application. In Kanchi, S., and Ahmed, S. (Eds.), *Green Metal Nanoparticles*. Hoboken, NJ: John Wiley & Sons, pp. 321–385.

Ahmad, N., and Sharma, S. (2012). Green synthesis of silver nanoparticles using extracts of *Ananas comosus*. *Green Sustainable Chemistry*, 2, 141–147.

Ahmad, N., Sharma, S., and Rai, R. (2012). Rapid green synthesis of silver and gold nanoparticles using peels of *Punica granatum*. *Advanced Materials Letters*, doi: 10.5185/amlett.2012.5357.

Almomani, F., Smith, D.W., and Gamal El-Din, M. (2008). Degradation of cyanobacteria toxin by advanced oxidation process. *Journal of Hazardous Materials*, 150, 238–249.

Amin, M. T., Alazba, A.A., and Manzoor, U. (2014). *Removal of Pollutants from Wastewater Using Different Types of Nanoparticles.* New Delhi: Hindawi Publishing Corporation.

Aromal, A., and Philip, D. (2012). Green synthesis of gold nanoparticles using *Trigonellafoenum-graecum* and its size-dependent catalytic activity. *Spectrochimica Acta A*, 97, 1–5.

Arunachalam, K., Annamalai, S., and Hari, S. (2013). One-step green synthesis and characterization of leaf extract-mediated biocompatible silver and gold nanoparticles from *Memecylon umbellatum. International Journal of Nanomedicine*, 1307.

Asha Rani, P.V., Mun, G.L.K., Hande, M.P., and Valiyaveettil, S. (2009). Cytotoxicity and genotoxicity of silver nanoparticles in human cells. *ACS Nano*, 3, 279–290.

Baiano, A. (2014). Recovery of biomolecules from food wastes: A review. *Molecules*, 19(9), 14821–14842.

Bankar, A., Joshi, B., Kumar, A.R., and Zinjarde, S. (2010a). Banana peel extract mediated novel route for the synthesis of silver nanoparticles. *Colloids and Surfaces A: Physicochemical Engineering Aspects*, 368, 58–63.

Bankar, A., Joshi, B., Kumar, A.R., and Zinjarde, S. (2010b). Banana peeled extract mediated novel route for the synthesis of palladium nanoparticles. *Materials Letters*, 64, 1951–1953.

Cai, W., Gao, T., Hong, H., and Sun, J. (2008). Applications of gold nanoparticles in cancer nanotechnology. *Nanotechnology, Science and Application*, 1, 17–32.

Chandran, S.P., Chaudhary, M., Pasricha, R., Ahmad, A., and Sastry, M. (2006). Synthesis of gold nanotriangles and silver nanoparticles using *Aloe-vera* plant extract. *Biotechnology Programme*, 22, 577–583.

Chen, X., and Schluesener, H.J. (2008). Nanosilver: A nano-product in medical applications. *Toxicology Letters*, 176, 1–12.

Cheong, S., Watt, J.D., and Tilley, R.D. (2010). Shape control of platinum and palladium nanoparticles for catalysis. *Nanoscale*, 2, 2045–2053.

Coccia, F., Tonucci,, L., Bosco, D., Bressan, M., and d'Alessandro, N. (2012). One pot synthesis of lignin-stabilized platinum and palladium nanoparticles and their catalytic behaviours in oxidation and reduction reactions. *Green Chemistry*, 14, 1073–1078.

Dang, H., Fawcett, D., and Poinern, G.E.J. (2017). Biogenic synthesis of silver nanoparticles from waste banana plant stems and their antibacterial activity against *Escherichia coli* and *Staphylococcus Epidermis*. *International Journal of Research in Medical Sciences*, 5(9), 3769–3775.

Deepa, Ameen, F., Amirul Islam, M., and Dhanker, R. (2022). Green synthesis of silver nanoparticles from vegetable waste of pea *Pisum sativum* and bottle gourd Lagenaria siceraria: Characterization and antibacterial properties. *Frontiers In Environmental Science*, 10, 941554.

Dhanker, R., Hussain, T., Tyagi, P., Singh, K., and Kamble, S. (2021). The Emerging Trend of Bio-Engineering Approaches for Microbial Nanomaterial Synthesis and Its Applications. *Frontiers In Microbiology*, 12, 638003.

Dhanker, R., Kumar, R., and Hwang, J. (2013). How effective are *Mesocyclops aspericornis* (Copepoda: Cyclopoida) in controlling mosquito immatures in the environment with an application of phytochemicals?. *Hydrobiologia*, 716(1), 147–162.

Espitia, P.J.P., Soares, N.F.F., dosReis Coimbrav J.S., de Andrade, N.J., Cruz, R.S., and Medeiros, E.A.A. (2012). Zinc oxide nanoparticles: Synthesis, antimicrobial activity and food packaging applications. *Food Bioprocess Technology*, 5, 1447–1464.

Ganapathi, R.K., Ashok, C.H., Venkateswara, R.K., Shilpa, C.C.H., and Akshaykranth, A. (2015). Eco-friendly synthesis of MgO nanoparticles from orange fruit waste. *International Journal of Advanced Research in Physical Science*, 2, 1–6.

Ghodake, G., Deshpande, N., Lee, Y., and Jin, E. (2010). Pear fruit extract-assisted room-temperature biosynthesis of gold nanoplates. *Colloids and Surfaces B: Biointerfaces*, 75, 584–589.

Ghosh, P., Han, G., De, M., Kim, C.K., and Rotello, V.M. (2008). Gold nanoparticles in delivery applications. *Advanced Drug Delivery Reviews*, 60, 1307–1315.

Ghosh, P.R., Fawcett, D., Sharma, S.B., and Poinern, G.E.J. (2017). Production of high-value nanoparticles via biogenic processes using aquacultural and horticultural food waste, *Multidisciplinary Digital Publishing Institute*, doi: 10.3390/ma10080852.Ghosh, P.R., Sharma, S.B., Haigh, Y.T., Barbara Evers, A.L., and Ho, G. (2015). An overview of food loss and waste: Why does it matter?. *COSMOS*, 11, 1–15.

Gustavsson, J., Cederberg, C., Sonesson, U., van Otterdijk, R., and Meybeck, A. (2011). *Global Food Losses and Food Waste: Extent, Causes and Prevention*. Rome, Italy: Food and Agriculture Organization of the United Nations.

Haytham, M.M.I. (2015). Green synthesis and characterization of silver nanoparticles using banana peel extract and their antimicrobial activity against representative microorganisms. *Journal of Radiation Research and Applied Sciences*, 8(3), 265–275, doi: 10.1016/j.jrras.2015.01.007.

Jain, D., Daima, H.K., Kachhwaha, S., and Kothari, S.L. (2009). Synthesis of plant-mediated silver nanoparticles using papaya fruit extract and evaluation of their antimicrobial activities. *Digest Journal of Nanomaterials and Biostructure*, 4(4), 557–563.

Jha, A., Prasad, K., Kumar, V., and Prasad, K. (2009). Biosynthesis of silver nanoparticles using Eclipta leaf. *Biotechnology Progress*, 25(5), 1476–1479.

Kantor, L.S., Lipton, K., Manchester, A., and Oliveira, V. (1997). Estimating and addressing America's food losses. *Food Review*, 20, 2–12.

Kasthuri, J., Kathiravan, K., and Rajendiran, N. (2009). Phyllanthin assisted biosynthesis of silver and gold nanoparticles: A novel biological approach. *Journal of Nanoparticle Research*, 11, 1075–1085.

Kaviya, S., Santhanalakshmi, J., Viswanathan, B., Muthumary, J., and Srinivasan, K. (2011). Biosynthesis of silver nanoparticles using *Citrus sinensis* peel extract and its antibacterial activity. *Spectrochimica Acta A*, 79, 594–598.

Kim, H., Mosaddik, A., Gyawall, R., Ahn, K.S., and Cho, S.K. (2012). Induction of apoptosis by ethanolic extract of mango peel and comparative analysis of the chemical constitutes of mango peel and flesh. *Food Chemistry*, 133, 416–422.

Kokila, T., Ramesh, P.S. and Getha, D. (2015). Biosynthesis of silver nanoparticles from Cavendish banana peel extract and its antibacterial and free radical scavenging assay: A novel biological approach. *Applied Nanoscience*, 5(8), 911–920.

Kokura, S., Handa, O., Takagi, T., Ishikawa, T., Naito, Y., and Yoshikawa, T. (2010). Silver nanoparticles as a safe preservative for use in cosmetics. *Nanomedicine*, 6, 570–574.

Konwarha, R., Gogoia, B., Philip, R., Laskarb, M.A., and Karaka, N. (2011). Biomimetic preparation of polymer supported free radical scavenging, cytocompatible and antimicrobial "green" silver nanoparticles using aqueous extract of *Citrus sinensis* peel. *Colloids and Surfaces B Biointerfaces*, 84, 338–345.

Kora, A.J., Sashidhar, R.B., and Arunachalam, J. (2010). Gum Kondagogu (*Cochlospermumgossypium*): A template for green synthesis and stabilization of silver nanoparticles with antibacterial application. *Carbohydrate Polymers*, 82, 670–679.

Kouvaris, P., Delimitis, A., Zaspalis, V., Papadopoulos, D., Tsipas, S.A., and Michailidis, N. (2012). Green synthesis and characterization of silver nanoparticles produced using *Arbutus unedo* leaf extract. *Materials Letters*, 76, 18–20.

Krishnaswamy, K., Vali, H., and Orsat, V. (2014). Value-adding to grape waste: Green synthesis of gold nanoparticles. *Journal of Food Engineering*, 142, 210–220.

Kumar, V., and Yadav, S.K. (2009). Plant-mediated synthesis of silver and gold nanoparticles and their applications. *Journal of Chemical Technology and Biotechnology*, 84, 151–157.

Kumar, V., Verma, S., Choudhury, S., Tyagi, M., Chatterjee, S., and Variyar, P.S. (2015). Biocompatible silver nanoparticles from vegetable waste: Its characterization and bio-efficacy. *International Journal of Nano and Material Sciences*, 4, 70–86.

Lakshmipathy, R., Palakshi Reddy, B., Sarada, N.C., Chidambaram, K., and Khadeer Pasha, S. (2015). Watermelon rind-mediated green synthesis of noble palladium nanoparticles: Catalytic application. *Applied Nanoscience*, 5, 223–228.

Li, W.R., Xie, X.B., Shi, Q.S., Duan, S.S., Ouyang, Y.S., and Chen, Y.B. (2011). Antibacterial effect of silver nanoparticles on *Staphylococcus aureus*. *Biometals*, 24, 135–141.

Lin, X., Wu, M., Wu, B., Kuga, S., Endo, T., and Huang, Y. (2011). Platinum nanoparticles using wood nanomaterials: Eco-friendly synthesis, shape control and catalytic activity for p-nitrophenol reduction. *Green Chemistry*, 13, 283–287.

Liu, X.Q., Dai, Q., Austin, L., Coutts, J., Knowles, G., Zou, J.H., Chen, H., and Huo, Q. (2008). A one-step homogeneous immunoassay for cancer biomarker detection using gold nanoparticle probes coupled with dynamic light scattering. *Journal of American Chemical Society*, 130, 2780–2782.

Loehr, R. (2012). *Agricultural Waste Management: Problems, Processes, and Approaches*. Amsterdam: Elsevier.

Lunge, S., Singh, S., and Sinha, A. (2014). Magnetic iron oxide (Fe_3O_4) nanoparticles from tea waste for arsenic removal. *Journal of Magnetism and Magnetic Materials*, 356, 21–31.

Madhumitha, G., Rajakumar, G., Roopan, S.M., Rahuman, A.A., Priya, K.M., Khan, F.R., Khanna, V.G., ... Jayaseelan, C. (2012). Acaricidal, insecticidal, and larvicidal efficacy of fruit peel aqueous extract of *Annona squamosa* and its compounds against blood-feeding parasites. *Parasitology Research Paper*, 111, 2189–2199.

Mikami, Y., Dhakshinamoorthy, A., Alvaro, M., and Garcia, H. (2013).Catalytic activity of unsupported gold nanoparticles. *Catalysis Science and Technology Journal*, 3, 58–69.

Morales-Sanchez, E., Guajardo-Pacheco, J., Noriega-Trevino, M., Quintero-Gonzalez, C., Compean-Jasso, M., Lopez-Salinas, F., Gonzalez-Hernandez, J., and Ruiz, F. (2011). Synthesis of silver nanoparticles using albumin as reducing agent. *Materials Sciences and Applications*, 2, 578–581.

Mourdikoudis, S., Pallares, R., and Thanh, N. (2018). Characterization techniques for nanoparticles: Comparison and complementarity upon studying nanoparticle properties. *Nanoscale*, 10(27), 12871–12934.

Nadagouda, M.N., and Varma, R.S. (2008). Green synthesis of silver and palladium nanoparticles at room temperature using coffee and tea extract. *Green Chemistry*, 10, 859–862.

Narayanamma, A., Rani, M.E., and Raju, K.M. (2016). Natural synthesis of silver nanoparticles by banana peel extract and as an antibacterial agent. *International Journal of Science and Research (IJSR)*, 5(8). Online.

Nisha, M.H., Tamileswari, R., and Jesurani, Sr. S. (2015a). Analysis of anti-bacterial activity of silver nanoparticles from pomegranate (*Punicagranatum*) seed and peel extracts. *International Journal of Engineering Research and Technology (IJERT)*, 4(4), 1044–1048.

Nisha, M.H., Tamileswari, R., Jesurani, Sr. S., Kanagesan, S., Hashim, M., Catherine, S., and Alexander, P. (2015b). Green synthesis of silver nanoparticles from pomegranate (*Punica Granatum)* leaves and analysis of anti-bacterial. *International Journal of Advanced Technology in Engineering and Science*, 3(6), 2348–7550. Online.

Patra, J.K., Das, G., and Shin, H.S. (2019). Facile green biosynthesis of silver nanoparticles using *Pisum sativum* L. outer peel aqueous extract and its anti-diabetic, cytotoxicity, antioxidant, and antibacterial activity. *Dovepress*, 6679–6690, doi: https://doi.org/10.2147/IJN.S212614.

Perez, J., Ba, L., and Escolano, C. (2005). Roadmap report on nanoparticles. Barcelona, Spain: Willems and van den Wildenberg (in Spanish).

Prasad, R., Kumar, V., and Prasad, K.S. (2014). Nanotechnology in sustainable agriculture: Present concerns and future aspects. *African Journal of Biotechnology*, 13(6), 705–713.

Puvanakrishnan, P., Park, J., Chatterjee, D., Krishnan S., and Tunnell, J.W. (2012). In vivo tumor targeting of gold nanoparticles: Effect of particle type and dosing strategy. *International Journal of Nanomedicine*, 7, 1251–1258.

Sahu, N., Soni, D., Chandrashekhar, B., Sarangi, B., Satpute, D., and Pandey, R. (2012). Synthesis and characterization of silver nanoparticles using *Cynodon dactylon* leaves and assessment of their antibacterial activity. *Bioprocess and Biosystems Engineering*, 36(7), 999–1004.

Shah, M., Fawcett, D., Sharma, S., Tripathy, S., and Poinern, G.E.J. (2015). Green synthesis of metallic nanoparticles via biological entities. *Materials*, 8, 7278–7308.

Shah, M., Poinern, G.E.J., and Fawcett, D. (2016). Biosynthesis of silver nanoparticles using indigenous *Xanthorrhoea Glauca* leaf extract and their antibacterial activity against *Escherichia coli* and *Staphylococcus epidermis*. *International Journal of Research in Medical Sciences*, 4, 2886–2892.

Sharma, K., Kaushik, S., and Jyoti, A. (2016). Green synthesis of silver nanoparticles by using waste vegetable peel and its antibacterial activities. *Journal of Pharmaceutical Science & Research*, 8(5): 313–316.

Song, J.Y., and Kim, B.S. (2009a). Rapid biological synthesis of silver nanoparticles using plant leaf extracts. *Bioprocess and Biosystems Engineering*, 32, 79–84.

Song, J.Y. and Kim, B.S. (2009b). Biological synthesis of metal nanoparticles. In Hou, C.T. and Shaw, J.-F. (Eds.), *Biocatalysis and Agricultural Biotechnology*, Boca Raton, FL: CRC Press, pp. 399–407.

Sotiriou, G.A., and Pratsinis, S.E. (2011). Engineering nanosilver as an antibacterial, biosensor and bioimaging material. *Current Opinion in Chemical Engineering*, 1, 3–10.

Torres-Chavolla, E., Ranasinghe, R.J., and Alocilja, E.C. (2010).Characterization and functionalization of biogenic gold nanoparticles for biosensing enhancement. *IEEE Transactions on Nanobiotechnology*, 9, 533–538.

Weiss, J., Takhistov, P., and Julian Mcclements, D. (2006). Functional materials in food nanotechnology. *Journal of Food Science*, 71, 107–116.

Yadav, S.G., Patil, S.H., Patel, P., Nair, V., Khan, S., Kakkar, S., and Gupta, A.D. (2018). Green synthesis of silver nanoparticles from plant sources and evaluation of their antimicrobial activity. *IJSRSET*, 5(4). Online.

Yang, N., Wei-Hong, L., and Hao, L. (2014). Biosynthesis of Au nanoparticles using agricultural waste mango peel extract and its in vitro cytotoxic effect on two normal cells. *Materials Letters*, 134, 67–70.

19 Artificial Intelligence-Based Smart Waste Management for the Circular Economy

Alaa El Din Mahmoud[1,2*] *and Omnya Desokey*[1]
[1]Environmental Sciences Department, Faculty of Science, Alexandria University, Alexandria, Egypt
[2]Green Technology Group, Faculty of Science, Alexandria University, Alexandria, Egypt
*Corresponding author Email: alaa-mahmoud@alexu.edu.eg

CONTENTS

19.1 Introduction 341
19.2 The Importance of Artificial Intelligence in Municipal Waste Management 343
19.3 The Role of Artificial Intelligence in the Stages of Waste Management 345
19.3.1 Dustbins 345
19.3.1.1 The Overflow Issue 346
19.3.1.2 Wrong Sorting Issue 346
19.3.2 Recycling 347
19.3.3 Energy Recovery 349
19.3.4 Landfill Disposal 350
19.4 The Circular Economy and Artificial Intelligence 351
19.4.1 Plastics 352
19.4.2 Aluminum 353
19.5 Artificial Intelligence and Sustainable Development Goals 353
19.6 Challenges of Artificial Intelligence 354
19.7 Conclusion 354
Acknowledgments 355
References 355

19.1 INTRODUCTION

Solid wastes are unwanted materials generated from anthropogenic processes (domestic, commercial, industrial, and healthcare) which become valueless to the persons or the organizations that generated them (Roy and Bandyopadhyay, 2008; Chandra and Kumar, 2017; Breitenmoser et al., 2019; Sunayana et al., 2020; Kumar et al., 2021). The two main factors contributing to the increasing amount of solid waste are population growth and the high gross domestic product (GDP) (Costa et al., 2018; Chavan et al., 2019; Khandelwal et al., 2019; Alam et al., 2021). The amounts of waste have increased globally because of the economic growth, rising populations, and widespread urbanization (Wainaina et al., 2019). Therefore, this has a negative impact on the public health and the natural environment because most of the waste finds its way into rivers, ponds, low-lying land, oceans, etc. without any treatment, resulting in odors, air, and water pollution, and greenhouse gas emissions (e.g. carbon dioxide), etc. (Khapre et al., 2019; Kumar et al., 2019; Manwatkar et al.,

DOI: 10.1201/9781003239956-21

2021). This problem can be mitigated by changing our mindset and acknowledging now that "waste is wealth" (Dhar et al., 2017).

The European Union (EU) defined municipal solid waste, as "the mixed waste which are collected separately from the households, which may contain paper, cardboard, glass, metals, biowaste, plastic, wood, textile, waste electrical and electronics, waste batteries, and huge waste, including mattresses and furniture" (Sarc et al., 2019). It is expected that the levels of the generated municipal solid waste (MSW) will double by 2025 (Hoornweg, 2012) and they are expected to increase by 3.4 million tonnes by 2050 (Das et al., 2021). Figure 19.1 illustrates the waste production globally by each region.

The solid waste in developing countries is mostly thrown in low-lying areas or landfilled beside the slums, and this waste may also contain hazardous materials (Manjunath et al., 2019a, 2019b; Pearse et al., 2017; Fungaro et al., 2021). This harms public health and the natural environment. It causes air pollution when it is burned and water pollution by leaching into the groundwater. Therefore, the contaminated water that people drink is an environment for insects or vermin to breed and for scavenger animals (Mahmoud 2020; Mahmoud et al., 2021). The literature shows that the incidence of diarrhea is twice as high, and the respiratory infections are six times as high in areas with uncollected waste, which is higher than in areas where the collection is done frequently (Hoornweg 2012). As well as this, the rise in infection due to the contagious CORONA-19 virus increased the municipal solid waste through the addition of packaging, face masks, PPE (personal protective equipment), and hazardous materials (Khan, Tirth et al., 2021; Khan, Abutaleb et al., 2021; Mousazadeh et al., 2021).

Therefore, dealing with municipal solid waste management is an important aspect in preserving people's health around the world (Goyal et al., 2014; Dhar et al., 2016; Kumar, N.S., Das et al., 2016; Kumar, S., Dhar et al., 2016; Kumar, S., Nimchuk et al., 2016). However, managing municipal solid waste through the collection, transportation, sorting, and treatment of waste while monitoring it is still hazardous to the health of the workers who are exposed to the handling of the solid waste (Kumar et al., 2011; Sil et al., 2012; Das et al., 2021). Also it is an extensive service

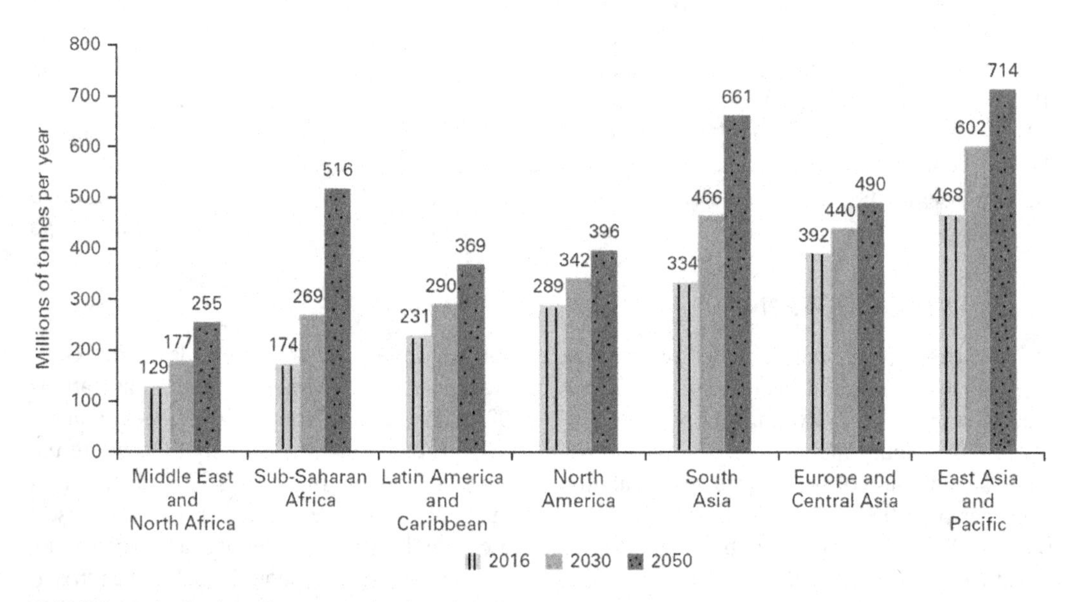

FIGURE 19.1 Projected waste production globally, by each region.

Source: Adapted from http://datatopics.worldbank.org/what-waste/trends_in_solid_waste_management.html).

that needs capacities in procurement, professional labor management, finance and the strong social contract between the municipality and the community, and other challenges for municipal services (Hoornweg, 2012).

Controlling and reducing the quantity of solid waste can be done through thinking before we buy new stuff whether we need it or not (ibid.). Then if we don't need this stuff or we can replace it with a more sustainable version, we should reject it, so, at the end, we will find that we have reduced the waste generated on a personal basis and maintained the balance at which the waste is managed (Costa et al., 2018).

Reuse is the next best choice in the waste management hierarchy. We can reuse any used materials several times, for example, we refill the empty glass bottles of juice with water and reuse the old clothes for cleaning, and so on. The reuse of different materials reduces the number of virgin materials which have become increasingly scarce. Furthermore, resource extraction and processing have contributed to greenhouse gas emissions and biodiversity loss, which should force us to shift toward more sustainable consumption (Heikkilä, 2020; Argya et al., 2021), then this is followed by recycling, recovery, and dumping in a landfill. According to Das et al. (2021), just 19% of the MSW is recovered for recycling and 70% is disposed of as landfill. European Union targets are to increase the reuse and recycling of MSW to a minimum of 65% by weight by 2035 (Sarc et al., 2019).

The aims of the waste management hierarchy are these elements (rethink, refuse, reduce, reuse, recycle, recover, etc.) and the circular economy is the most efficient and comprehensive usage of resources generally (Sara et al., 2019; Chowdhary et al., 2020a; Manjunatha et al., 2020; Rena et al., 2022). This system also contributes to improving public health and the environment, increasing the recycling rates, protecting the purity of recovered materials, improving the working conditions of the workers as well as producing high-quality secondary raw materials from waste (Wilts et al., 2021). Therefore, to achieve these goals, the new approaches of Industry 4.0 should be developed and implemented in the waste management field (Sarc et al., 2019).

We should take into consideration some factors of the design of a waste management system that vary from time to time and place to place:

1. *The nature of the waste*: the waste in developed countries has a low density which means it has a high content of packaging (paper, cardboard, plastic, glass, metal) while the waste content in developing countries is highly dense because of the high usage of food in general, which has high moisture content. The wastes with high moisture content are not suitable for incineration. Recycling and reusing reduce the proportions of combustible material (i.e., paper, plastic) before reaching the treatment stage (Senserna and Bandyopadhyay, 2008). According to Das et al. (2021), some economic entities and developed countries generate half of the total MSW produced globally. Some 46% of the global waste is organic but reaches 64% in developing countries.
2. *Access to waste collection points*: Public awareness and attitudes to waste are very important. The readiness to pay for waste management services and recycling and collection services and the political and socio-political willingness are vital. Institutional issues may limit the technology options. The policy of the government regarding the role of the private sector should be taken into account.

19.2 THE IMPORTANCE OF ARTIFICIAL INTELLIGENCE IN MUNICIPAL WASTE MANAGEMENT

Nowadays, there is a new discipline, known as Environmental Informatics, which combines research fields such as Artificial Intelligence, Geographical Information Systems (GIS), Modeling and Simulation, User Interfaces, etc. The rapid developments in information technologies and the increasing speed of the hardware have made interdisciplinary research links between environmental and computer scientists possible and very fruitful (Nair et al., 2016). This fruitful endeavor catalyzes the integration of data, information, and knowledge from various sources in the environmental sector. Hence, it helps in important and difficult tasks such as getting more insights into the environmental problems which are difficult to study through experimentation, for instance, biosphere destruction, air pollution, water contamination, and climate change. Information technologies have played an important role in the planning, prediction, supervision, and control of environmental processes at many different scales and within various periods (Cortes et al., 2000).

"Digitalization" generally describes the merging and inclusion of digital technologies into everyday life. This integration is called "Industry 4.0" because it embodies the fourth industrial revolution (Sarc et al., 2019). Artificial intelligence is an important element of Industry 4.0. Artificial intelligence (AI) is a branch of computer science, consisting of two words: artificial which means something that is made by humans and is not naturally produced, and intelligence which refers to the ability to think on its own, hence this makes artificial intelligence "thinking power made by humans". We create a computer or a robot or simply some machines to think and behave like humans, able to make decisions on their own. We are trying to simulate the human brain but it is impossible to recreate human consciousness (Dhankar et al., 2020).

There are three basic elements of AI as illustrated in Figure 19.2. First, the human element is the algorithm's language where 0s and 1s are hard to understand. It takes direction from the user, processes the data, and then provides accurate results, so the human element is the link and interpreting the results (ibid.). Second, the knowledge element is where AI analyzes the existing data within it, processes them, and compares large chunks of information to provide optimum results, so the more data that are fed in, the more efficient will be the result to find a solution to the present problems (ibid.). Third, the algorithm set where programmers and data scientists use various mathematical tools such as statistics, probability, calculus, and algebra to feed the machines with data. The AI interfaces with large chunks of data, it requires a certain set of instructions or algorithms to process and perform an operation on those data (ibid.).

Artificial intelligence depends largely on the quality and quantity of the available information used to make an appropriate decision to help conserve the environment. However, problems arise when the quantities of available information are huge and non-uniform. Computers are a very important part of AI, they are central in modern environmental conservation due to their role in monitoring, data analysis, communication, information storage and retrieval (Cortes et al., 2000).

Artificial intelligence goes through several stages of planning, reasoning, analyzing data, predicting outcomes and acting accordingly. It also involves the use of statistics and probability and various other mathematical tools. Artificial intelligence includes several other fields such as neural networks, deep learning, statistics, machine learning, which are proving to be successful in various domains, for example, security, research, robotics, voice recognition, transportation, and many more. Artificial intelligence is proving successful not only in reducing human workload but also in opening up new fields of research (Dhankar et al., 2020).

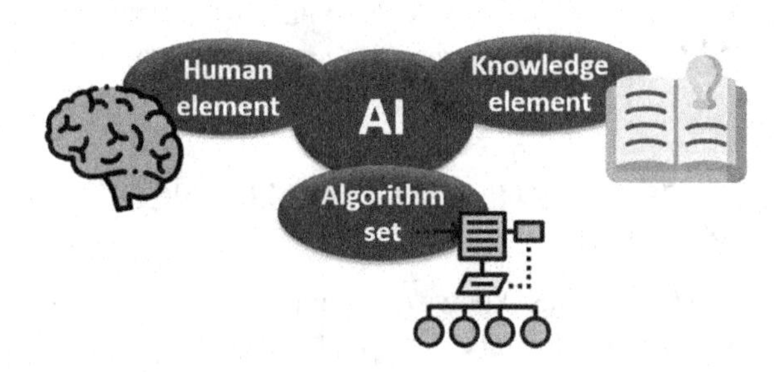

FIGURE 19.2 Basic elements of artificial intelligence (AI).

TABLE 19.1
Artificial Intelligence Techniques

Problem diagnosis technique	Decision support systems (DSS)	Environmental decision support systems (EDSS)
• Depend on the available knowledge to suggest strategies to recover/repair	• Recommend appropriate action plans based on evaluating alternatives to compare their relative costs and benefits	• Depend on knowledge-based systems (KBS)
• Try to recognize the different symptoms to develop/confirm hypotheses		• KBS is developed constantly to help us to make the right decision • Useful in the environmental management process, which consists of four spans and considers the ideal decision tool parameters

Artificial intelligence is going to "empower and improve every business, government organization, philanthropy there's no institution in the world that cannot be improved by machine learning" (Dauvergne, 2020). Therefore, artificial intelligence is very effective in the waste management process, as well as at every stage of it to make it more efficient in reducing our waste through recycling, the circular economy, energy recovery, etc. Artificial intelligence predicts consumer demand, its audit system can aid in managing supply chain logistics to reduce expenditures, speed up delivery, and ensure low carbon emissions of transportation (ibid.). This can be done through its techniques as illustrated in Table 19.1.

Ideal decision tools for valid recommendations on land, water, and environmental management must include quantitative and analytical components; they must span and integrate the physical, biological, socio-economic, and policy elements of decision-making. AI uses data interpretation and data mining techniques by screening the data to reveal patterns and identify potential problems or to find out similarities between current and past situations and learn new situations. Thus, our understanding of the relevant factors and their relationships can be improved and we can discover new ones (Cortes et al., 2000).

19.3 THE ROLE OF ARTIFICIAL INTELLIGENCE IN THE STAGES OF WASTE MANAGEMENT

19.3.1 Dustbins

The garbage bin in public places is one of the waste management issues, when it overflows before the beginning of the next cleaning process. This leads to various hazards such as bad odor and aesthetic issues which may be the main source for spreading various diseases. Nearly 235 million people who have their livelihood close to these garbage bins are victims of breathing illnesses like asthma due to inhaling polluted air. Furthermore, it is the root reason for the breeding of mosquitoes and houseflies, which are the main source of serious diseases like malaria. In order to reduce the human intervention and exposure to waste, robotic dustbins have been created which also makes the task of sorting and recycling very easy (Gupta et al., 2019). Sorting the waste is very important and makes the next steps of collection and recycling easier. According to Haridas and Ganesh (2020), waste is divided into two main groups: biodegradable (such as plants, fruits, vegetables, animals, flowers and paper); and non-biodegradable (such as rubber, plastic, metals, and so on). Every type of waste should be thrown separately in its correct dustbin but it is very difficult to monitor and make sure that the waste is placed in the right dustbin by the municipality. Therefore, artificial intelligence helps to overcome this issue.

19.3.1.1 The Overflow Issue

The IoT (Internet of Things) integrated with trash cans can sense and send the data about the trash volume to the servers through the internet. The dustbins' sensors send the trash volume data to the servers, which is essential for tracking and initiating the cleaning process accordingly on a routine. This guarantees that the trash is cleared well before it overflows and also prevents unnecessary trash pickup when the trash cans are not filled. The proposed system is performed in the city of Copenhagen, using Open Data that are the start of Big Data analytics, which have been improved with cyber systems. The collected data are used to statistically analyze the rate of these dustbins filling up (Gupta et al., 2019). As shown in Figure 19.3, the Smart Dustbin is a cylindrical structure connected to a piston that is used to compress the garbage. The trash bin plate is attached to the cylinder and the leaf switch is suspended upside down through the side hole. The leaf switch level is placed at a point lower than the maximum level. This is essential for the precautionary measures of the garbage overflow, in case of fault from the cleaning team. The compressing plate can reach down to press the switch. Once the threshold level is reached, the garbage will be refused entry to the trash can in order to avoid overflow. It saves manual effort, time and cost and reduces the overflowing trash cans by a factor of 4 (Gupta et al., 2019).

19.3.1.2 Wrong Sorting Issue

The trash boat is a robot trash bin that is empowered by AI. When someone throws something into its tank, sensors and machine learning detect the type of waste with 90% accuracy, even fluid can be extracted and placed in the right tank. It is useful for shopping malls, stadiums and airports and places that produce a lot of waste. In the future, the user will know if the discarded item has been recycled by installing LEDs in the system (Ahmed and Asadullah, 2020).

Artificial intelligence helps the dustbin to detect the waste item category through an image classification algorithm which is applied by using Tensor Flow, a free and Open Source software library for machine learning and artificial intelligence. Images are taken by using a camera and then these images are compared with the images with the dataset to find the right match for the waste collection, so AI helps in separating the waste into different categories and differentiates the biodegradable and non-biodegradable waste. This will reduce the manual sorting and number of scavengers, moreover

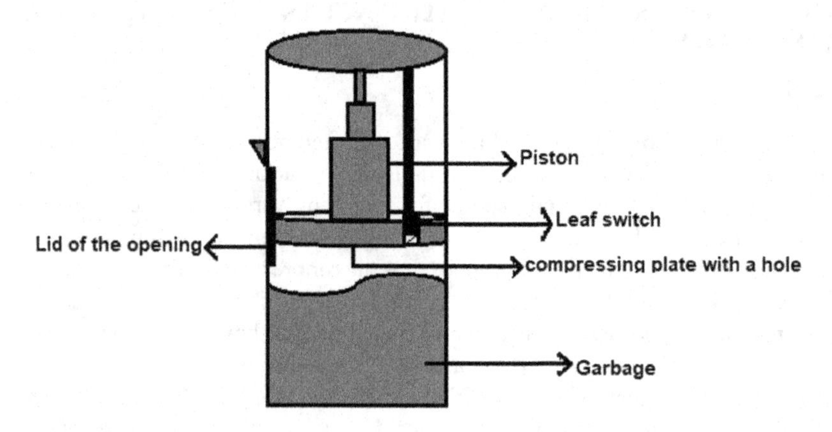

FIGURE 19.3 Smart dustbin design.

Source: Adopted from Kumar, N.S. et al. (2016).

the waste collection and sorting will be more accurate. In addition, the chances of finding non-biodegradable waste in the landfill will be decreased and then this will be followed by a reduction in pollution (Haridas and Ganesh. 2020).

19.3.2 Recycling

Recycling reduces our dependency on raw materials as well as reducing waste and subsequent landfills; it converts the useless waste materials into valuable recycled products (Gupta et al., 2019). This involves many steps, including collection in many different ways whether from curbside, drop-off recycling centers, retail locations, or deposit programs, processing the recyclables through going to a Material Recovery Facility (MRF) to be separated, cleaned, and turned into raw materials for further manufacturing (such as turning used cans into raw materials of aluminum, etc.). Of all the steps, waste separation is the most important component of recycling to remove waste contamination, which could decrease the benefit of recycling, while at the same time expanding and improving the recycling rate (Ozdemir et al., 2021).

It is also important to create ecologically sustainable development worldwide by ensuring that waste is properly disposed of and recycled to reduce the environmental impact and minimize the impact caused by incorrect disposal of garbage, especially domestic stuff (i.e., paper, plastic, glass and trash) (Costa et al., 2018).

Many strategies are used to separate materials in a waste stream, whether by its size, weight or thickness, such as moving beds, drums, and screens, besides air separators or by using magnets, such as an eddy current separator to separate ferrous and non-ferrous metals (Ozdemir et al., 2021). Costa et al. (2018) proposed using an automated system that is based on neural network techniques for the correct waste separation into recycling categories, which is divided into four classes: glass, paper, metal, and plastic, as well as the positive environmental, economic and social effects. Sensors can facilitate the recognition of optical, chemical, or physical properties of single particles/materials, and the mechanical system of separation for determining the type of wastes. The received sensor information affects the performance of waste separation which can happen manually in the home and be picked through any collection programs or else is automatically separated in the MRF. Sensor-based sorting techniques have been widely used lately as an alternative to manual sorting of solid waste which is difficult and unhygienic. There are several methods which involve robots and the concepts of machine learning. The source segregation mode of waste processing shows the possibility of waste recovery has the potential to achieve more than 25% with steadily increasing positives. Conventionally, processing of fully mixed waste is labor-intensive. Automating this process is an excellent use of the technology and skill in order to prevent human interference and the corresponding vulnerability of humans to respiratory and skin diseases. These lie on the higher end of financial investments with a waste recovery rate of mere 15% (Gupta et al., 2019). Additionally, one person can sort 20–40 options per minute, depending on the size of the object, though the manufacturers claim that previous models can sort 40–60 options per minute, and newer models can sort 60–80 options per minute. Expert opinion says that there can be at least one robot to perform the work of two people in different shifts (Ahmed and Asadullah, 2020).

The incorporation of recognition algorithms with artificial intelligence can be an effective solution in an automated process of waste segregation. A machine learning algorithm means that the machine has the capability of learning on its own, based on the pre-loaded data and pre-set examples. This is known to be the goal-driven aspect of machine learning. The use of adaptive and interactive modelling systems (AIMS) is to demonstrate the sensor data flow and to output a highly efficient description of object features that completely define the segregation strategies (Figure 19.4). The system is designed to solve the challenges of real-time data acquisition from the sensors and send the information sensed and learnt about the machine processing speeds (Gupta et al., 2019).

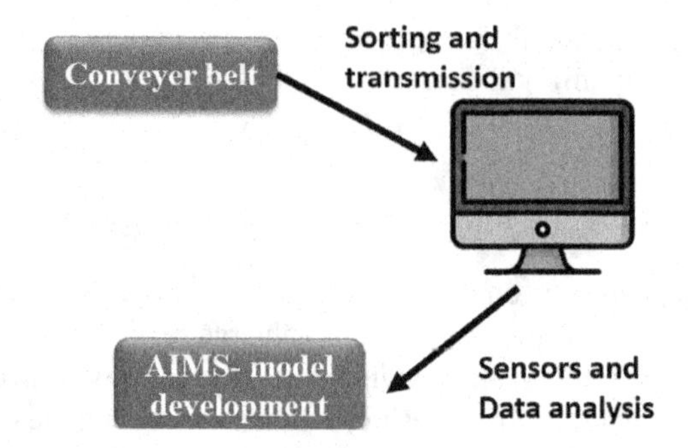

FIGURE 19.4 The scheme of the stream of waste segregation using the interactive modeling systems (AIMS).

FIGURE 19.5 Some examples of images of recycled objects of database for classifying.

Figure 19.5 illustrates the mode of learning by the system. The machine is trained to detect glass, metallic things and plastic, using a variety of possible shapes, deformation levels, diverse sizes and colors, as well as different levels of contamination. Each substance is made to pass through a systematically placed array of unique sensors. Their corresponding responses are recorded and a combined database is established which possesses the accumulated sensor data and the interpreted information. After feeding the machine with an adequate number of examples, the example-driven learning algorithm is used to create a classification function. The type of the learning algorithm and the related parameters are decided by the system. This is essential for the desired accuracy, speed and the stability of the output (ibid.).

There are other learning algorithms as well as AIMS. Radio frequency identification (RFID) and sensor network (SN) have been used to provide a new way to optimize waste management systems and minimize the impact of waste disposal. Two methods were proposed to estimate household waste volume based on the image analysis contents of the open container lid from RFID tags. For instance, the label is used to connect each bin to the address of its house and to improve the quality of eclectic collection by tracking the waste flow of a city, where each bit of waste is detected from information stored on an RFID tag that is associated with the waste, and during the processing step the RFID tags are read to provide some relevant information. Swedberg (2008) tracked the consumers by identifying them using an RFID tag associated with their recycle bin. Consumers are also financially rewarded, based on the weight of their recycling packages or how much they recycled.

Another method has been developed that is capable of selecting the best recycling container when considering the waste discarded by the user, and a model was presented using Ontology Web Language (OWL) to sort the smart waste items for better recycling of materials and to present information about the amount of valuable recyclable materials contained in each bundle of waste (Costa

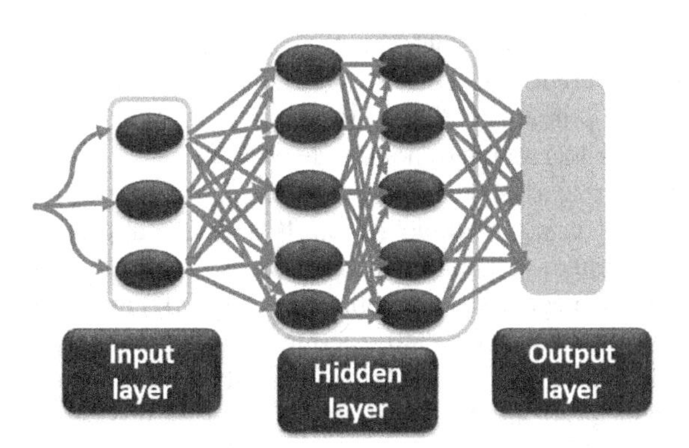

FIGURE 19.6 An example of CNN machine learning algorithm.

et al., 2018). Costa et al. created a database of images of recycled objects in six classes with about 400–500 images with a total of about 2,400 images. According to the authors, the data acquisition process involved using a white poster-board as a background, however, the lighting and pose for each photo are not the same, which causes variation in the database. Furthermore, using a convolutional neural network (CNN), the garbage is classified into six different recycling categories (i.e., metal, paper, glass, plastic, trash, and cardboard) with an accuracy rate ranging between 63% and 22%.

CNN is a general process which uses image recognition to classify and detect objects and minimize pre-processing data requirements to build a model (Figure 19.6). It is made up of neurons which receive multiple input signals and support the output by measuring the sum of the input data (Nasr et al. 2017). The convolutional layer is responsible for most of the calculations and merges the data with filters, an activation layer uses a rectifier function to correct the non-linearity of the image, the pooling layer limits the search for images with features such as dimensions. Finally, the fully connected layer converts the assembly into a column and transmits it to the neural network for processing (Ozdemir et al., 2021).

A pre-trained model CNN is called VGG16 that trains approximately 1.2 million images from the ImageNet Dataset. This model has 16 layers and can classify images into 1,000 object categories reaching 93% accuracy in its best scenario However, CNN requires better computational resources. ImageNet is a large collection of hierarchical labeled images (Costa et al., 2018).

Most modern systems use "vision" to collect visual data with high-quality 3D cameras. Cameras determine different colors, shapes, sizes, and object dimensions. Many advanced filter devices have multiple sensors and cameras. The data collected from the devices can be analyzed to maximize line performance. Automated editors can identify specific objects and see how they fit into the system. They compare new and historical data, in addition to data from different parts of the machine. Robots are not generally considered an alternative to visual filtering, but the system can work collaboratively to achieve high-quality production, Robots have simpler requirements than filters designed for operating conditions, including belt speed. They usually use suction cups to pick up objects, but some types use moving cones or finger-shaped cones (Ahmed and Asadullah, 2020). In 2025, the recycling and reuse rate must be 55%, and the targets for 2030 and 2035 are 60% and 65% respectively (Heikkilä, 2020).

19.3.3 Energy Recovery

Technologies have been developed to help in generating huge quantities of energy and in reducing the quantity of waste for its safe disposal (Ghosh, 2008). In general recovery, technologies can be divided into two groups: material recovery and energy recovery. Material recovery includes mechanical recycling, chemical or feedstock recycling and biological or organic material recovery/composting. Energy is recovered in the form of heat, steam, or electricity generation using waste materials as a substitute for primary fossil fuel resources. Energy can also be recovered from the organic fraction, whether biodegradable or non-biodegradable.

Some of the recovery techniques such as biochemical conversion are based on enzymatic decomposition of organic matter by using microbial action. It is usually applied to organic biodegradable material with a moisture content (Chandra and Chowdhary 2015; Chowdhary et al., 2020b). The technological options in this category are anaerobic digestion (AD) or what is called biomethanation and landfill gas recovery, on the other hand, thermo-chemical conversion involves the thermal decomposition of organic matter with low moisture content of organic non-biodegradable material. Incineration, pyrolysis and gasification are the methods used in this category (Ghosh, 2008). The quantity of waste and its physical, chemical characteristics (quality) are the parameters which affect the potential of recovery of energy from MSW (ibid.).

Refuse-derived fuel (RDF) separates MSW into noncombustible and combustible elements which are called RDF material and can be used in boilers. These feed the processing lines, which include primary and secondary trommel screens, magnetic separation, eddy current separation, recovery systems and a shredder. RDF has been efficiently burned in a variety of stokers boilers (Senserna and Bandyopadhyay, 2008).

19.3.4 Landfill Disposal

Garbage has become a major problem worldwide due to uncontrolled disposal of household (domestic) waste from citizens' home and industries without an effective and efficient waste management program, that can result in health risks and have a negative impact on the environment (Costa et al., 2018). To remedy this, artificial intelligence or technology represented in GIS and the analytical hierarchy process (AHP) are used to select the best location for MSW landfill sites that abide by the guidelines of landfill standards and environmental and socio-economic criteria (Chabok et al., 2020). Moreover, the chemical, biological and physical reactions that occur in landfills during waste decomposition result in extremely contaminated landfill leachate and landfill gases. Landfill leachate has negative impacts on soil and water resources (Abunama et al., 2018).

The Fuzzy-AHP method provides a flexible approach to deal with complex quantitative, qualitative, and contradictory issues, and, if combined with GIS, it can provide a powerful tool for locating MSW landfills. The transportation network and residential areas factors are the most important in determining the appropriate location for landfills, also the environmental, social, and economic factors in addition to government regulations and guidelines. When it comes to the environmental criteria, we should keep in mind that landfill sites must not be near any surface water body, especially rivers and groundwater, land cover, agricultural lands, and be some distance away from protected areas, sensitive ecosystem areas, and geological faults. Landfill causes harmful leaks. They must also be far from residential, commercial, and industrial areas, airports, archeological sites, railways, power transmission lines, protected areas, and gas, water and oil pipes. These sites were considered as social criteria because MSW landfill near these areas can have many negative effects, such as sanitary problems, or problems with public health, and land use. Moreover, access and proximity to roads and transportation links are a significant economic goal in locating landfills to reduce the cost of transportation of MSW and minimize the problems of public health, traffic, land pollution and reduce the impact on the tourism industry (Chabok et al., 2020).

Leachate is produced as a result of both waste moisture content and rainfall percolation into the landfill body; it has organic and inorganic contents as well as heavy metals so leachate management is very important when designing the landfill, taking into consideration the waste moisture content and composition, the landfill age and type, and the meteorological conditions, such as the level of precipitation. In addition, the spatial and temporal variations in these factors also affect the leachate and its quantity. Therefore, the prediction of leachate amounts is significant in sustainable waste management and leachate treatment processes, before they are discharged into the surrounding environment. The artificial intelligence (AI) models, such as the artificial neural network (ANN), can improve the modeling of various environmental processes, such as hydrological processes, and achieve accurate forecasting results. An ANN-based leachate model is able to mimic the empirical non-linear relationships between inputs and outputs and predict the leachate flow rate (Abunama et al., 2018).

19.4 THE CIRCULAR ECONOMY AND ARTIFICIAL INTELLIGENCE

The circular economy (CE) is a restorative and regenerative economic system. The linear economy, on the other hand, uses raw materials that are processed, transformed, used, and finally discarded into nature in the form of solid, liquid, and gaseous wastes causing environmental harm (Figure 19.7). Therefore, the linear economy is not sustainable in the long term and the principle of the circular economy keeps products and materials in use as long as they provide value. CE reduces and designs out waste and pollution, promotes recycling and re-use of materials, and increases the development of AI (Industry 4.0) collection and processing techniques for urban waste. Additionally, it decreases the need for virgin materials, lowers the emissions and the amount of disposed waste to the landfills and has a positive impact on society by creating jobs and boosting innovation (Heikkilä, 2020).

It is estimated that the levels of current global MSW generation are approximately 1.3 billion tonnes per year and they are expected to increase to approximately 2.2 billion tonnes per year by 2025 due to the linear economy (Ramadoss et al., 2018).

The CE model offers a solution to a cleaner and more sustainable future. In the case of MSW, this requires recycling and re-use of materials, especially dry mixed recyclable (DMR) waste, which contains paper, cardboard, metal (including aluminum) tins and cans and plastic. Plastics and aluminum make up a major proportion of MSW and have a significant role due to their high processing emissions and increasing virgin material scarcity (ibid.).

The circular economy makes the recycling, reuse, upcycling and recovery processes more efficient, but all the actions above are a kind of adaptation with the huge amount of waste which is generated to reduce it and its harm. Therefore, we need to think sustainably and use the circular economy concept in the product design phase before production, so we should be thinking about the waste after using the product and its fate. Therefore, we should use biodegradable, useful, and eco-friendly materials.

Industry 4.0 contributes to transforming the ways we produce products and track their lifecycles due to the digitalization of manufacturing, so it offers new opportunities to optimize material flows and enable the reverse material flow, which plays a significant role in the transition toward CE.

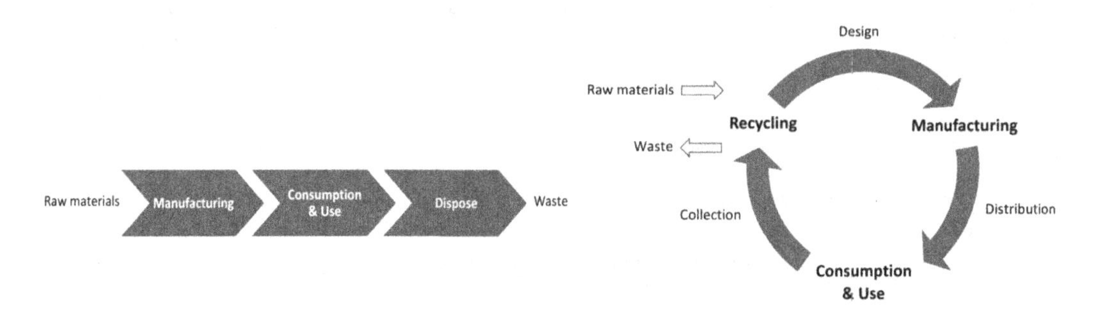

FIGURE 19.7 The linear economy versus the circular economy.

Source: Adopted from Heikkilä (2020).

Industry 4.0 consists of eight advanced technologies: the Internet of Things (IoT), augmented reality, additive manufacturing (AM), big data, cloud computing, simulation, autonomous robots, and cyber-security (Heikkilä, 2020).

The transformation from the linear economy to the circular requires a huge collective effort from governments, policy-makers and industries. CE workflow infrastructure would also generate more challenges such as asset management, a new supply chain, and establishing logistics with unfamiliar waste products and design for manufacturing service and quality control. For example, an electronic printed circuit board (PCB) system might have some parts which outlast other components, in this case, the surviving elements have to be designed for disassembly and reuse for future applications. On the other hand, in a conventional linear economy, industry operates those components at very low cost but they don't take in consideration any re-usability workflow and detachable provision for certain long-life chipsets or devices due to additional costs that would be incurred during the production (Ramadoss et al., 2018).

Digital transformation is forcing many companies to rethink their business and operating model. Industry 4.0 has reshaped many industries, companies and the waste industry by opening new ways to eliminate and reduce waste from different sectors (Heikkilä, 2020).

Sorting is a significant stage of waste management which identifies the secondary raw materials from the waste. There are two types of sorting: manual and automated. Automated sorting is represented by bag openers, trommels, ballistic separators, optical sorters, magnets, eddy current separators and pneumatic collection systems. We need both kinds of sorting because they complement each other, some sorting tasks still have to be carried out manually (manual picking), such as oversized, or recovered material or waste streams that still contain a significant number of recyclables that are incinerated or disposed of. In addition, current European municipal waste sorting plants are highly automated. Digitalization is very important in waste management, especially the sorting of municipal waste that will be a decisive next step in the transformation to a circular economy (Wilts et al., 2021).

Gundupalli et al. (2017) analyzed the world's first recycling robot-driven waste sorting device with AI (the ZRR2 system). These robots are provided with computer vision and deep learning algorithms, they separate precisely selected waste. They increase the efficiency and reduce the cost of waste separation. A single ZRR robot arm is capable of handling up to four different categories with a success rate of up to 98%. The ZRR robot was used for the first time to sort municipal household waste streams. In particular, the ZRR robot must be able to identify and sort the following materials: metals (ferrous and non-ferrous), wood, hard plastics (mixed plastics and hoses and pipes), hard materials (mixed materials, concrete, bricks, limestone, asphalt, etc.), cardboard, in addition to plastic bags by color, and must be trained to recognize and sort new materials: textiles, cardboard (OCC), high-density polyethylene (HDPE), low-density polyethylene (LDPE), and polypropylene (PP). As well, the technology must also adapt the catching system to the new waste conditions. The robot also works under stress conditions that could interfere with waste detection: a fast conveyor belt, multilayer distribution of the waste over the conveyor belt, the presence of dust, alkalis and organic waste (Wilts et al., 2021).

19.4.1 Plastics

In 2015, globally, about 50–60% of the total plastic waste was dumped in landfills, 20–30% was incinerated, and about 10–20% was recycled. Plastic pollutes our water bodies, destroying wildlife, entering our food chain, and ultimately damaging our well-being. North America is the country with the highest plastic production (Chidepatil et al., 2020).

There are three types of plastics: recyclable, non-recyclable and complex, and about 40–48% of annually produced plastics are non-recyclable. About 19–30% are potentially recyclable and nearly 20–27% fall into the complex or unknown category (ibid.).

By 2050, the EU targets are to recirculate 75% of steel, 50% of aluminum and 56% of plastics from the amounts that have already been produced, because it's known that recirculating materials reduces CO_2 emissions and saves energy, compared to new production, but we need to increase the value of secondary materials to be capable of substituting primary materials, while, at the same time, the design and dismantling should be enhanced too. The cost of used unrecycled materials should be raised to decrease its consumption (Heikkilä, 2020). These three factors (price, quality, and quantity) are important drivers of the demand-side for recycled plastic polymers (Chidepatil et al., 2020).

Recycling of plastics is challenging because of the material age, contamination, collectives and mixing of different types of plastics, so volumes of secondary plastics production are just 10% of the total demand in the EU. However, the potential of recirculating plastics is high, most of them can be recycled and used multiple times in addition to the development of technology to mark different plastic types and automate the plastics sorting and processing (Heikkilä, 2020). AI segregates plastic based on physico-chemical parameters, such as color, polymer type, density, etc. (Chidepatil et al., 2020).

The first step toward circular economy plastic is to follow a process that we term the total costing method (TCM), where we total the cost of each step of the circular economy, starting from the waste collection point to the final destination to recognize the actual potential for creating value. Logistics and processing are two stages which affect costs. Depending on how much the process adds or removes value, we have five types of processes, namely, upgrading, closed-loop recycling, downgrading, waste-to-energy, and dumping (ibid.). AI performs plastic segregation efficiently, and this helps in different processes of plastic.

19.4.2 Aluminum

Aluminum is a material that can be recycled multiple times without the material losing its original properties. The use of aluminum has continued to grow, and it is expected to increase in demand up to 40% in Europe by 2050. Some of this growth is generated because aluminum is replacing other materials on different markets: automotive, construction and packaging being the biggest industries. Most aluminum is produced with coal-fired electricity, which causes high CO_2 emissions. On the other hand, remelting existing aluminum requires just 5% of the energy of new aluminum, compared to producing new material which creates 18 tons of CO_2 per ton of aluminum. That could be cut by up to 98% with more efficient reuse of the material. Improved recycling practices and an advanced sorting technology are required to recycle better, and reuse aluminum products. They need better design and end-of-life treatment, additionally a higher amount of aluminum needs to be collected (Heikkilä, 2020).

European Aluminum's Vision 2050 has set a target to make aluminums' value chain decarbonized, circular and energy-efficient in Europe by 2050. Achieving aluminum's full potential for the CE will also contribute to the UN's Sustainable Development Goals (SDGs), in particular, SDG 12, "Responsible consumption and production" (ibid.).

19.5 ARTIFICIAL INTELLIGENCE AND SUSTAINABLE DEVELOPMENT GOALS

Artificial intelligence in waste management contributes widely to achieve 10 out of the 17 UN SDGs, which were set up in 2015 and are intended to be achieved by the year 2030. The SDGs are a collection of 17 global goals designed to achieve a better and more sustainable future for all.

Using AI in smart dustbins can save public health because the open dumping spreads diseases, so that SDG 3: "good health and well-being" will be achieved. The efficient sorting and recycling by AI reduce the amount of disposed waste which causes land and water pollution, so AI is helping in achieving SDG 6: "clean water and sanitation" and SDG 15: "life on land." When the amount of waste increases and the landfills can no longer accommodate it, people throw the rubbish in the oceans or incinerate it, causing water and atmospheric pollution. Polluting the atmosphere with greenhouse gases (GHGs) leads to climate change. Therefore, AI will make the waste management process healthier and will help to achieve SDG 13: "climate change" and SDG 14: "life below water."

Using recyclable materials from waste reduces our consumption of raw materials and energy. Adopting the concept of the circular economy with the aid of AI will help us to use other clean sources of energy, RDF and waste reduction. Therefore, the cost of raw materials and energy will decrease. The mechanical processes will be faster, thus SDG 7: "affordable and clean energy," SDG 8: "decent work and economic growth," SDG 9: "industry innovation and infrastructure," SDG 11: "sustainable cities and communities" and SDG 12: "responsible consumption and production" will be achieved.

19.6 CHALLENGES OF ARTIFICIAL INTELLIGENCE

- Robots and software are replacing workers, so the numbers of low educated people unemployed will increase (Dauvergne, 2020).
- The infrastructure to support AI is raising the demand for energy as well as accelerating the mining of cobalt, and rare earth elements in countries with distressing human rights records (ibid.).
- Automated marketing and advertising are increasing the consumption of goods. Therefore, the amount of consumed resources and waste will rise too, so the bad consequences of this will reflect on poor neighborhoods, fragile ecosystems, and future generations (ibid.).
- Privacy and security challenge because of inadequate software protection, and insufficient authorization (Ramadoss et al., 2018).
- AI serves business interests much more than the planetary interest because it accelerates the concentration of wealth and power rather than advancing global environmental sustainability, as a result of the efficiency and productivity gains from AI (Dauvergne, 2020).
- Data mining and data management are a challenge because of the huge amount of data required (Ramadoss et al., 2018).

19.7 CONCLUSION

Artificial intelligence (AI) helps us to deal with waste management efficiently because MSW has significant negative impacts on human health and the environment. Societies must reduce their generation of wastes from the source as the role of the consumers is significant. At the end of a product lifecycle, individuals have the purchasing power and can contribute to recycling by making the decision to properly dispose of a used plastic bottle rather than, for example, throwing it into the sea. Therefore, we need to raise environmental awareness, that means being aware of the environment and making choices that benefit Mother Earth, rather than damaging it, for a better future for the upcoming generations, starting at the individual level and spreading through society and organizations around the community, as argued by Dr. Dabas (2020).

Green technology is the application of environmental science for sustainable development. Therefore, integration of information technology and environmental awareness and environmental conservation, for example, paper-less offices has become a reality which directly benefits the environment as it decreases the number of trees cut to make paper.

The "robotic revolution" gives operators and companies more accuracy and flexibility, and more companies will adopt the robotic recycling as part of their processes in the next 10 years. Furthermore, the circular economy is considered a crucial solution to the problem of MSW, but the products and the sorting centers need to be redesigned, and the waste processing and handling need to be fully automated; artificial intelligence can help companies improve and accelerate the design and material selection process, based on circular design principles, in order to develop and design new circular products and businesses, as well as solve the solid waste problem. With the help of AI, a better reversed logistic infrastructure can be built, and the loops can be closed. We need stricter legislation to put pressure on more effective waste management and recycling processes.

ACKNOWLEDGMENTS

We thank Dr. Anthony Halog, School of Earth and Environmental Sciences at the University of Queensland in Australia, for his valuable comments.

REFERENCES

Abunama, T., Othman, F., Ansari, M., and El-Shafie, A. 2018. Leachate generation rate modeling using artificial intelligence algorithms aided by input optimization method for an MSW landfill. *Springer Link*, 26, 3368–3381, https://doi.org/10.1007/s11356-018-3749-5

Ahmed, A.A.A., and Asadullah, A. 2020. Artificial intelligence and machine learning in waste management and recycling. *Engineering International*, 8(1), 43–52. https://doi.org/10.18034/ei.v8i1.498

Alam, P., Mazhar, M.A., Khan, A.H., Khan, N.A., and Mahmoud, A.E.D. 2021. Seasonal characterization of municipal solid waste in the city of Jammu, India. *Materials Science and Engineering*, 1058(1), 012061.

Arya, S., Rautela, R., Chavan, D., and Kumar, S. 2021. Evaluation of soil contamination due to crude E-waste recycling activities in the capital city of India. *Process Safety and Environmental Protection*, 152, 641–653. https://doi.org/10.1016/j.psep.2021.07.001

Breitenmoser, L., Gross, T., Huesch, R., Rau, J., Dhar, H., Kumar, S., Hugi, C., and Wintgens, T. 2019. Anaerobic digestion of biowastes in India: Opportunities, challenges and research needs. *Journal of Environmental Management*, 15, 396–412. https://doi.org/10.1016/j.jenvman.2018.12.014

Chabok, M., Asakereh, A., Bahrami, H., and Jaafarzadeh, N.O. 2020. Selection of MSW landfill site by fuzzy-AHP approach combined with GIS: Case study in Ahvaz, Iran. *Environmental Monitoring and Assessment*, 192, 433.

Chandra, R., and Chowdhary, P. 2015. Properties of bacterial laccases and their application in bioremediation of industrial wastes. *Environmental Science: Processes and Impacts*, 17, 326–342.

Chandra, R., and Kumar, V. 2017. Phytoextraction of heavy metals by potential native plants and their microscopic observation of root growing on stabilised distillery sludge as a prospective tool for in-situ phytoremediation of industrial waste. *Environmental Science and Pollution Research*, 24, 2605–2619. https://doi.org/10.1007/s11356-016-8022-1

Chavan, D., Lakshmikanthan, P., Mandal, P., Kumar, S. and Kumar, R. 2019. Determination of ignition temperature of municipal solid waste for understanding surface and sub-surface landfill fire. *Waste Management*, 97, 123–130. https://doi.org/10.1016/j.wasman.2019.08.002

Chidepatil, A., Bindra, P., Kulkarni, D., Qazi, M., Kshirsagar, M., and Sankaran, K. 2020. From trash to cash: How blockchain and multi-sensor-driven artificial intelligence can transform circular economy of plastic waste. *MDPI*, 10, 23. doi:10.3390/admsci10020023.

Chowdhary, P., Raj, A., Verma, D., and Akhter, Y. (Eds.) 2020a. *Contaminants and Clean Technologies*, Boca Raton, FL: CRC Press.

Chowdhary, P., Raj, A., Verma, D., and Akhter, Y. (Eds.) 2020b. *Microorganisms for Sustainable Environment and Health*, Oxford: Elsevier.

Cortes, U., Sànchez-Marrèè, M., Ceccaroni, L., Roda, R., and Poch, M. 2000. *Artificial Intelligence and Environmental Decision Support Systems*, Dordrecht: Kluwer Academic, pp. 77–91.

Costa, B.S., Bernardes, A.C., Pereira, J.V., Zampa, V.H., Pereira, V.A., Matos, G.F., … and Silva, A.F. 2018. Artificial intelligence in automated sorting in trash recycling. In *Anais do XV Encontro Nacional de Inteligência Artificial e Computacional*, Lisbon: SBC, pp. 198–205.

Das, A., Islam, N., Billah, M., and Sarker, A. 2021. COVID-19 and municipal solid waste (MSW) management: A review. *Springer Link*, 28, 28993–2900.

Dauvergne, P. 2020. Is artificial intelligence greening global supply chains? Exposing the political economy of environmental costs. *Review of International Political Economy*, 696–718.

Dhankar, M., and Walia, N. 2020. Emerging trends in big data, IoT and cyber security. *An Introduction to Artificial Intelligence*, 105–108.

Dhar, H., Kumar, P., Kumar, S., Mukherjee, S.N., and Vaidya, A.N. 2016. Effect of organic loading rate during anaerobic digestion of municipal solid waste. *Bioresource Technology*, 217, 56–71. https://doi.org/10.1016/j.biortech.2015.12.004

Dhar, H., Kumar, S., and Kumar, R. 2017. A review on organic waste to energy systems in India. *Bioresource Technology*, 245, 1229–1237. https://doi.org/10.1016/j.biortech.2017.08.159

Dr. Dabas. 2020. Emerging trends in big data, IoT and cyber security, *Role of Information Technology and Media in Environment Awareness*, 225–230.

Fungaro, D.A., Silva, K.C., and Mahmoud, A.E.D. 2021. Aluminium tertiary industry waste and ashes samples for development of zeolitic material synthesis, *Journal of Applied Materials and Technology*, 2(2), 66–73.

Goyal., D., Kumar, S., and Sil, A. 2014. Municipal solid waste: Zero tolerance management strategy, *International Journal of Environmental Technology, and Management*, 17, 113–121. www.inderscienceonline.com/doi/abs/10.1504/IJETM.2014.061793

Gundupalli, S.P., Hait, S., and Thakur, A. 2017. A review on automated sorting of source-separated municipal solid waste for recycling. *Waste Management*, 60, 56–74.

Gupta, P.K., Shree, V. Hiremath, L., and Rajendran, S. 2019. The use of modern technology in smart waste management and recycling: Artificial intelligence and machine learning. *Springer Link*, 173, 188.

Haridas, K., and Ganesh, P. 2020. Artificial intelligence based smart dustbin collector for sustainable environment. *Journal of Xi'an Shiyou University, Natural Science Edition*, 16(7), 164–168.

Heikkilä, S. 2020. Circular value creation in DMR/MSW waste management ecosystem via smart robot, Master's thesis, University Lappeenranta-Lahti University of Technology, pp. 8–79.

Hoornweg, D., and Bhada-Tata, P. 2012. What a waste: A global review of solid waste management. Urban development series; Knowledge Papers no. 15. Washington, DC: World Bank. https://openknowledge.worldbank.org/handle/10986/17388.

Khanik, A.H., Abutaleb, A., Khan, N.A., Mahmoud, A.E.D., Khursheed, A., and Kumar, M. 2021. Co-occurring indicator pathogens for SARS-CoV-2: A review with emphasis on exposure rates and treatment technologies. *Case Studies in Chemical and Environmental Engineering*, 100113.

Khan, A.H., Tirth, V., Fawzy, M., Mahmoud, A.E.D., Khan, N.A., Ahmed, S., Ali, S.S., … Das, G. 2021. COVID-19 transmission, vulnerability, persistence and nanotherapy: A review. *Environmental Chemistry Letters*, 1–15.

Khandelwal., H., Thalla, A.K., Kumar, S., and Kumar, R. 2019. Life cycle assessment of municipal solid waste management options for India. *Bioresource Technology*, 288, 121515. https://doi.org/10.1016/j.biortech.2019.121515

Khapre, A., Kumar, S., and Rajasekaran, C. 2019. Phytocapping: An alternate cover option for municipal solid waste landfills. *Environmental Technology (UK)*, 40(17), 2242–2249.

Kumar, N.S., Vuayalakshmi, B., Prarthana, R.J., and Shankar, A. 2016. IOT based smart garbage alert system using Arduino UNO. In *2016 IEEE Region 10 Conference (TENCON)*, 1028–1034.

Kumar, S., Chiemchaisri, C., and Mudhoo, A. 2011. Bioreactor landfill technology in municipal solid waste treatment. *Critical Reviews in Biotechnology*, 42, 77–97.

Kumar, S., Das, A., Srinivas, G.L.K., Dhar, H., Ojha, V.K., and Wong, J.W.C. 2016. Effect of calcium chloride on abating inhibition due to volatile fatty acids during the start-up period in anaerobic digestion of municipal solid waste. *Environmental Technology*, 37, 1501–1509.

Kumar, S., Dhar, H., Nair, V.V., Bhattacharyya, J.K., Vaidya, A.N., and Akolkar, A.B. 2016. Characterization of municipal solid waste in high-altitude sub-tropical regions. *Environmental Technology*, 37, 2627–2637. https://doi.org/10.1080/09593330.2016.1158322

Kumar, S., Dhar, H., Nair, V.V., Rena Govani, J., Arya, S., Bhattarcharya, J.K., Vaidya, A.N., and Akolkar, A.B. 2019. Environmental quality assessment of solid waste dumpsites in high altitude sub-tropical regions. *Journal of Environmental Environment Management*, 252, 109681. https://doi.org/10.1016/j.jenvman.2019.109681

Kumar, S., Nimchuk, N., Kumar, R., Zeitsman, J., Tara, R., Spiegelman, C., and Kenney, M. 2016. Specific model for the estimation of methane emission form municipal solid waste landfills in India. *Bioresource Technology*, 216, 981–987. https://doi.org/10.1016/j.biortech.2016.06.050

Kumar, V., Ferreira, L.F.R., Sonkar, M., and Singh, J. 2021. Phytoextraction of heavy metals and ultrastructural changes of *Ricinus communis* L. grown on complex organometallic sludge discharged from alcohol distillery. *Environmental Technology & Innovation*, 22, 101382. https://doi.org/10.1016/j.eti.2021.101382

Mahmoud, A.E.D. 2020. Graphene-based nanomaterials for the removal of organic pollutants: Insights into linear versus nonlinear mathematical models. *Journal of Environmental Management*, 270, 110911.

Mahmoud, A.E.D., Fawzy, M., Hosny, G., and Obaid, A. 2021. Equilibrium, kinetic, and diffusion models of chromium (VI) removal using *Phragmites australis* and *Ziziphus spina-christi* biomass. *International Journal of Environmental Science and Technology*, 18(8), 2125–2136.

Manjunatha, G.S., Chavan, D., Lakshmikanthan, P., Singh, L., Kumar, S., and Kumar, R. 2020. Specific heat and thermal conductivity of municipal solid waste and its effect on landfill fires. *Waste Management*, 116: 120–130. https://doi.org/10.1016/j.wasman.2020.07.033

Manjunatha, G.S., Chavan, D., Lakshmikanthan, P., Swamy, R., and Kumar, S. 2019a. Estimation of heat generation from municipal solid waste landfills in India. *Science of the Total Environment*, 707, 135610. https://doi.org/10.1016/j.scitotenv.2019.135610

Manjunatha, G.S., Chavan, D., Lakshmikanthan, P., Swamya, R. and Kumar, S. 2019b. Estimation of heat generation and consequent temperature rise from nutrients like carbohydrates, proteins and fats in municipal solid waste landfills in India, *Science of the Total Environment*, 2019, 135610. https://doi.org/10.1016/j.scitotenv.2019.135610

Manwatkar, P., Dhote, L., Pandey, R.A., Middey, A., and Kumar, S. 2021. Combustion of distillery sludge mixed with coal in a drop tube furnace and emission characteristics. *Energy*, 221, 119871. https://doi.org/10.1016/j.energy.2021.119871

Mousazadeh, M., Naghdali, Z., Goldar, Z.M., Hashemi, M., Mahmoud, A.E.D., Al-Qodah, Z., Sandoval., M.A., Hashim, K.S., … Emamjomeh, M.M. 2021. A review of deciphering the successes and learning from the failures in preventive and health policies to stop the COVID-19 pandemic. *Environmental and Health Management of Novel Coronavirus Disease (COVID-19)*, 269–303.

Nair, V.V., Dhar, H., Kumar, S., Thalla, A.K., Mukherjee, S.N., and Wong, J.W.C., 2016. Artificial neural network based biomethan to evaluate methane yield from biogas in a laboratory-scale anaerobic bioreactor. *Bioresource Technology,* 217, 90–99. https://doi.org/10.1016/j.biortech.2016.03.046

Nasr, M., Mahmoud, A.E.D., Fawzy, M., and Radwan, A. 2017. Artificial intelligence modeling of cadmium (II) biosorption using rice straw. *Applied Water Science*, 7(2), 823–831.

Ozdemir, M.E., Ali, Z., Subeshan, B., and Asmatulu, E. 2021. Applying machine learning approach in recycling. *Journal of Material Cycles and Waste Management*, 23, 855–871. https://doi.org/10.1007/s10163-021-01182-y

Pearse, L.F., Hettiaratchi, J.P.A., and Kumar, S. 2017. Towards developing a representative biochemical methane potential (BMP) assay for landfill waste-a review. *Bioresource Technology,* 254, 312–324. https://doi.org/10.1016/j.biortech.2018.01.069

Ramadoss, T., Alam, H., and Seeram, R. 2018. Intelligence and internet of things enabled circular economy. *The International Journal of Engineering and Science (IJES)*, 7, 55–63. doi: 10.9790/1813-0709035563.

Rena, Y., Patel, S., Killedar, D.J., Kumar, S., and Kumar, R. 2022. Eco-innovations and sustainability in solid waste management: An Indian upfront in technological, organizational, start-ups and financial framework. *Journal of Environmental Management*, 302, 113953. https://doi.org/10.1016/j.jenvman.2021.113953

Sarc, R., Curtis, A., Kandlbauer, L., Khodier, K., Lorber, K.E., and Pomberger, R. 2019. *Digitalisation and Intelligent Robotics in Value Chain of Circular Economy Oriented Waste Management: A Review*, Oxford: Elsevier, pp. 476–492.

Senserna, R., and Bandyopadhyay, G. 2008. *Energy Recovery from Municipal Solid Waste*. Kolkata: West Bengal State Council of Science and Technology.

Sil, A., Wakadikar, K., Kumar, S., Babu, S., Sivagami, S.P.M., Tandon, S., Kumar, R., and Hettiaratchi, J.P.A. 2012. Toxicity characteristics of drilling mud and its effect on aquatic fish populations. *Journal of Hazardous, Toxic, and Radioactive Waste*, 15, 51–57. https://doi.org/10.1061/(ASCE)HZ.2153-5515.0000092

Sunayana, K.S., and Kumar, R. 2020. Forecasting of municipal solid waste generation using non-linear autoregressive (NAR) neural models. *Waste Management*, 121, 206–214. https://doi.org/10.1016/j.wasman.2020.12.011

Wainaina, S., Awasthi, M.K., Chen, H., Singh, E., Kumar, A., Ravindran, B., Sarsaiya, S., … Taherzadeh, M.J., 2019. Resource recovery and circular economy from organic solid waste using aerobic and anaerobic digestion technologies. *Bioresource Technology*, 301, 122778. https://doi.org/10.1016/j.biortech.2020.122778

Wilts, H., Garcia, B.R., Garlito, R.G., Gómez, L.S., and Prieto, E.G. 2021. Artificial intelligence in the sorting of municipal waste as an enabler of the circular economy. *Resources*, 10, 28. https://doi.org/10.3390/resources10040028

Index

Note: Figures are indicated with page numbers in *italics*. Tables are indicated with page numbers in **bold** type.

A

activated sludge process (ASP) 59
adsorption 41–42
aerobic bacteria 119
air contaminants 125
alcoholysis 192
anaerobic bacteria 119
anaphylactic shock 105
antibiotics 69–72, 71, 72, 78. See also antimicrobial resistance (AMR)
antimicrobial resistance (AMR): Advanced and Conventional Treatment Strategies for the Removal of Antibiotic Resistance Bacteria and Genetic Elements 75; AMR surveillance and related tools 78; Bacteria with High Levels of Antibiotic Resistance 78; GLASS-Focused Surveillance and Special Survey Activities 79; as health threat 76–77; removal strategies 74–75; sources 71–73; Sources and pathways for antibiotic contaminants in the environment 72; spread of 73–74; surveillance of 78–79; Timeline of Antibiotics: Year of Production, Name of Antibiotic, and Class 71; Wastewater treatment plants (WWTPs) as the transition point of AMR 74
Arthrobacter aurescens 166, 168
artificial intelligence-based waste management: Artificial Intelligence Techniques 345; Basic elements of artificial intelligence (AI) 344; and circular economy 351–53; Example of CNN machine learning algorithm 349; importance in municipal waste management 343–45; Linear economy versus the circular economy 351; overview of solid wastes 341–43; Projected waste production globally, by each region 342; role in stages of waste management 345–51; Scheme of the stream of waste segregation using the interactive modeling systems (AIMS) 348; Smart dustbin design 346; Some examples of images of recycled objects of database for classifying 348; sustainable development goals and challenges 353–54
assimilation 140
avoidance/excluders 246–47

B

Bacillus cereus 164
Bacillus subtilis 164, 170, 172, 317
bacterial resistance 70
Baekeland, Leo 185
Bakelite 185
Berger, Jacques Branden 185
bilge water 93
bio-based biodegradable plastics 188–89
biodegradability 192
BIOLEN IG30 171, 172
biopiling 157
bioplastics 182–85, 186
biopolymers 187
bioremediation applications. See microbial bioformulation technology
Bio-Save 170
biosurfactants 194
black carbon 90–91
black water 92–93
Burkholderia cepacian 70

C

Capsicum annuum 165
carbamate 158
cell electroporation 319
Central Asian and European Surveillance of Antimicrobial Resistance (CAESAR) network 79
chemical pollutants 158
chlorinated pesticides 51
circular economy (CE) 351–52
coagulation flocculation process 59
compostability 192
constructed wetlands 41
copper-based MOFs 220
coral reefs 84, 92, 166
Covid-19/CORONA-19 77, 87, 342
cylindraceous candida 184

D

DDT 59–60
de Chardonnet, Hilaire Bernigaud 185
Desmococcus olivaceus. See microalgae
Dickson, James Tennant 185
diclofenac (DIC) 11, 24–26, 28
digitalization 344
dispersive liquid-liquid micro extraction (DLLME) 28
distillery industry (DI) 130–32, 131
Dow Chemical 185
DuPont 185

E

electrical conductivity (EC) 314
Electro-Pure process 309
electrospinning method 299
emerging environmental pollutants (EPs): in aquatic environment 7; challenges and limitations of 14–15; characteristics and behavior of 4, 6–7; Concentration of Pesticide Concentration Surface/ Groundwater 8–9; Environmental Contaminants 5–6; health risks of 12–14; Management concept of EPs in the environment 13; Occurrence of Different Endocrine Disrupters (EDCs) in Water 9–10; overview 3–7; Pharmaceutical Drugs' Concentration in Water in Different Areas Around the Globe 11; Possible sources of emerging pollutants (EPs) in the environment 4; transport and bioaccumulation of 12

endocrine disruptors (ED) 25–26, 29–30
endosulfan 58–59
energy recovery 349–50
Enretech-1 171, 172
enterococci 70, 73
environmental contaminants management: categories of contaminants 125–26; Classification of pesticides 135; Different ex situ strategies 121; Different in situ strategies 120; Different Microbes Able to Degrade Pharmaceutical Wastewater 128; Different Microorganisms Having Hydrocarbon Degradation Potential 137–38; Different Microorganisms Suitable for PC's Wastewater Management 129; Different Microorganisms Used in Heavy Metals (HM) Degradation 134; Different types of microbial strategies used to contaminants management 121; environmental contaminants (ECs) 125–26; Important methods in rDNA technology 123; Major heavy metals and their sources 133; Major water pollutants 126; microbes in clean- up of environmental pollutants 119–25; microbes in water pollutants management 126–41; Microbial Enzymes Involved in Polymer Degradation 139; Microorganisms Involved in Organophosphorus Chemicals Degradation 135–36; Microorganisms Involved in Textile Industry Contaminants Management 128; Microorganisms Used to Decolorize Spent Wash and to Degrade Melanoidin 132; Overview of the major environmental contaminants and the microbes used for remediation 119; Process of microbial degradation of plastics 139; Several Fungal Species Having Lignocellulolytic Activity 125; Steps involved in biofilm formation 140; Use of Different Beneficial Microbes for Explosives Degradation 131
Environmental Protection Agency (EPA) 12, 54, 60, 93, 136
enzymes, inactivation of 319–20
EOS 100 171
Escherichia coli 70, 73, 78, 92
European Antimicrobial Resistance Surveillance Network (EARS-Net) 79
European Union (EU) 91, 108, 238, 342, 343
eutrophication 92, 126, 186, 236
ex situ 120
explosives industry 130, 131

F

family rearranging (FR) 124
fermentation 319
fire retardants 26
food additives 26
Food Allergen Labeling and Consumer Protection Act of 2004 (FALCPA) 107
food allergens: Allergenic Foods That Are Priority in Labeling Laws in Selected Countries 108; challenges and future prospects 108–9; Different Factors Influencing the Food Allergenicity in Animal Models 103; factors influencing food allergenicity 103–4; food allergens and common features 101–3; food hypersensitivity 100–101; Major Food Allergens From Offending Food Sources Which Induce Food Allergies in Children and Adults 102; management of 105–7; prevention 107–8
food and nutrition: active packaging 263–64; anti-microbial food packaging 264–65; Classification of Nanomaterials and Their Common Application in Food Systems 260; developing food nanocomposite polymers 259–62; improvement of mechanical, thermal, and barrier properties 266–67; nanocarriers for food ingredients 262–63; nanoparticles as sensors in smart packaging 267; nanosolutions against food fraud 268; overview of nanotechnology 257–59
food fraud 268
food processing industries 126–27
fossil-based biodegradable plastics 189
fragmentation 140
fuel oil 86
fungi 136–38
Fusarium solani 164

G

gasoline 85
gene regulating networks (GRN) 123
genetically engineered microorganisms (GEMs) 118
genetically modified organisms (GMOs) 118
genome rearranging (GR) 125
Global Antimicrobial Resistance Surveillance System (GLASS) 79
glycolysis 192
Great Pacific Garbage Patch (GPGP) 186
grey water 92–93

H

halophytes 246
heavy fuel oil (HFO) 86–87, 90–91
histamines 104
hypoallergenic food production 108

I

ibuprofen (IBU) 11, 24–25, 26, 30
immunotherapy 105
in situ 120
Industry 4.0 344
Inipol EAP22 171
insect pest management: in agriculture 273–74; Chemical and Physical Synthesis of AgNPs 276; Chemical methods of CRF preparation 282; Different approaches to synthesis of Ag nanoparticles (NPs) 277; Favourable development of nanopesticides' formulation 284; List of Chemical and Biological Nanoparticles Against Pests 281; nanomaterials 276–80; nanoscale components and their use 275; novel nano-insecticides 286–87; and pesticide nanoformulations 287; Physical methods of CRF preparation 281; Several Examples of Polymers Often Used in the Nanoparticles' Production 286
Isochrysis 165
isotatic polypropylene 185

J

Joule, James P. 309

K

ketoprofen (KET) 11, 24–26, 30
kitchen waste: applications of metal nanoparticles (MNPs) 335–36; Different types of nanoparticles and applications 336; Factors Affecting Biofabrication of Nanoparticles 333; and green synthesis of metal nanoparticles 330–33; List of Nanoparticles, Their Morphology and Sources 334; Processes involved in metal nanoparticles formation 332; silver and gold nanoparticles produced by 333–35; Types of nanoparticles generated by kitchen food waste 333
Klebsiella pneumoniae 73, 76, 78

L

Lancet Commission on Pollution and Health 157
lindane 57–58
liquid-based bioformulations 163–64

M

marine pollution: Classification of Plastic Litter According to Their Dimensions 85; Cruise liner dumping untreated sewage and wastewater into the ocean 94; cruise ship 88; Cruise ship pollution compared to car pollution in European cities in 2017 89; EU sulphur standards for marine and road fuels 88; Fractions of Crude Oil Refinery 86; Main physical, chemical and biological processes that command the fate of oil pollution 87; major sea pollutants 84–88; Nitrogen dioxide concentrations over France 90; Nitrogen dioxide concentrations over Italy 89; sewage and fossil fuel 89–94; Top 15 countries: share of fuel use by fuel type in 2019 92; Top 15 countries using fuel in the Arctic in 2019 91
melamine formaldehyde 185
membrane bioreactor (MBR) 40
mercury 218
metabolic engineering (ME) 120–22
metagenomics 79
metal ion removal 298–99
metal ion solution 330
metal matrix nanocomposites 298
metal organic frameworks (MOFs) 215–16
metallophytes 246
methyl parathion 58
microalgae: Algal Biomass Production by *Desmococcus Olivaceus* Grown in Open Raceway Pond Amended with Chrome Sludge from Electroplating Industry 209; Comparison of Physico-Chemical Parameters of Raw and Algae-Treated Electroplating Industrial Chrome Sludge in an Open Raceway Pond 208; effect of heavy metals contamination on environment 202; Flow chart showing effluent/wastewater treatment using microalgae and the biomass harvested and utilized in various applications of biofuel 210; Growth and Sludge Reduction by *Desmococcus Olivaceus* Grown in a Pilot Tank Amended with Chrome Sludge from Electroplating Industry 209; Heavy metals reduction and algal biomass production by *Desmococcus Olivaceus* grown in chrome sludge of electroplating industry in an open raceway pond 207; phycoremediation of chrome sludge 205–10; Phycoremediation of Heavy Metals by Microalgae in Wastewaters/ Effluents Treatment 205; phycoremediation of heavy metals using 204–5; phycoremediation techniques and mechanisms 203; raceway pond study 205–6; Toxicological properties of heavy metals in the environment 202
microbes, role of. See environmental contaminants management; plastics degradation
microbial adherence 138–40
microbial bioformulation technology: Advantages and Limitations of Biofonnulations 169; Advantages and Limitations of Bioremediation Approaches 161; available commercially 170–72; benefits and limitations 169–70; Bioformulation of Various Bioagents for the Bioremediation of Various Pollutants 166–67; in bioremediation applications 164–69; Commercially Available Bioformulations on the Market 170–71; environmental pollution and remediation 157–59; General Types of Pollutants Found in Wastewater From Different Industries 159; Main sources of environmental pollution 158; in non-bioremediation applications 164–65; Properties of Different Liquid-Based Bioformulations 164; Properties of Different Solid-Based Bioformulations 163; Proposed flow chart of customization of a bioformulation modified from the agricultural waste management method 173; Schematic diagram of the concept of bioformulation 162; Summary of conventional and recent remediation technologies 156; types 162–64; Various Remediation Technologies and Their Specific Use on Target Contaminants 160
microorganisms, inactivation of 317
microplastics/macroplastics 184, 186
mine tailings (MTs): advantages and limitations of 249; composition and treatment of 244–49; Contaminants in Mine Tailings and Plants Species with Their Phytoremediation Mechanisms 250–51; extraction and impact of 243; Phytoremediation approaches in polluted soil 247
mineralization 141
MOF-based sensors 217
moving bed biofilm reactor (MBBR) 60
municipal solid waste 342
Mycostop® 170

N

nanocarriers, five classes of 262–63
nano-emulsions (NEs) 284–85
nanofibers (NF) 259, 261
nanofungicides. See insect pest management
nanoherbicides. See insect pest management
nanolayers (NLs) 261
nanomaterials, categories of 259
nanoparticles (NPs) 261–62
nanopesticides. See insect pest management
nanotechnology, role of. See food and nutrition; insect pest management
Natta, Giulio 185

nickel-based MOF 220
nitrocellulose 185
nitrogen dioxide 88, 89, 90
Nobel Prize 185
non-biodegradable plastics 188
non-steroidal anti-inflammatory drugs: Action mechanism of NSAIDs 26; adsorption and advanced oxidation processes 41–43; analyzing in wastewater 25–29; constructed wetlands as treatment option 41; Details of NSAIDs Discussed 25; Details of WWTPs and Treatment Technologies 34–35; occurrence in Indian WWTPs 29, 34–36; presence in treated wastewater 23–24; removal 37–41; Schematic representation of DLLME 28; Schematic representation of SPE procedure 27; Summary of Target NSAIDs Mean and Range of Concentrations in WWTPs in the Indian Context 30–33
NSAIDs. See non-steroidal anti-inflammatory drugs

O

ohmic heating: advantages, disadvantages, and limitations 320–21; applications 316–20; Basic ohmic heating process 311; benefits 313; current status globally 324; effects on food quality 322–24; Electrical Conductivity (EC) of Common Foods 312; history of 309–10; and ideal food products 312–13; mathematical modelling 321–22; Ohmic Heating in Food Applications 316; parameters and factors that influence 314–16; principles 309–10; process and components of 310–11; Some Companies Producing an Ohmic Heating System 324
Oil Buster™ 171
Organisation for Economic Co-operation and Development (OECD) 60
organochlorine pesticides 50–51, 52
organophosphate pesticides 50–51

P

pasteurization 316
personal care products (PCPs) 25
pesticides: case study of Silk City, Bhagalpur, Bihar, India 56–60; Chronology of Pesticide Development 52; in Ganga River 56–60; Organochlorine Pesticides in the Ganga River System 54; overview 49–51; presence in river systems 53–55; regulatory measures for emerging contaminants 60–62; and water quality 51–52
petrochemical industries (PI) 129–30, 129
pharmaceutical industry 127, 128
pharmaceuticals 25
phycoremediation. See microalgae
phytoaccumulation 156, 248, 250–51
phytodegradation 248
phytoextraction 248, 249, 250–51
phytoremediation. See mine tailings (MTs)
phytosiderophores 245
phytostabilization 156, 248, 251
phytovolatilization 247–48
Piper nigrum 165
plasmids 123–24
plastics degradation: classification of 187–90; current research on 194–95; Factors affecting rate of biodegradation 191; future prospects 195; microbial mechanisms for 190–94; Options for plastic waste treatment 187; origins and pollution 185–87; overview of bioplastics 182–85; Plastic biodegradation: mechanism under aerobic and anaerobic conditions 191; Process of microbial degradation of plastics 139; role of microbes in 190–94; Role of microbes in degradation of plastics and directions toward greener bioplastic 183
polycaprolactone 190
polycyclic aromatic hydrocarbons (PAHs) 26, 136
polyethylene (PE) 183
polyethylene succinate 189–90
poly-ethylenetrephthalate (PET) 183
polyhydroxyalkanoates 189
polylactic acid (PLA) 189, 192
polymer/ layer silicate (PLS) nanocomposites 298
polymer nanocomposites. See wastewater treatment
polymer/ceramic nanocomposites 298
polymers 285
polypropylene (PP) 183, 352
polystyrene (PS) 141, 183, 185
polyurethane (PUR) 183
polyvinyl chloride (PVC) 167, 183, 185
PRP® Powder 171, 172
Pseudomonas aeruginosa 70, 128, 134
Pseudomonas cepacia 166, 167
Pseudomonas fluorescens 164, 165, 185
Pseudomonas putida 39, 129, 136, 161
pump-and-treat system 156, 159
pyrolyse 192

R

rayon 185
recycling 347–49
recycling routes 186–87
refuse-derived fuel (RDF) 350
RemActiv™ 171, 172
RootShield Plus+ Granules 170
Rotterdam Convention 61

S

Salmonella enterica 73
Schönbein, Christian 185
Serenade ASO 170, 172
soil contaminants 126
soil flushing 160
soil washing 156, 159, 160
solid-based bioformulations 162–63
solvolysis 192
Sonata 170
Staphylococcus aureus (MRSA) 70, 78, 128
starch 188, 323
Stenotrophomonas maltophilia 70
sterilization 316
Stockholm Convention 61–62
Streptomyces rochei 166, 168

T

textile industry (TI) 127–29, 128
thermal desorption 156
tolerance/accumulators 246–47

toxic heavy metals: Cu-MOF-based electrochemical senor for detection of Hg+2 ions in tap water and tuna fish 219; Fabrication of the MWCNTs–COOH/ UiO-66-NH2/ MWCNTs–COOH/ GCE for electrochemical sensing of Cd2+ and Pb2+ 219; and health concerns 215; Ln-MOF-based sensor for recognition of Pb+2 and Cd+2 ions 218; Major heavy metals and their sources 133; metal organic frameworks (MOFs) 215–16; Nickel-based MOF functionalized with ferrocene for sensing of copper, lead and cadmium metal ions 220; Toxicological properties of heavy metals in the environment 202; types of MOF composites 216–20; Various MOF-Based Sensors for Electrochemical Sensing of Heavy Metal Ions 221–22
transposons 124
Trichoderma harzanium 165, 170

U

United Nations Joint Group of Experts on the Scientific Aspects of Marine Pollution (GESAMP) 84

V

vermiwash and vermicompost: advantages over synthetic inorganic chemicals 236; as biopesticide/pest control agent 233–34; composition and role of vermiwash 232–33; effects of chemical fertilizers, pesticides, and weedicides 230; Functions of Different Vermiwash Components in Plants 233; Growth-Promoting Effects of Vermiwash on Crop Plants 236; overview 230–31; as plant growth promoters 234–35; preparation of vermiwash 231–32; Set-up of ERF method for the collection of vermiwash 231; in soil health and crop productivity 236–37; Use of Different Products of Vermicomposting for Pest Control 234

W

waste management system design 343
wastewater treatment: applications of polymer-based nanocomposites in 298–302; classification of nanocomposites 297–98; Illustration of the synthesis of superhydrophobic polymer-coated tetrapodal magnetic nanocomposite as efficient sorbent for the removal of oil from wastewater 301; Metal ion removal applications of poly(MMA-co-MA)/ Fe3O4 Magnetic material from wastewater 299; Methylene blue removal applications of carbon nanotubes-based polymer nanocomposites 300; Possible applications of polymer nanocomposites 296; Removal of Various Pollutants from Wastewater Bodies Using Polymer-Based Nanocomposites 302; Sources of natural polymer, their composites, and applications 297; sources of natural polymers 296–97
water absorption capacity (WAC) 322
water contaminants 126
water solubility index (WSI) 322
Whinfield, Rex 185
Wyeth, Nathaniel 185

Z

Ziegler, Karl 185

Printed in the United States
by Baker & Taylor Publisher Services